The Soybean

Botany, Production and Uses

The Soybean

Botany, Production and Uses

Edited by

Guriqbal Singh

Department of Plant Breeding and Genetics
Punjab Agricultural University
Ludhiana, India

www.cabi.org

CABI is a trading name of CAB International

CABI Head Office
Nosworthy Way
Wallingford
Oxfordshire OX10 8DE
UK

Tel: +44 (0)1491 832111
Fax: +44 (0)1491 833508
E-mail: cabi@cabi.org
Website: www.cabi.org

CABI North American Office
875 Massachusetts Avenue
7th Floor
Cambridge, MA 02139
USA

Tel: +1 617 395 4056
Fax: +1 617 354 6875
E-mail: cabi-nao@cabi.org

A catalogue record for this book is available from the British Library, London, UK.

Library of Congress Cataloging-in-Publication Data

The soybean : botany, production and uses / edited by Guriqbal Singh
p. cm.
Includes bibliographical references and index.
ISBN 978-1-84593-644-0 (alk. paper)
1. Soybean. 2. Soybean–Processing. 3. Soybean products. I. Singh, Guriqbal, agronomist II. Title.

SB205.S7S534 2010
633.3'4–dc22

2010004392

ISBN-13: 978 1 84593 644 0

Commissioning editor: Meredith Carroll
Production editor: Fiona Chippendale

Typeset by AMA DataSet Ltd, UK.
Printed and bound in the UK by CPI Antony Rowe.

Contents

About the Editor

Dr Guriqbal Singh is currently working as a Senior Agronomist (Pulses) (equivalent to Professor) in the Department of Plant Breeding and Genetics, Punjab Agricultural University, Ludhiana, India. He received his BSc Agri. (Hons) and MSc Agronomy degrees from the Punjab Agricultural University, Ludhiana, India, where he was a scholarship holder. He received his PhD from the University of Wales, Bangor, UK, on a Commonwealth Scholarship, where he studied the effects of herbicides on biological nitrogen fixation in peas (*Pisum sativum*).

During 2005–2006 Dr Singh worked with the International Centre for Agricultural Research in the Dry Areas (ICARDA) at its Regional Office for Central Asia and the Caucasus, Tashkent, Uzbekistan. Here, he worked as the Technical Coordinator for an Asian Development Bank-funded project on 'Improving rural livelihoods through efficient water and soil fertility management in Central Asia'.

He has approximately 20 years' experience in working on various grain legumes at the Punjab Agricultural University. His main areas of research include conservation agriculture, planting method and planting time, weed management, nutrient management, plant population and planting geometry and water management. He has provided very useful recommendations to farmers for raising the productivity of their grain legume crops and reducing the costs of cultivation. He is actively involved in teaching agronomy courses to students, as well as in extension education programmes for farmers.

Dr Singh has published 68 research articles in journals of national and international repute, 47 abstracts in conference proceedings, 70 extension articles, 10 book chapters and three bulletins. He is a senior editor of two books, *Recent Advances in Agronomy* (2002) and *Pulses* (2005). He has participated in approximately 20 national and international conferences and workshops. He is a life member of many professional societies, such as the Indian Society of Pulses Research and Development, the Indian Society of Agronomy, the Indian Society of Soybean Research and Development, the Indian Society of Weed Science and the Indian Ecological Society. He is also a Fellow of the Indian Society of Pulses Research and Development.

In 1998 Dr. Singh was awarded the Dr P.S. Deshmukh Young Agronomist Award by the Indian Society of Agronomy for his significant research contributions. In 2009 he was honoured by the Indian Society of Pulses Research and Development (ISPRD) with the ISPRD Recognition Award for Crop Production for his outstanding contributions in pulses agronomy.

Contributors

Mr Navneet Aggarwal, Assistant Agronomist (Pulses), Department of Plant Breeding and Genetics, Punjab Agricultural University, Ludhiana 141004, India. E-mail: navneetpulsespau@yahoo.com

Dr Nawab Ali, Ex-Deputy Director General (Engineering), Indian Council of Agricultural Research, KAB-II, Pusa, New Delhi 110012, India. Present address: House Number SDX-40, Minal Residency, JK Road, Bhopal 462 023, India. E-mail: alinawab11@gmail.com

Dr Kavita Bisht, Lecturer, SAP Kanya Mahavidyalaya, Kichha, 263148 (Kumaon University, Nainital), India. E-mail: bishkavita@gmail.com

Professor Ru-Zhen Chang, Professor, The National Key Facility for Crop Gene Resources and Genetic Improvement (NFCRI)/Key Lab of Germplasm Utilization (MOA), Institute of Crop Science, Chinese Academy of Agricultural Sciences, 100081 Beijing, PR China. E-mail: dadousoybean@yahoo.com.cn

Dr G.S. Chauhan, Ex-Director, Directorate of Soybean Research, Khandwa Road, Indore 452017, Madhya Pradesh, India. E-mail: gschauhan_46@yahoo.co.in

Dr Jonas N. Chianu, Principal Agricultural Economist, Agriculture 2 Division (OSAN.2), Agriculture & Agro Industry Department (OSAN), African Development Bank, Agence Temporaire de Relocalisation, 15 Avenue du Ghana, Angle Rues Hedi Nouira & Pierre de Coubertin, B.P. 323 Tunis, 1002 Tunis Belvedere, Tunisia. E-mail: jchianu@yahoo.com, j.chianu@afdb.org

Dr Dennis B. Egli, Professor, Department of Plant and Soil Sciences, University of Kentucky, 1405 Veterans Drive, Lexington, KY 40546-0091, USA. E-mail: degli@uky.edu

Professor B. Fawole, Professor of Nematology, Crop Protection and Environmental Biology Department, Faculty of Agriculture and Forestry, University of Ibadan, Ibadan, Nigeria. E-mail: bamidelefawole@yahoo.com

Dr Prabal K. Ghosh, Senior Scientist, Food Development Centre, 810 Phillips Street, Portage la Prairie, Manitoba, Canada, R1N 3J9. E-mail: prabal.ghosh@gov.mb.ca

Dr Glen L. Hartman, Research Plant Pathologist, USDA Agricultural Research Service, and Professor, Department of Crop Sciences, National Soybean Research Center, University of Illinois, Urbana, Illinois, USA. E-mail: ghartman@illinois.edu

Mr Curtis B. Hill, Principal Research Specialist, Department of Crop Sciences, National Soybean Research Center, University of Illinois, Urbana, Illinois, USA. E-mail: curthill@illinois.edu

Dr Digvir S. Jayas, Vice-President (Research) and Distinguished Professor, Department of Biosystems Engineering, University of Manitoba, Winnipeg, Manitoba, Canada, R3T 5V6. E-mail: Digvir_Jayas@umanitoba.ca

Mr Kevin D. Johnson, Doctorial Candidate, Department of Entomology, Iowa State University, 117 Insectary, Ames, IA 50011, USA. E-mail: john2057@iastate.edu

Dr Vineet Kumar, Senior Scientist, Biochemistry (Plant Sciences), Directorate of Soybean Research, Khandwa Road, Indore 452017, Madhya Pradesh, India. E-mail: vineetksahni@yahoo.com

Dr Saratha Kumudini, Assistant Professor, Department of Plant and Soil Sciences, University of Kentucky, 1405 Veterans Drive, Lexington, KY 40546-0091, USA. E-mail: s.kumudini@uky.edu

Professor David L. McNeil, Director, Tasmanian Institute of Agricultural Research, Chair of Agricultural Science, University of Tasmania, Private Bag 54, Hobart, 7001, Tasmania, Australia. E-mail: david.mcneil@utas.edu.au

Dr J.S. Mishra, Principal Scientist (Agronomy), Directorate of Sorghum Research, Rajendranagar, Hyderabad 500030, Andhra Pradesh, India. E-mail: jsmishra31@gmail.com

Dr S.K. Mishra, Head, Germplasm Evaluation Division, National Bureau of Plant Genetic Resources, Pusa Campus, New Delhi 110012, India. E-mail: skmishra_gene@rediffmail.com

Dr Ephraim M. Nkonya, Senior Research Fellow, International Food Policy Research Institute (IFPRI), Washington, DC, USA. E-mail: e.nkonya@cgiar.org

Dr Matthew E. O'Neal, Assistant Professor, Department of Entomology, Iowa State University, 117 Insectary, Ames, IA 50011, USA. E-mail: oneal@iastate.edu

Dr Edward O. Oyekanmi, Research Fellow-Nematologist, Nematology Unit, International Institute of Tropical Agriculture, Oyo Road, Ibadan, Nigeria; Crop Protection and Environmental Biology Department, Faculty of Agriculture and Forestry, University of Ibadan, Ibadan, Nigeria; and Biological Sciences Department, Wesley University of Science and Technology, Ondo, Nigeria. E-mail: eoyekanmi@yahoo.com, eoyekanmi@cgiar.org

Dr Dilip R. Panthee, Assistant Professor, Department of Horticultural Science, North Carolina State University, Mountain Horticultural Crops Research and Extension Center, Mills River, NC 28759, USA. E-mail: dilip_panthee@ncsu.edu

Dr Li-Juan Qiu, Professor, The National Key Facility for Crop Gene Resources and Genetic Improvement (NFCRI)/Key Lab of Germplasm Utilization (MOA), Institute of Crop Science, Chinese Academy of Agricultural Sciences, 100081 Beijing, PR China. E-mail: qiu_lijuan@263.net

Dr Rita S. Raghuvanshi, Dean, College of Home Science, GB Pant University of Agriculture and Technology, Pantnagar 263145, Uttarakhand, India. E-mail: rita_raghuvanshi@yahoo.com, deanhsc1@gmail.com

Dr Hari Ram, Wheat Agronomist, Department of Plant Breeding and Genetics, Punjab Agricultural University, Ludhiana 141004, India. E-mail: hr_saharan@yahoo.com

Dr Anita Rani, Principal Scientist (Plant Breeding), Directorate of Soybean Research, Khandwa Road, Indore 452017, Madhya Pradesh, India. E-mail: anitavks@yahoo.co.in

Dr K. Sammi Reddy, Senior Scientist, Indian Institute of Soil Science, Nabi Bagh, Berasia Road, Bhopal 462038, Madhya Pradesh, India. E-mail: ksr@iiss.ernet.in

Dr S. Shanmugasundaram, Agricultural Consultant and Ex-Deputy Director General Research, AVRDC – The World Vegetable Center. Present address: 27 Bayard Road, Somerset, NJ 08873, USA. E-mail: sundar19392004@yahoo.com

Dr B.G. Shivakumar, Senior Scientist, Division of Agronomy, Indian Agricultural Research Institute, New Delhi 110012, India. E-mail: bgskumar@yahoo.com

Dr Guriqbal Singh, Senior Agronomist (Pulses), Department of Plant Breeding and Genetics, Punjab Agricultural University, Ludhiana 141004, India. E-mail: singhguriqbal@rediffmail.com

Dr A. Subba Rao, Director, Indian Institute of Soil Science, Nabi Bagh, Berasia Road, Bhopal 462038, Madhya Pradesh, India. E-mail: asrao_iiss@yahoo.co.in

Dr V.D. Verma, Officer In-charge, National Bureau of Plant Genetic Resources, Regional Station, Phagli, Shimla 171004, Himachal Pradesh, India. E-mail: headnbpgr@dataone.in

Ms Miao-Rong Yan, Principal Research Assistant, Legume Unit, AVRDC – The World Vegetable Center, P.O. Box 42, Shanhua, Tainan, Taiwan 74199. E-mail: miao-rong.yan@worldveg.org

Dr Edilegnaw W. Zegeye, Senior Lecturer, Department of Agricultural Economics, School of Agricultural Sciences and Agribusiness, University of KwaZulu-Natal, P Bag X01 Scottsville 3209, Pietermaritzburg, South Africa. E-mail: walee@ukzn.ac.za

Preface

Soybean (*Glycine max* (L.) Merrill), with its countless and varied uses, is an important crop at the global level. Its seeds are rich in oil (approximately 20%) and protein (approximately 40%). In 2007, the global area, production and productivity of soybean were 90.1 million ha, 220.5 million t and 2.44 t ha^{-1}, respectively. The USA, Brazil, Argentina, China and India are the major soybean-producing countries.

Soybean is grown in various sequential and inter/mixed cropping systems. Many biotic and abiotic stresses limit soybean production in different parts of the world. Much research has been carried out worldwide on breeding, production, protection, processing and utilization aspects of soybean. Vast information on all of these aspects is available in different journals, research reports, magazines and leaflets. There has been, however, a dire need to compile this scattered information in one place in the form of a book.

A humble request was made to experts in their fields to contribute chapters to this book. This book consequently contains 20 chapters written by eminent researchers from different countries including Australia, Canada, China, India, Nigeria, South Africa, Taiwan and the USA. With the combined wisdom of so many reputable scientists from different parts of the world, I hope this book will achieve the status of the world soybean reference book.

The book has been divided into six sections. The first section on 'History and Importance' includes two chapters: the first on the origin and history of soybean and the second on the role of soybean in agriculture. The second section on 'Botany, Genetics and Physiology' includes four chapters on soybean growth and development, soybean genetic resources, varietal improvement in soybean and soybean yield physiology. The 'Production' section, third in the series, includes six chapters on agro-techniques for soybean production, nutrient management, water management, weed management, biological nitrogen fixation and storage of soybean. Three chapters relating to diseases, insect pests and nematodes are included in a 'Protection' section. The section on 'Utilization' includes four chapters on soybean processing and utilization, the nutritional value of soybean, uses of soybean and vegetable soybean. The last section on 'Marketing and Trade' includes a final chapter on global soybean marketing and trade. The chapters, therefore, address advanced and diverse topics covering almost all aspects of soybean. Each chapter has a good number of references at the end to enable the interested reader to go to the original source.

I hope that this book will be useful to researchers, teachers, students, extension personnel and others with an interest in soybean.

I would like to thank all of the contributors for their wonderful cooperation. All permissions to reproduce copyright materials are gratefully acknowledged. Thanks are also due to the CABI for publishing the book so well.

Guriqbal Singh
20 December 2009
Ludhiana, India

1 The Origin and History of Soybean

Li-Juan Qiu and Ru-Zhen Chang
The National Key Facility for Crop Gene Resources and Genetic Improvement/ Key Lab of Germplasm Utilization (MOA), Institute of Crop Science, Chinese Academy of Agricultural Sciences, Beijing, PR China

1.1 Introduction

The protein content in soybean (*Glycine max* (L.) Merrill) seed is approximately 40% and the oil content is approximately 20%. This crop has the highest protein content and the highest gross output of vegetable oil among the cultivated crops in the world. In 2007, the total cultivated area of soybean in the world was 90.19 million ha and the total production was 220.5 million t (FAO, 2009).

The origin of soybean cultivation is China. China was the world's largest soybean producer and exporter during the first half of the 20th century. In the 1950s soybean production developed rapidly in the USA, and the USA is now the largest soybean-producing country in the world. In the 1970s soybean production developed in Brazil, and this country is now the second largest soybean-producing country. Since then, soybean production developed rapidly in Argentina, now the third major soybean-producing country. These soybean-producing countries use machines in commercial production and the commodity rate of soybean is high. They are therefore not only large producers of soybean, but also large exporters. Soybean production in India is developing rapidly and the cultivated area of soybean is about the same as in China, but the yield per unit area is still relatively low.

The cultivated area of soybean in China in 2007 was 8.90 million ha, the total production was 13.80 million t and the yield per unit area was 1550 kg ha^{-1} (FAO, 2009). Compared with the large soybean-producing countries, the yield of soybean in China is obviously low. The main reasons for this are that the scale of soybean cultivated by farmers is small and, therefore, advanced cultural practices have not been adopted. Along with economic developments and improvements in people's living standards, the demand for soybean in China is increasing rapidly and the domestic production of soybean cannot meet these demands. China began importing soybean in 1996 and is now the largest soybean importer in the world.

The Chinese people are accustomed to eating soybean. Traditional soybean products such as bean curd (tofu), soybean milk, dried rolls of bean milk cream, soy sauce and so on are favoured foods to Chinese people. The demand for soybean products and the amount of processed soybean products is continuously increasing. The raw materials used for processing soybean products are non-transgenic soybean produced in China. Most of the commercially imported soybean contains transgenics and is used for extracting oil. The refined oil is used as edible oil; soybean cake is used as feed.

Because China has a long history of growing soybean and a rich array of soybean germplasm has been bred through long-term natural and artificial selection, this provides a rich base for the selection and breeding of soybean varieties and for making a great contribution to soybean production and breeding in the world. China has made extensive improvements in soybean varieties and the high-yield culture techniques of soybean continue to improve. There is still, however, great potential for further improvements in soybean yield.

1.2 The Origin

The evidences of origin of soybean in China

Scholars generally agree that the origin of soybean cultivation is in China. First, the annual wild soybean (*G. soja*), the kindred ancestor of the current cultivated soybean (*G. max*), is found throughout China. The distribution of *G. soja* is limited to China, Japan, Korea and the Far East area of Russia in East Asia, but its distribution in China is the most extensive, its numbers the largest and its diversity of types the richest. Second, China has the earliest written records of soybean cultivation. According to historical records, the Emperor Xuanyuan Huangdi studied the climatic changes in the four seasons and cultivated five kinds of crops: panicgrass (*Panicum antidotale*), broomcorn millet (*P. miliaceum*), beans, wheat (*Triticum aestivum*) and rice (*Oryza sativa*). This was about 4500 years ago. 'Shu', the name of soybean expressed in Chinese characters in ancient times, can be found in many ancient Chinese books. An initial word expressing soybean appeared in inscriptions on unearthed bones and tortoise shells of the Yin and Shang Dynasties (3700 years ago). Third, soybean has been found in unearthed artefacts. Carbonized soybean seeds were found during the excavation of the 2600-year-old Dahaimeng site in Yongji County, Jilin Province. The remains of soybean seeds have also been excavated from the site of a primitive society in Damudan Tun Village, Ningan County, Heilongjiang Province (3000 years old). In the site of the Eastern Zhou Dynasty in Niucun Village, Houma City, Shanxi Province, the remains of soybean seeds have been excavated from a cellar for storing foods. These remains are estimated to be 2590 years old, based on carbon dating. These soybean seeds are yellow in colour and are now preserved in the Beijing

Museum of Natural History. The remains of soybean seeds have also been excavated from the Shaogou Han Dynasty Tomb in Luoyang City, Henan Province, from the Mawangdui Han Dynasty Tomb in Changsha City, Hunan Province, and from the No. 168 Han Dynasty Tomb in Fenghuangshan Mountain, Jiangling County, Hubei Province. Finally, soybeans cultivated in different countries in the world were introduced directly or indirectly from China. The pronunciation of the word of soybean in many countries is about the same as the Chinese 'Shu'; for instance, it is pronounced 'soya' in England, 'soy' in the USA, 'Соя' in Russia and 'ダイズ' in Japan.

While the origin of soybean cultivation may be China, scholars have different viewpoints on the original areas of soybean domestication. The evidence for each theory of origin is summarized and discussed below.

The theory that soybean originated from northeast China

Fukuda (1933), a Japanese scholar, thought that the origin of soybean is northeast China, based on the observations that semi-natural wild soybeans are extensively distributed in northeast China but not in other regions, that there are many soybean varieties in this region and that many of them possess 'original' characteristics. In addition, a record has been found in the ancient Chinese prose Guanzi-Jiepian of Qi Huangong obtaining 'Shu' (soybean) from Shanrong when he sent a punitive expedition against the Shanrong nationality in the north of his state, and since then soybean has been cultivated extensively. According to the dissemination of soybean from Shanrong and the carbonized soybean seeds excavated from Jilin Province, Li (1987, 1994) thought that the origin of soybean should be limited from the northeastern Hebei Province to southeastern areas of northeast China.

Fukuda (1933) stated that the extensive distribution of semi-natural wild soybeans in northeast China, while only few are known from other areas, might well be influenced by differences in efforts to investigate and collect materials. In fact, many small black soybean germplasm have 'primitive' traits, and these are extensively distributed in the lower and middle reaches of the Yellow River, especially in North Shaanxi and North Shanxi provinces. Therefore, their distribution area is much larger than northeast China alone. Maliao Dou and Nidou (*G. max* L.), which are closely related to semi-natural soybean, are even distributed as far south as the Yangtze River valley. As for the large number of soybean varieties, Fukuda indicated that the number of soybean varieties in the Yangtze River valley is also very large and that Shanxi and Shaanxi provinces alone already have 3000 accessions of germplasm resources of soybean, which is far more than from northeast China. Next to these spring-type soybeans, the number of varieties of summer-planting types of soybean in the Yangtze River valley is also very large.

As mentioned above, Qi Huangong (685–643 BC), a powerful leader of feudal lords in the Spring and Autumn Periods (770–476 BC), obtained 'Shu' from Shanrong. However, this record of soybean is 400 years younger than that in the records of the Western Zhou Dynasty. Other records also indicate that soybean from Shanrong was distributed across ancient China, but a plausible explanation is that soybean from Shanrong was a good soybean variety that also had good adaptability, which made it appropriate for extensive distribution.

The theory that soybean originated from the Huanghuai region of China

Among the eight independent origin regions of major cultivated crops in the world defined by Vavilov (1982), the largest is formed by the central part of China and the mountainous areas of western China and their adjacent low-lying lands. Vavilov pointed out that the most important characters of origin of cultivated crops in China are the large amount of cultivated crops and three species of cereal crops. The most important indigenous species of the temperate region are buckwheat (*Fagopyrum esculentum* Moench), soybean and various pulse crops. Several thousand genetic types with obvious differences can be identified for soybeans, persimmons and citruses. For Vavilov it was clear that soybean is a temperate-zone species and that its origin is the mountainous areas in central and western China and their adjacent low-lying lands (i.e. the middle reach of the Yellow River).

Johnson (1980) stated in *The Encyclopedia Americana* that 'The ancient Chinese documents consider that soybean had been cultivated widely due to its high nutritional value before recorded in detail in literature. Soybean was treated as an important crop as early as 2000 BC and it is one of the five cereals of the base of Chinese civilization.' Cuzin (1976) wrote in *Bolshaya Sovetskaya Entsiklopediya* that 'The origin of soybean is China. China began growing soybean 5000 years ago and this crop is introduced from China to the south and south east Asia, and then it is introduced into Europe in 18th century.'

Hymowitz (1970), a scholar from the USA, thought that the origin of soybean was the eastern part of northern China, which he referred to as the winter wheat (*T. aestivum*)–sorghum (*Sorghum bicolor*) growing region (i.e. the lower reaches the Yellow River). He thought that wild soybean was cultivated in this region and that wild soybean can be seen everywhere. He also made researches on the ancient Chinese character 'Shu' (菽). The earliest ancient Chinese character 'Shu' was '尗', in which the horizontal stroke in the centre means the ground, the vertical strokes on and above the horizontal stroke are the stem and root of soybean and the dots mean the root nodules. The ancient Chinese character 'Shu' (菽) can be traced back to the 11th century BC.

The blooming dates of wild soybean and cultivated soybean are the same at 35°N, but differ when going further north or south. Therefore, 35°N is the turning point of the photoperiodic response of soybean and cultivated soybean varieties may have been derived from local wild soybean at around 35°N. In addition, the protein content of cultivated soybean is close to that of wild soybean at 34–35°N. The original cultivated soybean seems to be evolved from wild soybean in the Yellow River valley.

Wang (1985) studied the origin of soybean by using ancient Chinese literature, inscriptions on bones and tortoise shells of the Shang Dynasty and so on, and concluded that the earliest region for cultivating soybean was around the middle or downstream of the Yellow River. Chang (1989) stated that the origin of soybean is the Yellow River valley, based on his study of the relationship between the origin of agriculture and the origin of soybean, in which he unearthed cultural relics in archaeological studies and combined the evolutionary process of soybean with plant ecology and botany. As a professional researcher in agricultural heritage, Guo (1993) systematically collected literature related to soybean in past dynasties of China and wrote the book *The History of Soybean Cultivation in China*. He analysed the arguments related to the origin of soybean in ancient literatures and thought that the origin of cultivated soybean in China is northern China, but that the exact origins of soybean remain unknown. He gave various possible locations, including northeast China, north China, the central Shaanxi plain and the Yellow River valley. He thought that these arguments are not conclusive in pinpointing the exact location and that, therefore, further research is needed.

The relationship between the origin of agriculture and the origin of soybean is an important argument with regard to the origin of soybean in the Yellow River valley. The ancient Chinese civilization originated along the middle reaches of the Yellow River and is closely related to the occurrence and development of dryland farming in northern China. In the Neolithic Age, as we can deduce from painted pottery, the sites that mankind inhabited were concentrated in the foot hills or loess platform near the Yellow River. The Yangshao and Banpo cultures were located in these areas and panicgrass (*P. antidotale*), broomcorn millet (*P. miliaceum*), bast-fibre plant (*Linum usitatissimum* and *Cannabis sativa*) and other dry crop seeds have been unearthed from these sites. The civilization in the Yangtze River valley is closely related to the appearance and development of wet farming. Rice seeds have been unearthed from the sites of the Hemudu culture, Qujialing culture and other sites of culture. All of these facts relate the dry crop soybean to the origin of dry farming in northern China.

Scientific investigations on wild soybean in the Yellow River valley have found an abundance of wild soybean in this area and much variation in the seeds of wild soybean. Large-seeded wild soybean has been found in investigations made along the Yellow River in Shanxi Province. For example, one wild soybean accession collected on the banks of the Yellow River in Yongji County had a 100-seed weight of 4–5 g, clear differentiation of the main stem and large leaves. In the investigations, wild soybeans with yellow, green, brown and black seed coats were collected, and also no-bloom seeds. Thus, in this region the wild soybean has extensive variation, which is a prerequisite for domestication of wild soybean into cultivated soybean.

The theory that soybean cultivation originated in south China

Based on the wide distribution of wild soybean in southern China, primitive soybean varieties such as Nidou, Maliao Dou, Xiao Huangdou and others cultivated in south China, the abundance of initial varieties of soybean in south China and the short-day character, which is considered to be the initial physiological state of soybean, Wang (1947) thought that the origin of soybean could be south China. After analysis of the photoperiod of wild soybean of China, Wang *et al.* (1973) found that the wild soybeans in the Yangtze River valley and its southern areas had the strongest initial short-day character, and they therefore considered south China to be the centre of the origin of soybean.

Gai *et al.* (2000) studied the origin and evolution of soybean by comparing biological characteristics. They thought that the key is the typical sample (including the size of the sample) obtained; the selected characteristics of the typical sample should reflect the whole process of evolution. They therefore thought that the agronomic characteristics related to yield and quality – which are the objectives of improvements by humans and which are affected greatly by current artificial selection – cannot be used to trace the conditions in ancient times. In a study that focused on 11 morphological characteristics of wild soybeans that are less influenced by artificial selection, they compared 250 accessions of typical cultivated and wild soybeans with local cultivated soybean varieties, representing the six geographical and season-ecological populations. They also used isoenzymes and restriction fragment length polymorphism (RFLP) markers of chloroplast and mitochondrial DNA in the study. The results showed that cultivated soybeans in southern China, especially the later-maturing types, are much closer to wild soybeans in genetic terms than cultivated soybeans in northeast China or the Huanghuai region. Therefore, the wild soybean in south China might be the common ancestor of cultivated soybeans, from which the various early-maturing types have been derived during the process of their dissemination to the north. Their further analysis of single sequence repeat (SSR) data and botanical traits confirmed genetic differentiation related to the geographic region of the sources, and genetic diversity of the south China population was higher than that of both northern and Huanghuai populations (Ding *et al.*, 2008).

The theory of multiple origins

Lü (1978) provided three arguments as to why soybean might not have originated from one region, but from several regions. First, both south and north China have regions with early developed cultures, and he thought it natural that the ancients in different areas used local wild soybean as food. Therefore, it is not unlikely that they would have domesticated wild soybeans into cultivated ones. Second, there is evidence from the occurrence of wild soybean and cultivated soybean in the same regions, and both species have similarities in morphological characters. Third, the characters of strong and weak short-day wild soybean enabled its cultivation in different regions across China. In Lü's opinion, the geographical distribution of the short-day character of wild soybean indicates the possibility of multiple origins of cultivated soybean.

1.3 Evolution

Classification and distribution of perennial species

The genus *Glycine* is thought to be of ancient polyploid origin due to the high chromosome number of the majority of the species (n = 20) compared to closely related genera (mostly n = 10 or 11, one with n = 14; Goldblatt, 1981). Additional lines of evidence exist, including cytogenetic studies in haploid *G. max* (Crane *et al.*, 1982), supporting this hypothesis of polyploid origin. Schuelter *et al.* (2004) found that the *Glycine* genome has gone through two major rounds of duplication, the first estimated at 41.6 million years ago and another at 14.5 million years ago. Van *et al.* (2008) looked at evolutionary events, revealing that the recent divergence of two soybean homoeologous regions occurred at 60 and 12 million years ago, respectively. The type of polyploidy was tested and discussed by Doyle *et al.* (2003). Clarindo *et al.* (2007) found that the karyograms support soybean's tetraploid nature (4× = 40), specifically for the presence of chromosomes with identical morphology, and suggested that chromosome rearrangements may have occurred during the speciation of *G. max*.

The genus *Glycine* Willd. is divided into two subgenera, *Glycine* (perennials) and *Soja* (Moench) F.J. Herm. (annuals). A list of species of the genus *Glycine* is presented in Table 1.1.

The perennial species are extremely diverse in morphology, cytology and genome composition. They grown in very diverse climatic and soil conditions and have a wide geographic distribution. The species have been screened for many physiological and biochemical traits as well as for sources of resistance to economic pathogens. Some perennial *Glycine* species are sources of resistance to soybean cyst nematode and a source of lack of Bowman-Birk protease inhibitor (Hymowitz, 2004).

Distribution of annual wild soybean

Taxonomically, both the annual wild soybean (*G. soja* Sieb. & Zucc.) and the cultivated soybean (*G. max* (L.) Merrill) are subgenera of *Soja*. The wild soybean was named *G. ussuriensis* by Regel and Maack, and this name was commonly used until 1979. Vordcourt advocated *G. soja* as the scientific name for the annual wild soybean to conform to the formal procedure as the name *G. soja* is older (Hymowitz and Newell, 1981).

The distribution of wild soybean in China is extensive. Results of investigations have shown that the distribution of the wild soybean is from Mohe in Heilongjiang at 53°N in north China to Guangdong's Shaoguan region at 24°N in south China; from Gansu's Jingtai County at about 104°E in northwest China to Tibet's Chayu County at about 97°E in southwest China; and from the banks of the Wusulijiang River at 135°E in northeast China to the north part of Taiwan Province in the southeast. With regard to altitude, the upper limit in northeast China is about 1300 m, while it is 1500–1700 m in

Table 1.1. Species in the genus *Glycine*, together with the 2n chromosome number, genome symbol and geographical distribution (reprinted with permission from Hymowitz, 2004).

Species	2n	Genome[a]	Distribution
Subgenus *Glycine*			
1. *G. albicans* Tind. & Craven	40	I_1	Australia
2. *G. aphyonota* B. Pfeil	40	?	Australia
3. *G. arenaria* Tind.	40	HH	Australia
4. *G. argyrea* Tind.	40	A_2A_2	Australia
5. *G. canescens* F. J. Herm.	40	AA	Australia
6. *G. clandestina* Wendl	40	A_1A_1	Australia
7. *G. curvata* Tind.	40	C_1C_1	Australia
8. *G. cyrtoloba* Tind.	40	CC	Australia
9. *G. dolichocarpa* Tateishi & Ohash	80	?	(Taiwan)
10. *G. falcata* Benth.	40	FF	Australia
11. *G. hirticaulis* Tind. & Craven	40	H_1H_1	Australia
	80	?	Australia
12. *G. lactorirens* Tind & Craven	40	I_1I_1	Australia
13. *G. latifolia* (Benth.) Newell & Hymowitz	40	B_1B_1	Australia
14. *G. latrobeana* (Meissn) Benth.	40	A_3A_3	Australia
15. *G. microphylla* (Benth.) Tind.	40	BB	Australia
16. *G. peratosa* B. Pferl & Tind.	40	?	Australia
17. *G. pindanica* Tind. & B. Craven	40	H_2H_2	Australia
18. *G. pullenii* B. Pfeil. Tind. & Craven	40	?	Australia
19. *G. rubiginosa* Tind. & B. Pfeil	40	?	Australia
20. *G. stenophita* B. Pferl & Tind.	40	B_3B_3	Australia
21. *G. tabacina* (Labill.) Benth.	40	B_2B_2	Australia
	80	Complex[b]	Australia
22. *G. tomentella* Hayata	38	EE	South Pacific Islands, south China, Australia
	40	DD	Australia, Papua New Guinea, south China
	78	Complex[c]	Australia, Papua New Guinea
	80	Complex[d]	Australia, Papua New Guinea, south China
Subgenus *Soja* (Moench) F. J. Herm.			
23. *G. soja* Sieb. & Zucc.	40	GG	China, Russia, Japan, Korea (wild soybean)
24. *G. max* (L.) Merr.	40	GG	Cultigen (soybean)

[a]Genomically similar species carry the same symbols.
[b]Allopolyploid (A and B genomes) and segmental allopolyploid (B genomes).
[c]Allopolyploid (D and E, A and E genomes or any other unknown combination).
[d]Allopolyploid (A and D genomes or any other unknown combination).

the Yellow River and Yangtze River valleys. The uppermost limit of the distribution of wild soybean is 2650 m in Yunnan's Ninglang County. In an analysis of four natural distributed wild soybean populations from northeast China, the results indicated that genetic patches were on average approximately 20 m^2 in size, while the effective neighbourhood sizes varied between 10 and 15 m^2 (Jin *et al.*, 2006).

Biology of domestication of wild soybean to cultivated soybean

The wild soybean was domesticated by ancient people under certain agricultural conditions. The first piece of evidence is that the number of chromosomes of both the cultivated soybean and the wild soybean is 2n = 40. The chromosome set is GG. If we cross the cultivated soybean with the wild soybean, the fertility and seed-setting percentage of the F_1 generation are normal and there is no obvious difference as compared to crosses within cultivated soybeans. This shows that there is no isolation between the cultivated soybean and the wild soybean and that they are (at the very least) close relatives.

The second piece of evidence is that when the cultivated soybean is crossed with the wild soybean, the seed size, plant height, lodging and other traits are inherited as quantitative traits, with some intermediate types occurring, which show that the two groups accumulated minor variants of the underlying genes. Third, new variations always occur while growing soybean in the field. For instance, early-maturing variants have been identified in late-maturing varieties. Investigations on wild soybean have found that early-maturing variants with large seeds and thick stems are also minor variations of quantitative traits. The fourth piece of evidence is that among the rich germplasm resources in China there are wild soybeans, semi-wild soybeans and highly evolved cultivated soybeans. All of these germplasm resources of soybean, with different degrees of evolution, adapt to different natural environments, cultural conditions and utilization requirements. The evolution of soybean is clearly a continuous accumulation of minor variations and a continuing process from quantitative variation to qualitative variation.

More evidence comes from molecular data. Within the wild species of subgenus *Glycine*, considerable differences in repeat size occur in several species, but no variation of ribosomal DNA-RFLP has been found in >40 accessions of the two species between the cultivated soybean and its wild progenitor, *G. soja* (Doyle and Beachy, 1985). Both *G. max* and *G. soja* are close in their genome structure, detected by simple repetitive sequences (Yanagisawa *et al.*, 1994). The subgenus *Soja*, comprised of two highly variable species (*G. max* and *G. soja*), was confirmed by RFLP of chloroplast DNA variation (Shoemaker *et al.*, 1986; Close *et al.*, 1989; Abe *et al.*, 1999), genomic DNA variation (Keim *et al.*, 1989), random amplified polymorphic DNA data (Chen and Nelson, 2004) and single nucleotide polymorphisms of *GmHs1*$^{pro\text{-}1}$ (Yuan *et al.*, 2008), as well as by SSR data (Powell *et al.*, 1996).

What about the process of the first domestication of soybean? To start with, variation occurred in the ancestor of cultivated soybean, the wild soybean, through natural selection under the natural environment. For instance, in investigations on wild soybean, variants with large seeds, clear differentiations of the main stem and early flowering have been found and these variants have been differentiated from the typical wild soybean. Under artificial cultural conditions, according to the demands of usage, these minor variations have probably been accumulated by people through continuous selection. Artificial selection has further promoted the differentiation of these traits into the soybean types we currently know.

People mainly use soybean seeds. In long-term production activity and use, people have focused on the selection of large seeds. Selection of one trait inevitably results in corresponding changes in other traits. While the seed of soybean has been enlarged, correspondingly the pod has been enlarged, the plant height reduced and the stem thickened. Reductions in plant height have been favourable for the development from vine type to vertical type. The vertical-type plant is easy to manage and good management conditions have promoted the selection of strong and lodging-resistant types, thus promoting further the evolution of traits.

Shu *et al.* (1986) have made comparative studies on the traits of wild, semi-wild and cultivated soybeans. The results show that from the wild soybean to the cultivated soybean, the most significant change of traits is in seed size. The 100-seed weight has increased from 1.61 g for the wild soybean to 15.14 g for the cultivated soybean: a 9.4-fold increase. But as the number of seeds per plant has decreased by 8.24 times, the seed weight produced per plant is only 32% more. The pod size and leaf area have increased by 4.7 and 2.6 times, respectively. The plant height has decreased by 2.6 times and the number of branches has decreased by 2.73 times, but the seed number per pod is practically unchanged. The reproductive period has been lengthened, which is favourable to the accumulation of dry matter, thus enlarging the seed (Table 1.2).

The genetic differentiation and diversity from wild soybean to cultivated soybean have been observed at DNA sequence level, including the soybean Kunitz trypsin inhibitor (*SKTI*) gene (Wang *et al.*, 2005, 2008b) and acyl coenzyme A-dependent diacylglycerol acyltransferase (*GmDGAT*) (Wang *et al.*, 2006), 11S globulin molecular (Zakharova *et al.*, 1989) and glycinin subunit genes (Wang *et al.*, 2008a). Variations in storage proteins (Natarajan *et al.*, 2006), major seed allergens (Xu *et al.*, 2007) and Kunitz trypsin inhibitors (Natarajan *et al.*, 2007) in wild (*G. soja*) and cultivated (*G. max*) soybean seeds has been observed using proteomic analysis. Most of the above results appear to indicate higher genetic diversity in the wild soybean than in the cultivated soybean.

Many elite traits in wild soybean (*G. soja* Sieb. & Zucc.) have been identified, such as tolerance to salt (Luo *et al.*, 2005; Yang *et al.*, 2007), chilling stress and dehydration stress (Chen *et al.*, 2006), a high lutein content (Kanamaru *et al.*, 2006) and so forth, which can be used in breeding programmes. The wild soybean also can be used for producing fertile hybrids between domestic and wild soybeans (Singh, 2007).

Table 1.2. Comparison of the evolution of wild and semi-wild soybeans (reprinted with permission from Shu *et al.*, 1986).

				Comparison of three types (ratio)[a]		
Trait	Wild $\bar{x}$	Semi-wild $\bar{x}$	Cultivated $\bar{x}$	Wild	Semi-wild	Cultivated
Leaf area (cm^2)	28.9	56.45	85.62	1.00	2.00	3.00
Leaf length (cm)	7.45	10.28	11.42	1.00	1.38	1.53
Leaf width (cm)	3.63	5.43	7.27	1.00	1.50	2.00
Plant height (cm)	248.87	162.41	95.81	2.60	1.70	1.00
No. of nodes	35.41	25.90	18.88	1.87	1.37	1.00
No. of branches	18.60	12.19	6.81	2.73	1.79	1.00
Pod size (cm × cm)	2.14 × 0.45	3.21 × 0.67	4.53 × 0.99	1.00	2.00	4.70
Pod no. per plant	1253.89	525.19	168.84	7.43	3.11	1.00
Seeds per plant	2304.61	930.58	279.86	8.24	3.33	1.00
Seed weight per plant (g)	34.32	44.73	45.27	1.00	1.30	1.32
Seed no. per pod	1.89	1.83	1.81	1.04	1.01	1.00
100-seed weight (g)	1.61	5.37	15.14	1.00	3.34	9.40
Emergence: flowering (days)	92.67	62.45	68.19	1.48	1.00	1.09
Flowering: maturation (days)	60.70	76.01	74.92	1.00	1.25	1.23
Growing period (days)	158.16	144.06	140.20	1.13	1.03	1.00

[a]The ratio is calculated relative to the smallest value of the three groups.

1.4 Distribution

Data pertaining to area, production and yield of soybean in major soybean-growing countries are presented in Table 1.3.

Asia

Asia has the longest history of growing soybean, and the cultivated area of soybean in China is the largest in the world. Soybean is also cultivated in Japan, the Republic of Korea (South Korea), the Democratic People's Republic of Korea (North Korea), Indonesia, Thailand, Vietnam and other countries. Most of the varieties of soybean in Japan are large-seed types and are used as vegetable soybean, which is called *edamame* in Japan. The 100-fresh-seed weight is >70 g and the 100-dry-seed weight is >30 g. The 100-seed weight of another kind of small-seed soybean is <10 g and is used for the production of *nattō*. Most of the varieties of soybean in Korea are

Table 1.3. The area, production and yield of soybean in selected countries in 2006 and 2007 (adapted from FAO, 2009).

	Area (ha)		Production (t)		Yield (kg ha^{-1})	
Country	2006	2007	2006	2007	2006	2007
United States of America	30,190,680	25,960,000	83,510,000	72,860,400	2,766	2,806
Brazil	22,047,349	20,565,300	52,464,640	57,857,200	2,379	2,813
Argentina	15,130,038	15,981,264	40,537,364	47,482,784	2,679	2,971
China	9,100,085	8,900,068	15,500,187	13,800,147	1,703	1,550
India	8,334,000	8,880,000	8,857,000	10,968,000	1,062	1,235
Paraguay	2,200,000	2,429,000	3,800,000	5,856,000	1,727	2,410
Canada	1,201,200	1,171,500	3,465,500	2,695,700	2,885	2,301
Bolivia	950,118	958,279	1,618,966	1,595,947	1,703	1,665
Ukraine	714,800	583,100	889,600	722,600	1,244	1,239
Russian Federation	810,130	709,900	806,570	651,840	995	918
Indonesia	580,534	459,116	747,611	592,634	1,287	1,290
Uruguay	309,100	366,535	631,900	779,920	2,044	2,127
Nigeria	630,000	638,000	605,000	604,000	960	946
Italy	177,909	132,604	551,292	442,151	3,098	3,334
South Africa	240,570	183,000	424,000	205,000	1,762	1,120
Serbia, Republic of	156,680	146,988	429,639	303,950	2,742	2,067
Korea, Democratic People's Republic of	300,000	300,000	345,000	345,000	1,150	1,15
Romania	177,481	109,314	344,909	136,094	1,943	1,244
Vietnam	185,600	190,100	258,100	275,500	1,390	1,449
Iran, Islamic Republic of	81,775	110,000	184,967	260,000	2,261	2,363
Japan	142,100	138,300	229,200	226,700	1,612	1,639
Thailand	137,640	128,872	214,773	203,973	1,560	1,582
Croatia	62,810	46,506	174,214	90,637	2,773	1,948
Korea, Republic of	90,248	76,267	156,404	114,245	1,733	1,497
Uganda	145,000	147,000	175,000	176,000	1,206	1,197
France	45,263	37,000	122,995	102,000	2,717	2,756
Myanmar	122,000	123,000	120,000	122,000	983	991

medium- and small-seed types; the 100-seed weight is <15 g and these soybeans are used for the production of bean sprouts.

In Indonesia *danbei*, a food produced with fermented soybean, is popular. India and Nepal produce *kinema* with fermented soybean. Bean curd is a popular food in China, Japan and Korea. Soybean used for production of bean curd should have a high protein content; in particular, the content of water-soluble protein should be high.

Americas

Hymowitz (1984) pointed out that soybean was cultivated in the USA as early as 1765, when Samuel Bowen, a sailor from the East India Company, brought soybean from China to Savennah (Georgia). Benjamin Franklin was the second to introduce soybean to the USA by transporting it from France to Philadelphia in 1770 when he was an ambassador to France. During 1927–1931, the USA twice sent scientists to China, Korea and Japan to collect soybean germplasm, and these scientists collected several thousand accessions of soybean from these countries. Some of the germplasms have become the primary parents of soybean breeding in the USA. The earliest report on the introduction of soybean to South America was in 1882.

The USA in North America and Brazil and Argentina in South America rank in the first three positions for the cultivated area and production of soybean. According to the latest statistics available (FAO, 2009), in 2007 the cultivated area of soybean in the USA was 25.96 million ha and the production of soybean was 72.86 million t. In the same year, the cultivated area of soybean in Brazil was 20.56 million ha and the total production of soybean was 57.85 million t. The cultivated area of soybean in Argentina was 15.98 million ha and the total production of soybean was 47.48 million t. The area used for soybean cultivation in South America increased by about 50% during the period 2000–2006, and production is now higher than that of North America.

The rest of the world

Due to the climatic reasons the cultivated area of soybean in Europe is not very large. Ukraine, Russia, Italy, Romania, Serbia, Croatia and France are all soybean-producing countries, and the production of soybean in these countries is >100,000 t. The yield of soybean per unit area in Italy is >3 t ha^{-1}. A small amount of soybean is produced in Australia (50,000 t $year^{-1}$).

The cultivated area of soybean in Africa is not large. Nigeria has a large area under soybean, followed by South Africa, Uganda, Zimbabwe, Congo, Zambia and others. Africa has great potential in the development of soybean and needs support and help from the major soybean-producing countries. Supports should be given not only in the introduction of varieties and cultural techniques, but also in the processing and utilization of soybean and in the production of soybean foods suitable for consumption by the local people.

1.5 Soybean in China

Distribution

Three soybean growing regions in China can be distinguished according to the cropping system. Within these, ten subregions can be identified according to the climatic conditions and geographical features:

- The north spring-sowing soybean subregion (the north region):
  1. The northeast spring-sowing soybean subregion (the northeast region).
  2. The north plateau spring-sowing soybean subregion (the north plateau subregion).
  3. The northwest spring-sowing soybean subregion (the northwest subregion).
- The Huang Huai Hai valleys summer-sowing soybean subregion (the Huang Huai Hai subregion):
  4. The central Hebei-Shanxi summer- and spring-sowing soybean subregion (the central Hebei-Shanxi subregion).
  5. The Huang Huai Hai valleys summer-sowing soybean subregion (the Huang Huai Hai subregion).
- The south China multiple-sowing soybean region (the south region):
  6. The Yangtze River valley spring- and summer-sowing soybean subregion (the Yangtze River valley subregion).
  7. The southeast autumn- and spring-sowing soybean subregion (the southeast subregion).
  8. The central-south spring-, summer-, and autumn-sowing soybean subregion (the central-south subregion).
  9. The southwest plateau summer-sowing soybean subregion (the southwest subregion).
  10. The south China multiple-sowing soybean subregion (the south China subregion).

The northeast spring-sowing soybean region is the largest soybean-producing region in China. This region includes Heilongjiang, Jilin, Liaoning and the greater part of Inner Mongolia. Soybean is sown here in spring (from the last ten days of April to the first ten days of May) and harvested in autumn (from the middle ten days to the last ten days of September) (i.e. one crop year^{-1}). The cultivated area in 2006 was 4.863 million ha, which accounted for 52.4% of the total area of soybean in China. The total production of soybean was 8.548 million t, which accounted for 53.5% of the total production of the country. The cultivated area and the total production of soybean in Heilongjiang province ranked the first (37.0% and 37.3%, respectively).

The Huang Huai Hai summer-sowing soybean region is the second-largest soybean-producing region. This region includes Shandong, Henan, the central-south part of Hebei Province, the north part of Jiangsu and Anhui provinces, the central-south part of Shanxi Province and the Shaanxi plain area. Soybean is sowed from the middle ten days to the last ten days of June as the second crop after the harvest of winter wheat. Soybean is harvested from the last ten days of September to the first ten days of October before winter wheat sowing. The cultivated area in 2006 was 2.735 million ha (29.5% of the total area of soybean in China), producing 4.254 million t (26.6% of China).

The south China multiple-sowing soybean region includes the provinces south of the Yangtze River. This region has spring-, summer- and autumn-sowing soybeans. The spring soybean is sown in the Yangtze River valley from March to the first ten days of April and harvested from the first ten days to the middle ten days of July. Late rice or winter wheat are planted after the soybean harvest or summer soybean is sown after the harvest of winter rapeseed (*Brassica* species). The summer soybean is sown from the last ten days of May to the first ten days of June and harvested in October. The autumn soybean is sown after the harvest of early rice from the last ten days of July to early August and harvested in the first ten days of November. The cultivated area of soybean in the south region in 2006 was 1.502 million ha, producing 2.812 million t.

Data pertaining to area, production and yield of soybean in major soybean-growing provinces in China are presented in Table 1.4.

Table 1.4. The cultivated area, total production and yield of soybean in provinces with cultivated areas >0.15 million ha in 2006 (National Statistical Bureau in China, 2009, personal communication).

Province	Cultivated area (10,000 ha)	Total production (10,000 t)	Yield (kg ha^{-1})
The whole country	928.01	1596.7	1721
Heilongjiang	343.68	596.0	1734
Anhui	96.30	125.0	1298
Inner Mongolia	75.45	104.5	1385
Henan	51.63	64.9	1257
Jilin	44.84	121.4	2707
Shaanxi	32.03	42.3	1321
Hebei	23.80	44.6	1874
Shanxi	22.61	27.7	1225
Shandong	22.40	62.1	2772
Liaoning	22.31	32.9	1475
Jiangsu	21.43	53.7	2506
Guangxi	20.55	30.4	1479
Sichuan	20.00	39.3	1965
Hunan	18.35	42.6	2322
Hubei	17.16	38.5	2244

Utilization

The consumption of soybean in China was about 44 million t in 2006, but the country produced 15.50 million t only. That year China imported 28 million t of soybean from America. Soybean used directly for food was about 8.5 million t, mainly in the form of traditional soybean products such as bean curd, soybean milk, soybean paste and bean curd stick. Less than 0.4 million t was used for the production of modern processed products such as soybean milk powder and soybean protein powder. Soybean used for foods is domestically produced non-transgenic soybean. The largest amount of soybean was used to extract oil – about 34.70 million t, mainly from imported transgenic soybean. The amount of domestically produced soybeans used for oil was only 6.3 million t. Approximately 0.9 million t was used as seed soybean. Nearly 0.3 million t of non-transgenic soybean was exported.

Breeding and cultivation

Soybean varietal improvement in China began in 1913 when the Gongzhuliang Agricultural Experimental Station was established in Jilin Province and began collecting local soybean varieties. The variety 'Huanbaozhu' was bred in 1923 through pure line selection, and then 'Fengdihuang', 'Xiaojinhuang No. 1' and other varieties were released. In the 1930s 'Mancangjin', 'Mandijin', 'Yuanbaojin' and other varieties were bred using sexual hybridization. 'Mancangjin' and 'Xiaojinhuang No. 1' were the main cultivars in northeast China in the 1950s. The cultivated area of 'Mancangjin' reached 1 million ha.

Since 1949, along with the development of soybean production, research on soybean was also strengthened and developed. Two large-scale collection activities were carried out in 1956 and 1980, in which 15,000 and 10,000 accessions of soybean germplasm were collected, respectively. Currently, a large number of accessions of soybean germplasm, including improved varieties, native varieties and annual wild soybeans, are stored in the National Gene Bank. In addition, 2500 accessions of introduced varieties are stored. This abundant germplasm resource of soybean provides the basis for soybean breeding.

China has bred >1200 soybean varieties since 1949 and there are >100 scientific research institutions engaged in soybean breeding. In recent years, seed companies have also started carrying out research on soybean breeding. The main method for the improvement of soybean varieties is cross-breeding and the main selection approaches are the pedigree and single-seed descent methods. Regional experiments on soybean varieties are undertaken in each province and the varieties are spread across the province through regional experiments. The regional experiments on soybean varieties at state level can be divided into 13 experimental groups according to the cultivation and ecological regions of soybean. The varieties are released and used in production after being approved by the National Crops Varieties Examination

Committee. These regional experiments at state and provincial levels have formed a system that has guaranteed the popularization and utilization of improved varieties.

The improved varieties have high yield capacity, high disease resistance and abiotic stress tolerance, and the quality of soybeans has also improved greatly. 'Yuejin No. 5', 'Hefeng No. 25', 'Zhonghuang 13', 'Suinong 14', 'Heihe 27', 'Yudou 22' and other good varieties exhibit high yield capacity and high stress tolerance. The accumulated cultivated area of 'Hefeng No. 25' has reached >12 million ha since its release in 1984, while the accumulated cultivated area of 'Suinong 14', 'Yuejin No. 5' and 'Zhonghuang 13' has also reached 0.6–0.7 million ha. Most of the improved varieties are resistant to soybean mosaic virus and the spring soybean varieties in northeast China are also resistant to soybean frogeye leaf spot. Varieties with drought and salt tolerance include 'Jindou 21' and 'Zhonghuang No. 10'. Varieties with a protein content >45% or an oil content >23% include 'Yudou 12' and 'Chuandou No. 4' (50.6–50.7% protein) and 'Jihuang 13' and 'Jiyu 67' (23.6–24.1% oil). Variety 'Wandou 12' combines 45.12% protein with a 22.98% oil content. Breeding of varieties with high isoflavones content and the absence of lipoxygenase and non-trypsin inhibitor has also generated improved varieties.

As for cultural practices, the sowing method and field management have changed towards more intensive farming. Levels of mechanized farming increase year by year; machines are extensively used in sowing and harvesting and the yield of soybean has been improved. Model cultural techniques are studied and practised in various soybean-cultivating regions. For instance, the 'ridge three' cultural technique is practised in the northeast region, in which, under the conditions of mechanized farming, the three basic measures – deep loosening, layer by layer fertilizing and precision sowing – are adopted in combination with chemical weeding and disease and insect pest control. This has increased the soybean yield by 15–20%. Dwarf varieties and close planting have also resulted in yield increases of ≥15%.

Industry

The industrial chain of soybean in China has developed along with the economy of the country. Soybean products have gradually changed from crude oil and bean meal to high-value-added products. The traditional soybean processed products are mainly bean curd, bean curd stick and bean curd cheese. The processing of these products has moved to commercial soybean product processing factories. The production of products has increased and the quality of products improved. The production of modern soybean processed products such as soybean milk flour, isolated soy protein, concentrated soy protein and structural protein has also continuously increased, along with the production of functional soybean foods such as phospholipid, saponins, isoflavones, oligosaccharides and edible fibre.

Between 2001 and 2006, the soybean-processing capacity of China doubled to >76 million t. In 2005, there were 95 enterprises with a daily processing capacity of >1000 t. At present, five of the 11 enterprises with daily soybean-processing capability >5000 t in the world are in China. Most soybean-processing enterprises are located in coastal areas and the soybeans they process are mainly imported. The production of isolated and concentrated soy protein is increasing steadily; the process capacity is about 150,000 t $year^{-1}$. The change of the structure of processed soybean products has also promoted the development of emulsified and soluble soybean products such as meat, milk, flour, instant and fast-frozen food, candy and drinks. Commercial production is realized in the extraction of high-value-added phospholipids, saponins and isoflavones and these products can be purchased as functional health foods.

1.6 Development of Soybean Production and Processing Globally

Development of soybean production

In ancient times, the production of soybean in China was mainly concentrated around the middle and lower reaches of the Yellow River. It expanded to the south and north during the Qin and Han Dynasties and then gradually across the whole country. The northeast of China has been the major soybean-producing region since the 19th century. The peak of soybean production was in the 1930s, with a total production of soybean in China of 11.30 million t in 1936. However, production fell and it took half a century for the total production of soybean to reach that level again (11.6 million t in 1986). From 1986 the production of soybean in China increased year by year, reaching 15.5 million t in 2006, although dropping to 13.8 million t in 2007.

Soybean was introduced to Japan through Korea about 2000 years ago, although soybean cultivated in the south of Japan was separately brought by merchant ship from east China. Japan's current demand for soybean is nearly 5 million t, of which 0.2 million t is produced in Japan. About four-fifths of soybean consumption is used for extracting oil and the amount of soybean used for foods (bean curd, soybean milk, *nattō* and so on) is nearly 1 million t. The cultivated area of soybean in the Democratic People's Republic of Korea is 0.3 million ha and the total production of soybean is only 0.35 million t. The cultivated area of soybean in the Republic of Korea is <0.1 million ha, which produces 0.15 million t of soybean. The production of soybean in India is developing rapidly and the cultivated area is currently about the same as that in China. The total production of soybean is nearly 9 million t, which makes India the fifth largest soybean-producing country in the world.

The production of soybean in the USA has developed rapidly since the 1950s. The main producing areas are Iowa, Illinois, Indiana, Missouri, Ohio and some other states in central and western USA. In recent years, the cultivated area of soybean in the USA has grown to >28 million ha, which produces >80 million t.

In 1961, the cultivated areas of soybean in Brazil and Argentina were only 240,000 and 1000 ha, respectively, and the production of soybean was 0.27 million t and 1000 t. Since then, production has expanded amazingly. In 2000, the cultivated areas of soybean in the two countries were 13.64 and 8.64 million ha, respectively, producing 32.73 and 20.21 million t. Even more recently, the production of soybean in Brazil and Argentina reached more than 60 and 40 million t, respectively. According to predictions made by the US Department of Agriculture, the production of soybean in Argentina will reach as much as 48.5 million t in 2008.

Processing and utilization

Soybean has many uses. It is mainly pressed to extract soybean oil, after which a soybean meal remains, which is a rich source of protein. Soybean oil can be used for the production of edible oils such as kitchen oil, salad oil and others through refining and deep processing. Soybean oil is also used for the production of printing ink and biodiesel. Soybean meal is mainly used for the production of compound feed. It is the main protein source in feed for livestock farming. The native soybean meal produced under low-temperature conditions is mainly used for the production of isolated soy protein, concentrated protein and structural protein. These proteins are added to various foods in the food-processing industry for the production of soybean protein-rich foods. For instance, wheat flour is supplemented with a certain amount of soybean protein for the production of bread and cake. Soybean protein supplementation improves the water absorption of meat and the palatability of sausages. Soybean protein can be used to process protein fibre, which can be blended with cotton, wool or chemical fibres. The texture of the resulting fabric is soft and of high quality.

Many soybean food products, including the traditional non-fermented soybean products such as bean curd, soybean milk and bean curd stick, can be processed by using soybean as a raw material. In China, soybean is used to produce Bei bean curd (the coagulating agent is $MgCl_2$), Nan bean curd (the coagulating agent is $CaCl_2$), lactone bean curd (the coagulating agent is gluconolactone) and others through soaking, grinding, boiling and adding different coagulating agents. The fermented soybean products are soy paste, fermented soybeans, soybean cheese, soybean sauce and others. Small seed soybean sprouts are used for making dishes or soup. Soybean sprout soup is common in Korea, while soybean sauce soup is often eaten in Japan.

Along with the depth of research on the nutritional elements of soybean, soybean functional foods such as soy peptide, isoflavones, saponins, phosphatides, sterol, oligosaccharide and edible fibre have been developed. Lactoserum waste water is produced during the processing of bean curd and other products and 2–5 t of lactoserum waste water can be produced from 1 t of soybean. Soybean protein content in lactoserum is 8.2%. Through filtration of lactoserum waste water by using dynamic membranes, 85–93% of the protein can be recovered. The lactoserum protein is a natural surface active agent and can be used for cosmetics. Lactoserum protein is easily digested and assimilated and has a high metabolic rate and biological value. Lactoserum waste water can be used for the extraction of oligosaccharides, which can promote intestinal peristaltics and ease constipation. It also promotes the growth of *Bifidobacterium* and improves the structure of the intestinal bacterial flora.

Isoflavones also can be extracted from lactoserum waste water. Soybean isoflavone consists of flavone glycoside (97–98%) and aglycones (2–3%). Aglycones have biological activity. Isoflavone glycoside is separated from aglycones by the actions of different isoflavone-glucosidases, and the genistein with biological activity is then released. Genistein can attenuate postmenopausal osteoporosis in humans. Isoflavones have inhibitory effects on the early transformation and proliferation of cancer cells. They can effectively inhibit the angiogenesis of a cancer structure and thus block the supply of nutrients to cancer cells. Therefore, isoflavone is of therapeutic use in breast cancer, colon cancer, lung cancer, prostate cancer, leukaemia and others.

Phosphatide, sterone and vitamin E can be extracted from the residues that remain after soybean extraction. The main contents of soybean phosphatide are phosphatidylcholine, phosphatidylethanolamine, phosphatidylinositol, phosphatidylserine and phosphatidic acid. Soybean phosphatide is a natural emulsifier and can be used to supplement the nutrient requirements of the human body; therefore, it is used extensively in the production of candies, biscuits, chocolate, artificial cream and other food products. Soybean phosphatide is a by-product of oil extraction, but as its source is rich and the price is cheap, it has broad prospects for applications in food, medicine and animal production.

Soybean polypeptide is a hydrolyzed product of protein through special treatment. Generally, it consists of peptides of 3–6 amino acids. Soybean polypeptide has a high nutritional value, high digestibility coefficient and low antigenicity, and the results of experiments show that its digestibility coefficient is much better than that of protein or amino acids. Soybean polypeptide can be used as a raw material for or additive to health foods. It has a therapeutic effect on high blood pressure and cardiovascular and cerebrovascular diseases, and is safe and reliable. Soybean polypeptide also decreases the deposition of subcutaneous fat and increases fat burning and it is, therefore, a safe food for people who want to lose weight. Soybean polypeptide also has an antioxidant effect, and it has been claimed that the muscle cells of athletes recover faster when they imbibe a polypeptide-containing drink (Wang *et al.*, 2004).

References

Abe, J., Hasegawa, A. and Fukushi, H. (1999) Introgression between wild and cultivated soybeans of Japan revealed by RFLP analysis for chloroplast DNAs. *Economic Botany* 53, 285–291.

Chang, R.Z. (1989) Studies on the origin of cultivated soybean. *Oil Crop of China* 1, 1–6.

Chen, Y.W. and Nelson, R.L. (2004) Genetic variation and relationships among cultivated, wild, and semiwild soybean. *Crop Science* 44, 316–325.

Chen, Y.Y., Chen, P.Y. and de los Reyes, G. (2006) Differential responses of the cultivated and wild species of soybean to dehydration stress. *Crop Science* 46, 2041–2046.

Clarindo, W.R., de Carvalho, C.R. and Alves, B.M.G. (2007) Mitotic evidence for the tetraploid nature of *Glycine max* provided by high quality karyograms. *Plant Systematics and Evolution* 265, 101–107.

Close, P.S., Shoemaker, R.C. and Keim, P. (1989) Distribution of restriction site polymorphism within the chloroplast genome of the genus *Glycine*, subgenus *Soja*. *Theoretical and Applied Genetics* 77, 768–776.

Crane, C.F., Bwversdorf, W.D. and Bingham, E.T. (1982) Chromosome pairing and associations at meiosis in haploid soybean (*Glycine max*). *Canadian Journal of Genetics and Cytology* 24, 293–300.

Cuzin, V.F. (1976) *Soviet Great Encyclopedia*, Volume 24, Book 1, 3rd edn. Moscow Publishing Company 'Soviet Encyclopedia', Moscow, pp. 281–351.

Ding, Y.L., Zhao, T.J. and Gai, J.Y. (2008) Genetic diversity and ecological differentiation of Chinese annual wild soybean (*Glycine soja*). *Biodiversity Science* 16, 133–142.

Doyle, J.J. and Beachy, R.N. (1985) Ribosomal gene variation in soybean (*Glycine*) and its relatives. *Theoretical and Applied Genetics* 70, 369–376.

Doyle, J.J., Doyle, J.L. and Harbison, C. (2003) Chloroplast-expressed glutamine synthetase in *Glycine* and related Leguminosae: Phylogeny, gene duplication, and ancient polyploidy. *Systematic Botany* 28, 567–577.

FAO (2009) *FAOSTAT*. Food and Agriculture Organization of the United Nations, Rome, Italy. Available at: http://faostat.fao.org (accessed 3 December 2009).

Fukuda, Y. (1933) Cytogenetical studies on the wild and cultivated Manchurian soybeans (*Glycine* L.). *Japanese Journal of Botany* 6, 489–506.

Gai, J.Y., Xu, D.H., Gao, Z., Shimamoto, Y., Abe, J., Fukushi, H. and Kitajiama, S. (2000) Studies on the evolutionary relationship among eco-types of *G. max* and *G. soja* in China. *Acta Agronomica Sinica* 5, 513–520.

Goldblatt, P. (1981) Cytology and the phylogeny of Leguminosae. In: Polhill, R.M. and Raven, P.H. (eds) *Advances in Legume Systematics*, Part 2. Royal Botanic Gardens, Kew, UK, pp. 427–463.

Guo, W.T. (1993) *The History of Soybean Cultivation*. Hehai University Press, Nanjing, China.

Hymowitz, T. (1970) On the domestication of the soybeans. *Economic Botany* 23, 408–421.

Hymowitz, T. (1984) Dorsett-Morse soybean collection trip to East Asia: 50 retrospective. *Economic Botany* 38, 378–388.

Hymowitz, T. (2004) Diversity within the perennial *Glycine* spp. In: *Proceedings of World Soybean Research Conference VII*, pp. 518–521.

Hymowitz, T. and Newell, C.A. (1981) Taxonomy of the genus *Glycine*, domestication and uses of soybeans. *Economic Botany* 35, 272–288.

Jin, Y., He, T.H. and Lu, B.R. (2006) Genetic spatial clustering: Significant implications for conservation of wild soybean (*Glycine soja*: Fabaceae). *Genetica* 28, 1–3.

Johnson, H. (1980) *The Encyclopedia Americana*, International Edition. Americana Corporation, Volume 25, pp. 348–351.

Kanamaru, K., Wang, S.D. and Abe, J. (2006) Identification and characterization of wild soybean (*Glycine soja* Sieb. et Zecc.) strains with high lutein content. *Breeding Science* 56, 3231–3234.

Keim, P., Shoemaker, R.C. and Palmer, R.G. (1989) Restriction fragment length polymorphism diversity in soybean. *Theoretical and Applied Genetics* 77, 786–792.

Li, F.S. (1987) Discussion about the earliest cultivation area of soybean in China. *China Seeds* 1, 3–4.

Li, F.S. (1994) Study on origin and evolution of soybean. *Soybean Science* 13, 61–66.

Lü, S.L. (1978) Discussion on the original region of cultivated soybean in China. *Scientia Agricultura Sinica* 4, 90–94.

Luo, Q., Yu, B. and Liu, Y. (2005) Differential sensitivity to chloride and sodium ions in seedlings of *Glycine max* and *G. soja* under NaCl stress. *Journal of Plant Physiology* 162, 1003–1012.

Natarajan, S.S., Xu, C. and Bae, H. (2006) Characterization of storage proteins in wild (*Glycine soja*) and cultivated (*Glycine max*) soybean seeds using proteomic analysis. *Journal of Agricultural and Food Chemistry* 54, 3114–3120.

Natarajan, S.S., Xu, C.P. and Bae, H. (2007) Proteomic and genomic characterization of Kunitz trypsin inhibitors in wild and cultivated soybean genotypes. *Journal of Plant Physiology* 164, 756–763.

Powell, W., Morgante, M., Doyle, J.J., McNicol, J.W., Tingey, S.V. and Rafalski, A.J. (1996) Genepool variation in genus *Glycine* subgenus *Soja* revealed by polymorphic nuclear and chloroplast microsatellites. *Genetics* 144, 793–806.

Schuelter, J.A., Dixon, P., Granger, C., Grant, D., Clark, L., Doyle, J. and Shoemaker, R. (2004) Mining the EST databases to resolve evolutionary events in major plant species. *Genome* 47, 868–876.

Shoemaker, R.C., Hatfield, P.M. and Palmer, R.G. (1986) Chloroplast DNA variation in the genus *Glycine* subgenus *Soja*. *The Journal of Heredity* 77, 26–30.

Shu, S.Z., Li, F.S. and Chang, R.Z. (1986) Preliminary study on evolution of main traits in soybean. *Acta Agronomica Sinica* 4, 255–259.

Singh, R.J. (2007) Methods for producing fertile crosses between wild and domestic soybean species. US Patent. Application number 11/417,369. Publication number US 2007/0261139.

Van, K., Kim, D.H. and Cai, C.M. (2008) Sequence level analysis of recently duplicated regions in soybean [*Glycine max* (L.) Merr.] genome. *DNA Research* 15, 93–102.

Vavilov, N.I. (1982) *The World Origin Centers of Main Cultivated Crops* (Translated by Dong, Y.C.). Agricultural Press, Beijing, China.

Wang, C.M., Wu, X.L. and Jia, F.X. (2008a) Genetic variations of glycinin subunit genes among cultivated and wild type soybean species. *Progress in Natural Science* 18, 33–41.

Wang, H.W., Zhang, J.S. and Gai, J.Y. (2006) Cloning and comparative analysis of the gene encoding diacylglycerol acyltransferase from wild type and cultivated soybean. *Theoretical and Applied Genetics* 112, 1086–1097.

Wang, J.L. (1947) Evaluation of soybean traits. *Agricultural Journal* 5, 6–11.

Wang, J.L., Meng, Q.X. and Zhu, Q.C. (1973) Study on photoperiodic response of wild soybean originating from different locations of China. *Genetics Communication* 3, 1–8.

Wang, K.J., Takahata, Y. and Kono, Y. (2005) Entity evidence for differentiation between Tia and Tib types of soybean Kunitz trypsin inhibitor: Detection of a novel transitional variant type between Tia and Tib in wild soybean (*Glycine soja* Sieb. & Zucc.). *Theoretical and Applied Genetics* 112, 66–71.

Wang, K.J., Takahata, Y. and Kono, Y. (2008) Allelic differentiation of Kunitz trypsin inhibitor in wild soybean (*Glycine soja*). *Theoretical and Applied Genetics* 117, 565–573.

Wang, L.Z. (1985) The origin, evolution and distribution of soybean. *Soybean Science* 1, 1–6.
Wang, Q.R., Li, S.F., Yang, Z.Y., Zhou, L.L., Gao, H., Xu, B.H., Fang, Z.L., Gong, S.L., Gao, J., Wu, Y.M., Cai, M.Y., Shi, W.Y. and Ge, C.H. (2004) The effects of soybean peptides supplementation on serum biomarkers in distance runners. *Chinese Journal of Sports Medicine* 1, 33–37.
Xu, C.P., Caperna, T.J. and Garrett, W.M. (2007) Proteomic analysis of the distribution of the major seed allergens in wild, landrace, ancestral, and modern soybean genotypes. *Journal of the Science of Food and Agriculture* 87, 2511–2518.
Yanagisawa, T., Hayashi, M. and Hirai, A. (1994) DNA fingerprinting in soybean [*Glycine max* (L.) Merrill] with oligonucleotide probes for simple repetitive sequences. *Euphytica* 80, 129–136.
Yang, Y., Yan, C.Q. and Cao, B.H. (2007) Some photosynthetic responses to salinity resistance are transferred into the somatic hybrid descendants from the wild soybean *Glycine cyrtoloba* ACC547. *Physiologia Plantarum* 129, 658–669.
Yuan, C.P., Zhou, G.A. and Li, Y.H. (2008) Cloning and sequence diversity analysis of $GmHs1^{pro\text{-}1}$ in Chinese domesticated and wild soybeans. *Molecular Breeding* 22, 1380–3743.
Zakharova, E.S., Epishin, S.M. and Vinetski, Y.P. (1989) An attempt to elucidate the origin of cultivated soybean via comparison of nucleotide sequences encoding glycinin B_4 polypeptide of cultivated soybean, *Glycine max*, and its presumed wild progenitor, *Glycine soja*. *Theoretical and Applied Genetics* 78, 852–856.

2 The Role of Soybean in Agriculture

Guriqbal Singh[1] and B.G. Shivakumar[2]
[1]*Department of Plant Breeding and Genetics, Punjab Agricultural University, Ludhiana, India;* [2]*Division of Agronomy, Indian Agricultural Research Institute, New Delhi, India*

2.1 Introduction

Soybean (*Glycine max* (L.) Merrill) is one of the most valued oilseed crops in the world. As per the latest data available, soybean accounts for 36.65 million t of oil (FAO, 2009), putting it far ahead of all other field crops raised for oil extraction (Table 2.1). Although its cultivation dates back to >5000 years ago in China, it came to prominence only during the last 200 years. It has been cultivated for varying purposes during different periods of history in different parts of the world. Its earlier uses have varied from a green manure crop to a forage crop and a nitrogen-fixing crop due to its ability to fix substantial quantities of atmospheric nitrogen in association with nodule-forming bacteria (*Bradyrhizobium*).

The face of the crop changed irreversibly with the increasing demand for oil during the early 20th century, especially during the two World Wars (1914–1918 and 1939–1945) and with the simultaneous realization of soybean's potential as a source of vegetable oil. Its success in the USA during this period led to its introduction to South America, especially Argentina and Brazil, and in the process helped expand its cultivation to an unprecedented level. At present, soybean is cultivated mainly for oil extraction and for the protein-rich de-oiled cake, which is a very important by-product of this crop with great commercial value.

In India, soybean was introduced as an oilseed crop in 1960s and has borne great success. The area has increased from meagre 0.03 million ha in 1970 to >9.6 million ha in 2008 (DSR, 2009). As per 2007 data, soybean has become a leading oilseed crop in India, leaving behind the traditional oilseed crops such as groundnut (*Arachis hypogaea*) and rapeseed mustard (*Brassica* species) (FAO, 2009). Besides being a major source of edible oil, soybean is established as a major foreign-exchange earner due to export of de-oiled cake.

Table 2.1. Pattern of edible oil production from global field crops during 2007 (adapted from FAO, 2009).

Crop	Edible oil production (million t)
Soybean (*Glycine max*)	36.65
Rapeseed (*Brassica* spp.)	17.24
Sunflower (*Helianthus annuus*)	11.60
Groundnut (*Arachis hypogaea*)	5.52
Cotton seed (*Gossypium* spp.)	5.18
Maize (*Zea mays*)	2.19
Sesame (*Sesamum indicum*)	0.90
Safflower (*Carthamus tinctorius*)	0.15

Besides its stated purpose as oilseed crop, soybean has several significant beneficial features. Its role in improving soil properties through its deep and proliferated tap-root system, residue incorporation by way of shedding leaves as well as green manuring crop, soil and moisture conservation due to its thick and dense foliage, contribution to soil nitrogen enrichment through biological nitrogen fixation (BNF) and improvement in the soil biological health have been recognized from the beginning. Indeed, it is one of the leading crops grown under rainfed conditions, exploiting the limited moisture available for agriculture depending on the rainfall pattern in a given locality. Its moisture stress tolerance, contribution to soil fertility, lesser pest and disease incidence and leguminous nature have made soybean suitable for many mixed and sequential cropping systems.

2.2 Soybean-based Cropping System

Cropping system refers to the spatial and temporal arrangement of different crops to exploit natural resources and enhance productivity per unit area and time. The spatial arrangement of crops helps in the effective utilization of land, soil moisture, nutrients and solar radiation. This is brought about by choosing appropriate crops of varying morpho-physiological nature and planning their planting geometry to reduce mutual competition for resources and enhance complementarities to increase overall productivity. In general, this is achieved by mixed and intercropping systems.

On the other hand, the temporal arrangement aims at growing crops one after another in sequence to exploit the congenial conditions of different seasons. This is mostly achieved by sequential cropping systems. Thus, both mixed/intercropping and sequential cropping systems are practised to enhance the production of different crops to meet the ever-growing demand for them. Soybean, being a leguminous crop, has been an important component in both the cropping systems in different crop rotations. However, the exact nature of mixed/intercropping or sequential cropping systems varies from location to location and situation to situation, depending upon the prevalent agroclimatic conditions and growers' needs. There are a large number of cropping systems under both mixed/intercropping and sequential cropping systems in major soybean-growing areas worldwide.

Mixed/intercropping system

Soybean is often cultivated in mixed and intercropping systems when it is grown under rainfed or dryland conditions. In mixed cropping systems, the seeds are mixed in a desired ratio before sowing and then sown either in rows or broadcast. There may not be any defined ratio of seed mixing. The plant population of the crops grown together will be mixed and uneven. In intercropping systems, the seeds of different crops are sown in defined row proportions. The ratio depends on the component crops. The major objective of the mixed and intercropping systems is to ensure against the crop failure due to uncertainties of weather or other factors that are detrimental to achieving a good harvest. The assumption is that if one crop fails, the other crop will provide some yield and total crop failure is thereby avoided. Furthermore, the combined yield/net income of both the crops will be more than the yield/net income of either of the sole crops. The selection of component crops depends on their morphological features, rooting pattern, growth phenology and type of economic yield. A number of mixed/intercropping systems are followed in different soybean-growing areas. Inclusion of a cereal in soybean-based cropping systems is a very common feature in many areas. Some of the common mixed cropping/intercropping systems involving soybean are listed in Table 2.2.

Table 2.2. Common mixed/intercropping systems involving soybean.

Country	Mixed/intercropping system	Reference(s)
Argentina	Soybean + sunflower	Cerrudo *et al.* (2009)
China	Soybean + tea (*Camellia sinensis*)	Long *et al.* (2008)
	Soybean + maize	Ming and Ming (2004); Nian and Cheng (2009)
	Soybean + sugarcane (*Saccharum officinarum*)	Wen *et al.* (2004); Nian and Cheng (2009)
India	Soybean + maize	Bhatnagar and Joshi (1999)
	Soybean + cotton	Bhatnagar and Joshi (1999)
	Soybean + pigeon pea (*Cajanus cajan*)	Bhatnagar and Joshi (1999)
	Soybean + sorghum (*Sorghum bicolor*)	Bhatnagar and Joshi (1999)
	Soybean + pearl millet (*Pennisetum typhoides*)	Bhatnagar and Joshi (1999)
	Soybean + sunflower	Bhatnagar and Joshi (1999)
	Soybean + finger millet (*Eleusine coracana*)	Bhatnagar and Joshi (1999)
	Soybean + groundnut	Bhatnagar and Joshi (1999)
Iran	Soybean + maize	Mohammad (2009)
USA	Soybean + maize	Kanwar *et al.* (2005)

Sequential cropping system

Soybean has become an important component crop in many traditional sequential cropping systems worldwide. The timing of soybean in sequential cropping systems is mainly determined by the climatic conditions suitable for luxuriant growth and reduced occurrence of insect pests and diseases, leading to bumper productivity. Sowing coinciding with the beginning of the rainy season and maturity with the start of winter has been found suitable in many soybean-growing areas the world over. However, its cultivation has also been observed in other seasons where the distinction between the seasons, especially with regard to the length of the photoperiod, is not very rigid (Bhatnagar and Joshi, 1999). Other important factors that determine the success of soybean in the sequential cropping systems are the growing period available, soil moisture availability and restrictions of other production parameters such as government legislation, market trends, demand for the produce and so on.

When soybean is cultivated in a double-cropping system, a cereal either preceding or succeeding is quite common. In addition, the introduction of either an oilseed or fibre or another legume crop is not uncommon. Sequential cropping under rainfed situations is often limited due to a shortage of moisture. However, under irrigated or assured rainfall conditions, such difficulties are easily overcome and double- or even triple-cropping involving soybean is possible. Furthermore, sequential cropping is very common in tropical and subtropical conditions rather than in temperate conditions due to the short growing period available in the latter situation. Some of the common sequential cropping systems involving soybean, followed in different agroclimatic situations, are listed in Table 2.3.

2.3 Biological Nitrogen Fixation by Soybean

One of the major features of soybean that makes it an attractive crop in many cropping systems is its efficient BNF in association with *Bradyrhizobium* in the root nodules; soybean thereby requires low nitrogen supplies in the form of chemical fertilizers for meeting its own nitrogen requirement. The quantum of nitrogen fixed varies with the climatic conditions experienced during growing period, soil conditions, agronomic practices followed, genotype and so on. There is a wide variation in the proportion of nitrogen derived from nitrogen fixation and the quantum of nitrogen fixed by soybean (Table 2.4), and this is clearly due to the diverse conditions under which it is cultivated in different countries. Furthermore, the methods used to quantify the exact amount of nitrogen fixed by the crop are not always accurate. The nitrogen fixation is often quantified by indirect methods rather than the most accurate methods such as ^{15}N techniques. Each method has its own merits and limitations (Unkovich and Pate, 2000; Herridge *et al.*, 2008). Nonetheless, all of these reports bring to the fore the undisputed fact that the BNF is of great importance in soybean.

Table 2.3. Common sequential cropping systems involving soybean.

Country	Sequential cropping system	Reference
Argentina	Wheat (*Triticum aestivum*)–soybean	Monzon *et al.* (2007)
	Groundnut–maize–soybean	Meriles *et al.* (2006)
Australia	Soybean–cotton	Rochester *et al.* (1998, 2001)
	Soybean–maize	Peoples *et al.* (2008)
Brazil	Soybean–wheat	Borkert (1990); Alves *et al.* (2003)
	Oats (*Avena sativa*)–soybean	Alves *et al.* (2003)
China	Wheat–maize–soybean	Zhu and Yu (2007)
	Maize–soybean relay cropping	Zhu *et al.* (2007)
	Cotton–soybean	Jun *et al.* (2005)
India	Central Zone Soybean–rapeseed/mustard, soybean–chickpea (*Cicer arietinum*), soybean–lentil (*Lens culinaris*), soybean–wheat, soybean–potato (*Solanum tuberosum*), soybean–safflower Southern Zone Wheat–soybean–finger millet–French bean (*Phaseolus vulgaris*), oats–cowpea (*Vigna sinensis*)–barley (*Hordeum vulgare*)–soybean, soybean–wheat–groundnut, soybean–tobacco (*Nicotiana* spp.) North Zone Soybean–wheat, soybean–chickpea, soybean–safflower, soybean–mustard North Hill Zone Soybean–wheat, soybean–barley	Bhatnagar and Joshi (1999); Reddy *et al.* (2007); Nemade *et al.* (2008); Singh *et al.* (2008); Vyas *et al.* (2008)
Nigeria	Soybean–maize	Sanginga (2003); Okogun *et al.* (2005)
USA	Maize–soybean	Zhu and Fox (2003); Jagadamma *et al.* (2008)
	Wheat–soybean	Cordell *et al.* (2006)
	Soybean–maize	Gentry *et al.* (2001); Varvel and Wilhelm (2003)
	Soybean–sorghum	Varvel and Wilhelm (2003)
	Maize–wheat–soybean	Hong *et al.* (2004)

How much nitrogen soybean fixes is an important question, but it is very difficult to give the 'correct' answer because nitrogen fixation is influenced by many diverse factors. Unkovich and Pate (2000) compiled information from the literature and reported that nitrogen in soybean shoot could be 0–450 kg ha^{-1}, with the percentage of soybean nitrogen derived from the atmosphere (%Ndfa) being 0–95%. These authors further suggested that under irrigated conditions, average atmospheric nitrogen

Table 2.4. Proportion of nitrogen derived from nitrogen fixation (P_{fix}) and nitrogen fixed by soybean in different countries.

	P_{fix} (%)		Nitrogen fixed (kg ha^{-1})		
Country	Mean	Range	Mean	Range	Reference
Australia	83	73–90	371	246–488	Rochester *et al.* (1998)
	72	–	–	290–364	Chapman and Myers (1987)
	95	–	–	289–312	Chapman and Myers (1987)
	–	81–87	–	435–488	Rochester *et al.* (2001)
	53	13–73	178	42–311	Peoples *et al.* (1995)
	–	0–83	–	0–233	Herridge and Holland (1992)
	–	49–83	–	41–117	Peoples *et al.* (2008)
Brazil	–	70–85	–	70–250	Alves *et al.* (2003)
India	–	13–81	–	7–53	Singh *et al.* (2004)
Nigeria	–	6–81	–	–	Okogun *et al.* (2005)
	–	26–64	–	24–168	Sanginga *et al.* (1997)
	–	84–87	–	14–188	Eaglesham *et al.* (1982)
	–	–	–	38–126	Sanginga (2003)
	–	39–54	–	51–78	Osunde *et al.* (2003)
Switzerland	–	24–54	–	150–260	Oberson *et al.* (2007)
Tanzania	–	36–78	–	23–83	Okogun and Sanginga (2003)
USA	57	0–96	84	0–185	Russelle and Birr (2004)

fixation by soybean is around 175 kg N ha^{-1} for shoots (about 248 kg including roots), whereas under rainfed conditions it is around 100 kg N ha^{-1} (142 kg including roots). About 50–60% of the nitrogen demand of soybean crop is met by BNF (Salvagiotti *et al.*, 2008). The total nitrogen fixed by soybean annually in four major soybean-producing countries (USA, Brazil, Argentina and China) is estimated to be 16.44 Tg (Herridge *et al.*, 2008), with an average %Ndfa of 68%.

Earlier studies, based on root excavations, suggested that root nitrogen represents only a small proportion of the total plant nitrogen (Bergersen *et al.*, 1989). However, recent studies with ^{15}N feeding have clearly indicated the much higher amounts of nitrogen in roots (Rochester *et al.*, 1998; Unkovich and Pate, 2000; Herridge *et al.*, 2008; Peoples *et al.*, 2008). Nodulated roots and rhizodeposition of nitrogen during the growth of a legume crop may account for 30–50% of the total nitrogen in the crop (Peoples *et al.*, 2008). This clearly shows that the earlier estimates of nitrogen fixation by soybean published in the literature were underestimates.

The BNF efficiency depends on: (i) climatic factors (temperature and photoperiod); (ii) the interaction between environmental factors and the soybean plant, such as the efficiency of a soybean cultivar in fixing atmospheric nitrogen, soil fertility conditions and macro- and micronutrient supply; and (iii) bacterial strain competitiveness, the amount and the quality of the inoculant, the care in the inoculation process and the absence of antagonistic agrochemicals on the seed (Campo and Hungria, 2004). In Brazil, several efficient imported *Bradyrhizobium* strains have been found to be unable to compete with native soil microflora and other previously introduced *Bradyrhizobium* strains when they were introduced for the first time; after some acclimatization, however, these strains became much more efficient (Alves *et al.*, 2003). Alves *et al.* (2003) further reported that the selection of an appropriate strain of *Bradyrhizobium* is a prerequisite for increasing BNF efficiency. Various other factors may also influence nitrogen fixation in soybean, as described below.

Edaphic factors

Soil is the most important factor influencing the rate and amount of nitrogen fixation in soybean. The physical, chemical and biological characteristics of soil have a profound influence on BNF activity. Among the physical properties of soil, the type, texture and structure, having an effect on water-holding capacity, groundwater table and so on, affect the nitrogen-fixing microbes and thereby the amount of nitrogen fixed. In general, loamy and clay soils favour better nitrogen fixation than sandy soils. This is attributed to the poor microbial activity and lesser water-holding capacity of the latter types of soil. On the other hand, soils with water pooling on the surface for unusually longer periods after rains or heavy irrigations or shallow ground water affect the aeration in the rhizosphere and consequently the microbes involved in nitrogen fixation (Puiatti and Sodek, 1999). Soils rich in available nitrogen tend to subdue the activity of nitrogen fixation (Bo *et al.*, 1997). In addition, soils inherently high in salts or acidity leading to either unusually high or low pH affect nitrogen fixation. Lack of organic matter in the soil is often a major factor, resulting in little or no microbial activity and rendering BNF less effective. Waluyo *et al.* (2004) reported that under acidic soil conditions, calcium and phosphorus were limiting factors for BNF.

Crop factors

Among crop factors, the genetic constitution of the crop, its compatibility with nitrogen-fixing microbes, crop duration, different phenological stages and yield potential have all been found to affect the quantum of BNF (Nicolas *et al.*, 2006; Abaidoo *et al.*, 2007). Reports indicate the variability of varieties in influencing the nitrogen fixation activity of microbes (Farnia *et al.*, 2005). Furthermore, the duration of different phenological stages and total crop duration also have an important role (Shiraiwa *et al.*, 1994; Botha *et al.*, 1996). Since BNF starts only after the initial seedling growth stage, the time taken for initiating the vegetative phase will determine the time period available for BNF. Likewise, BNF tends to decline with the onset of podding and the grain-filling stage. Thus, the different phenological phases decide the total amount of nitrogen fixed. Finally, high-yielding varieties requiring rapid translocation of photosynthates as well as longer time periods tend to affect the rate and amount of nitrogen fixed by the crop (Pandey, 1996; Chechetka *et al.*, 1998).

Climatic factors

Climatic factors (i.e. temperature and rainfall) affect the activity of the microbes involved in BNF. Very high or very low ambient temperatures affect soil temperatures. In spite of soil's great buffering capacity, there seems to be sufficient fluctuation in the soil temperature to affect the efficacy of nitrogen-fixing microbes (Shiraiwa *et al.*, 2006). Rainfall, in terms of both quantity and distribution, affects the normal functioning of the crop as well as of the microbes. Heavy downpours resulting in waterlogging and long dry spells leading to moisture stress equally influence the efficiency of BNF activity and thus affect the amount of nitrogen fixed (Sung, 1993; Sridhara *et al.*, 1995; Jung *et al.*, 2008).

Management factors

Various agronomic practices (i.e. time of sowing, depth of sowing, cropping practices such as sole or intercropping, tillage operations, seed inoculation, irrigation method and frequency, use of plant protection chemicals, intercultivation practices and so on) have a profound influence on microbial activity, rhizosphere aeration and crop performance. These, in turn, influence the rate of nitrogen fixation. Seed inoculation with efficient strains of *Bradyrhizobium*, a starter dose of nitrogen through fertilizers, light irrigations to avoid waterlogging and avoiding the use of plant protection chemicals that harm the microbes, positively influence the BNF and lead to greater amounts of nitrogen fixation. On the other hand, untimely sowing, a poor or uneven plant stand, lack of seed inoculation, heavy doses of nitrogen fertilizers and so on, result in shy nodulation and a lower amount of nitrogen fixation. In an experiment to study the impact of different residue man agement and tillage practices, no significant difference in soybean yield or nitrogen accumulation was observed, but BNF was higher in zero tillage as compared to conventional tillage (Alves *et al.*, 2002). Hughes and Herridge (1989) reported higher number of nodules, nodule dry weight, nitrogen fixation and nitrogen balance in soil in a no-tilled condition than in a tilled one. Tillage stimulates mineralization of organic matter in the soil; this results in the availability of high levels of nitrate, which may depress nodulation and nitrogen fixation. It implies that no-tillage conditions are preferred to repeated tillage operations as far as BNF is concerned.

2.4 Effects of Soybean on Soil Properties

Cultivation of soybean has been observed to influence the different properties of soil. This may be attributed to two major factors: (i) a deep and well-proliferated tap-root system; and (ii) the addition of a large quantity of biomass by way of roots left in the soil after the harvest of the aboveground portion and the addition of litter through leaf-shedding prior to harvest in many cultivated varieties. Even assuming a modest 1:3 root to shoot ratio, a crop yielding 2 t of grain at a harvest index of 0.33 tends to add about 2 t of root biomass underground and another 1 t of biomass in the form of litter. The addition of this much biomass by a crop growing for around 4 months will obviously influence soil properties. The effects of soybean cultivation on different soil properties are described below.

Chemical properties

The important chemical properties influenced by soybean are related to the status of different essential nutrients. As the crop is capable of fixing atmospheric nitrogen in association with *Bradyrhizobium*, it is often observed to influence the nitrogen balance in the soil. In addition, as soybean is usually grown with a large dose of phosphorus and potassium, the status of these elements in the soil also tends to be affected by the cultivation of soybean. The addition of a sizeable quantity of crop residues is likely to influence many other chemical properties of the soil. Gawande *et al.* (2007) reported significant nitrogen, phosphorus and potassium build-up in soil in soybean-based cropping systems. Shoko and Tagwira (2007) also observed significant improvements in chemical properties with the introduction of soybean to sugarcane production systems.

Organic carbon

In a soybean–wheat cropping system, over a period of 30 years, soil organic carbon has been found to increase by 29% to 104% with the use of different combinations of nitrogen, phosphorus, potassium and farmyard manure (FYM) (Bhattacharyya *et al.*, 2008). Even in the unfertilized control, soil organic carbon increased by 9% over 30 years, possibly due to carbon addition through the roots and crop residues, as each year about 124 kg ha^{-1} leaf-fall biomass of soybean and 85 kg ha^{-1} stubble biomass of wheat was added to the soil under this treatment. The organic carbon content in soil has been found to be higher in a maize + soybean–wheat cropping system than in a maize–wheat system (Sharma and Behera, 2009).

Nitrogen

Nitrogen is one of the most dynamic nutrients present in the soil. Due to several chemical reactions associated with this nutrient, it is difficult to quantify the effect of soybean on this element. However, several studies (Bhatia *et al.*, 2001; Joon *et al.*, 2005; Ho *et al.*, 2008) have conclusively proven the positive effect of soybean cultivation on nitrogen availability after the harvest of soybean. Furthermore, the increased availability of nitrogen in the soil has also been linked to mineralization of soybean crop residues left in the soil due to leaf shedding and root biomass (Toomsan *et al.*, 1995).

The incorporation of soybean residues has been found to increase the nitrogen content in the soil (Galal and Thabet, 2002). When soybean was grown as a sole crop or intercropped with maize it added residues to the tune of 4.84 and 1.75–2.36 t ha^{-1}, respectively, with corresponding nitrogen addition of 60.0 and 21.7–29.3 kg ha^{-1} (Sharma and Behera, 2009). After two years, the apparent nitrogen balance was positive (+59.4 kg ha^{-1}) in the case of a maize + soybean–wheat cropping system and negative (–6.1 kg ha^{-1}) in the case of a maize–wheat system (Sharma and Behera, 2009). In many areas, soybean biomass is removed from the field after harvest. Under such situations, only the residual biomass of soybean is the source of nitrogen. In India, the residual biomass of soybean, comprising leaf-fall, root, nodules and rhizodeposition, contributes nitrogen in the ranges of 7.02–16.94, 11.65–28.83, 3.31–8.91, and 11.30–23.80 kg N ha^{-1}, respectively (Singh *et al.*, 2004).

In a soybean–wheat cropping system, the total soil nitrogen has been found to increase by 51–86% with the application of nitrogen, phosphorus, potassium and FYM over a 30-year period, compared with a 23% increase in an unfertilized control (Bhattacharyya *et al.*, 2008). Ramesh and Reddy (2004) reported that soybean-based cropping systems enrich the soil nitrogen, especially with the application of recommended doses of nitrogen.

Phosphorus

Phosphorus is another element that has been found to be influenced by the cultivation of soybean. However, whether it is positive or negative balance depends on the initial available status in soil and phosphorus added in the form of fertilizers. Since a sizeable quantity of applied phosphorus in addition to a lot of crop residue remains in the soil, this element is likely to become enriched in areas where regular fertilization has been in vogue. Shoko and Tagwira (2007) observed a significant increase in soil phosphorus with soybean introduction as a break crop in sugarcane production systems. In a long-term trial with soybean–wheat, the available phosphorus content was increased by 25–50% with the application of phosphorus fertilizer and FYM (Bhattacharyya *et al.*, 2008).

Potassium

The extent of the effect of soybean cultivation on potassium dynamics in soil is variable and complicated. Since there is equilibrium in the different types of potassium (i.e. fixed and available), it is often difficult to study the minor fluctuations consequent to soybean cultivation. However, application of suboptimal doses of potassium over a period of time may deplete the total soil potassium (Bhattacharyya *et al.*, 2006). The available soil potassium content may not increase, despite the application of potassium fertilizer, due to its removal by not only soybean but also by other crops in the rotation (Kundu *et al.*, 2007; Bhattacharyya *et al.*, 2008). Gawande *et al.* (2007) observed a higher build-up of potassium after a soybean–sorghum cropping sequence as compared to other grain legume-based cropping systems, indicating a positive effect of soybean on potassium build-up in soil.

Other parameters

Other important chemical parameters that are likely to be influenced following cultivation of soybean are pH and electrical conductivity. A substantial quantity of biomass is added to the soil with every crop. This is likely to increase the organic matter content. Furthermore, with this, the increase in organic carbon content and other ions, cation exchange capacity and electrical conductivity are also likely to be affected.

Physical properties

The deep and well-spread tap-root system of soybean, resulting in a tilling effect to the soil, coupled with the addition of a large quantity of biomass through the roots and leaf shedding, tends to improve the physical properties of the soil. The impact is more pronounced in the lighter soils and under those conditions where the inherent organic matter content is less and its degradation rate slow. In tropical conditions, however, the impact is less owing to the faster degradation of organic matter due to high temperatures. The effect could mostly be on the soil bulk density, aeration, infiltration and water-holding capacity (Bhattacharyya *et al.*, 2008). Soil strength, measured as cone penetrometer readings during the cotton growing phase, has been found to be significantly lower when cotton was sown after soybean as compared with after continuous cotton (Rochester *et al.*, 2001).

Biological properties

The biological properties of soil have been reported to be influenced by soybean cultivation (Adeboye and Iwuafor, 2007; Bhattacharyya *et al.*, 2008). The micro- and macro-flora and fauna of the soil are benefitted by soybean cultivation. This is possibly due to the rooting pattern, leading to good aeration in the rhizosphere, and the addition of crop biomass, enriching the organic matter content that forms the basis for biological activity in the soil. In a cropping system experiment, higher nitrate-nitrogen, hydrolysable-nitrogen, protease and microbial activities were detected in samples collected at maize flowering and harvest from a soybean–maize rotation than from maize monoculture, indicating the existence of larger labile nitrogen pools and higher capacity for nitrogen mineralization in soils under maize rotated with soybean than under maize grown in monoculture (Conti *et al.*, 1998). In another experiment, Long *et al.* (2008) reported that soybean intercropped with tea significantly improved the microecology of the tea plantation by reducing weed growth and pest and disease occurrence, and increased tea yield and improved the economics of the tea plantation.

Allelopathic effects

Long-term continuous cultivation of soybean results in a decline in soybean yield (Liu and Herbert, 2002; Kelley *et al.*, 2003). This may be due to many reasons, including root diseases, insect pests and nematodes, imbalance of the soil environment or deterioration of soil properties, change of rhizosphere microbes or toxicity of residual and root exudates. Decomposed root material, due to the allelopathy phenomenon, may decrease the germination, growth and yield of soybean (Liu and Herbert, 2002). Crop rotations, rather than continuous soybean cultivation, help in enhancing soybean yields (Kelley *et al.*, 2003). In a long-term (30-year) study with a soybean–wheat cropping system, no allelopathic-related yield reduction in soybean was observed (Kundu *et al.*, 2007; Bhattacharyya *et al.*, 2008). Soybean may have allelopathic effects on weeds and, therefore, could help in weed management. For example, Rose *et al.* (1984) reported that soybean root exudates reduced the dry weight of velvetleaf (*Abutilon theophrasti*), but did not inhibit foxtail millet (*Setaria italica*).

The growth of double-crop soybean following winter wheat is adversely affected due to wheat straw leachate (Hariston *et al.*, 1987). Supplemental application of nitrogen, however, overcomes depressed growth and yield of soybean.

2.5 Effects of Soybean on Diseases, Insect Pests and Weeds

As with other crops, soybean has its own set of diseases and insect pests specific to it. However, barring a few diseases and insect pests, the frequency of occurrence and extent of damage is comparatively less than for other crops. It has been observed that most of the diseases and insect pests have their own hot spots, beyond which their prevalence and damage is by and large limited. Prominent diseases such as yellow mosaic virus, soybean mosaic virus and soybean rust and insect pests such as stem borers, defoliators, leaf miners, pod borers and sap feeders are observed in this crop.

Diseases

Soybean per se may not have a great effect on the proliferation of diseases of other crops. However, it has been found to check the spread of several diseases by affecting their life cycle when used as a component crop in cropping systems (Gil *et al.*, 2008; Qun *et al.*, 2008).

Insect pests

Many insect pests are crop-specific and, in the absence of such a specific host, there is less likelihood of them occurring on a regular basis and attaining threshold level. Soybean, if used appropriately as an inter/mixed crop or in a sequential cropping system, has a positive role in preventing the insect pest cycles of other crops. The effects of intercropping castor (*Ricinus communis*) and black soybean on the biological prevention and control of major pests have been investigated by Hua *et al.* (2003). They observed that intercropping castor and black soybean had a significant effect on the prevention and control of *A. glycines* and *L. glycinivorella*, but the effect depended on the cultivar. When castor and black soybean 1, which had weak resistance to pests, were intercropped, the number of aphids per plant and the rate of pest damage decreased by 61.5% and 16.2%, respectively, compared with single cropping. When castor and black soybean 2, which had high resistance to pests, were intercropped, the number of aphids per plant decreased by 32.4% compared with single cropping. When castor and the two black soybean cultivars were intercropped with the same row arrangement (2:4), the effect on the prevention and control of *A. glycines* and *L. glycinivorella* was better. In another study, a significant difference in the weight of maize earworm larvae, which were allowed to feed on different genotypes of fodder soybean, was observed, indicating the positive effect of fodder soybean in controlling the earworm larvae in maize (Javaid *et al.*, 2006).

Weeds

Maize grown after soybean has been found to have lesser *Striga hermonthica* parasitism as compared with maize grown after sorghum (Carsky *et al.*, 2000). However, more studies are required.

2.6 Residual Effects of Soybean on Succeeding Crops

Due to its deep and well-proliferated tap-root system conferring many positive features on the soil, capability to fix atmospheric nitrogen and provision of a large quantity of root biomass as well as litter by way of leaf shedding, soybean is able to leave behind many beneficial residual effects for the succeeding crop in sequential cropping systems. Because soybean is partially independent of soil nitrogen, relying mostly on nitrogen from biological fixation, soil nitrogen may be left to be used for further cropping, which may explain the benefits observed in the field in crop rotation systems with soybean.

Productivity

Soybean has several residual effects on the succeeding crop. These include improved soil conditions, increased soil fertility, higher moisture conservation and a reduced incidence of insect pests and diseases. Many positive effects of soybean on companion crops in intercropping and succeeding crops in rotation have been observed. Intercropping of soybean has improved soil fertility and promoted larch (*Larix decidua*) growth in north-east China (Wang *et al.*, 2006).

Increased productivity of the succeeding crop represents the sum total of all beneficial effects of previous soybean. Rotation between cotton and soybean has considerable economic, ecological and social benefits (Jun *et al.*, 2005). After an experiment to study the residual effect of soybean on the succeeding crop of maize, Osunde *et al.* (2003) reported significantly greater plant height, shoot biomass, grain yield and nitrogen uptake of maize in plots previously sown to soybean than in previously fallow plots. A late-maturing genotype exhibited better residual effects than a medium-maturing genotype. Grain yields of maize, when rotated with soybean, were observed to be greater than with continuous maize, indicating a positive influence of including soybean in the system (Carsky *et al.*, 1997; Lamb *et al.*, 1998; Gentry *et al.*, 2001). In Brazil, yields of winter cereal are higher after soybean than after maize or a fallow, which may be due to nitrogen input through BNF by soybean (Alves *et al.*, 2003). In a wheat–soybean–wheat sequence, it was noted that wheat planted after soybean produced a higher grain yield than that grown before soybean (Kumbhar *et al.*, 2007).

In the case of soybean, 90–100% of its leaves are shed at physiological maturity, containing about 110 kg N ha^{-1}. This source of nitrogen might be one of the factors responsible for the increase in maize yield that followed soybean (20–24%) compared with a continuous maize plot (Okogun *et al.*, 2007). Alves *et al.* (2002) opined that the benefit to the subsequent crop was due to the release of nitrogen from extremely labile soybean residues of low carbon to nitrogen ratio and not because of a net gain of nitrogen from BNF. The grain yield of wheat has been found to be higher after maize + soybean intercropping than after sole maize (Sharma and Behera, 2009), possibly due to the incorporation of soybean residue, having 21.7–29.3 kg N ha^{-1}.

Nitrogen economy

Soybean fixes a lot of atmospheric nitrogen with the symbiotic relationship with *Bradyrhizobium*. A substantial portion of this is used by the growing soybean, but some is left unused in the soil and some in the nodules. Once soybean is harvested, this leftover nitrogen is available to the next crop. The extent of this leftover nitrogen depends on the efficiency of BNF and utilization by the crop. Furthermore, it is also influenced by the prevailing environmental conditions and the ability of the succeeding crop to utilize nitrogen. Some estimates of nitrogen balance/residual fixed nitrogen by soybean are given in Table 2.5.

In long-term studies with soybean–maize and soybean–sorghum cropping systems, it was found that maize and sorghum obtained 65 and 80 kg N ha^{-1} from soybean, respectively (Varvel and Wilhelm, 2003). Wheat planted after soybean required 21 kg N/ha^{-1} less than wheat planted after

Table 2.5. Estimates of nitrogen balance/residual fixed nitrogen by soybean.

Country	Succeeding crop	Nitrogen balance/residual fixed nitrogen (kg ha^{-1})	Reference
Australia	Cotton	+73 to +284	Rochester *et al.* (1998)
Australia	Pasture	+55 to +110	Hughes and Herridge (1989)
Australia	Soybean	–75 to +109	Hughes and Herridge (1989)
Australia	Cotton	+243 to +266	Rochester *et al.* (2001)
Australia	–	–69 to +45	Herridge and Holland (1992)
Nigeria	Maize	+10 to +24	Sanginga (2003)
Switzerland	–	–140 to –300	Oberson *et al.* (2007)

grain sorghum (Staggenborg *et al.*, 2003). At maturity, there may be 37 kg N ha^{-1} in the roots and stem bases of soybean and 30–68 kg N ha^{-1} in above-ground plant parts (excluding seed) (Chapman and Myers, 1987), which is all available for use by the succeeding crop. In this study, soybean was found to offer the possibility of a marginal reduction in the nitrogen fertilizer need of the succeeding rice (*Oryza sativa*) crop, with greater benefits when residues were incorporated. Maize crop sown after soybean required about half of the nitrogen fertilizer required by continuously cropped maize, as soybean added the equivalent of 150 kg fertilizer N ha^{-1} (Omay *et al.*, 1998). Fortuna *et al.* (2008) observed that soybean coupled with tillage reduced the fertilizer nitrogen requirement of maize in maize–soybean rotation.

2.7 How to Improve Contributions of Soybean in Agriculture?

BNF is beneficial to the environment (Jensen and Hauggaard-Nielsen, 2003) and is an important source of nitrogen in agriculture, which needs to be further enhanced by various means including agronomic, microbiological and plant breeding (Hardarson and Atkins, 2003). However, the breeding approach for enhanced atmospheric nitrogen fixation has not been a success in soybean (Herridge *et al.*, 2001).

Soybean has great potential. It has the capability to fix a large quantity of atmospheric nitrogen. Nitrogen fixation has a large role in meeting the nitrogen requirements of the crop as well as, to some extent, those of the succeeding crop in the system. However, the present BNF realizations are quite low and there are prospects for improving this. It requires an understanding of the production factors and ironing out the areas that are pulling down the potential of this crop.

Very high seed yields (6000–8600 kg ha^{-1}) of soybean have been reported from various research studies and national soybean contests in the USA (Cooper, 2003). Soybean seed is rich in protein, and therefore a very high nitrogen requirement is expected for obtaining such high yields. Although some amount of this nitrogen requirement may be met through soil nitrogen and fertilizer nitrogen, the role of BNF becomes more important due to the high cost involved with fertilizer nitrogen, along with other environmental considerations of fertilizer production and use.

Bradyrhizobium inoculation of soybean seed increases the seed yield of soybean substantially and the response is further enhanced with the use of *Bradyrhizobium* inoculation plus 26 kg P ha^{-1} (Ndakidemi *et al.*, 2006). An increase in nodulation, seed yield and seed nitrogen content is possible only when seed is inoculated with effective and adequate number of rhizobia (Albareda *et al.*, 2009). Substantial losses of viability of inocula (94.0–99.9%) may occur between inoculation and sowing (Brockwell *et al.*, 1988); therefore, the inoculation method should be such that there are no or minimum losses of rhizobia. It has now been amply established that there are wide variations in the ability of *Bradyrhizobium* strains to affect BNF. Screening for *Bradyrhizobium* strains appropriate for different soil conditions, agroclimatic conditions and cropping systems can improve BNF with no cost involved. Furthermore, screening of native strains for their competitiveness is another way of knowing if microbes are a limiting factor. The use of biotechnological tools for incorporating the genes suitable for enhancing BNF in *Bradyrhizobium* strains needs to be attempted.

Inappropriate production practices often hamper the potential of BNF in the system. Genotypes/breeding lines of soybean differ in %Ndfa and the amount of nitrogen fixed (Osunde *et al.*, 2003; Sanginga, 2003; Houngnandan *et al.*, 2008). Genotypes with a high BNF potential along with high seed yields therefore need to be grown. Timely and appropriate methods of sowing, the application of a starter dose of nitrogen and sufficient quantity of other nutrients, irrigation practices devoid of waterlogging or moisture stress, the use of plant protection measures less detrimental to microbes in the soil and so on can pave way for improved BNF.

No-till sowing should be encouraged for enhanced nitrogen fixation. In the Pampas region of Argentina, most of the soybean area (approximately 90% of a total 12×10^6 ha) is under no-till cultivation (Austin *et al.*, 2006). In Brazil, no-till is followed in almost 50% of soybean-based crop rotations and the yields are similar to those of the conventional tillage system (Alves *et al.*, 2003). Under moderate levels of soil nitrate, the number and dry weight of nodules, amount of nitrogen fixed, proportion of nitrogen fixed and nitrogen balance by soybean are higher in no-tillage system than in cultivated ones (Herridge and Holland, 1992).

Most soils used for soybean cultivation have 5–10 µg N g^{-1} in 0–30 cm depth. However, higher soil NO_3^- levels (30 µg N g^{-1} in 0–30 cm depth) delay nodule initiation, retard nodule development, reduce the extent of nodulation and consequently impair nitrogen fixation (Herridge *et al.*, 1984). Therefore, for obtaining high levels of nitrogen fixation, soils high in soil nitrate should not be selected for soybean cultivation or soil nitrate should be exhausted prior to sowing by including a nitrogen-demanding crop in the cropping system. In areas where soybean nodulation is poor due to high temperature, thermotolerant bradyrhizobial strains should be selected and used (Rahmani *et al.*, 2009).

Nutrition should be optimum and balanced. Phosphorus deficiency decreases plant growth, photosynthetic rate, nodule dry weight and BNF in soybean (Chaudhary *et al.*, 2008). An adequate supply of phosphorus will enhance BNF by stimulating plant growth. An optimum dose of nutrients should not be supplied to soybean only, but to all of the crops in the cropping system through chemical fertilizers and organic manures such as poultry manure, vermicompost and FYM so that optimum nutrient status is maintained in the soil (Behera *et al.*, 2007; Behera, 2009), subsequently ensuring higher rates of BNF.

2.8 Conclusions

Soybean is an important oilseed crop in the world. It can be grown in various inter/mixed and sequential cropping systems. BNF is an important source of nitrogen for the soybean crop. Nitrogen fixed by the soybean crop, its roots, leaves shed during crop growth and residue at harvest contribute greatly to improving the chemical, physical and biological properties of soil. Improvements in such soil properties help in obtaining high yields of the succeeding crops in the rotation. Furthermore, the nitrogen requirements of the succeeding crops following soybean are reduced, thereby reducing the costs of cultivation for raising the crops and consequently increasing the net income to farmers. There are many means that can help in further improving BNF in soybean and, therefore, the soybean crop has the potential to play an even greater role in sustainable agriculture.

References

Abaidoo, R.C., Keyser, H.H., Singleton, P.W., Dashiell, K.E. and Sanginga, N. (2007) Population size, distribution, and symbiotic characteristics of indigenous *Bradyrhizobium* spp. that nodulate TGx soybean genotypes in Africa. *Applied Soil Ecology* 35, 57–67.

Adeboye, M.K. and Iwuafor, E.N. (2007) Soil microbial biomass and water-soluble organic carbon in crop rotation systems in a moist tropical savanna Alfisol. *Nigerian Journal of Soil and Environmental Research* 7, 45–52.

Albareda, M., Rodríguez-Navarro, D.N. and Temprano, F.J. (2009) Soybean inoculation: Dose, N fertilizer supplementation and rhizobia persistence in soil. *Field Crops Research* 113, 352–356.

Alves, B.J.R., Zotarelli, L., Boddey, R.M. and Urquiaga, S. (2002) Soybean benefit to a subsequent wheat crop in a cropping system under zero tillage. In: *Nuclear Techniques in Integrated Plant Nutrient, Water and Soil Management. Proceedings of an International Symposium*, Vienna, Austria, 16–20 October 2000, pp. 87–93.

Alves, B.J.R., Boddey, R.M. and Urquiaga, S. (2003) The success of BNF in soybean in Brazil. *Plant and Soil* 252, 1–9.

Austin, A.T., Piñeiro, G. and Gonzalez-Polo, M. (2006) More or less: Agricultural impacts on the N cycle in Argentina. *Biogeochemistry* 79, 45–60.

Behera, U.K. (2009) Organic manuring for soil biological health and productivity of a wheat–soybean cropping system in the Vertisols of central India. *Crop & Pasture Science* 60, 382–393.

Behera, U.K., Sharma, A.R. and Pandey, H.N. (2007) Sustaining productivity of wheat–soybean cropping systems through integrated nutrient management practices on the Vertisols of central India. *Plant and Soil* 297, 185–199.

Bergersen, F.J., Brockwell, J., Gault, R.R., Morthorpe, L., Peoples, M.B. and Turner, G.L. (1989) Effects of available soil N and rates of inoculation on nitrogen fixation by irrigated soybeans and evaluation of $\delta^{15}N$ methods for measurement. *Australian Journal of Agricultural Research* 40, 763–780.

Bhatia, C.R., Nichterlein, K. and Maluszynski, M. (2001) Mutations affecting nodulation in grain legumes and their potential in sustainable cropping systems. *Euphytica* 120, 415–432.

Bhatnagar, P.S. and Joshi, O.P. (1999) Soybean in cropping systems in India. In: *Integrated Crop Management*, Vol. III. Food and Agriculture Organization of the United Nations, Rome, Italy, pp. 11–14.

Bhattacharyya, R., Prakash, V., Kundu, S., Ghosh, B.N., Srivastva, A.K. and Gupta, H.S. (2006) Potassium balance as influenced by farmyard manure application under continuous soybean–wheat cropping system in a Typic Haplaquept. *Geoderma* 137, 155–160.

Bhattacharyya, R., Kundu, S., Prakash, V. and Gupta, H.S. (2008) Sustainability under combined application of mineral and organic fertilizers in a rainfed soybean–wheat system of the Indian Himalayas. *European Journal of Agronomy* 28, 33–46.

Bo, G.Y., Jing, C. and Stulen, I. (1997) Effects of nitrogen application at different growth stages on growth, nodulation and yield of soyabeans. *Soybean Science* 16, 125–130.

Borkert, C.M. (1990) Potash fertilization for soybean–wheat double-cropping in southern Brazil. *Better Crops International* 6, 10–11.

Botha, A.D.P., Pretorius, D.C., Bienkowski, J.F. and Coetzer, R.L.J. (1996) Quantification of the uptake of atmospheric and soil N by soyabeans [*Glycine max* (L.) Merrill] from ^{15}N-enriched soil. *South African Journal of Plant and Soil* 13, 139–146.

Brockwell, J., Gault, R.R., Herridge, D.F., Morthorpe, L.J. and Roughley, R.J. (1988) Studies on alternative means of legume inoculants: microbiological and agronomic appraisals of commercial procedures for inoculating soybeans with *Bradyrhizobium japonicum*. *Australian Journal of Agricultural Research* 39, 965–972.

Campo, R.J. and Hungria, M. (2004) Sources of nitrogen to reach high soybean yields. In: *Proceedings of VII World Soybean Research Conference, IV International Soybean Processing and Utilization Conference, III Congresso Brasileiro de Soja Brazilian Soybean Congress*, Foz do Iguassu, PR, Brazil, 29 February–5 March 2004, pp. 1275–1280.

Carsky, R.J., Abaidoo, R., Dashiell, K.and Sanginga, N. (1997) Effect of soybean on subsequent maize grain yield in Guinea savanna of West Africa. *African Crop Science Journal* 5, 31–38.

Carsky, R.J., Berner, D.K., Oyewole, B.D., Dashiell, K. and Schulz, S. (2000) Reduction of *Striga hermonthica* parasitism on maize using soyabean rotation. *International Journal of Pest Management* 46, 115–120.

Cerrudo, A., Sadras, V., Echarte, L., Monzón, J., Razalli, R. and Andrade, F. (2009) Sunflower–soybean intercropping: Identification and evaluation of soybean varieties with contrasting photomorphogenic response. In: Qiu, L., Guan, R., Jin, J., Song, Q., Guo, S., Li, W., Wang, Y., Han, T., Liu, X., Yu, D., Jiang, L. and

Peng, D. (eds) *Developing a Global Soy Blueprint for a Safe Secure and Sustainable Supply. Abstracts for oral presentations and posters of the World Soybean Research Conference VIII*, Beijing, China, August 10–15, 2009. The Chinese Academy of Agricultural Sciences, Beijing, China, p. 25.

Chapman, A.L. and Myers, R.J.K. (1987) Nitrogen contributed by grain legumes to rice grown in rotation on the Cunnunurra soils of the Ord Irrigation Area, Western Australia. *Australian Journal of Experimental Agriculture* 27, 155–163.

Chaudhary, M.I., Adu-Gyamfi, J.J., Saneoka, H., Nguyen, N.T., Suwa, R., Kanai, S., El-Shemy, H.A., Lightfoot, D.A. and Fujita, K. (2008) The effect of phosphorus deficiency on nutrient uptake, nitrogen fixation and photosynthetic rate in mashbean, mungbean and soybean. *Acta Physiologiae Plantarum* 30, 537–544.

Chechetka, S.A., Piskorskaya, V.P., Bruskova, R.K., Troitskaya, G.N. and Izmailov, S.F. (1998) Partitioning of ^{14}C-photoassimilates in soybean plants assimilating symbiotic and nitrate nitrogen. *Russian Journal of Plant Physiology* 45, 203–208.

Conti, M.E., Palma, R.M., Arrigo, N.M., Zourarakis, D.P. and Cappelletti, C.A. (1998) Long-term rotation effect of soybean with no-till maize on soil N availability indices and microbial activity in the Argentine Pampa. *Soil and Tillage Research* 49, 267–270.

Cooper, R.L. (2003) A delayed flowering barrier to higher soybean yields. *Field Crops Research* 82, 27–35.

Cordell, M.L., Brye, K.R., Longer, D.E. and Gbur, E.E. (2006) Residue management practice effects on soybean establishment and growth in a young wheat soybean double-cropping system. *Journal of Sustainable Agriculture* 29, 97–120.

DSR (2009) *Director's Report and Summary Tables of Experiments (2008–2009)*. All India Coordinated Research Project on Soybean, Directorate of Soybean Research, Indore, India.

Eaglesham, A.R., Ayanaba, A., Ranga Rao, V. and Eskew, D.L. (1982) Mineral N effects on cowpea and soyabean crops in a Nigerian soil. II. Amounts of N fixed and accrual to the soil. *Plant and Soil* 68, 183–192.

FAO (2009) *FAOSTAT*. Food and Agriculture Organization of the United Nations, Rome, Italy. Available at: http://faostat.fao.org (accessed 8 September 2009).

Farnia, A., Noormohammadi, G. and Cyadat, A. (2005) Evaluation of soybean cultivars growth analysis effected from *Rhizobium japonicum* strains. *Journal of New Agricultural Science* 1, 26–34.

Fortuna, A., Blevins, R.L., Frye, W.W., Grove, J. and Cornelius, P. (2008) Sustaining soil quality with legumes in no-tillage systems. *Communications in Soil Science and Plant Analysis* 39, 1680–1699.

Galal, Y.G.M. and Thabet, E.M.A. (2002) Effect of soybean residues, *Azospirillum* and fertilizer-N on nitrogen accumulation and biological fixation in two wheat cultivars. *Egyptian Journal of Microbiology* 37, 15–27.

Gawande, C.G., More, S.D. and Magar, S.M. (2007) Soil fertility dynamics as influenced by cropping system and fertilizer levels. *Asian Journal of Soil Science* 2, 48–50.

Gentry, L.E., Below, F.E., David, M.B. and Bergerou, J.A. (2001) Source of the soybean N credit in maize production. *Plant and Soil* 236, 175–184.

Gil, S.V., Meriles, J.M., Haro, R., Casini, C. and March, G.J. (2008) Crop rotation and tillage systems as a proactive strategy in the control of peanut fungal soilborne diseases. *BioControl* 53, 685–698.

Hardarson, G. and Atkins, C. (2003) Optimising biological N_2 fixation by legumes in farming systems. *Plant and Soil* 252, 41–54.

Hariston, J.E., Sanford, J.O., Pope, D.F. and Horneck, D.A. (1987) Soybean–wheat doublecropping: Implications from straw management and supplemental nitrogen. *Agronomy Journal* 79, 281–286.

Herridge, D.F. and Holland, J.F. (1992) Production of summer crops in northern New South Wales. I. Effects of tillage and double cropping on growth, grain and N yields of six crops. *Australian Journal of Agricultural Research* 43, 105–122.

Herridge, D.F., Roughley, R.J. and Brockwell, J. (1984) Effect of rhizobia and soil nitrate on the establishment and functioning of the soybean symbiosis in the field. *Australian Journal of Agricultural Research* 35, 149–161.

Herridge, D.F., Turpin, J.E. and Robertson, M.J. (2001) Improving nitrogen fixation of crop legumes through breeding and agronomic management: Analysis with simulation modelling. *Australian Journal of Experimental Agriculture* 41, 391–401.

Herridge, D.F., Peoples, M.B. and Boddey, R.M. (2008) Global inputs of biological nitrogen fixation in agricultural systems. *Plant and Soil* 311, 1–18.

Ho, S.J., Hee, L.S., Son, C.Y., Eun, L.J., Keun, L.C. and Up, K.W. (2008) Estimation of soybean N fraction derived from N sources by ^{15}N in soybean cultivation with rye as green manure. *Korean Journal of Crop Science* 53, 50–57.

Hong, N., Crozier, C.R., White, J.G., Weisz, R., Heiniger, R.W. and Sripada, R.P. (2004) Remote sensing for precision N management in a corn–wheat–soybean rotation. In: *Proceedings of the 7th International Conference on Precision Agriculture and Other Precision Resources Management,* Hyatt Regency, Minneapolis, MN, USA, 25–28 July 2004, pp. 1351–1365.

Houngnandan, P., Yemadje, R.G.H., Oikeh, S.D., Djidohokpin, C.F., Boeckx, P. and Cleemput, O.V. (2008) Improved estimation of biological nitrogen fixation of soybean cultivars (*Glycine max* L. Merril) using ^{15}N natural abundance technique. *Biology and Fertility of Soils* 45, 175–183.

Hua, D.A., Ming, M.J., He, L.X., Xin, W.J. and Ming, Z.Q. (2003) Studies on effects of intercropping castor and black soybean on biological prevention and control of major pests of black soybean. *Journal of Jilin Agricultural University* 25, 598–601.

Hughes, R.M. and Herridge, D.F. (1989) Effect of tillage on yield, nodulation and nitrogen fixation of soybean in far north-coastal New South Wales. *Australian Journal of Experimental Agriculture* 29, 671–677.

Jagadamma, S., Lal, R., Hoeft, R.G., Nafziger, E.D. and Adee, E.A. (2008) Nitrogen fertilization and cropping system impacts on soil properties and their relationship to crop yield in the central Corn Belt, USA. *Soil and Tillage Research* 98, 120–129.

Javaid, I., Dadson, R.B., Hashem, F.M. and Joshi, J. (2006) Antibiosis of forage soybean as an ecological alternative for the control of corn earworm. *Agronomy for Sustainable Development* 26, 55–59.

Jensen, E.S. and Hauggaard-Nielsen, H. (2003) How can increased use of biological N_2 fixation in agriculture benefit the environment? *Plant and Soil* 252, 177–186.

Joon, P.S., Han, K.W., Eun, L.J., Up, K.Y., Chul, S.J., Hwan, R.Y. and Chun, S.R. (2005) Nitrogen balance and biological nitrogen fixation of soybean in soybean–barley cropping system. *Korean Journal of Crop Science* 50, 1–4.

Jun, W.J., Li, L.S., Tong, L.G., Yong, Z. and Xin, K. (2005) Studies on the pattern of cultivation of spring-sown soybean in cotton fields in northern Xinjiang. *Xinjiang Agricultural Sciences* 42, 192–194.

Jung, G., Matsunami, T., Oki, Y. and Kokubun, M. (2008) Effects of waterlogging on nitrogen fixation and photosynthesis in supernodulating soybean cultivar Kanto 100. *Plant Production Science* 11, 291–297.

Kanwar, R.S., Cruse, R.M., Ghaffarzadeh, M., Bakhsh, A., Karlen, D.L. and Bailey, T.B. (2005) Corn–soybean and alternative cropping systems effects on NO_3-N leaching losses in subsurface drainage water. *Applied Engineering in Agriculture* 21, 181–188.

Kelley, K.W., Long Jr., J.H. and Todd, T.C. (2003) Long-term crop rotations affect soybean yield, seed weight, and soil chemical properties. *Field Crops Research* 83, 41–50.

Kumbhar, A.M., Buriro, U.A., Kumbhar, M.B., Oad, F.C., Jamro, G.H. and Chachar, Q.I. (2007) Yield of wheat, cotton, berseem and soybean under different crop sequences and fertility regimes. *Asian Journal of Plant Sciences* 6, 143–147.

Kundu, S., Bhattacharyya, R., Prakash, V., Gupta, H.S., Pathak, H. and Ladha, J.K. (2007) Long-term yield trend and sustainability of rainfed soybean–wheat system through farmyard manure application in a sandy loam soil of the Indian Himalayas. *Biology and Fertility of Soils* 43, 271–280.

Lamb, J.A., Dowdy, R.H., Anderson, J.L. and Allmaras, R.R. (1998) Water quality in an irrigated sandy soil: Ridge tillage in rotated corn and soybean compared with full-width tillage in continuous corn. *Soil and Tillage Research* 48, 167–177.

Liu, X. and Herbert, S.J. (2002) Fifteen years of research examining cultivation of continuous soybean in northeast China: A review. *Field Crops Research* 79, 1–7.

Long, L.J., Feng, T.P., Na, C., Chi, T.J., Rong, W.X., Hai, N., Hong, L. and Long, Y.X. (2008) Effects of tea intercropping with soybean. *Scientia Agricultura Sinica* 41, 2040–2047.

Meriles, J.M., Gil, V.S., Haro, R.J., March, G.J. and Guzman, C.A. (2006) Glyphosate and previous crop residue effect on deleterious and beneficial soil-borne fungi from a peanut–corn–soybean rotations. *Journal of Phytopathology* 154, 309–316.

Ming, L.W. and Ming, Y.A. (2004) Study on optimization combination for intercropping spring corn with spring soybean. *Acta Agriculturae Zhejiangensis* 16, 131–135.

Mohammad, M.R. (2009) The comparison of yield in pure soybean (*Glycine max*) and intercropping. In: Qiu, L., Guan, R., Jin, J., Song, Q., Guo, S., Li, W., Wang, Y., Han, T., Liu, X., Yu, D., Jiang, L. and Peng, D. (eds) *Developing a Global Soy Blueprint for a Safe Secure and Sustainable Supply. Abstracts for oral presentations and posters of the World Soybean Research Conference VIII*, Beijing, China, 10–15 August 2009. The Chinese Academy of Agricultural Sciences, Beijing, China, p. 236.

Monzon, J.P., Sadras, V.O., Abbate, P.A. and Caviglia, O.P. (2007) Modelling management strategies for wheat–soybean double crops in the southeastern Pampas. *Field Crops Research* 101, 44–52.

Ndakidemi, P.A., Dakora, F.D., Nkonya, E.M., Ringo, D. and Mansoor, H. (2006) Yield and economic benefits of common bean (*Phaseolus vulgaris*) and soybean (*Glycine max*) inoculation in northern Tanzania. *Australian Journal of Experimental Agriculture* 46, 571–577.

Nemade, S.M., Solanke, A.V., Mohod, N.B., Shinde, R.H. and Mohod, A.B. (2008) Economics of soybean potato cropping sequence with integrated nutrient management systems. *Journal of Maharashtra Agricultural Universities* 33, 252–253.

Nian, H. and Cheng, Y.B. (2009) Soybean multiple cropping patterns in South China. In: Qiu, L., Guan, R., Jin, J., Song, Q., Guo, S., Li, W., Wang, Y., Han, T., Liu, X., Yu, D., Jiang, L. and Peng, D. (eds) *Developing a Global Soy Blueprint for a Safe Secure and Sustainable Supply. Abstracts for oral presentations and posters of the World Soybean Research Conference VIII*, Beijing, China, August 10–15, 2009. The Chinese Academy of Agricultural Sciences, Beijing, China, p. 27.

Nicolas, M.F., Hungria, M. and Arias, C.A.A. (2006) Identification of quantitative trait loci controlling nodulation and shoot mass in progenies from two Brazilian soybean cultivars. *Field Crops Research* 95, 355–366.

Oberson, A., Nanzer, S., Bosshard, C., Dubois, D., Mäder, P. and Frossard, E. (2007) Symbiotic N_2 fixation by soybean in organic and conventional cropping systems estimated by ^{15}N dilution and ^{15}N natural abundance. *Plant and Soil* 290, 69–83.

Okogun, J.A. and Sanginga, N. (2003) Can introduced and indigenous rhizobial strains compete for nodule formation by promiscuous soybean in the moist savanna agroecological zone of Nigeria? *Biology and Fertility of Soils* 38, 26–31.

Okogun, J.A., Sanginga, N., Abaidoo, R., Dashiell, K.E. and Diels, J. (2005) On-farm evaluation of biological nitrogen fixation potential and grain yield of *Lablab* and two soybean varieties in the northern Guinea savanna of Nigeria. *Nutrient Cycling in Agroecosystems* 73, 267–275.

Okogun, J.A., Sanginga, N. and Abaidoo, R.C. (2007) Evaluation of maize yield in an on-farm maize–soybean and maize–*Lablab* crop rotation systems in the northern Guinea savanna of Nigeria. *Pakistan Journal of Biological Sciences* 10, 3905–3909.

Omay, A.B., Rice, C.W., Maddux, L.D. and Gordon, W.B. (1998) Corn yield and nitrogen uptake in monoculture and in rotation with soybean. *Soil Science Society of America Journal* 62, 1596–1603.

Osunde, A.O., Bala, A., Gwam, M.S., Tsado, P.A., Sanginga, N. and Okogun, J.A. (2003) Residual benefits of promiscuous soybean to maize (*Zea mays* L.) grown on farmers' fields around Minna in the southern Guinea savanna zone of Nigeria. *Agriculture, Ecosystems and Environment* 100, 209–220.

Pandey, M. (1996) Effect of photosynthate availability on N_2-fixation in soybean nodules. *Indian Journal of Plant Physiology* 1, 18–20.

Peoples, M.B., Gault, R.R., Lean, B., Sykes, J.D. and Brockwell, J. (1995) Nitrogen fixation by soybean in commercial irrigated crops of central and southern New South Wales. *Soil Biology and Biochemistry* 27, 553–561.

Peoples, M.B., McLennan, P.D. and Brockwell, J. (2008) Hydrogen emission from nodulated soybeans [*Glycine max* (L.) Merr.] and consequences for the productivity of a subsequent maize (*Zea mays* L.) crop. *Plant and Soil* 307, 67–82.

Puiatti, M. and Sodek, L. (1999) Waterlogging affects nitrogen transport in the xylem of soybean. *Plant Physiology and Biochemistry, Paris.* 37, 767–773.

Qun, Z.Y, You, H.W., Zhi, W.B., Dong, Z.F. and Li, Y. (2008) Effect of chilli pepper intercropping system on nutrient utilization, main diseases and pests and yield of chilli pepper. *Journal of Wuhan Botanical Research* 26, 412–416.

Rahmani, H.A., Saleh-rastin, N., Khavazi, K., Asgharzadeh, A., Fewer, D., Kiani, S. and Lindström, K. (2009) Selection of thermotolerant bradyrhizobial strains for nodulation of soybean (*Glycine max* L.) in semi-arid regions of Iran. *World Journal of Microbiology and Biotechnology* 25, 591–600.

Ramesh, P. and Reddy, K.S. (2004) Productivity and nutrient balance of soybean (*Glycine max*)-based cropping systems as influenced by nitrogen levels in deep Vertisols of central India. *Indian Journal of Agronomy* 49, 140–142.

Reddy, S.V.K., Krishna, S.K., Rao, J.A.V.P., Kumar, P.H. and Krishnamurthy, V. (2007) Effect of application of biofertilizers to soybean (*Glycine max*) and nitrogen to tobacco (*Nicotiana tabacum*) in soybean–tobacco cropping system. *Indian Journal of Agronomy* 52, 294–299.

Rochester, I.J., Peoples, M.B., Constable, G.A. and Gault, R.R. (1998) Faba beans and other legumes add nitrogen to irrigated cotton cropping systems. *Australian Journal of Experimental Agriculture* 38, 253–260.

Rochester, I.J., Peoples, M.B., Hulugalle, N.R., Gault, R.R. and Constable, G.A. (2001) Using legumes to enhance nitrogen fertility and improve soil condition in cotton cropping systems. *Field Crops Research* 70, 27–41.

Rose, S.J., Burnside, O.C., Specht, J.E. and Swisher, B.A. (1984) Competition and allelopathy between soybeans and weeds. *Agronomy Journal* 76, 523–528.

Russelle, M.P. and Birr, A.S. (2004) Large-scale assessment of symbiotic dinitrogen fixation by crops: Soybean and alfalfa in the Mississippi River Basin. *Agronomy Journal* 96, 1754–1760.

Salvagiotti, F., Cassman, K.G., Specht, J.E., Walters, D.T., Weiss, A. and Dobermann, A. (2008) Nitrogen uptake, fixation and response to fertilizer N in soybeans: A review. *Field Crops Research* 108, 1–13.

Sanginga, N. (2003) Role of biological nitrogen fixation in legume based cropping systems: A case study of West Africa farming systems. *Plant and Soil* 252, 25–39.

Sanginga, N., Dashiell, K., Okogun, J.A. and Thottappilly, G. (1997) Nitrogen fixation and N contribution in promiscuous soyabeans in southern Guinea savanna of Nigeria. *Plant and Soil* 195, 257–266.

Sharma, A.R. and Behera, U.K. (2009) Recycling of legume residues for nitrogen economy and higher productivity in maize (*Zea mays*)–wheat (*Triticum aestivum*) cropping system. *Nutrient Cycling in Agroecosystems* 83, 197–210.

Shiraiwa, T., Sinclair, T.R. and Hashikawa, U. (1994) Variability in nitrogen fixation activity among soybean cultivars grown under field conditions. *Japanese Journal of Crop Science* 63, 111–117.

Shiraiwa, T., Sakashita, M., Yagi, Y. and Horie, T. (2006) Nitrogen fixation and seed yield in soybean under moderate high-temperature stress. *Plant Production Science* 9, 165–167.

Shoko, M.D. and Tagwira, F. (2007) Benefits of soyabeans as a breakcrop in sugarcane production systems in the South Eastern Lowveld of Zimbabwe. *Sugar Journal* 70, 18–22.

Singh, A.B., Saha, J.K. and Ghosh, P.K. (2008) Effect of nutrient management practices on soybean (*Glycine max*)–chickpea (*Cicer arietinum*) cropping system for improving seed yield, quality and soil biological health under rainfed condition. *Indian Journal of Agricultural Sciences* 78, 485–489.

Singh, M., Kundu, S., Biswas, A.K., Saha, J.K., Tripathi, A.K. and Acharya, C.L. (2004) Quantification of N_2 fixation and annual N benefit from N_2 fixation in soybean accrued to the soil under soybean–wheat continuous rotation. *Journal of Plant Nutrition and Soil Science* 167, 577–583.

Sridhara, S., Thimmegowda, S. and Prasad, T.G. (1995) Effect of water regimes and moisture stress at different growth stages on nodule dynamics, nitrogenase activity and nitrogen fixation in soybean [*Glycine max* (L.) Merrill]. *Journal of Agronomy and Crop Science* 174, 111–115.

Staggenborg, S.A., Whitney, D.A., Fjell, D.L. and Shroyer, J.P. (2003) Seeding and nitrogen rates required to optimize winter wheat yields following grain sorghum and soybean. *Agronomy Journal* 95, 253–259.

Sung, F.J.M. (1993) Waterlogging effect on nodule nitrogenase and leaf nitrate reductase activities in soybean. *Field Crops Research* 35, 183–189.

Toomsan, B., McDonagh, J.F., Limpinuntana, V. and Giller, K.E. (1995) Nitrogen fixation by groundnut and soyabean and residual nitrogen benefits to rice in farmers' fields in Northeast Thailand. *Plant and Soil* 175, 45–56.

Unkovich, M.J. and Pate, J.S. (2000) An appraisal of recent field measurements of symbiotic N_2 fixation by annual legumes. *Field Crops Research* 65, 211–228.

Varvel, G.E. and Wilhelm, W.W. (2003) Soybean nitrogen contribution to corn and sorghum in Western Corn Belt rotations. *Agronomy Journal* 95, 1220–1225.

Vyas, A.K., Billore, S.D., Chauhan, G.S. and Pandya, N. (2008) Agro-economic analysis of soybean-based cropping systems. *Indian Journal of Fertilisers* 4, 41–42 & 45–51.

Waluyo, S.H., An, L.T. and Mannetje, L. (2004) Effect of phosphate on nodule primordia of soybean (*Glycine max* Merrill) in acid soils in rhizotron experiments. *Indonesian Journal of Agricultural Science* 5, 37–44.

Wang, F., Sun, Jia-bao and Zu, Yuan-gang (2006) The influence of larch–soybean intercropping on the growth characteristics of Xingan larch plantations in northeast China. *Eurasian Journal of Forest Research* 9, 45–50.

Wen, W.C., Chang, Y.H., Kuan, C.X., Yong, L.J., Xin, L., Jun, Z. and Guang, W.J. (2004) Effect of soybean intercropping on the growth and yield of sugarcane at seedling stage. *South-west China Journal of Agricultural Sciences* 17, 645–650.

Zhu, W. and Yu, Y.W. (2007) New system of soybean planting in hilly soils of South China. *Crop Research, Hisar* 34, 35–38.

Zhu, W., Yu, Y.W. and Lin, W.Q. (2007) Effects of shading in maize/soybean relay-cropping system on the photosynthetic characteristics and yield of soybean. *Acta Agronomica Sinica* 33, 1502–1507.

Zhu, Y. and Fox, R.H. (2003) Corn–soybean rotation effects on nitrate leaching. *Agronomy Journal* 95, 1028–1033.

3 Soybean Growth and Development

Saratha Kumudini
Department of Plant and Soil Sciences, University of Kentucky, Lexington, Kentucky, USA

3.1 Introduction

The process of plant growth and development is important to the successful adaptation of a species to its geographic and climatic environment. Adaptation of a species to the growing season of a region ensures the species' reproductive success. In annual species, the seed must germinate, grow, flower, set seed and mature within the growing season or risk reproductive failure. The developmental process can also be important in improving a crop's yield potential. The yield potential of a crop can be improved by tailoring the occurrence of important developmental phases to coincide with the occurrence of favourable ambient conditions.

Advances in plant breeding have resulted in the development of crop species that are adapted to a number of geographic and climatic regions of the world, thereby extending their area of production from the initial region of adaptation. Geographical and historical evidence suggests that soybean (*Glycine max* (L.) Merrill) first emerged as a domesticate in the eastern part of northern China (*c.* 1500–100 BC) (Hymowitz, 2004). From there, the crop was introduced into other regions of Asia; soybean landraces have been found in Japan, Indonesia, the Philippines, Vietnam, Thailand, Malaysia, Myanmar, Nepal and north India. Interest in the crop grew in Europe and the USA in the early 1900s. In the USA, soybean was grown predominantly as a forage crop for many years, before it was grown for grain (Probst and Judd, 1973). Today, the crop is produced throughout the world including much of North America, South America and Asia. The USA and Brazil are the world's largest producers of soybean.

Soybean belongs to the family *Fabaceae*, genus *Glycine* and subgenus *Soja* (Moench) F.J. Herm. This subgenus is comprised of the annuals of the genus. The cultivated soybean has an erect, bushy and annual growth habit. The form and structure of a soybean plant varies vastly. This variation is partly due to selection pressures during the development of landraces in East Asia. In East Asia soybean has been grown for a variety of specialty uses such as for food, feed, medicinal, religious and ceremonial purposes (Hymowitz, 2004). Researchers have detailed much of the vegetative and reproductive morphological characteristics of the soybean plant (Carlson, 1973; Carlson and Lersten, 2004; Lersten and Carlson, 2004).

The growth and developmental processes of a crop can have important implications on the success of the plant as a crop species. Soybean production today includes production areas that are widely disparate from its region of origin. This crop's wide geographic adaptation, reproductive success and yield potential are at least in part due to the nature of its growth and development processes. The aim of this chapter is to give an overview of the vegetative and reproductive structures of soybean and then discuss the environmental and genetic factors that control soybean growth and development from seedling to maturity.

3.2 Vegetative and Reproductive Morphology

Leaves

The soybean plant has four different leaf structures: seed (cotyledon) leaves, primary (unifoliolate) leaves, trifoliolate leaves and prophylls. The pair of seed leaves are oppositely arranged and occur first on the plant. Next are a pair of ovate-shaped, oppositely arranged, primary (unifoliolate) leaves. The node refers to the part of the stem where the leaves attach. The first two leaf types occur on the first two nodes (Fig. 3.1). All subsequent nodes have the alternatively arranged, trifoliolate leaves. Individual leaflets have entire margins and range in shape from oblong to ovate to lanceolate. On occasion, the alternatively arranged leaves may have four to seven leaflets and lateral leaflets may fuse with the terminal leaflets. Pulvini are found at the point of attachment of the petiole to the stem of each primary and trifoliolate leaf (Fig. 3.1). Smaller pulvini occur at the base of each petiolule. Changes in pulvinar osmotic pressure allow for the diurnal movement of soybean leaves and leaflets (Lersten and Carlson, 2004). Prophylls are the fourth leaf type. They occur as small pairs of simple leaves, found at the base of lateral branches and the lower part of the pedicel of each flower (Hicks, 1978). Prophylls lack petioles and pulvini.

All commercial cultivars of soybean are pubescent. Trichomes can be found on leaves, stems, sepals and pods. There is genetic variation for trichome density, which includes glabrous genotypes, although these are not generally commercially viable as they are prone to heavy insect damage. Stomata are present on both the adaxial and abaxial leaf surfaces, with significantly more stomata on the abaxial surface (Carlson, 1973).

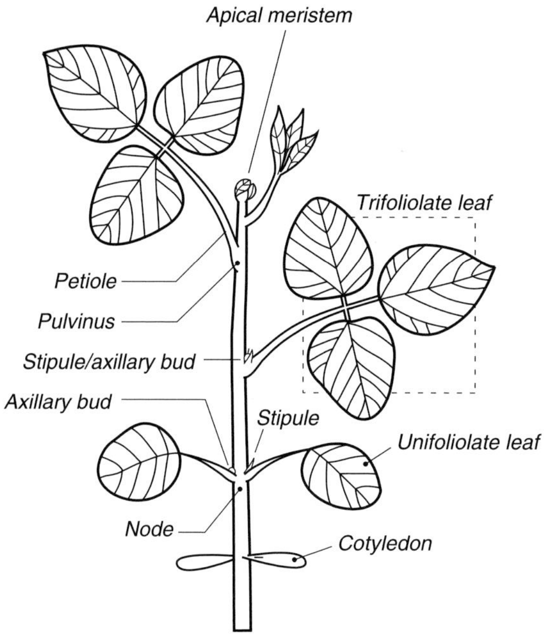

Fig. 3.1. Diagram of a young, vegetative soybean plant.

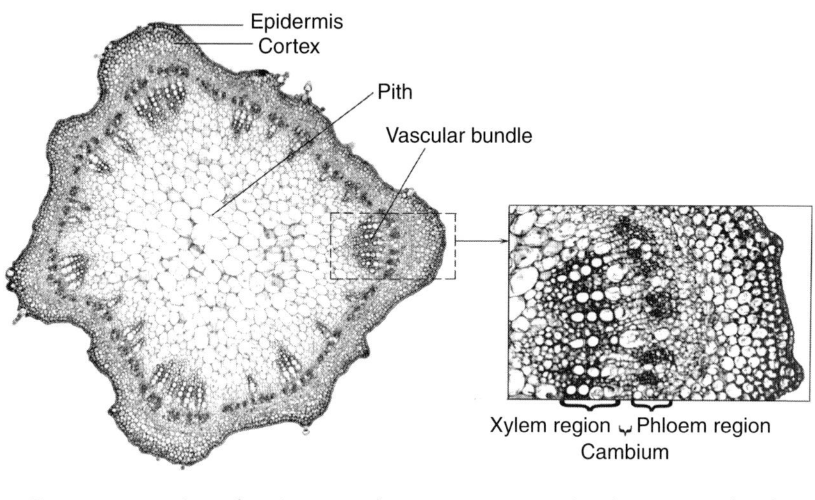

Fig. 3.2. Transverse section of an intact soybean stem (4×), with a large central pith, a circular arrangement of vascular bundles and a narrow cortex below the epidermis. The detailed vascular bundle (10×) shows the xylem region (primary and secondary xylem), the cambium and the phloem region (phloem and phloem fibres).

Stems

The mature primary stem consists of a central pith of thin-walled parenchyma cells lacking chloroplasts, a zone of vascular bundles arranged in a circular pattern and a cortex layer between the vascular bundles and the epidermis (Fig. 3.2). The trichomes on the epidermis resemble those in the leaf, as described above. Parenchyma cells extending from the pith separate the vascular bundles and merge with the cortex cells. Stem vascular bundles are of the collateral type, with xylem towards the pith, phloem towards the cortex and a strip of potential cambial cells in between (Fig. 3.2).

As the plant develops, the stem undergoes secondary growth. To accommodate the additional demands placed by the growing stem, additional vascular and supporting tissues are added: the cambial tissue between the primary xylem and phloem adds secondary xylem and phloem tissue. This occurs first only within the vascular bundles, but as the cambial tissue activity increases, meristematic tissues also begin to form between each vascular bundle to form a complete cylinder of meristematic tissue. The newly formed complete cylinder of cambial tissue then forms a complete ring of secondary xylem and phloem tissue. In regions of the soybean stem with considerable secondary growth, the pith cells collapse to form a hollow stem.

Roots

Soybean roots are composed of an outer layer of epidermis. Root hairs form from epidermal cells as early as 4 days after germination and about 1 cm from the tip of the primary root. Inside the epidermis is a large cortex area made of parenchyma cells. Food storage is apparently not a function of the cortex, since starch grains are rarely observed in the parenchyma cells (Carlson, 1973). The inner layer of the cortex differentiates into the endodermis. The central stele includes the pericycle, phloem and the metaxylem and protoxylem arranged in a characteristic tetrarch pattern. The pericycle makes up the outer region of the stele. Secondary growth due to activity of the vascular cambium produces a central rounded core of xylem surrounded by phloem, somewhat obscuring the tetrarch protoxylem pattern (Fig. 3.3).

The protoxylem points of the tetrarch are the points from which the branch roots arise. Therefore, the xylem pattern determines the number of rows of branch roots. Radial elongation of the pericycle is the first indication of the initiation of a branch root. The root apex of the branch forces its way through the endodermis, cortex and epidermis to finally break through into the soil. Secondary roots emerge acropetally as the primary root increases in length (Sun, 1955).

Nodules form on the roots of soybean plants as a consequence of a mutually beneficial relationship between the plant and *Bradyrhizobium japonicum*, a Gram-negative bacterium present in the soil. This bacterium multiplies within the soybean root nodules and obtains carbon-rich energy compounds from the plant. In exchange, the bacterium reduces atmospheric nitrogen into ammonia, which is subsequently utilized by the plant. Biological nitrogen fixation is restricted to these and other prokaryotes that possess the nitrogenase enzyme; in the absence of nitrogenase, reduction of atmospheric nitrogen would not be possible. The nitrogenase enzyme is irreversibly inactivated by oxygen and, therefore, biological nitrogen fixation requires anoxic or nearly anoxic conditions. In soybean, in order to

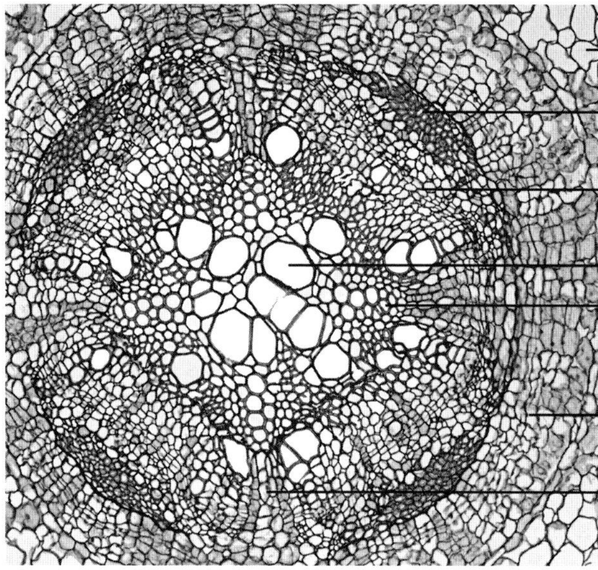

Fig. 3.3. Transverse section of a soybean root after initiation of secondary growth. Secondary growth has begun to obscure the tetrarch pattern formed by the protoxylem at the centre of the root.

protect nitrogenase from an oxic environment, nitrogen fixation occurs within the root nodules in specialized tissues of the soybean root with morphological and biochemical characteristics that limit the exposure of nitrogenase to oxygen (Lhuissier *et al.*, 2001; Timmers, 2008). Soybean root nodules appear as visible spherical swellings of the root. In the presence of the bacterium, several hundred nodules per plant may be found, to depths as deep as a metre below the surface (Grubinger *et al.*, 1982). There is a well-orchestrated and interactive play between the bacterium and the soybean root as they identify the presence of the other. Further to this, processes and changes occur that lead to the initiation and development of the soybean root nodule. These processes have been characterized and well detailed in a number of reviews (Lhuisssier *et al.*, 2001; Lersten and Carlson, 2004; Oldroyd and Downie, 2008; Timmers, 2008).

Flowers

The fully developed flower of a soybean can be described as typical of its *Papilionoideae* subfamily. Its five petals consist of one large posterior banner petal, two lateral wing petals and two anterior keel petals (Fig. 3.4). A ring of ten diadelphous stamen filaments (nine fused and one free stamen filament) surrounds the pistil. The single unicarpellate pistil has four ovules. The style curves back towards the free posterior stamen and terminates in a capitate stigma. Trichomes are present on the pistil, tubular calyx (with five

Fig. 3.4. The mature soybean flower with (a) a single banner petal, (b) two wing petals, (c) two keel petals and (d) ten stamens arranged around (e) the pistil and (f) a tubular calyx.

unequal sepal lobes), bracts and bracteoles of the flower, but are not present in the petals or stamens (Guard, 1931).

Soybean flowers develop from axillary buds on the main stem and branches. Once induced, the axillary buds develop into floral racemes typically with clusters of eight to 16 flowers, although clusters of two to 35 flowers have been reported (Piper and Morse, 1923; Guard, 1931). When inflorescences are initiated, there are marked changes in the development of the axillary buds. The two opposite prophyll primordia are initiated as in vegetative development, except that the typical distichous phyllotaxy of the leaf primordia is replaced with a spiral two-fifth phyllotaxy. The prophylls develop into the floral bracts. A knob-like primordium in the axil of the bract is the first sign of a developing flower. The first whorl of floral organs to be initiated is the sepals. Next, the calyx tube emerges from the base. The next whorl to appear are the petals, which develop very slowly and are soon surpassed by the outer and inner whorl of stamens, followed lastly by the development of the single free stamen. At about the same time as the last whorl of stamens, the carpel primordium also appears. Simultaneous ovule and carpel development occurs (Guard, 1931). Staminal tube, free stamen and style growth are synchronous so that the anthers are lifted to the stigma at maturity (Johns and Palmer, 1982). Therefore, at maturity, the pollen are shed directly on the stigma, resulting in a high percentage of self-fertilization (Williams, 1950). Natural crossing varies, but is at most 1% (Carlson, 1973).

Pods and seeds

The first cell division of the zygote typically occurs 32 h after fertilization (Carlson and Lersten, 2004). Within 7 days post-fertilization, the cotyledonary leaves are initiated. The initial position of the cotyledons is displaced 90° from the final position in the mature seed. As the seed develops, the cotyledons rotate and assume their final position at about 10–14 days after fertilization. The primordia of the two primary leaves are initiated at about this point. The primary leaves continue to enlarge until they reach their maximum dormant embryo size. As the primary leaves reach their maximum size, the leaf primordium for the first trifoliolate is initiated. During this period, assimilates accumulate in the cotyledons.

After fertilization, the flower style and stigma dry out while the calyx persists and the ovary starts developing into the fruit. The soybean pod consists of the two halves of the single carpel joined by a dorsal and ventral suture, which itself consists of the main and marginal veins of the former carpel. The wall of the young pod is composed of an epidermis with varying degrees of trichome density. Beneath the epidermis is a wide zone of parenchyma tissue in which the extensive vascular system is embedded, and an inner zone of parenchyma tissue that will form the membranous endocarp. As the pod matures, the outer epidermal cells develop thickened walls covered by a well-developed cuticle. Separation of the two halves of the pod at maturity is preceded by the appearance of clefts in the parenchyma of the dorsal and ventral sutures (Carlson and Lersten, 2004).

The mature soybean seed, generally oval in shape, consists of a seed coat surrounding a large embryo. The seed coat has a hilum that varies in shape and colour. At one end of the hilum is the micropyle, a tiny hole formed during seed development that becomes covered by a cuticle at maturity. The dormant embryo consists of the two fleshy cotyledons, two well-developed primary leaves enclosing a trifoliolate leaf primordium and a hypocotyl-radicle axis.

3.3 Growth and Development

Seed germination and seedling development

Under the appropriate temperature and moisture, the soybean seed will imbibe water and initiate epigeous germination. The imbibition of water changes the normally oval-shaped seed to a reniform (kidney) shape. Once both the seed coat and the embryo are fully imbibed, the radicle emerges by breaking the seed coat in the region of the micropyle. The radicle then grows rapidly down into the soil. The soybean stem originates from the embryo axis, which consists of the hypocotyl and the epicotyl. The hypocotyl is the base of the stem and lies below the epicotyl. The epicotyl consists of the cotyledons, the two primary leaves and the apical bud.

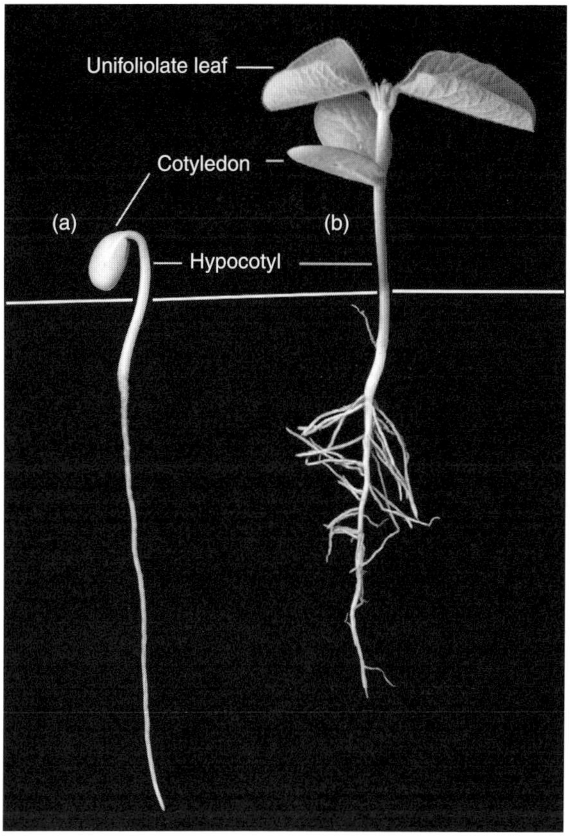

Fig. 3.5. Soybean seedlings at the (a) VE (emergence) and (b) VC growth stage. Note that the leaf edges of the unifoliolate leaves are no longer touching.

Fehr and Caviness (1977) developed a system to stage soybean development that has gained wide acceptance. The earliest vegetative stage they describe is denoted as VE (Fig. 3.5a) and corresponds to emergence. The emergence stage refers to the point at which the cotyledons are above the soil surface. During germination and emergence the hypocotyl elongates, elevating the cotyledons above the ground (Fig. 3.5a). Once at the soil surface, the cotyledons turn green and photosynthetic. Microbodies associated with the conversion of stored lipids to hexose sugars are found in the cotyledons of germinating seeds (Liu *et al.*, 1971). Therefore, the cotyledons provide nutrients through the mobilization of stored reserves as well as through the production of photosynthetic assimilates. The stem system above the hypocotyl emerges from the epicotyl and the axillary buds of the cotyledons. Following the elongation of the radicle, root branching also commences, signalling the development of the soybean root system. After the emergence of the cotyledons above the soil surface, the two embryonic primary leaves in the apical bud begin to expand. When the primary leaves have expanded to a point where their edges no longer touch, the plant is considered to be at the VC growth stage (Fig. 3.5b) (Fehr and Caviness,

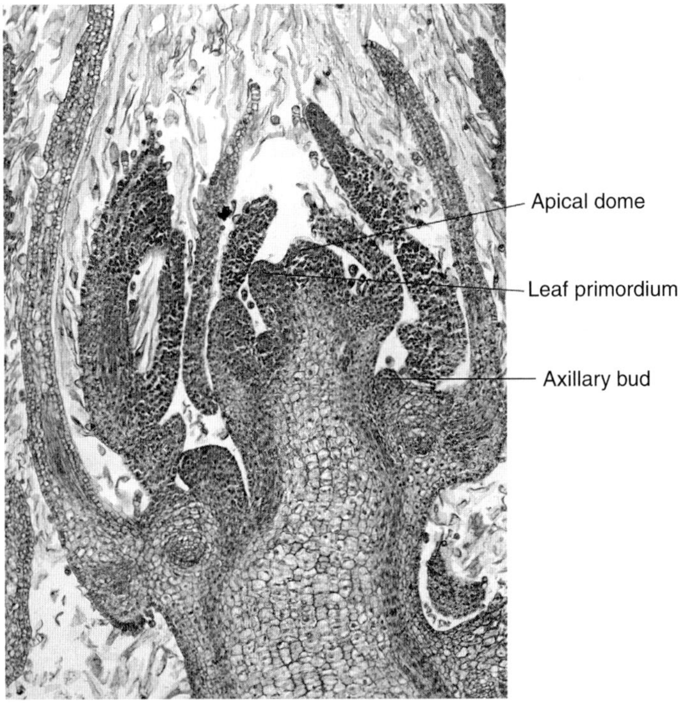

Fig. 3.6. Longitudinal section of the shoot apex of a young soybean plant (24 days after planting). Note the trifoliolate leaf primordium beside the apical dome.

1977). After the expansion of the primary leaves, the trifoliolate leaf primordium initiates the development of the first trifoliolate leaf. Stem growth continues as the stem apical bud initiates trifoliolate leaves in an alternate pattern along the main stem. Axillary buds form at the leaf axils (Fig. 3.6).

Vegetative development

Vegetative development in soybean is quantified based on main stem node accrual (Fehr and Caviness, 1977). All vegetative stages following VE and VC are designated as Vn, where n is determined by counting the number of nodes on the main stem, beginning with the unifoliolate nodes that have or have had a fully developed leaf. A leaf is considered to be fully developed when the leaf at the node directly above it has expanded sufficiently that the edges of the leaflets are no longer touching.

Vegetative development impacts the size and structure of the soybean canopy. The rate of development of main stem nodes, the final main stem node number and branching off of the main stem can all influence soybean morphology. Both environmental and genetic factors have been reported to regulate soybean vegetative development and thus influence canopy size and morphology.

The rate of node development may be estimated from the plastochron, since the node is the point of attachment of the leaf to the stem. The plastochron is the time interval between the initiation of two consecutive leaf primordia on the shoot apex. Based on observations of the longitudinal sections of the shoot apex, Sun (1957) described the initiation of leaf buds in soybean. He found that when an individual primordium reaches a height of about 80–90 μm, the next leaf primordium is initiated. The leaf primordia arise laterally on the shoot apex about 30–50 μm from the top (Sun, 1957). The enlarging leaf primordium appears as a bulge on the side of the apical dome (Fig. 3.6). When the leaf primordium is 140–200 μm high, cell divisions at two points on the adaxial margins result in the production of two papillate leaflet primordia. Miksche (1961) described soybean leaf initiation in terms of time, and observed that, on average, successive leaf primordia are generally initiated at 2-day intervals. Although the macroscopic appearance of leaves on the main stem follows the initiation of leaves on the apical primordium, there has not been a consensus as to whether the time of leaf initiation and the time of leaf appearance are related (Hunt and Chapleau, 1986; Hay and Kirby, 1991).

The rate of node development has been related to the rate of leaf appearance, the reciprocal of which is the phyllochron. The phyllochron is the interval between the macroscopic appearance of successive leaves on the main stem. Both the phyllochron and the rate of leaf appearance can be readily measured in a field setting, but the phyllochron is best determined under controlled-environment conditions and then tested under field conditions. In an attempt to calculate the soybean phyllochron, researchers have used a variety of methodologies. For example, both leaf appearance (measured as when a leaf reaches a certain leaf area) and nodal accrual (V-staging, as in Fehr and Caviness, 1977) have been used. These studies were sometimes carried out under field conditions and sometimes under controlled-environment conditions. Hesketh *et al.* (1973) calculated that the days per trifoliolate on the main stem was fairly similar between determinate and indeterminate cultivars ('Dare' and 'Wayne', respectively), but that it was longer for the first two nodes relative to subsequent nodes. They reported that the phyllochron ranged from about 2 to 8 days per trifoliolate depending on temperature – the cooler the mean temperature, the longer the time interval between the appearance of successive trifoliolates. Their findings were supported by Hofstra *et al.* (1977), who estimated the soybean phyllochron to range from approximately 2.5 to 6 days per trifoliolate depending on temperature. Consistent with Hesketh *et al.* (1973), Fehr and Caviness (1977), using node numbers (V-staging), also noted that node accrual was lower during early vegetative development. They reported that after the V5 growth stage, node accrual can range from 2 to 5 days, depending on temperature. Wilcox *et al.* (1995), using field node number data, reported an average phyllochron of 3.3 days that was consistent across soybean lines of varying maturity. Bastidas *et al.* (2008), using field node accrual data, reported rates of 3.7 to 4.1 days per node. As in other reports, they also noted that early soybean development did involve a lower rate of node accrual, but that this was only apparent until the V1 growth stage.

Although the rate of node accrual appears to be similar across genotypes, the size and structure of the soybean canopy can vary greatly depending on environmental and genetic factors. Two gene series known to impact soybean vegetative morphology and canopy size are the *Dt* and *E*-gene series.

The *Dt* genes control soybean growth habit and are known to have a significant impact on the final main stem node number (Bernard, 1972; Curtis *et al.*, 2000). There are two main types of growth habits (also referred to as stem types) in soybean: determinate and indeterminate. The indeterminate stem types trace their ancestry to northeastern China, whereas the determinate types can be found in south China as well as in Japan and Korea (Nagata, 1960). In North America, cultivars of maturity group (MG) 000 to IV, grown in the more northern regions of the continent, tend to be indeterminates. Cultivars of MG V and higher, grown in the southern USA, tend to be predominantly determinate stem types. These two major stem types in soybean are regulated by alleles at two *Dt* loci (Dt_1 and Dt_2) (Bernard, 1972; Thompson *et al.*, 1997).

In the determinate stem type (dt_1, dt_1, *Dt2*, *Dt2* or dt_1, dt_1, dt_2, dt_2), vegetative growth of the main stem stops shortly after flowering begins (Bernard, 1972). This has been attributed to the cessation of vegetative growth on the terminal bud when it becomes a terminal inflorescence. Cessation of vegetative growth on the main stem will cut short the continued accrual of nodes on the main stem, thereby reducing the final main stem node number of determinate genotypes.

The determinate phenotype is characterized by a long inflorescence at the apex. Determinate plants have both axillary racemes and a single terminal raceme. However, Bernard (1972) questioned the availability of botanical data to substantiate the existence of a terminal raceme in determinate lines. Determinate plants are characterized at maturity by a shorter and thicker stem with substantially fewer nodes on the main stem than equivalent indeterminate stem types when grown side by side. Although main stem vegetative growth of determinate lines ceases soon after flowering, these lines do continue to produce nodes on their branches, well into late reproductive development (Gai *et al.*, 1984; Egli *et al.*, 1985). Gai *et al.* (1984) argued that determinate lines have a greater propensity for branching than indeterminate lines. Vegetative growth does continue in determinate lines following the onset of flowering, but fewer nodes are produced on the main stem. If indeed the apical meristem of determinate lines becomes a terminal raceme then this somewhat abrupt main stem termination would be a consequence of loss of apical dominance, which would also explain the continuation of vegetative growth from branches lower in the canopy. Branches may arise from the axillary buds of the cotyledonary node or from nodes above the cotyledonary node. Environmental factors such as photoperiod, row spacing, plant populations and fertility have also been reported to affect the branching pattern of soybean (Carlson, 1973).

In plants with an indeterminate stem type (Dt_1, Dt_1, dt_2, dt_2), stem elongation and node production of the main stem continues after flowering begins in the axillary nodes. These plants have substantially more nodes on the main stem at maturity than their determinate counterparts when grown side by side. This phenotype has a rather long, tapering stem with thinner internodes near the apex of the plant (Bernard, 1972; Hartung *et al.*, 1981). In addition, some distinctly intermediate stem types, called semi-determinates (Dt_1, Dt_1, Dt_2, Dt_2), also occur (Bernard, 1972; Thompson *et al.*, 1997). In this stem type, again node production on the main stem does not cease at flowering. However, these phenotypes are not quite as long as their indeterminate counterparts (when grown side by side), but have just a few less nodes on the main stem than their indeterminate near-isogenic lines (NILs).

The alleles at the *Dt* loci have an important influence on the main stem node number (Bernard, 1972; Thompson *et al.*, 1997; Curtis *et al.*, 2000). Determinate plants are quite distinct from the other two growth habits, since they abruptly terminate main stem node production soon after flowering. However, it can be difficult to distinguish between semi-determinates and indeterminates (Bernard, 1972). Both semi-determinates and indeterminates can continue vegetative development on the main stem for some time after flowering, and consequently both stem termination types tend to have many nodes produced on the main stem after flowering begins.

Even within a growth habit, there is considerable variation in vegetative size and morphology due at least in part to the *E*-genes present in the genotype. The *E*-genes are a series of seven genes identified and studied due to their impact on time to flowering and maturity (Bernard, 1971; Buzzell, 1971; Buzzell and Voldeng, 1980; McBlain and Bernard, 1987; Cober and Voldeng, 2001). The *E*-gene series is both temperature- and photoperiod-sensitive (Cober *et al.*, 2001; Stewart *et al.*, 2003). There are two alleles at each *E*-gene locus; late flowering and maturity is a partially dominant trait and early flowering and maturity is a recessive trait (Cober *et al.*, 2001). The *E*-genes impact the final main stem node number because time to flowering can impact termination of node accrual on the main stem. Delayed flowering increases the main stem node number in all three stem termination types. Since the dominant *E*-gene allele delays flowering, especially under an extended photoperiod (Cober *et al.*, 2001; Stewart *et al.*, 2003; Kumudini *et al.*, 2007), the presence of dominant alleles increases the main stem node number (Curtis *et al.*, 2000). Curtis *et al.* (2000) studied a number of NILs with known *E*-gene and *Dt* composition and noted that the presence of either the dominant *E*-gene allele or the Dt_1 allele (indeterminate growth habit) will significantly increase the main stem node number of the NIL. Therefore, the alleles present at the *Dt* and *E*-gene loci, as well as photoperiod, can dramatically influence the main stem node number and the morphology of the soybean canopy.

Vegetative development is, in part, a function of the rate of development of main stem nodes, the final main stem node number and branching, all of which influence the canopy structure. Although the rate of node accrual is temperature dependent, it appears to be constant across stem types and maturity groups. The final number of nodes on the main stem is influenced by both temperature and photoperiod, as well as by the *Dt* and *E*-gene alleles. Branching is influenced by the *Dt* alleles as well as by photoperiod, row spacing, plant populations and fertility. Therefore, vegetative development in soybean is regulated by a number of environmental and genetic factors that contribute to the observed large variation in the size and structure of soybean canopies.

Reproductive development

Timely flowering and seed maturity ensures the soybean crop's geographic adaptation and reproductive success. The relationship reported between the duration of the seed-filling period and crop yield (Gay *et al.*, 1980; Smith and Nelson, 1986) also underlines the importance of the reproductive development phase for improving the yield potential of soybean.

In their classification of soybean development, Fehr and Caviness (1977) used the appearance of the first open flower on the main stem, termed the R1 growth stage, to signal the beginning of the reproductive phase of development. They categorized reproductive development based on flowering, pod development, seed development and plant maturation stages. The first two stages – R1 and R2 – refer to flowering stages. The next two stages – R3 and R4 – refer to pod development. Seed development begins when the pod nears its maximum size. The R5 and R6 stages refer to seed development phases, whereas the R7 and R8 stages refer to phases of plant maturation.

Flowering signals the beginning of reproductive development and involves the transition of a vegetative meristem to a reproductive floral meristem. The timing of floral development can be critical to the adaptation of a species to a geographic region. The floral initials in soybean can form on axillary buds or on the apical bud. In soybean, as in many other crop species, floral induction is controlled mainly by temperature, photoperiod and genetics (Borthwick and Parker, 1938; Thomas and Raper, 1983; Wilkerson *et al.*, 1989; Upadhyay *et al.*, 1994; Cober *et al.*, 2001; Stewart *et al.*, 2003). The genes known to be involved in flowering in soybean, as mentioned earlier, are the series of genes known as the *E*-gene series (Bernard, 1971; Buzzell, 1971; Buzzell and Voldeng, 1980; McBlain and Bernard, 1987; Cober and Voldeng, 2001). The *E*-genes have been studied extensively for their role in time to flowering. The ability of soybean to adapt to a wide range of latitudes is attributable, at least in part, to the *E*-genes.

Although the appearance of an open flower on the main stem is easily distinguished as the first sign of reproductive development, a number of phases precede the appearance of the first flower. The time from emergence to first open flower includes four phases of development that vary in photoperiod sensitivity (Wilkerson *et al.*, 1989; Adams *et al.*, 2001). These phases are: (i) the photoperiod-insensitive (juvenile) phase; (ii) the photoperiod-sensitive inductive phase; (iii) the photoperiod-sensitive post-inductive phase; and (iv) the photoperiod-insensitive floral development phase.

The early phase of development is termed the 'juvenile' phase. During this phase plants are not yet competent of perceiving the photoperiod flowering stimulus. This phase is regulated by temperature. There have been reports of genetic differences for the juvenile phase in soybean. Wilkerson *et al.* (1989) reported that most soybean cultivars are competent to receive the photoperiod stimulus soon after seed germination, and consequently lack a juvenile phase. Upadhyay *et al.* (1994) argued that the juvenile phase is present in soybean and its length is related to the *E*-gene alleles. Hartwig and Kiihl (1979) identified a recessive trait (later referred to as the 'long juvenile trait') in soybean PI 159925, which delayed the flowering response under short-day conditions. This recessive trait has been reported to prolong the juvenile phase of soybean under the inductive photoperiods tested by the researchers (Wilkerson *et al.*, 1989; Collinson *et al.*, 1993).

Once capable of being induced to flower, inductive processes must occur to commit the meristem to floral development. This phase is known as the photoperiod-sensitive inductive phase (Adams *et al.*, 2001). Commitment of the meristematic tissue to reproductive development is dependent on the number of photoperiod inductive cycles perceived. The minimum number of inductive cycles required in soybean is dependent on the photoperiod to which the plants are exposed and to plant genetics (Bothwick and Parker, 1938; Thomas and Raper, 1983; Wilkerson *et al.*, 1989; Upadhyay *et al.*, 1994). Upadhyay *et al.* (1994) reported that the *E*-genes, either individually or in positive epistatic combination, impact the number of cycles required to induce floral initials. Under long-day conditions, the dominant *E*-gene alleles were found to increase the duration of the photoperiod-sensitive inductive phase (i.e. require a greater number of cycles for floral induction). Under short-day conditions, floral induction has been seen as early as after just two long-night cycles (Bothwick and Parker, 1938; Thomas and Raper, 1983; Upadhyay *et al.*, 1994).

Photoperiod has also been shown to affect the early phases of floral development and hasten anthesis after floral induction (Adams *et al.*, 2001). This phase has been referred to as the photoperiod-sensitive post-inductive phase. Several researchers have reported that the photoperiod after floral induction can significantly impact the subsequent development of the flower bud (Bothwick and Parker, 1938; Johnson *et al.*, 1960; Thomas and Raper, 1983; Zhang *et al.*, 2001). Bothwick and Parker (1938) noted that after floral induction, continued long nights hastened the opening of the flowers. Thomas and Raper (1983) also noted that after the development of floral initials, anthesis occurred much later in plants exposed to 15- and 16-h photoperiods than in those grown under shorter photoperiods. Johnson *et al.* (1960) and Zhang *et al.* (2001) reported that both photoperiod and genetics affect the rate of floral development. Zhang *et al.* (2001) reported that long-day treatments delayed floral bud growth, and that this effect was more apparent for late-maturing soybean genotypes.

Many plant species have been shown to be insensitive to photoperiod during the final phase of flower development. Once the floral primordium has reached a certain developmental phase, the meristem is committed to flower production; after this time, photoperiod no longer has an impact on floral development (Adams *et al.*, 2001). This phase is known as the photoperiod-insensitive floral development phase. In their study of seven soybean genotypes, Wilkerson *et al.* (1989) observed that the last 6.3–8.7 days of flower development appear to be independent of photoperiod effects. Zhang *et al.* (2001) also reported that long-day treatments did not affect time to flowering when plants were treated late in floral development (8 days after floral bud initiation). Upadhyay *et al.* (1994) further suggested that despite the impact of photoperiod, *E*-gene alleles may have a pleiotropic effect on this phase.

Therefore, the time to first flower in soybean is dependent on four different phases of development that are regulated by either temperature or temperature and photoperiod. In addition, genetics may modify the response of the plant to these environmental conditions. Consequently, the impact of temperature, photoperiod or genetics on one or more of these four phases of development can have a great influence on the time it takes for a soybean plant to develop its first fully open flower.

The first flower initiates reproductive development, after which, pod extension and then seed filling begin. After the appearance of the first open flower on a main stem node, flowering continues on both main stem nodes and branch nodes. This period of flowering on the various nodes of a plant can occur over a relatively long period in both determinate and indeterminate stem types (Gai *et al.*, 1984). This results in the production of a variety of reproductive structures, at various stages of development, on a single soybean plant. A soybean plant during reproductive development may be observed to have flowers, pods and pods with developing seeds on different nodes on the same plant.

Environmental and genetic factors have been shown to regulate reproductive development in soybean, even after the development of the first open flower (Thomas and Raper, 1983; Morandi *et al.*, 1988; Asumadu *et al.*, 1998; Summerfield *et al.*, 1998; Kumudini *et al.*, 2007). Photoperiod, temperature and genetics work to regulate the post-flowering developmental period of both determinate and indeterminate soybeans. In cultivars of both stem habit, the period from first flower to last flower (or flower on stem apex) has been shown to be regulated by photoperiod and *E*-gene alleles (Thomas and Raper, 1983; Morandi *et al.*, 1988; Asumadu *et al.*, 1998). In a study of indeterminate soybean *E*-gene NILs, Asumadu *et al.* (1998) noted that both long days and dominant *E*-gene alleles tend to prolong the flowering duration. Using the same genetic material, Summerfield *et al.* (1998) reported that both flowering duration and reproductive duration (i.e. period from first flower to maturity) were regulated by photoperiod and *E*-gene alleles. These earlier studies were conducted using potted plants under generally controlled-environment conditions. Under field conditions, Kantolic and Slafer (2001, 2005, 2007) also observed a positive correlation between photoperiod and post-flowering reproductive development (R3–R6). They further noted that this response was greater for the later-maturing soybean cultivars that they tested. Since *E*-gene alleles have the potential to change

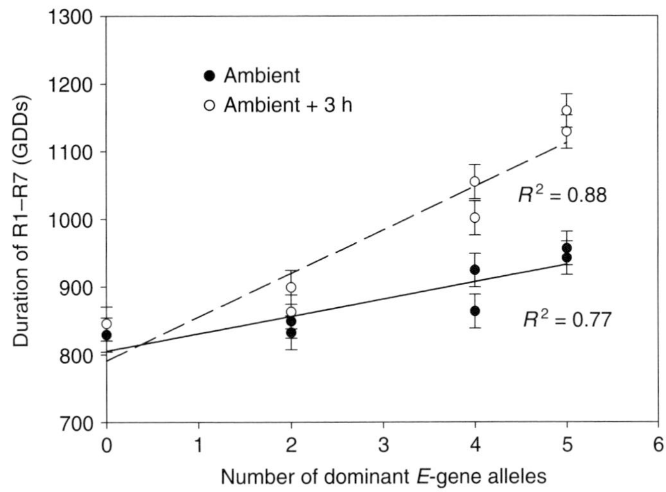

Fig. 3.7. Effect of photoperiod and *E*-gene alleles (number of dominant alleles, with E_1 counted as two dominant alleles) on the duration (growing-degree days; GDDs) of the reproductive phase (R1–R7) of seven *E*-gene near-isogenic lines grown in Lexington, KY, USA. The two photoperiods were ambient and ambient plus 3 h day-length extension. The bars represent standard error of the means (adapted from data in Kumudini *et al.*, 2007).

the maturity group to which a genotype belongs, it is likely that later-maturing cultivars have more dominant *E*-gene alleles (Kumudini *et al.*, 2007).

In an effort to determine the role of *E*-gene NILs and photoperiod on the post-flowering reproductive phase, Kumudini *et al.* (2007) conducted a field study with two post-flowering photoperiod treatments and seven *E*-gene NILs under two genetic backgrounds (to account for epistatic effects). The post-flowering photoperiod treatments were achieved by synchronizing flowering of the NILs and exposing the flowering plants to either the same ambient photoperiod or ambient photoperiod plus 3 h day-length extension. In this manner they were able to show that the duration of reproductive development (R1–R7) was extended by day-length extension, and that this response was dependent on the number and presence of dominant *E*-gene alleles (Fig. 3.7). The post-flowering phase during which soybeans are receptive to photoperiod has been estimated to be from first to last flower, a period that ends roughly around growth stage R5 (Asmadu *et al.*, 1998; Kantolic and Slafer, 2007).

The impact of genetics and photoperiod on the reproductive phase is considered to be of agronomic importance since a number of studies have suggested that the soybean reproductive phase, specifically the seed-filling phase (R4–R7) is critical for yield determination (Dunphy *et al.*, 1979; Gay *et al.*, 1980; Nelson, 1986; Smith and Nelson, 1986). Indeed, the results of controlled-environment studies have indicated that the presence of dominant *E*-gene alleles under long-day conditions extends the post-flowering phase and that this, in turn, resulted in an increase in soybean yield (Asmadu *et al.*, 1998; Ellis *et al.*, 2000). However, the evidence for an increase in seed yield has not been borne out when similar experiments have been conducted under field-growing conditions.

Under field conditions, Kantolic and Slafer (2001, 2005, 2007) showed that post-flowering day-length extension can extend the R3–R6 growth phase. They showed that extension of this phase increases seed number; a yield parameter that is important to yield determination in soybean. The problem, however, has been that under field conditions, the increase in seed number is often associated with reduced seed size, which may undermine the gains made in seed number. Despite the number of field experiments illustrating the impact of day length on the duration of the post-flowering growth phase, no evidence of its impact on seed yield has been reported. There is hope, however, that manipulation of day-length sensitivity through genetic means will allow for greater control of post-flowering reproductive development, potentially leading to future yield improvements.

Owing to the association between the duration of the seed-filling phase and soybean yield, soybean breeders have attempted to select for increased duration of the seed-filling phase. They have had difficulty, probably because the trait has low heritability and is strongly affected by the environment (Salado-Navarro *et al.*, 1985; Pfeiffer and Egli, 1988). Kumudini *et al.* (2007) suggested that a problem breeders may encounter when selecting for the duration of this phase is the hitherto ignored need to control for the impact of photoperiod on the duration of the seed-filling phase. In other words, to select for the genetic traits associated with the duration of this phase, all genotypes tested must be exposed to the same post-flowering photoperiod. It has been well documented in studies with different maturity groups as well as in studies with *E*-gene NILs that flowering occurs later in later-maturing lines (Egli, 1993, 1994; Cober *et al.*, 2001; Stewart *et al.*, 2003; Kumudini *et al.*, 2007). When a variety of maturity group cultivars are planted at the same time, time to first flower is positively related to the maturity of the genotype: later-maturing lines flower later than earlier-maturing lines (Fig. 3.8). This is significant because of the temporal changes that occur in photoperiod during the growing season. It has been shown that as later-maturing lines flower later during the growing season, they experience shorter ambient photoperiods post-flowering than earlier-maturing lines (Fig. 3.9) when planted in the spring in mid-temperate latitudes. Furthermore, this effect was observed whether planting was in early or late spring (Kumudini *et al.*, 2007).

The duration of reproductive development in soybean is controlled by both genetics and photoperiod (Kumudini *et al.*, 2007). Therefore, genotype selection for duration of reproductive development must be conducted under similar photoperiodic conditions – otherwise, the seasonal change in photoperiod during crop growth and maturation will confound the results and reduce the efficacy of selection for genes regulating the duration of the seed-filling phase (Kumudini *et al.*, 2007).

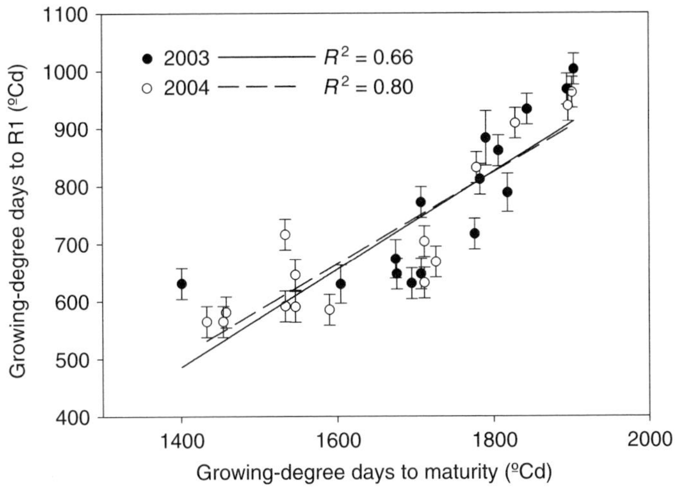

Fig. 3.8. Relationship between time to maturity (in growing-degree days) and time to first flower (in growing-degree days) of 15 *E*-gene near-isogenic lines grown in 2003 and 2004 in Lexington, KY, USA (data modified from Kumudini *et al.*, 2007).

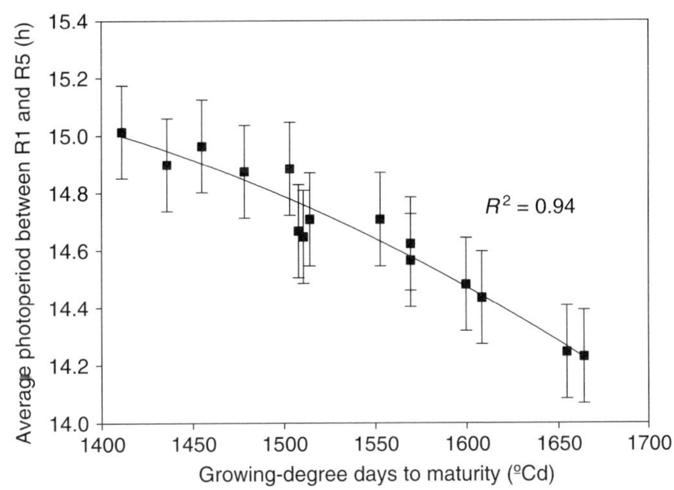

Fig. 3.9. Relationship between time to maturity (growing-degree days) and the average ambient photoperiod (includes civil twilight) experienced during the R1–R5 growth stage of 15 *E*-gene near-isogenic lines grown in Lexington, USA. The bars represent the standard error of the mean (modified from data in Kumudini *et al.*, 2007).

One way of ensuring that genotypes are exposed to the same photoperiod post-flowering is to synchronize flowering time. If all genotypes flower at the same time then they will experience the same ambient photoperiod and temperature conditions post-flowering. Kumudini *et al.* (2007) were able to synchronize the flowering of seven genotypes, ranging in maturity

(MG 00–IV), by using the Stewart *et al.* (2003) gene-based flowering model and historical weather data to stagger the planting date. This can be challenging for breeders when historical weather data do not fit well with ambient conditions during the trial, and if the *E*-gene composition of the genotypes being tested is unknown.

Research into the regulation of the post-flowering reproductive phase by *E*-gene alleles has helped to elucidate the mechanism by which soybeans are able to utilize environmental cues to adapt to the geographic region (Kumudini *et al.*, 2007). Once flowering has occurred, soybeans are receptive to the ambient photoperiod and respond accordingly. They either mature quickly, if they possess the recessive alleles, or they mature later, if they possess the dominant allele and the photoperiod is long (i.e. indicating a long period prior to killing frost). If they mature quickly, they have ensured seed maturation. If they mature later, they have taken full advantage of the longer duration of the lifecycle to fill the growing seeds.

Reproductive development in soybean is a dynamic process with the possibility of flowering, pod development and seed growth occurring simultaneously at various nodes on the plant. Environmental factors such as temperature and photoperiod, as well as genes such as the *E*-genes, work to regulate time to first flower . The *E*-genes also regulate the post-flowering reproductive phase and reveal a possible mechanism for the geographic adaptation of soybean to a wide range of latitudes worldwide (Kumudini *et al.*, 2007).

Maturity and senescence

In their classification of soybean reproductive development, Fehr and Caviness (1977) classified the final developmental phases of soybean as phases of plant maturation. The R7 growth stage is said to be reached when a normal pod on the main stem reaches its mature pod colour. The R8 growth stage is categorized as full maturity, when 95% of the pods on the main stem have reached their mature pod colour. Although these categories are useful in visual assessment and for comparative purposes, their utility is in their correlation with agronomically important events.

Agronomically important phases of maturation are physiological maturity and harvest maturity. Harvest maturity refers to when the crop is at a moisture level appropriate for field harvest operation. Physiological maturity generally refers to the point when soybean seeds no longer continue to grow and have reached their maximum dry weight (Crookston and Hill, 1978). At this point the maximum grain yield of the crop is attained. This phase of development is of agronomic importance and has been studied in a number of crop species. Daynard and Duncan (1969) reported that corn (*Zea mays* L.) kernels stop gaining dry weight after the vascular connections to the kernels are broken. The break in the vascular connection is due to the formation of a visually observable abscission layer, known as the 'black layer'. Therefore, physiological maturity in corn can be determined as the point when the black layer appears. No such easily determined, visual indicator of physiological maturity has been found for soybean. Howell *et al.* (1959) noted that soybean seeds reach maximum dry weight and low seed respiration rates when seed moisture concentration ranges from 50% to 60%. However, they did not report a corresponding visual indicator of physiological maturity. A number of researchers have attempted to find a non-destructive, observable cue that may be correlated to physiological maturity in soybean. Crookston and Hill (1978) looked at 11 visual indicators in an attempt to correlate one with physiological maturity. Of the 11 indicators selected, initiation of seed shrinkage and loss of green pigment from the pods were reliable indicators of physiological maturity. TeKrony *et al.* (1979) noted that low respiration rates, which could be used to estimate the time of physiological maturity, correspond to when seed moisture concentration dropped between 55% and 60%, consistent with the report by Howell *et al.* (1959). TeKrony *et al.* (1979) also conducted a greenhouse experiment in which they exposed soybeans to $^{14}CO_2$ during plant maturation and noted the accumulation of ^{14}C in the seeds of pods and seeds grouped by colour. Practically no ^{14}C was recovered from yellow seeds, regardless of the colour of the pod. Since most yellow seeds occurred in pods that had lost their green pigment, their results were consistent with those of Crookston and Hill (1978). In the interest of a consistent system, TeKrony *et al.* (1979) proposed the use of the R7 growth stage as the phenological stage that indicates physiological maturity. They argued that the R7 growth stage was appropriate, although on average it occurs a little before physiological maturity, and the R8 growth stage (full maturity) occurs on average 9–16 days after physiological maturity. Their main arguments for proposing the use of this phenological stage were: (i) it is the closest growth stage to when most seeds were yellow (and, therefore, the closest stage to physiological maturity in soybean); and (ii) it was impossible to detect significant differences in yield between plots harvested at growth stage R7 and those harvested at R8. Ghikpi and Crookston (1981) concurred. They also noted that R7 preceded physiological maturity, but found the variation in days to physiological maturity to be only 1–9 days after R7. The R7 growth stage has been widely accepted to be the growth stage that corresponds to physiological maturity in soybean.

Monocarpic plant species such as soybean complete their life cycle after a single reproductive phase. Soybean plants flower, set seeds and then pass through leaf and finally plant senescence as they mature. In monocarpic plants, a tight correlation between the initiation of leaf senescence and the development of reproductive organs has been observed. Among crop species, soybean specifically shows a marked correlation between leaf senescence and seed filling. Pod removal has been reported to delay the reduction of green leaf area in soybean (Crafts-Brandner *et al.*, 1984; Nooden and Leopold, 1988; Kumudini *et al.*, 2001). This relationship has generated suggestions of a cause and effect relationship between reproductive development and leaf senescence and speculation on various mechanisms of reproductive organ-induced leaf senescence (Sinclair and de Wit, 1976; Nooden, 1984).

Sinclair and de Wit (1976) hypothesized that nutrient withdrawal from the leaves to the developing pods and seeds results in a depletion of nitrogen from the leaves causing the plant to 'self-destruct'. They speculated that the particularly marked correlation between leaf senescence and seed filling in soybean is due to the high demand for nitrogen from the nitrogen-rich soybean seeds. Nooden (1984) alternatively postulated that soybean seeds exert a lethal senescence signal late in pod fill that targets mainly the leaf tissue. Nooden (1984, 1985) argued that this signal causes the senescence of leaf tissue. These studies postulated that without functional leaves the plants ultimately die, hence it is leaf senescence that leads to plant senescence.

Studies reporting a relationship between the stay-green trait and yield increase (Duvick, 1992; Kumudini *et al.*, 2001) have driven interest in the use of the stay-green trait as a potential tool for yield improvement. It has been shown that newer, high-yielding soybean cultivars maintain green leaf area longer during the seed-filling period than older, low-yielding cultivars (Kumudini *et al.*, 2001; Kumudini, 2002). Kumudini *et al.* (2001) speculated that the greater radiation interception due to delayed leaf senescence contributes to greater dry matter accumulation and higher yields in the modern genotypes. It is important, however, to distinguish between visual and functional stay-green, as only the latter can result in increased dry matter accumulation. In functional stay-green, the potential photosynthetic capacity of green leaves is maintained longer during the seed-filling period, whereas in visual stay-green sustained greenness of the leaves may or may not be associated with photosynthetic capacity. Differences in response to the advance in crop development between visual stay-green (i.e. leaf chlorophyll content) and functional stay-green (i.e. leaf photosynthetic rate) has been reported for two newer and older maize genotypes (Echarte *et al.*, 2008).

In order to take advantage of stay-green, there needs to genetic variation in stay-green that may be tapped to improve soybean yield. Genetic differences in rate of leaf senescence have been observed. Abu-Shakra *et al.* (1978) reported that genetic variation exists for the maintenance of carboxylation activity further into reproductive development. Begonia *et al.* (1987) observed that both genetic and environmental factors regulate leaf persistence. The complication in isolating genes that regulate leaf senescence in soybean may well lie in the nature of monocarpic senescence. The tight correlation between leaf senescence and seed fill has been postulated to be controlled by a coordinated signalling system (Biswal and Biswal, 1999). Nooden and Leopold (1988) postulated that the genes that control reproductive development also influence monocarpic senescence. Gan and Amasino (1997) termed the relationship between seed development and leaf senescence 'correlative control' of leaf senescence. They argued that leaf senescence allows the transport of assimilates from the source leaves to the seed sinks. Consequently, the two processes of seed filling and leaf senescence may be under correlative genetic control, making selection for stay-green a difficult task.

3.4 Summary

Soybeans undergo a complex system of growth and development, influenced by environmental and genetic factors. Temperature and photoperiod are two important known environmental factors that influence vegetative and reproductive development. Two important genetic factors known to impact soybean development and morphology are the *Dt* and *E*-genes. These genes impact both vegetative and reproductive development as well as plant morphology. The functions of the *E*-gene alleles are themselves modified by environmental factors such as photoperiod and temperature, adding further to the complexity of the system. The study of the influence of these *E*-genes on post-flowering development has revealed a possible mechanism for their wide range in geographic adaptation. Selection for duration of developmental phases and developmental processes such as an extended duration of the seed-filling phase and delayed leaf senescence can be complicated by environmental influences on the genes present and the possible correlative process that occur in monocarpic species. Through continued research on the processes regulating development, it may be possible to elucidate the mechanisms of this crop's wide geographic adaptation as well as to find new means to improve soybean yield potential.

References

Abu-Shakra, S.S., Phillips, D.A. and Huffaker, R.C. (1978) Nitrogen fixation and delayed leaf senescence in soybeans. *Science* 199, 973–975.

Adams, S.R., Pearson, S. and Hadley, P. (2001) Improving the quantitative flowering models through a better understanding of the phases of photoperiod sensitivity. *Journal of Experimental Botany* 52, 655–662.

Asumadu, H., Summerfield, R.J., Ellis, R.H. and Qi, A. (1998) Variation in the durations of the photoperiod-sensitive and photoperiod-insensitive phases of post-first flowering development in maturity isolines of soyabean *Glycine max* (L.) Merrill 'Clark'. *Annals of Botany* 82, 773–778.

Bastidas, A.M., Setryono, T., Dobermann, A., Cassman, K.G., Elmore, R.W. Graef, G.L. and Specht, J.E. (2008) Soybean sowing date: The vegetative, reproductive, and agronomic impacts. *Crop Science* 48, 727–740.

Begonia, G.B., Hesketh, J.D., Frederick, J.R., Finke, R.L. and Pettigrew, W.T. (1987) Factors affecting leaf area duration in soybean and maize. *Photosynthetica* 21, 285–295.

Bernard, R.L. (1971) Two major genes for time of flowering and maturity in soybeans. *Crop Science* 11, 242–244.

Bernard, R.L. (1972) Two genes affecting stem termination in soybeans. *Crop Science* 12, 235–239.

Biswal, B. and Biswal, U.C. (1999) Leaf senescence: Physiology and molecular biology. *Current Science* 77, 775–782.

Borthwick, H.A. and Parker, M.W. (1938) Influence of photoperiods upon the differentiation of meristems and the blossoming of Biloxi soy beans. *Botanical Gazette* 99, 825–839.

Buzzell, R.I. (1971) Inheritance of a soybean flowering response to fluorescent-daylength conditions. *Canadian Journal of Genetics and Cytology* 13, 703–707.

Buzzell, R.I. and Voldeng, H.D. (1980) Inheritance of insensitivity to long daylength. *Soybean Genetics Newsletter* 7, 26–29.

Carlson, J.B. (1973) Morphology. In: Caldwell, B.E. (ed.) *Soybeans: Improvement, Production, and Uses*. Agronomy Monograph 16. American Society of Agronomy, Madison, WI, USA, pp. 71–95.

Carlson, J.B. and Lersten, N.R. (2004) Reproductive morphology. In: Boerma, H.R. and Specht, J.E. (eds) *Soybeans: Improvement, Production, and Uses*, 3rd edn. Agronomy Monograph 16. American Society of Agronomy, Crop Science Society of America, and Soil Science Society of America, Madison, WI, USA, pp. 59–95.

Cober, E.R. and Voldeng, H.D. (2001) Low R:FR light quality delays flowering of E7E7 soybean lines. *Crop Science* 41, 1823–1826.

Cober, E.R., Stewart, D.W. and Voldeng, H.D. (2001) Photoperiod and temperature responses in early-maturing, near-isogenic soybean lines. *Crop Science* 41, 721–727.

Collinson, S.T., Summerfield R.J., Ellis R.H. and Roberts E.H. (1993) Durations of the photoperiod-sensitive and photoperiod-insensitive phases of development to flowering in four cultivars of soyabean (*Glycine max* L. Merril). *Annals of Botany* 71, 389–394.

Crafts-Brandner, S.J., Below, F.E., Harper, J.E. and Hageman, R.H. (1984) Effect of nodulation on assimilate remobilization in soybean. *Plant Physiology* 76, 452–455.

Crookston, K.R. and Hill, D.S. (1978) A visual indicator of the physiological maturity of soybean seed. *Crop Science* 18, 867–870.

Curtis, D.F., Tanner, J.W., Luzzi, B.M. and Hume, D.J. (2000) Agronomic and phenological differences of soybean isolines differing in maturity and growth habit. *Crop Science* 40, 1624–1629.

Daynard, T.B. and Duncan, W.G. (1969) The black layer and grain maturity in corn. *Crop Science* 9, 473–476.

Dunphy, E.J., Hanway, J.J. and Green, D.E. (1979) Soybean yields in relation to days between specific developmental stages. *Agronomy Journal* 71, 917–920.

Duvick, D.N. (1992) Genetic contributions to advances in yield of US maize. *Maydica* 37, 69–79.

Echarte, L., Rothstein, S. and Tollenaar, M. (2008) The response of leaf photosynthesis and dry matter accumulation to N supply in an older and a newer maize hybrid. *Crop Science* 48, 656–665.

Egli, D.B. (1993) Cultivar maturity and potential yield of soybean. *Field Crops Research* 32, 147–158.

Egli, D.B. (1994) Cultivar maturity and reproductive growth duration in soybean. *Journal of Agronomy and Crop Science* 173, 249–254.

Egli, D.B., Guffy, R.D. and Leggett, J.E. (1985) Partitioning of assimilate between vegetative and reproductive growth in soybean. *Agronomy Journal* 77, 917–922.

Ellis, R.H., Asumadu, H., Qi, A. and Summerfield, R.J. (2000) Effects of photoperiod and maturity genes on plant growth, partitioning, radiation use efficiency, and yield in soyabean [*Glycine max* L. Merrill] 'Clark'. *Annals of Botany* 85, 335–343.

Fehr, W.R. and Caviness, C.E. (1977) *Stages of Soybean Development*. Special Report No. 80, Cooperative Extension Service, Agriculture & Home Economics, Experimental Station, Iowa State University, Ames, IA, USA.

Gai, J., Palmer, R.G. and Fehr, W.R. (1984) Bloom and pod set in determinate and indeterminate soybeans grown in China. *Agronomy Journal* 76, 979–984.

Gan, S. and Amasino, R.M. (1997) Making sense of senescence: Molecular genetic regulation and manipulation of leaf senescence. *Plant Physiology* 113, 313–319.

Gay, S., Egli, D.B. and Reicosky, D.A. (1980) Physiological aspects of yield improvement in soybeans. *Agronomy Journal* 72, 387–391.

Ghikpi, P.I. and Crookston, R.K. (1981) A whole-plant indicator of soybean physiological maturity. *Crop Science* 21, 469–472.

Grubinger, V., Zobel, R., Vendeland, J. and Cortes, P. (1982) Nodule distribution on roots of field-grown soybeans in subsurface soil horizons. *Crop Science* 22, 153–155.

Guard, A.T. (1931) Development of floral organs of the soy bean. *Botanical Gazette* 91, 97–102.

Hartung, R.C., Specht, J.E. and Williams, J.H. (1981) Modification of soybean plant architecture by genes for stem growth habit and maturity. *Crop Science* 21, 51–56.

Hartwig, E.E. and Kiihl, R.A.S. (1979) Identification and utilization of a delayed flowering character in soybeans for short-day conditions. *Field Crops Research* 2, 145–151.

Hay, R.K.M. and Kirby, E.J.M. (1991) Convergence and synchrony: A review of the coordination of development in wheat. *Australian Journal of Agricultural Research* 42, 661–700.

Hesketh, J.D., Myhre, D.L. and Willey, C.R. (1973) Temperature control of time intervals between vegetative and reproductive events in soybeans. *Crop Science* 13, 250–254.

Hicks, D.R. (1978) Growth and development. In: Norman, G.A. (ed.) *Soybean Physiology, Agronomy and Utilization*. Academic Press, Inc., New York, USA, pp. 17–44.

Hofstra, G., Hesketh, J.D. and Myhre, D.L. (1977) A plastochron model for soybean leaf and stem growth. *Canadian Journal of Plant Science* 57, 167–175.

Howell, R.W., Collins, F.I. and Sedgwick, V.E. (1959) Respiration of soybean seeds as related to weathering losses during ripening. *Agronomy Journal* 51, 677–679.

Hunt, L.A. and Chapleau, A.M. (1986) Primordia and leaf production in winter wheat, triticale, and rye under field conditions. *Canadian Journal of Botany* 64, 1972–1976.

Hymowitz, T. (2004) Speciation and cytogenetics. In: Boerma, H.R. and Specht, J.E. (eds) *Soybeans: Improvement, Production, and Uses*, 3rd edn. Agronomy Monograph 16. American Society of Agronomy, Crop Science Society of America, and Soil Science Society of America, Madison, WI, USA, pp. 97–136.

Johns, C.W. and Palmer, R.G. (1982) Floral development of a flower-structure mutant in soybeans, *Glycine max* (L.) Merr. (Leguminosae). *American Journal of Botany* 69, 829–842.

Johnson, H.W., Borthwick, H.A. and Leffel, R.C. (1960) Effects of photoperiod and time of planting on rates of development of the soybean in various stages of the life cycle. *Botanical Gazette* 122, 77–95.

Kantolic, A.G. and Slafer, G.A. (2001) Photoperiod sensitivity after flowering and seed number determination in indeterminate soybean cultivars. *Field Crops Research* 72, 109–118.

Kantolic, A.G. and Slafer, G.A. (2005) Reproductive development and yield components in indeterminate soybean as affected by post-flowering photoperiod. *Field Crops Research* 93, 212–222.

Kantolic, A.G. and Slafer, G.A. (2007) Development and seed number in indeterminate soybean as affected by timing and duration of exposure to long photoperiods after flowering. *Annals of Botany* 99, 925–933.

Kumudini, S. (2002) Trials and tribulations: A review of the role of assimilate supply in soybean genetic yield improvement. *Field Crops Research* 75, 211–222.

Kumudini, S., Hume, D.J. and Chu, G. (2001) Genetic improvement in short season soybeans: I. Dry matter accumulation, partitioning and leaf area duration. *Crop Science* 41, 391–398.

Kumudini, S., Pallikonda, P. and Steele, C. (2007) Photoperiod and *E*-genes directly influence the duration of soybean reproductive development. *Crop Science* 47, 1510–1517.

Lersten, N.R. and Carlson, J.B. (2004) Vegetative morphology. In: Boerma, H.R. and Specht, J.E. (eds) *Soybeans: Improvement, Production, and Uses*, 3rd edn. Agronomy Monograph 16. American Society of Agronomy, Crop Science Society of America, and Soil Science Society of America, Madison, WI, USA, pp. 15–57.

Lhuissier, F.G.P., De Ruijter, N.C.A., Sieberer, B.J., Esseling, J.J. and Emons, A.M.C. (2001) Time course of cell biological events evoked in legume root hairs by *Rhizobium* Nod factors: State of the art. *Annals of Botany* 87, 289–302.

Liu, K.C., Pappelis, A.J. and Kaplan, H.M. (1971) Microbodies of soybean cotyledon mesophyll. *Transactions of the Illinois State Academy of Science* 64, 136–141.

McBlain, B.A. and Bernard, R.L. (1987) A new gene affecting the time of flowering and maturity in soybeans. *Journal of Heredity* 178, 68–70.

Miksche, J.P. (1961) Developmental vegetative morphology of *Glycine max*. *Agronomy Journal* 53, 121–128.

Morandi, E.N., Casano, L.M. and Regggiardo, L.M. (1988) Post-flowering photoperiodic effect on reproductive efficiency and seed growth in soybean. *Field Crops Research* 18, 227–241.

Nagata, T. (1960) Studies on the differentiation of soybean in Japan and the world. *Memoires of the Hyogo University of Agriculture (Agronomy Series)* 3, 63–102.

Nelson, R.L. (1986) Defining the seed-filling period in soybeans to predict yield. *Crop Science* 26, 132–135.

Nooden, L.D. (1984) Integration of soybean pod development and monocarpic senescence. *Physiologia Plantarum* 62, 273–284.

Nooden, L.D. (1985) Regulation of soybean senescence. In: Shibles, R. (ed.) *World Soybean Research Conference III: Proceedings*. Westview Press, Boulder, CO, USA, pp. 891–900.

Nooden, L.D. and Leopold, A.C. (1988) *Senescence and Aging in Plants*. Academic Press, New York, USA.

Oldroyd, G.E.D. and Downie, J.A. (2008) Coordinating nodule morphogenesis with rhizobial infection in legumes. *Annual Review of Plant Biology* 59, 519–546.

Pfeiffer, T.W. and Egli, D.B. (1988) Heritability of seed-filling period estimates in soybean. *Crop Science* 28, 921–925.

Piper, C.V. and Morse, W.J. (1923) *The Soybean*. McGraw-Hill, New York, USA.

Probst, A.H. and Judd, R.W. (1973) Origin, U.S. history and development, and world distribution. In: Caldwell, B.E. (ed.) *Soybeans: Improvement, Production, and Uses*. Agronomy Monograph 16. American Society of Agronomy, Madison, WI, USA, pp. 1–15.

Salado-Navarro, L.R., Hinson, K. and Sinclair, T.R. (1985) Nitrogen partitioning and dry matter allocation in soybeans with different protein concentrations. *Crop Science* 25, 451–455.

Sinclair, T.R. and de Wit, C.T. (1976) Analysis of the carbon and nitrogen limitations to soybean yield. *Agronomy Journal* 68, 319–324.

Smith, J.R. and Nelson, R.L. (1986) Selection for seed-filling period in soybean. *Crop Science* 26, 466–469.

Stewart, D.W., Cober, E.R. and Bernard, R.L. (2003) Modeling genetic effects on the photothermal response of soybean phenological development. *Agronomy Journal* 95, 65–70.

Summerfield, R.J., Asumadu, H., Ellis, R.H. and Qi, A. (1998) Characterization of the photoperiodic response of post-flowering development in maturity isolines of soyabean [*Glycine max* (L.) Merrill] 'Clark'. *Annals of Botany* 82, 765–771.

Sun, C.N. (1955) Growth and development of primary tissues in aerated and non-aerated roots of soybean. *Bulletin of the Torrey Botanical Club* 82, 491–502.

Sun, C.N. (1957) Histogensis of the leaf structure of the shoot apex in *Glycine max* (L.) Merrill. *Bulletin of the Torrey Botanical Club* 84, 163–174.

TeKrony, D.M., Egli, D.B., Balles, J., Pfeiffer, T. and Fellows, R.J. (1979) Physiological maturity in soybean. *Agronomy Journal* 71, 771–775.

Thomas, J.F. and Raper Jr, C.D. (1983) Photoperiod and temperature regulation of floral initiation and anthesis in soya bean. *Annals of Botany* 51, 481–489.

Thompson, J.A., Bernard, R.L. and Nelson, R.L. (1997) A third allele and the dt1 locus. *Crop Science* 37, 757–762.

Timmers, A.C.J. (2008) The role of the plant cytoskeleton in the interaction between legumes and rhizobia. *Journal of Microscopy* 231, 247–256.

Upadhyay, A.P., Summerfield, R.J., Ellis, R.H., Roberts, E.H. and Qi, A. (1994) Variation in the durations of the photoperiod-sensitive and photoperiod-insensitive phases of development to flowering among eight maturity isolines of soybean [*Glycine max* (L) Merrill]. *Annals of Botany* 74, 97–101.

Wilcox, J.A., Wiebold, W.J., Niblack, T.L. and Kephart, K.D. (1995) Growth and development of soybean ioslines that differ for maturity. *Agronomy Journal* 87, 932–935.

Wilkerson, G.G., Jones, J.W., Boote, K.J. and Buol, G.S. (1989) Photoperiodically sensitive interval in time to flower of soybean. *Crop Science* 29, 721–726.

Williams, L.F. (1950) Structure and genetic characteristics of the soybean. In: Markley, K.S. (ed.) *Soybeans and Soybean Products*, Volume 1. Interscience Publishers, New York, USA, pp. 111–134.

Zhang, L., Wang, R. and Hesketh, J.D. (2001) Effects of photoperiod on growth and development of soybean floral bud in different maturity. *Agronomy Journal* 93, 944–948.

4 Soybean Genetic Resources

S.K. Mishra[1] and V.D. Verma[2]
[1]*Germplasm Evaluation Division, National Bureau of Plant Genetic Resources, Pusa Campus, New Delhi, India;* [2]*National Bureau of Plant Genetic Resources, Regional Station, Phagli, Shimla, Himachal Pradesh, India*

4.1 Introduction

Soybean (*Glycine max* (L.) Merrill) is one of the most important crops in the world today. It is considered to be a miracle crop as it is extraordinarily rich in protein (~40%) and oil (~20%). It originated in China and has been cultivated for >5000 years (Qiu *et al.*, 1999). It is believed that with the development of sea and land trades, soybean moved out of China to nearby countries such as Burma (Myanmar), Japan, India, Indonesia, Malaysia, Nepal, the Philippines, Thailand and Vietnam between the first century AD and 1100 AD. However, it remained a minor crop everywhere except in China. With its introduction into the USA in the 18th century, and its systematic breeding in that country in the 1940s and 1950s, soybean was transferred from an inefficient fodder-type crop to a highly productive erect plant type, and the USA has been the largest producer of soybean in the world ever since (Hymowitz and Harlan, 1983). The Food and Agriculture Organization of the United Nations lists >85 countries that produce soybeans. The main soybean-producing countries, by weight, are the USA, Brazil, Argentina and China, which have together made up 80% of global production in the last several years. In addition, India, Paraguay, Canada, Bolivia and Indonesia are significant producers. This chapter discusses the genetic resources of soybean in general, with examples taken from India.

In India, work on soybean was undertaken at the Indian Agricultural Research Institute (IARI), New Delhi, on a small scale, with material built-up by the Plant Introduction Division during the 1950s. The Indian Council of Agricultural Research sanctioned the All India Coordinated Research Project on Soybean on April 1st, 1967, with the main centres located in Pantnagar, Jabalpur and Delhi. Initially, 1400 germplasm lines were assembled at the Uttar Pradesh Agriculture University, Pantnagar (now the G.B. Pant University of Agriculture and Technology), to start breeding work at Pantnagar and J.N. Krishi Vishwa Vidyalaya, Jabalpur, in collaboration with the University of Illinois, Urbana-Champaign, USA. Bragg, a yellow-seeded soybean, was the first variety released for commercial cultivation across the country, except in the southern parts. Several soybean varieties could be released by systematic testing of the germplasm at various centres. However, several other US-bred varieties (Hardee, Lee, Semmes, Clark-63, Monetta and Improved Pelican) were also later released as direct introductions. It was soon felt that some of the released varieties had become susceptible to diseases, insect pests and poor seed germinability. Rapid steps were taken to assemble soybean germplasm from various countries of the world. The National Bureau of Plant Genetic Resources (NBPGR) (formerly a plant introduction division of IARI, New Delhi) made efforts to collect germplasm from exotic and indigenous sources through direct introductions and explorations. This collected material was made available to various breeding centres at state agricultural universities and agriculture departments. Soybean breeders at Pantnagar, Uttar Pradesh (now in Uttarakhand), Jabalpur (Madhya Pradesh), Bangalore (Karnataka), IARI-Delhi, Kalyani (West Bengal), Amravati (Maharashtra) and Palampur (Himachal Pradesh) started using soybean germplasm through hybridization to increase yield levels and overcome the problems of susceptibility to diseases and insect pests, poor seed germinability, pod shattering and photoperiod insensitivity. Nearly 80 varieties have now been bred and released by Indian soybean breeders using exotic/indigenous germplasm through hybridization at various centres; furthermore, eight varieties have been released as direct introductions.

4.2 Taxonomy and Distribution

Taxonomy

Soybean belongs to the family Fabaceae (Leguminosae), subfamily Papilionoideae, tribe Phaseoleae and genus *Glycine*. The name was originally introduced by Linnaeus (1737) in the first edition of his *Genera Plantarum*. The generic name *Glycine* is derived from the Greek word 'glyks' (sweet). Linnaeus listed eight *Glycine* species, all of which were subsequently moved to other genera with the exception of *G. javanica*, which remained as the lectotype in the genus until 1966 (Hitchcock and Green, 1947). Now the Greek 'glyks' does not refer to any of the current *Glycine* species. Soybean has been known under various names, including *G. hispida*, *G. soja* and *G. max*. Kelsey and Dayton (1942) considered *G. soja* to be the approved botanical name, but the name *G. max*, proposed by Merrill (1917), is widely accepted as the valid designation.

According to recent taxonomical classification, soybean belongs to the genus *Glycine*, which has two subgenera: *Soja* and *Glycine*. Cultivated soybean (*G. max*) and its wild annual relative *G. soja* belong to the subgenus *Soja*. The subgenus *Glycine* contains 16 wild perennial species, mostly found

Table 4.1. Species in the genus *Glycine* and their geographical distribution (reprinted with permission from Hymowitz *et al.*, 1997).

Genus: *Glycine*	
Subgenus: *Soja* (annual)	
1. *G. max* (L.) Merr. (soybean)	Cultigen of eastern Asia, now grown worldwide
2. *G. soja* Sieb & Zucc. (wild soybean)	China, Japan, Korean peninsula, Russia
Subgenus: *Glycine* (perennial)	
3. *G. albicans* Tind. & Craven	Australia
4. *G. arenaria* Tindale	Australia
5. *G. argyrea* Tindale	Australia
6. *G. canescens* F.J.Herm	Australia
7. *G. clandestine* Wendl.	Australia
8. *G. curvata* Tindale	Australia
9. *G. crytoloba* Tindale	Australia
10. *G. falcate* Benth.	Australia
11. *G. hirticaulis* Tindale & Craven.	Australia
12. *G. lactovirens* Tindale & Craven.	Australia
13. *G. latifolia* (Benth.) Newell & Hymowitz	Australia
14. *G. latrobeana* (Meissn.) Benth.	Australia
15. *G. microphylla* (Benth.) Tindale	Australia
16. *G. pindanica* (Tind. & Craven.)	Australia
17. *G. tabacina* (Labill.) Benth.	Australia, China (Fujian, Taiwan), Japan (Ryukyu islands), South Pacific islands
18. *G. tomentella* Hayata	Australia, China
(2×)	(Fuzian, Taiwan)
(4×)	Philippines, Papua New Guinea

in Australia (Table 4.1). All of these species generally carry 2n = 40 chromosomes, except for *G. hirticaulis, G. tabacina* and *G. tomentella* (Vaughan and Hymowitz, 1983; Brown *et al.*, 1987; Hymowitz *et al.*, 1997). Biosystematics of the genus *Glycine* have been described by Hymowitz *et al.* (1997) (Table 4.2). Some of these wild perennial species also have polyploid cytotypes. *Glycine* is believed to be an ancient polyploid having × = 10; however, plants with 2n = 40 behave cytologically like diploids. The annual *Glycine* is derived from the perennial forms. Each subgenus has a different centre of diversity. The subgenus *Soja* is most diverse in the eastern half of north China, whereas maximum diversity for the subgenus *Glycine* occurs in Australia. The wild perennial *Glycine* species found outside of Australia were taken to other neighbouring regions by migratory birds via long distance dispersal (Hymowitz *et al.*, 1997). Over the last two decades, a large germplasm of 16 perennial species of *Glycine* has been assembled by the US Department of Agriculture (USDA). These collections are now maintained in Canberra, Australia, and are recognized by the International Plant Genetic Resources Institute as the world base collection for perennial *Glycine*.

Table 4.2. List of the genus *Glycine*, three-letter code, 2n, accession number and symbol (reprinted with permission from Hymowitz *et al.*, 1997).

Sr. no.	Soybean species	Code	2n	IL	PI	Genome[a]
Subgenus: *Glycine*						
1.	*G. albicans* Tind. & Craven	ALB	40	889	–	I I
2.	*G. arenaria* Tind.	ARE	40	689	505204	H H
3.	*G. argyrea* Tind.	ARG	40	768	505151	A2 A2
4.	*G. canescens* F.J.Herm.	CAN	40	434	440932	A A
5.	*G. clandestina* Wendl.	CLA	40	490	440958	A1 A1
6.	*G. curvata* Tind	CUR	40	791	505166	C1 C1
7.	*G. crytoloba* Tind.	CYR	40	481	440962	C C
8.	*G. falcate*	FLA	40	674	505179	F F
9.	*G. microphylla* (Benth.) Tind.	MIC	40	449	440956	B B
10.	*G. hirticaulis* Tind. & Craven.	HIR	40	1246	–	H1 H1
			80	943	–	–
11.	*G. lactovirens* Tind. & Craven.	LAC	40	1247	–	I1 I1
12.	*G. latifolia* (Benth.) Newell & Hymowitz	LAT	40	373	378709	B1 B1
13.	*G. latrobeana* (Meiss.) Benth.	LTR	40	659	483196	A3 A3
14.	*G. pindanica* (Tind. & Craven)	PIN	40	1251	–	H2 H2
15.	*G. tabacina* (Labill.) Benth.	TAB	40	370	373990	B2 B2
			80	–	–	Complex[b]
16.	*G. tomentella* Hayata	TOM	38	398	440998	E E
			40	709	5052	D D
			78	–	–	Complex[c]
			80	–	–	Complex[d]
Subgenus: *Soja* (Moench) F.J.Herm.						
17.	*G. soja* Sieb & Zucc.	SOJ	40	–	81762	G G
18.	*G. max* (L.) Merr.	MAX	40	cv Williams	82	G G

IL, a temporary number assigned at Urbana, Illinois, USA; PI, Plant Introduction Number assigned by the US Department of Agriculture.
[a]Genomically similar species carry the same letter symbols.
[b]Allopolyploids (A and B genome) and segmental allopolyploids (B genome).
[c]Allopolyploids (D and E, A and E or any other unknown combination).
[d]Allopolyploids (A and D or any other unknown combination).

Distribution

Soybean is believed to be of Chinese origin, having been derived from a slender, twig-like plant known as *G. ussuriensis* Regal & Maack. According to Fukuda (1933), Manchuria should be the centre of origin, since soybean exhibits wide genetic diversity in this area. Nagata (1959, 1960) suggested that the species originated in China proper, probably in the north and central regions. He based his conclusions partially on the distribution of *G. ussuriensis*, which is considered to be the progenitor of *G. max*, the cultivated form. Piper and Morse (1923) considered that the wild form *G. ussuriensis* was known to occur in China, Manchuria and Korea and stated that soybean is native of eastern Asia. According to Hymowitz (1970), *G. ussuriensis* grows wild in Korea, Taiwan and Japan throughout the Yangtze valley, the northern provinces of China and the adjacent areas of the former USSR. Based on cytogenetic evidence, Hymowitz (1970) concluded that *G. max* and *G. ussuriensis* are the same species and also stated that historical and geographical evidence points to the eastern half of northern China as the area where soybean was first domesticated around the 11th century BC.

Nagata (1959, 1960) suggested that the cultivated form of soybean was introduced into Korea from China and then disseminated to Japan between 200 BC and the third century AD. Morse (1950) presented a comprehensive review of the history of soybean production and mentioned that production was more or less localized in China until after the First Sino-Japanese War (1894–1895), when the Japanese began to import soybean cake for use as a fertilizer. The Russo-Japanese War (1904–1905) brought about a wider interest in soybean and its products. Shipments of soybeans and soybean products were made to Europe around 1908 and soybean attracted worldwide attention. Europeans had become aware of soybeans in 1712 through the writing of Engelbert Kaempfer, a German botanist who had spent 2 years (1691–1692) in Japan. Soybean seeds sent from China by missionaries were planted as early as 1740 in the Jardin des Plantes, Paris. Aiton (1814) indicated that soybean was first brought to England in 1790 and cultivated at the Royal Botanic Gardens, Kew, in that year. The greatest effort to expand soybean cultivation in Europe was from Frederich Haberlandt in Vienna, who grew 19 Chinese and Japanese varieties in 1873. Four of these varieties matured and seeds were distributed to various cooperators throughout Europe.

Piper and Morse (1923) gave an account of the early distribution of soybean in China, Korea, Japan and other Asiatic areas and in Australia, Africa and the Americas. According to them, not more than eight varieties of soybean were grown in the USA, prior to the numerous introductions by the USDA, until 1898. Hymowitz and Barnard (1991) made a detailed account of early introductions in the USA and mentioned that during the first two decades, new soybean accessions were introduced from India and China into the USA by USDA plant explorers Charles V. Piper and Frank N. Meyer, respectively. Both Piper and Meyer collected soybean as part of a plant exploration programme. However, in the 1920s, two major soybean exploration trips were undertaken by USDA scientists. From August 1924 through to December 1926, P.H. Dorsett collected soybean germplasm from northeast China. From March 1929 to February 1931, P.H. Dorsett and W.J. Morse collected soybean germplasm from Japan, Korea and China. Unfortunately, seed viability was lost due to lack of preservation facilities in the USA. In 1949, Martin G. Weiss of the USDA and Jackson L. Carter of the US Regional Soybean Germplasm Laboratory at Urbana, Illinois, initiated the development of a comprehensive germplasm collection.

Soybean was introduced to neighbouring countries (Japan, India, Nepal, Russia) from China around the first century AD. It appears that missionaries may have been the first to bring soybean to Europe early in the 18th century. Soybean was first introduced to the USA in 1765 (Hymowitz and Harlan, 1983) and was then spread to Canada and Latin America. Soybean production began only recently in Africa, during the second half of the 20th century.

Soybeans were taken to Brazil in 1822 by Gustavo Dutra. The real development of soybean in Brazil began in 1900, when the government of the state of São Paulo distributed about 200 kg of soybean seeds to around 70 farmers. A more organized planting by Japanese immigrants in the state of São Paulo occurred in 1908. By 1914, soybean had penetrated to Rio Grande do Sul, taken by professor F.G. Graig from the Technical University, which today is the Federal University of Rio Grande do Sul. In Argentina, the first soybean planting was made in 1862 in the Pampean plain. This marked a turning point in the basket of foods Argentina offers the world (Larreche and Brenta, 1999).

In India, soybean has been traditionally grown for many years on a small scale in Himachal Pradesh, the Kumaon hills of Uttaranchal, eastern Bengal, the Khasi Hills, Manipur, the Naga Hills and parts of central India covering Madhya Pradesh. Several attempts have been made to popularize soybean cultivation in India, including an initiative taken by Mahatma Gandhi in 1935. However, soybean did not initially find favour because of its late-maturing and fodder-type behaviour. It gained in popularity with introductions made from the USA during 1968, when the first soybean variety Bragg was released for commercial cultivation. Currently, about 80 varieties of soybean are being cultivated in various states of India.

4.3 Centres of Diversity

The Chinese characters for soybean appear many times in the ancient Chinese book *Shi Jing*, written during Zhou Dynasty (1000–200 BC). Later in the agricultural book of Guan Zi, written during the Han Dynasty (approximately 200 AD), soybean was classified into two types, small and large, based on the seed size. In the sixth century, soybean varieties such as 'Huang luo dou', 'Chang shao' and 'Niu jian' were recorded in the famous agricultural book *Qi Min Yao Su*. During the Song Dynasty in the tenth century, the book *Tu jing Ben Cao* described soybean varieties that differed in seed coat colour, maturity, seed size and shape. In the Ming Dynasty of the 16th century, the book *Tian Gong kai Wu* described the 'Gao jiang huang' that was planted after harvesting early rice in Yangtze valley and could mature in 90 days. This indicates soybean planting after rice in a cropping system.

Thousands of soybean landraces with great genetic diversity have been selected and preserved by Chinese farmers during a long history of cultivation. The Yellow River region of China is generally considered as the centre of origin of soybean, based on the existence of a great number of wild soybeans and the earliest record of soybean in China (Hymowitz and Kaizuma,

1981). Wild soybean (*G. soja* Sieb. & Zucc.) is widely distributed in nearly all provinces of China, Korea, Japan and parts of Russia (Hymowitz and Singh, 1987). Based on tremendous diversity in cultivated and wild soybean, the Chinese Academy of Agricultural Sciences, Beijing, has collected 23,000 accessions of *G. max*. In addition, 5300 accessions of *G. soja* have been conserved in a gene bank for long-term storage.

Thirteen wild perennial species of soybean collected by USDA explorers are indigenous to Australia (Hymowitz and Bernard, 1991). All carry 2n = 40 chromosomes. *G. tabacina* (Labill.) Benth., with 2n = 40 or 80 chromosomes, has been found in Australia, Taiwan, the South Pacific Islands (New Caledonia, Fizi, Tonga, Vanuatu, Niue) and the islands of the west central Pacific (Mariana, Ryukyu). All accessions of *G. tabacina* collected outside of Australia are tetraploid (2n = 80) and, even including Australia, the tetraploid predominates over the diploid form (Singh *et al.*, 1987, 1989; Hymowitz and Bernard, 1991), demonstrating that the complexes of *G. tabacina* and *G. tomentella* evolved through alloploidy in Australia. This clearly indicates that the wild perennial species of soybean have invaded Australia and associated areas, and the wild annual *G. soja* has invaded central and northern Asia. Since *G. soja* is the wild ancestor of soybean (Hymowitz and Newell, 1980) and all morphological and genetic variability exist in China in the form of landraces and primitive cultivars, this indicates that China is the centre of diversity.

4.4 Germplasm Collection and Introduction

Germplasm includes primitive cultivars, landraces, wild species closely related to cultivated crop plants, genetic stocks, inbred lines and hybrids. For crop improvement programmes, the diversity within the species is very important as a first-hand tool for easy hybridization. Wild species are equally important, but they are difficult to utilize in crossing programmes due to extremely low intersubgeneric crossibility. Breeders always look for high-yielding genetic stocks to increase production, wider adaptation and high nutritive value, along with resistance to biotic and abiotic stresses. Hence, germplasm is an essential basic raw material to meet current and future needs. Its assemblage is a continuous process, necessary for crop improvement programmes.

Collection of soybean germplasm is important work that is carried out in many countries the world over. It is estimated that worldwide there are >147,000 (Kolhe and Hussain, 2009) or 170,000 (Nelson, 2009) soybean accessions, with some accessions in duplication.

In 1944, the NBPGR (New Delhi) began efforts to increase its collection of soybean germplasm. A large number of accessions (2813) of soybean, including exotic, indigenous, wild perennial and wild annual sources, were maintained at the NBPGR Regional Station (Akola, Maharashtra) (Verma *et al.*, 1993). This collection includes exotic soybean germplasm introduced from 30 countries. The majority of accessions are from the USA (609), followed by Argentina (190), Germany (152), Taiwan (110), Australia (109) and Nigeria (75). Several high-yielding accessions with resistance to biotic and abiotic stress introduced from the Asian Vegetable Research and Development Center (AVRDC)-Taiwan, the USA and Argentina are performing well in various agroclimatic zones of India. The early introductions (Bragg, Monetta, Improved Pelican, Clark-63, Lee and Hardee) were high-yielding cultivars that gave 3–4 t ha^{-1} yield during the 1970s. Vegetable-type soybean varieties (Kim, Kenrich, Harasoy, Magna and Prize) were very bold seeded and suitable for green pod seeds as vegetables with almost no or a less beany flavour.

During 2000–2006, soybean germplasm was introduced to India from Australia (20), AVRDC-Taiwan (1366 *G. max*, five wild perennial species and 19 annual wild species), Nigeria (43), Sri Lanka (two), Thailand (six) and the USA (824 *G. max* and 59 wild perennial species) for high yield, low linolenic acid, vegetable type, low trypsin inhibitor, photoperiod insensitivity, resistance to rust, bacterial pustules, downy mildew, yellow mosaic virus, mungbean yellow mosaic virus, drought and heat and root nematodes, sensitivity to herbicides and seed shattering. These trait-specific accessions are being characterized, evaluated and maintained at the National Research Centre for Soybean, Indore (Madhya Pradesh). In addition, during 2000–2006 310 soybean accessions were introduced from the USA through NBPGR for the research and development programmes of private seed companies.

In China, >23,000 soybean accessions have been collected and preserved (Gai, 2009). The National Gene Bank of China has also conserved about 3000 foreign soybean accessions (Liu *et al.*, 2009), collected from 23 countries or regions, with most from the USA. Similarly, soybean germplasm is also collected and introduced by other countries.

4.5 Germplasm Evaluation and Documentation

The germplasm assembled from exotic and indigenous sources, including wild species, cannot be used by breeders and other researchers until it has been properly evaluated, characterized, classified and documented. Scientists evaluate and characterize soybean germplasm using the International Board for Plant Genetic Resources descriptors for 23 qualitative and quantitative characters, including oil content. Various important attributes, (e.g. flower and pubescence colour, seed colour, hilum colour, seeds per pod, 100-seed weight, lodging, pod shattering and oil content) and scores for major diseases (e.g. bacterial pustules, soybean mosaic and pod blight) are recorded.

Genetic variability

A wide range of genetic variability has been observed in soybean germplasm (Verma *et al.*, 1993), which provides a vast potential for exploiting various useful economic traits. These accessions vary in days to flowering (22–78 days), days to maturity (68–140 days), the number of leaflets (3–4), plant height (8.7–122.0 cm), number of seeds per pod (1.3–3.9), yield per plant (0.1–30.0 g) and oil content (13.0–24.7%). Variability in qualitative characters may be found in flower colour (purple or white), pubescence colour (tawny or grey), pubescence type (appressed, semi-appressed, erect or curly), pubescence density (normal, dense, sparse, semi-sparse or glabrous), leaf shape (normal, normal narrow, broad or small), leaf colour (light green, green or dark green), pod colour (light brown, brown or light black), seed coat colour (yellow, yellowish green, olive green, chocolate, light brown, brown, black, light grey, or black shedding to buff), reaction to bacterial pustules, soybean mosaic and pod blight (free, moderate or susceptible), lodging (free, moderate or susceptible), pod shattering (free, moderate or susceptible) and growth habit (determinate, semi-determinate or indeterminate) (Verma and Thomas, 1991). A large variation in seed weight in germplasm accessions at different locations in Pakistan has also been reported (Ashraf and Ghafoor, 2009).

Based on characterization and evaluation studies, various specific donors have been identified in soybean germplasm at the NBPGR (Regional Station, Akola, India) for early maturity, ideal plant height with plant yield, high numbers of pods per plant, four seeded pods, resistance to bacterial pustules (*Xanthomonas campestris* pv. *glycines*) (Verma, 1990), pod blight (*Colletotrichum dematium* (Pers, ex Fr) Grove *vartruncatum* (Schw.)), defoliation and pod shattering, high seed germinability, high yield, vegetable type and high oil content (Table 4.3). Glabrous and dense pubescence in soybean have specific significance in terms of insect pest infestation. Plants with glabrous or dense pubescence provide non-preferential (antixenosis)-type resistance. Glabrous pubescence genotypes include EC76736, EC85602-A, EC95272-A, EC95278, EC95296 and EC274472, whereas important dense pubescence accessions include EC251867, EC254683, EC274684, EC274755, EC280134, EC280146, EC287469 and EC287472.

Some germplasm lines and genotypes have been reported to be tolerant to waterlogging (e.g. VND 2, Nam Vang and ATF 15-1) (Tran *et al.*, 2009), drought (e.g. PK 34 and KB 79) (Arunkumar *et al.*, 2009) and rust (e.g. BR01-18437) (de Farias Neto *et al.*, 2009). Similarly, some germplasm lines and genotypes have been found to contain low linolenic acid and high oleic acid (e.g. EC-251405, EC-39160, EC-391181, IC-118429, IC-172659 and VLS-59) (Manjaya and Mondal, 2009).

Pod shattering is a serious problem in tropical climates, where day temperatures can be very high at the time of maturity. Yield losses due to pod shattering range from negligible to as high as 90% depending upon the time of harvesting, environmental conditions and genetic endowment of the variety. With this in mind, several pod-shattering-resistant accessions have been identified (Table 4.3). Among the wild annual and perennial species, reactions to disease and other useful traits have also been recorded.

Soybean seeds have a short life and the problem of poor stand establishment is pronounced, especially under tropical conditions, where seed

Table 4.3. Promising donor genotypes identified for various important economic characters in soybean germplasm (reprinted with permission from Verma *et al.*, 1993).

Character	Promising donor accession
Early maturity (68–72 days)	EC26295, EC30198, EC34160, EC34349, EC34383, EC39121, EC39156, EC39158
Plant height (<30 cm) with high yield	EC251372, EC251447, EC251484, EC251510, IC118034
Plant height (>30 cm) with high yield	EC242063, EC251329, EC251410, EC251527, EC251713, EC251820, IC96356, IC117930, IC118106, IC118264, IC118366, IC118414, IC118642
High no. pods per plant (>100)	EC172665, EC244707, EC250581, EC251528, EC274717, EC274755, EC309517
Four seeded pods	EC241110, EC241780, EC245482, EC245985, EC251529
Resistance to bacterial pustules	IC118005, IC118008
Resistance to pod blight	EC2578, EC14475, EC24046, EC75193, EC232082, EC251362, EC251526
Defoliator resistance	EC106991, EC106992
Multiple resistance	EC06991, EC106998, EC127501, EC251342, EC251500, EC260565, EC271621, EC280130, EC291393
Shattering resistance	EC251393, EC274674, EC280127, EC287401, EC291402, EC309508, EC309530
Vegetable type	EC250575, EC301881, EC301884, P-1366
High yield (>25 g)	EC241771, EC251372, EC251379, EC287401, EC287465, EC287465, EC287478, EC291391, EC291400, EC291402, EC309517, EC309535, EC309539
High oil (>23%)	EC76736, EC102612, EC113767, EC251298, EC251387, EC251432
High protein (>50%)	*G. soja*
Resistance to Bihar hairy caterpillar	*G. soja*

deterioration is accelerated by high temperatures and humidity. Musgrave *et al.* (1980) suggested methanol stress as a test of seed vigour in soybean. Kueneman (1982) also described the use of the methanol stress test in soybean for detecting high seed longevity. To select soybean genotypes for high seed germinability through the methanol stress test, 135 genetically diverse genotypes were tested after 3 weeks of storage (Verma, 1992). From each accession 100 seeds were put into 20% methanol (V/V) solution for 2 h, then soaked in water for 5 min and finally placed for germination in two replications under laboratory condition. The mean germination under methanol stress was 90–94% in IC202, IC574, IC81820, EC251418, EC251369, Yavatmal local, G2554, EC19729 and G2680, followed by 80–88% in EC24142, EC99991, EC172654, IC41680, EC39495, EC241771, EC244673, IC24020, EC172587, EC172638 and EC251459 (Verma, 1992). Out of 21 accessions selected for high seed germinability 13 were black seeded and seven yellow seeded. This indicates that black-seeded soybean gives predominantly higher germination than yellow-seeded soybean. Accessions that showed high seed germinability under methanol stress were mostly small to medium seeded (8–11 g 100^{-1} seeds).

Germplasm documentation

Soybean germplasm has been documented at various centres around the world. For example, a catalogue on soybean was published by the NBPGR (Regional Station, Akola, India) in 1983 for 439 indigenous and 1569 exotic accessions for 18 descriptors (Bhatia *et al.*, 1983). Later on, information on passport data and 23 descriptors were documented, including oil content, in 2737 accessions of soybean to allow easy access when querying the database on interrelated traits (Verma *et al.*, 1993). The retrieval of information on important traits is always helpful when selecting accessions for specific traits or combinations of traits in ongoing research programmes. The NBPGR has brought out an 'Inventory of Soybean Genetic Resources in India', which gives the details of soybean germplasm introduced into the country from 1944 to 1981, either through the NBPGR's own initiatives or on the request of scientists.

4.6 Germplasm Registration

A system of value-added germplasm registration has been developed at the NBPGR (New Delhi, India), wherein suitable candidate lines are registered and one set of their seed sample or other propagating material is kept for long-term storage. The seed sample of soybean's unique germplasm to be registered should contain fresh, dry, physiologically mature seeds that are free from infestation and sufficient in quantity (i.e. 4000–6000 seeds).

4.7 Germplasm Conservation

Soybean seeds are inherently short-lived. They deteriorate more rapidly than the seeds of rice (*Oryza sativa*), maize (*Zea mays*), sorghum (*Sorghum bicolor*), wheat (*Triticum aestivum*) and many other crops under the same conditions of production, harvesting, drying and storage (Delouche, 1982). Thus, it is imperative to store soybean germplasm at low temperature and humidity to maintain seed viability for present and future uses.

Medium- and long-term storage facilities have been created in various countries for the safe conservation of germplasm. Seeds are stored under a controlled environment to maintain seed viability for longer durations. Conditions are generally maintained at 5–7°C and 35% relative humidity for medium-term storage and –20°C for long-term storage. In 2009, the

National Gene Bank of India had successfully conserved 3427 accessions of soybean at −18°C, representing four species, 120 released cultivars and two registered genetic stocks (Radhamani, 2009).

4.8 Utilization of Germplasm

Plant breeders now have greater access to a large range of genetic diversity through national and international networking, subject to some technical constraints. However, these vast germplasm collections are not being used by breeders, resulting in a narrowing of the genetic base. Breeders have utilized limited soybean breeding material, staying with materials that are familiar and reasonably adapted to the environment as opposed to alien materials, which requires a lengthy programme of pre-adaptation. A thorough evaluation of soybean germplasm is an essential prerequisite for its utilization in crop improvement.

Harlan and de Wet (1971) developed a unified system for determining the total available gene pool of a cultivated plant and assigned taxa to one of the three gene pools – primary, secondary or tertiary. Hymowitz and Bernard (1991) explained that the primary gene pool (GP-1) for soybean consists of its domesticated and wild form, *G. soja*. Among forms in this gene pool, crossing is easy and hybrids are generally fertile with good chromosome pairing. At present, soybean does not have a secondary gene pool (GP-2) (Hymowitz and Bernard, 1991). GP-2 forms include those biological species that can exchange genes with the domesticate. Gene transfer is possible but difficult. Hybrids tend to be sterile and chromosomes pair poorly or not at all. The 15 wild perennial *Glycine* species comprise the tertiary gene pool (GP-3). Crosses can be made with soybean, but hybrids tend to die or to be completely sterile. Gene transfer is only possible utilizing extreme techniques, such as embryo culture, doubling of chromosome number or using bridge species to obtain some fertility (Hymowitz and Bernard, 1991). GP-3 is the outer limit of the conventional potential gene pool of the crop. There are no reports in the literature concerning the successful recovery of fertile diploid plants of crosses between soybean and wild perennial *Glycine* species. However, a soybean cultivar has been successfully backcrossed (BC1) to an amphiploid (soybean × *G. tomentella*) (Singh *et al.*, 1990).

Systematic efforts to utilize soybean germplasm for varietal development in various countries have long been proceeding, and consequently many varieties have been developed with specific characteristics such as high yield, disease resistance, good quality oil and so on. In India, two advanced breeding lines (NRC-64 and NRC-67) exhibit comparatively higher oleic acid, while VLS-59 exhibits lower linolenic acid. These sources can be exploited for developing varieties with improved oxidative stability of soybean oil. EC39490 and AGS-2, which exhibit oleic acid of >45%, can be employed for developing varieties with improved oxidative stability of soybean oil. Shilajeet, NRC-7, LSb-1, JS-335 and Sapori Midori are suitable for consumption at green pod stage. Hardee, Pb-1, KHSb-2 and Shilajeet are suitable for food uses. In India alone, nearly 80 improved varieties of soybean have been bred and released for cultivation since the mid-1960s.

In China, seven foreign soybean cultivars have been immediately used in production and 134 cultivars have been bred using foreign germplasm, accounting for a planting area of 25% of the total since 1980 (Liu *et al.*, 2009). Since 2001, 35 germplasm lines have been introduced to China from Ukraine by the Jilin Academy of Agricultural Sciences. These have been used in breeding programme, and consequently seven superior lines are expected to be released (Wang and Yang, 2009).

4.9 Future Perspectives

Establishing soybean germplasm resources with vast genetic diversity is the right approach to achieving major research priorities in soybean. Large collections of exotic and indigenous soybean germplasm have been built and maintained in different countries, but there is still poor representation of wild annual and perennial species. The low productivity in many countries (nearly 1.0 t ha^{-1}) as compared to in the USA (2.8 t ha^{-1}) and the world average (2.4 t ha^{-1}) represents a great challenge. Crops have the potential to exhibit better productivity and production in the coming years with the provision of research back-up, technology transfer and policy support from governments. By creating awareness of the health benefits of soybean as a functional food and using it in daily meals, the widespread energy–protein malnutrition can be addressed to a great extent. Taking these points into account, research into the following major areas of soybean production are thus imperative:

- **Poor seed germinability/viability**: there is a need to develop and introduce soybean varieties with high seed germinability and long viability (i.e. >75% germination after 20 months of storage under ambient conditions).
- **Photo-thermo insensitivity**: since farmers in some parts of the world (e.g. central parts of India) are growing two crops of soybean in a year, there is a need to develop material that is suitable for a winter/spring crop.
- **Earliness for specific situations**: early-maturing cultivars fit well in relay and catch cropping systems. There is a need to develop narrow-leaf and early types for intercropping with crops such as maize (*Zea mays*), sesame (*Sesamum indicum*) and cotton (*Gossypium* species).
- **Pod shattering**: pod shattering is an important factor that reduces soybean yield by 10–70%. Therefore, high-yielding cultivars with non-shattering pod characteristics should be developed.
- **High yield and high protein**: low soybean yields in many countries in comparison to the USA and Brazil reveal that there is still scope for increased productivity. Similarly a gain of 2–3% in the oil content of

seeds would have an impact on the total oil yield. Thus, there is a need to develop high-yielding genotypes with >23% oil content.

- **Cold tolerance**: the introduction of cold-tolerant genotypes will aid soybean cultivation in hilly areas.
- **Oil quality**: soybean oil quality could be improved by reducing the proportion of linolenic acid in the oil. This 18:3 (i.e. 18 carbon atoms and three double bonds) fatty acid accounts for 7–8% of the total oil content and is responsible for lowered stability and poor flavour. The level of linolenic plus linoleic acid in soybean oil is inversely proportionate to the level of oleic acid in soybean. Interspecific hybridization cannot be used to reduce the linolenic acid concentration in soybean because cultivated species are lower in linolenic acid than wild ones (Howell *et al.*, 1972). Therefore, the available soybean germplasm should be screened for linoleic, linolenic and oleic acid levels.
- **Kunitz trypsin inhibitor and lipoxygenase**: soybean Kurtz trypsin inhibitor is one of the most important antinutritional factors. The oxidation of fatty acids causes a grass-beany and bitter flavour in soybean products. There is a need to develop genotypes of soybean lacking in Kurtz trypsin inhibitor and lipoxygenase.
- **Vegetable type**: in some countries, people prefer to consume the green seeds of immature pods as a vegetable. Vegetable soybeans are nutritionally similar or even superior to green peas (Rao *et al.*, 1999). There is a need to search out bold, green seeded accessions that are sweet in taste without a beany flavour.
- **Insect pest resistance**: soybean pests such as girdle beetle, leaf miner, stem borer and Bihar hairy caterpillar are becoming a major problem. A large number of fungi (e.g. *Pythium*, *Phytophthora*, *Colletotrichum*, *Aspergillus*, *Fusarium*, *Macrophomina* and *Monilia*) and a few bacterial species of *Pseudomonas*, *Bacillus* and so on are responsible for seed and seedling rot in soybean. In addition, bacterial pustules, bacterial blight, rust, charcoal rot, phyllody, soybean yellow mosaic virus, soybean mosaic, frog-eye leaf spot, *Alternaria* leaf blight and bud blight are prevalent in some countries. Soybean genotypes with a wide array of resistance against major insect pests should be searched out.

References

Aiton, W.T. (1814) *Epitome of Hortus kewensis*, 2nd edn. Richard and Arthur Taylor, London, UK.

Arunkumar, N.P., Swamy, M., Chandrappa, M. and Manja, N.C. (2009) Identification of drought tolerant genotypes of soybean (*Glycine max* L.). In: Qiu, L., Guan, R., Jin, J., Song, Q., Guo, S., Li, W., Wang, Y., Han, T., Liu, X., Yu, D., Jiang, L. and Peng, D. (eds) *Developing a Global Soy Blueprint for a Safe Secure and Sustainable Supply. Abstracts for oral presentations and posters of the World Soybean Research Conference VIII*, Beijing, China, 10–15 August 2009. The Chinese Academy of Agricultural Sciences, Beijing, China, p. 6.

Ashraf, M. and Ghafoor, A. (2009) Collection and evaluation of germplasm for soybean development in Pakistan. In: Qiu, L., Guan, R., Jin, J., Song, Q., Guo, S., Li, W., Wang, Y., Han, T., Liu, X., Yu, D., Jiang, L. and Peng, D. (eds) *Developing a Global Soy Blueprint for a Safe Secure and Sustainable Supply. Abstracts for oral presentations and posters of the World Soybean Research Conference VIII*, Beijing, China, 10–15 August 2009. The Chinese Academy of Agricultural Sciences, Beijing, China, p. 5.

Bhatia, H.P., Patel, D.P. and Kale, R.K. (1983) *Catalogue on Soybean Genetic Resources*. National Bureau of Plant Genetic Resources, Regional Station, Akola, India.

Brown, A.H.D., Grace, J.P. and Speer, S.S. (1987) Designation of a "core" collection of perennial *Glycine*. *Soybean Genetics Newsletter* 14, 59–70.

de Farias Neto, A.L., de Souza, P.I.M., Moreira, C.T., Santin, M.R., da Silva, S.A. and Dianese, A.D.C. (2009) Evaluation of soybean (*Glycine max* L.) germplasm for resistance to soybean rust (*Phakopsora pachyrhizi*) in Brazil. In: Qiu, L., Guan, R., Jin, J., Song, Q., Guo, S., Li, W., Wang, Y., Han, T., Liu, X., Yu, D., Jiang, L. and Peng, D. (eds) *Developing a Global Soy Blueprint for a Safe Secure and Sustainable Supply. Abstracts for oral presentations and posters of the World Soybean Research Conference VIII*, Beijing, China, 10–15 August 2009. The Chinese Academy of Agricultural Sciences, Beijing, China, pp. 55–56.

Delouche, J.C. (1982) Physiological changes during storage that affect soybean and stand establishment. In: Sinclair, J.B. and Jackobs J.A. (eds) *Soybean Seed Quality and Stand Establishment*. Proceedings of a Conference for Scientists of Asia, Colombo, Sri Lanka, 25–31 January 1981, INTSOY-SERIES 22. University of Illinois, Urbana-Champaign, IL, USA, pp. 57–66.

Fukuda, Y. (1933) Cytogenetical studies on the wild and cultivated Manchurian soybeans (*Glycine* L.). *Japanese Journal of Botany* 6, 489–506.

Gai, J.Y. (2009) Conservation and sustainable utilization of genetic diversity in soybean. In: Qiu, L., Guan, R., Jin, J., Song, Q., Guo, S., Li, W., Wang, Y., Han, T., Liu, X., Yu, D., Jiang, L. and Peng, D. (eds) *Developing a Global Soy Blueprint for a Safe Secure and Sustainable Supply. Abstracts for oral presentations and posters of the World Soybean Research Conference VIII*, Beijing, China, 10–15 August 2009. The Chinese Academy of Agricultural Sciences, Beijing, China, pp. 169–170.

Harlan, J.R. and de Wet, J.M.J. (1971) Toward a rational classification of cultivated plants. *Taxon* 20, 509–517.

Hitchcock, A.S. and Green, M.L. (1947) Species lectotypicae generum Linnaei. *Britonica* 6, 114–118.

Howell, R.W., Brim, C.A. and Rinne, R.W. (1972) The plant geneticist's contribution toward changing lipid and amino acid composition of soybeans. *Journal of American Oil Chemists' Society* 49, 30–32.

Hymowitz, T. (1970) On the domestication of the soybean. *Economic Botany* 24, 408–421.

Hymowitz, T. and Barnard, R.L. (1991) Origin of soybean and germplasm introduction and development in North America. In: Shands, H.L. and Weisner, I.E. (eds) *Use of Plant Introductions in Cultivars Development. Part 1*. Special Publication 17. Crop Science Society of America, Madison, WI, USA, pp. 147–164.

Hymowitz, T. and Harlan, J.R. (1983) Introduction of soybeans to North America by Samuel Bowen in 1765. *Economic Botany* 37, 371–379.

Hymowitz, T. and Kaizuma, N. (1981) Soybean protein electrophoresis profile from 15 Asian countries or regions: Hypotheses on paths of dissemination of the soybean from China. *Economic Botany* 35, 10–23.

Hymowitz, T. and Newell, C.A. (1980) Taxonomy, speciation, domestication, dissemination, germplasm resources and variation in the genus *Glycine*. In: Summerfield, R.J. and Bunting, A.H. (eds) *Advances in Legume Science*. Royal Botanical Gardens, Kew, UK, pp. 251–264.

Hymowitz, T. and Singh, R.J. (1987) Taxonomy and speciation. In: Wilcox, J.R. (ed.) *Soybeans: Improvement, Production, and Uses*, 2nd edn. Agronomy Monograph 16. American Society of Agronomy, Crop Science Society of America, and Soil Science Society of America, Madison, WI, USA, pp. 23–48.

Hymowitz, T., Singh, R.J. and Kollipara, K.P. (1997) Biosystematics of the genus *Glycine*, 1996. *Soybean Genetics Newsletter* 24, 119–120.

Kelsey, H.P. and Dayton, W.A. (1942) *Standardized Plant Names*. J. Horace McFarland Co., Harrisburg, PA, USA.

Kolhe, S. and Hussain, S.M. (2009) On-line soybean germplasm information system. In: Qiu, L., Guan, R., Jin, J., Song, Q., Guo, S., Li, W., Wang, Y., Han, T., Liu, X., Yu, D., Jiang, L. and Peng, D. (eds) *Developing a Global Soy Blueprint for a Safe Secure and Sustainable Supply. Abstracts for oral presentations and posters of the World Soybean Research Conference VIII*, Beijing, China, 10–15 August 2009. The Chinese Academy of Agricultural Sciences, Beijing, China, pp. 48–49.

Kueneman, E.A. (1982) Genetic differences in soybean seed quality: Screening methods for cultivar improvement. In: Sinclair, J.B. and Jackobs, J.A. (eds) *Soybean Seed Quality and Stand Establishment*. Proceedings of a Conference for Scientists of Asia, Colombo, Sri Lanka, 25–31 January 1981, INTSOY-SERIES 22. University of Illinois, Urbana-Champaign, IL, USA, pp. 31–41.

Larreche, H.J. and Brenta, L.M.F. (1999) State of the soybean industry in Argentina. In: *Proceedings of World Soybean Research Conference VI*, Chicago, IL, USA, 4–7 August 1999, pp. 5–11.

Linnaeus, C. (1737) *Genera Plantarum*. Conrad Wishoff, Leiden, The Netherlands.

Liu, Z.X., Qiu, L.J., Chang, R.Z., Guan, R.X., Li, Y.H. and Jin, L.G. (2009) Introduction, evaluation and utilization of soybean germplasm from foreign countries. In: Qiu, L., Guan, R., Jin, J., Song, Q., Guo, S., Li, W., Wang, Y., Han, T., Liu, X., Yu, D., Jiang, L. and Peng, D. (eds) *Developing a Global Soy Blueprint for a Safe Secure and Sustainable Supply. Abstracts for oral presentations and posters of the World Soybean Research Conference VIII*, Beijing, China, 10–15 August 2009. The Chinese Academy of Agricultural Sciences, Beijing, China, pp. 170–171.

Manjaya, J. and Mondal, S. (2009) Identification of low linolenic acid genotypes and its validation by linked molecular markers in soybean [*Glycine max* (L.) Merrill]. In: Qiu, L., Guan, R., Jin, J., Song, Q., Guo, S., Li, W., Wang, Y., Han, T., Liu, X., Yu, D., Jiang, L. and Peng, D. (eds) *Developing a Global Soy Blueprint for a Safe Secure and Sustainable Supply. Abstracts for oral presentations and posters of the World Soybean Research Conference VIII*, Beijing, China, 10–15 August 2009. The Chinese Academy of Agricultural Sciences, Beijing, China, p. 7.

Merrill, E.D. (1917) *An Interpretation of Rumphius's Herbarium Amboinense*. Bureau of Printing, Manila, The Philippines.

Morse, W.J. (1950) History of soybean production. In: Markley, K.S. (ed.) *Soybeans and Soybean Products*. I. Interscience Publisher Inc., New York, USA, pp. 3–59.

Musgrave, M.E., Priestley, D.A. and Leopold, A.C. (1980) Methanol stress as a test of seed vigour. *Crop Science* 20, 626–630.

Nagata, T. (1959) Studies on the differentiation of soybeans in the world with special regard to that of Southern Asia. Proceedings of Crop Science Society of Japan, No. 28, pp. 79–82.

Nagata, T. (1960) Studies on the differentiation of soybeans in Japan and the world. *Memoirs Hyogo University of Agriculture* 3, 63–102.

Nelson, R. (2009) Collection, conservation, and evaluation of soybean germplasm. In: Qiu, L., Guan, R., Jin, J., Song, Q., Guo, S., Li, W., Wang, Y., Han, T., Liu, X., Yu, D., Jiang, L. and Peng, D. (eds) *Developing a Global Soy Blueprint for a Safe Secure and Sustainable Supply. Abstracts for oral presentations and posters of the World Soybean Research Conference VIII*, Beijing, China, 10–15 August 2009. The Chinese Academy of Agricultural Sciences, Beijing, China, pp. 5–6.

Piper, C.V. and Morse, W.J. (1923) *The Soybeans*. McGraw Hill, New York, USA.

Qiu, L., Chang, R., Jianying, S., Xianghua, L., Zhanglin, C. and Zenglu, L. (1999) The history and use of primitive varieties in Chinese soybean breeding. In: *Proceedings of World Soybean Research Conference VI*, Chicago, IL, USA, 4–7 August 1999, pp. 165–172.

Radhamani, J. (2009) Conservation and management of soybean (*Glycine max* L.) genetic resources at National Gene Bank, New Delhi, India. In: Qiu, L., Guan, R., Jin, J., Song, Q., Guo, S., Li, W., Wang, Y., Han, T., Liu, X., Yu, D., Jiang, L. and Peng, D. (eds) *Developing a Global Soy Blueprint for a Safe Secure and Sustainable Supply. Abstracts for oral presentations and posters of the World Soybean Research Conference VIII*, Beijing, China, 10–15 August 2009. The Chinese Academy of Agricultural Sciences, Beijing, China, p. 6.

Rao, M.S.S., Bhagsari, A.S. and Mohamed, A.L. (1999) Soybeans: A potential vegetable crop. In: *Proceedings of World Soybean Research Conference VI*, Chicago, IL, USA, 4–7 August 1999, pp. 172–176.

Singh, R.J., Kollipara, K.P. and Hymowitz, T. (1987) Polyploid complexes of *Glycine tabacina* (Labill.) Benth. and *G. tomentella* Hayata revealed by cytogenetics analysis. *Genome* 29, 490–497.

Singh, R.J., Kollipara, K.P. and Hymowitz, T. (1989) Ancestors of 80 and 78 chromosome *Glycine tomentella* Hayata (Leguminosae). *Genome* 32, 796–801.

Singh, R.J., Kollipara, K.P. and Hymowitz, T. (1990) Backcross derived progeny from soybean and *Glycine tomentella* Hayata intersubgeric hybrids. *Crop Science* 18, 991–996.

Tran, C.H.T., Van Toai, T.T., Nguyen, N.H.T., Nguyen, H. and Shannon, G. (2009) Tolerance of soybean (*Glycine max* L.) germplasm from Southeast Asia to soil waterlogging. In: Qiu, L., Guan, R., Jin, J., Song, Q., Guo, S., Li, W., Wang, Y., Han, T., Liu, X., Yu, D., Jiang, L. and Peng, D. (eds) *Developing a Global Soy Blueprint for a Safe Secure and Sustainable Supply. Abstracts for oral presentations and posters of the World Soybean Research Conference VIII*, Beijing, China, 10–15 August 2009. The Chinese Academy of Agricultural Sciences, Beijing, China, p. 6.

Vaughan, D.A. and Hymowitz, T. (1983) Progress in wild perennial soybean characterization. *Plant Genetic Resources Newsletter* 56, 7–12.

Verma, V.D. (1990) Evaluation of soybean (*Glycine max*) germplasm for bacterial pustules caused by *Xanthomonas campestris* pv *glycines*. *Indian Journal of Agricultural Sciences* 60, 832–833.

Verma, V.D. (1992) Screening of soybean (*Glycine max*) germplasm lines for high seed germinability through methanol stress test. *Indian Journal of Agricultural Sciences* 62, 698–699.

Verma, V.D. and Thomas, T. (1991) Soybean genetic resources and their evaluation. *Indian Journal of Plant Genetic Resources* 4, 29–33.

Verma, V.D., Patel, D.P., Loknathan, T.R., Singh, B., Sapra, R.L. and Rana, R.S. (1993) *Evaluation of Soybean Germplasm*. National Bureau of Plant Genetic Resources, Regional Station, Akola, India.

Wang, Y. and Yang, G.Y. (2009) Utilization of soybean germplasm introduced from Ukraine. In: Qiu, L., Guan, R., Jin, J., Song, Q., Guo, S., Li, W., Wang, Y., Han, T., Liu, X., Yu, D., Jiang, L. and Peng, D. (eds) *Developing a Global Soy Blueprint for a Safe Secure and Sustainable Supply. Abstracts for oral presentations and posters of the World Soybean Research Conference VIII*, Beijing, China, 10–15 August 2009. The Chinese Academy of Agricultural Sciences, Beijing, China, p. 122.

5 Varietal Improvement in Soybean

Dilip R. Panthee
Department of Horticultural Science, North Carolina State University, Mountain Horticultural Crops Research and Extension Center, Mills River, North Carolina, USA

5.1 Introduction

Soybean (*Glycine max* (L.) Merrill) (2n = 2X = 40) is one of the most important legumes in the world. It is grown on an estimated area of 91 million ha, globally producing 222 million t year^{-1} (Soytech Inc., 2007). Most soybean is grown in North America, South America and Asia. The major soybean-producing countries by area are the USA (28.2%), Brazil (23.7%), Argentina (18.5%), China (9.7%), India (9.1%), Paraguay (2.6%) and others (8.2%). Production-wise, the USA produces the most soybean (32.2%), followed by Brazil (27.5%), Argentina (21.2%), China (7.0%), India (3.6%), Paraguay (2.8%) and others (5.8%) (Soytech Inc., 2007). Disparity between the total area and total production in different countries is associated with average yield. For example, the average yield of soybean in Brazil is 2.84 t ha^{-1}, compared to 2.80 t ha^{-1} in Argentina, 2.79 t ha^{-1} in the USA, 2.58 t ha^{-1} in Paraguay, 1.77 t ha^{-1} in China and 0.95 t ha^{-1} in India. It is estimated that about 50% of the yield improvement is attributed to improved genetics and the other 50% to cultural practices including fertilizers, plant protection and irrigation. The differences in average yield emphasize the importance of plant breeding in improving the yield potential of soybean varieties in developing countries such as India, where the average yield is one third that of developed countries. In the USA, it is estimated that the yield is improving at the rate of 25 kg ha^{-1} year^{-1} (Orf *et al.*, 2004).

Soybean is mainly grown for protein and oil, which make up about 40% and 20% on a dry weight basis, respectively. In North America and Europe, soybean is regarded as a major oilseed crop and soybean meal is used as a major source of protein in animal feed. In other parts of the world it is consumed by humans in various forms such as tofu, sprout, *nattō*, soy milk, soy nuts, cooking oil and soy paste. Soybean oil also has several industrial applications such as in lubricants, cosmetics and toner ink. The possibility of developing soybean as a source of biofuel has increased the interest of researchers, industry people and policy-makers in this crop. Because of its contribution to the world economy and multiple uses, there is also growing interest in improving various traits of soybean, including seed composition, agronomic traits and disease resistance, so that the market value of the crop is further improved. Plant breeding will play a pivotal role in bringing about all of these improvements. This chapter discusses the components that are required in variety improvement programmes.

5.2 Wild Relatives and Genetic Resources

Genetic resources are the basis for crop improvement, and naturally available genetic variation has been the basis of crop improvement in the past. In fact, soybean was a wild plant until 3000 BC. Chinese farmers used their skills in agriculture to convert the wild plant to a domesticated crop through a series of selections (Hymowitz, 2004). Today, cultivated as well as wild relatives of soybean are being used in crop improvement programmes. Annual cultivated relatives or germplasms are created through recombination in breeding programmes. However, there are several wild relatives, which have important genes conferring disease resistance, seed quality and other agronomic traits.

The genus *Glycine* has two subgenera: *Soja* and *Glycine* (Hymowitz, 2004). Species of the subgenera *Glycine* are perennial and those under *Soja* are annuals. There are only two species under the subgenera *Soja*: *Glycine max* and *G. soja*. Thirty perennial wild relatives of soybean have been identified so far. Some of the perennials have been reported to be cross-compatible with cultivated soybean (*G. max*), whereas the compatibility of others is unknown. This has provided an enormous genetic pool for the improvement of traits of interest in soybean.

Annual relatives

One of the most important annual wild relatives of soybean is *G. soja* (Fig. 5.1). It has been used extensively in several breeding programmes throughout the world to improve numerous traits (LeRoy *et al.*, 1991; Rebetzke *et al.*, 1997; Concibido *et al.*, 2003). It is a source of protein (Zakharova *et al.*, 1989; Weng *et al.*, 1995; Sebolt *et al.*, 2000), disease resistance (Hegstad *et al.*, 1998; Wang *et al.*, 2001) and stress tolerance (Goldman *et al.*, 1989). Because of cross-compatibility and diversity at molecular levels, it has been used not only for transferring agronomically useful genes but also for constructing the molecular genetic linkage map of soybean (Gardner *et al.*, 2001; Weng *et al.*, 2001). There are still several useful genes in *G. soja* that could be transferred to *G. max* to improve the traits of interest.

Fig. 5.1. Annual and perennial species of soybean: (from left to right) *Glycine max, G. soja* and *G. tomentella*. While *G. soja* is a source of seed storage protein, *G. tomentella* is the widely used source of resistance to Asian soybean rust (caused by *Phakopsora pachyrhizi*).

Perennial relatives

Out of 30 perennial wild relatives of soybean, fewer have been utilized in soybean breeding programmes (Bodanese-Zanettini *et al.*, 1996). One of the widely utilized perennial relatives is *G. tomentella* (Fig. 5.1), which has been used as a source of rust resistance genes in *G. max* (Singh *et al.*, 1993; Patzoldt *et al.*, 2007). *G. tomentella* has been reported to have an aneuploidy (2n = 2X = 38) as well as tetraploidy (2n = 4X = 80) genome. In the past, this species has been used in a number of experiments in Australia, China and India to transfer soybean rust resistance genes. In the USA, there has been growing interest in this species after confirmation of the incidence of soybean rust in 2004 (Stokstad, 2004). Regarding ploidy levels, *G. tabacina* has been reported to have tetraploidy (2n = 4X = 80) in addition to normal diploid (2n = 2X = 40) plants, and the rest of the species is diploid (Hymowitz, 2004). Perennial species under the subgenus *Glycine* are *G. ablicans* Tind. & Craven, *G. aphyonota* B. Pfeil, *G. arenaria* Tind., *G. argyrea* Tind., *G. canescens* F.J. Herm, *G. clandestina* Wendl, *G. curvata* Tind., *G. cyrtoloba* Tind., *G. dolichocarpa* Tateishi & Ohashi, *G. falcata* Benth., *G. hirticaulis* Tind. & Craven, *G. lactovirens* Tind. & Craven, *G. latifolia* (Benth.) Newell & Hymowitz, *G. latrobeana* (Meissn.) Benth, *G. microphylla* (Benth.) Tind., *G. peratso* B. Pfeil & Tind., *G. pindanica* Tind. & Craven, *G. pullenii* B. Pfeil, Tind. & Craven, *G. rubiginosa* Tind. & B. Pfeil, *G. stenophita* B. Pfeil & Tind., *G. tabacina* (Labill.) Benth and *G. tomentella* Hayata. This is an enormous resource from which genes of interest can be transferred into *G. max* to improve the crop.

5.3 Mode of Reproduction

The soybean flower is a complete flower comprised of a calyx, corolla, androecium and gynoecium (Carlson and Lersten, 2004). As in other

Papilionaceous flowers, the calyx has five unequal sepals and a corolla with five petals. Among the five petals, the outer one is a standard protecting two lateral wings and two anterior keels each. The androecium consists of ten stamens; filaments of nine are fused together and elevated as a single structure, while one posterior stamen is separate. The gynoecium, consisting of a stigma, style and ovary with one to four ovules, is at the centre. Anthesis, the opening of the flower, occurs only after the maturation of pollen and eventually pollination is complete. This flower structure makes the soybean flower a cleistogamy and soybean a self-pollinated plant, with >99% self-pollination.

Flower clusters appear on a node of the stem. The first node with a flower cluster is on the fifth or sixth node. Depending upon growth habit – determinate or indeterminate – flower buds will appear as terminal or auxiliary buds. In soybean with a determinate type of growth habit, a terminal bud stops the further vegetative growth of the plant. In soybean with an indeterminate type of growth habit, on the other hand, vegetative and reproductive growth continue simultaneously for some time. The inflorescence of soybean, called a raceme, initially contains about 3–35 single flower buds. However, up to 90% abortion has been reported, leaving only a few flowers per node, which can eventually develop into pods. Soybean cultivars with many flowers per node tend towards higher flower drop, whereas those with fewer flowers per node have less (Carlson and Lersten, 2004). Flower drop takes place at different stages right from flower bud initiation to pod formation. However, it is most common within a week of flower initiation. Flower drop is caused by drought and heat stress, but unfertilized flowers will drop even under favourable growing conditions.

5.4 Crossing Methods

Selection of parents

Careful consideration needs to be given to selecting the parents. Parent selection depends on breeding objectives and available genetic resources. Crosses between multiple parents often have to be made to introgress the genes conferring a trait. This is necessary especially when the target trait is controlled by multiple genes – called a quantitative trait. Generally, one of the parents should be well adapted to the existing climatic and environmental conditions, while the other may be exotic, containing gene(s) conferring a trait(s) of interest (Orf *et al.*, 2004). As a general rule, parent selection is based on either performance of the genotypes or progeny. In either case, parents should have complementary attributes that are suitable for the target environment (Witcombe and Virk, 2001). Traits to improve could be disease resistance, stress tolerance or seed quality, in addition to the seed yield. It should be noted that crosses using successful parents are likely to produce even better progeny by accumulating the favourable alleles from both parents. Comparative evaluation of lines or breeding materials provides the performance of the lines that can be selected for parents. If parents are to be selected for developing disease resistance, test-crosses can be performed to make sure that the resistance gene is present in the lines that are going to be used as parents.

Crossing block

Selected parents are planted in a separate block so that controlled cultural practices can be provided, and vegetative and reproductive growth stages can be monitored. Each parent is planted in a row of 1.2–1.8 m on at least four different planting dates, with an interval of a week. This is important to synchronize the flowering times. While the days to flower of the parents may be known, parents are sometimes selected without prior information on flowering date. Furthermore, the growing environment, particularly photoperiod – which is beyond the control of soybean breeder – affects the flowering date significantly. Therefore, planting at different dates is necessary.

Emasculation

The removal of anthers from a flower, called emasculation, is necessary for hybridization between selected parents. Female parents are emasculated by removing all ten anthers carefully in the morning until 10:00–11:00. The calyx can be removed by first holding on to the tip of the sepals and pulling slightly out from one side of the flower bud. The flower bud is flat when it is ready to emasculate, and removal of the calyx from the other side becomes easier after removing it from one side. The flower bud, white or purple, will now be visible, with the corolla. Because the anthers and corolla are attached at the base, holding the bud firmly at the base of the flower with forceps and pulling up gently will remove the entire corolla and anthers. It will now be possible to see the slightly curved, hairy style with swollen base at the centre.

Care should be taken to ensure the stigma is not damaged during the emasculation process. The use of magnifying glass is helpful to see all of the structures. A skilled person can emasculate about 25–30 flowers h^{-1}. After emasculation, the flowers should be tagged properly and kept out of direct sunlight to avoid desiccation. While the female parent can be traced on the basis of the parent row tag, it is always good to clearly write both the male and female parents on a tag.

Pollination

The transfer of pollen grains from anther to stigma, called pollination, can be done in the morning. However, some soybean breeders pollinate even in the afternoon (Fehr, 1987). Soybean pollens are collected just after a burst of anthers. Visual observation of the flower bud is very important to identify this stage. A slight emergence of petals from the flower bud indicates that the anthers have burst and pollens are ready to collect. A slight extrusion of white or purple petals through the calyx indicates that pollens are at the appropriate stage for harvesting. Some soybean breeders prefer to collect pollens in a vial before emasculation and keep them in a desiccator. This helps to remove any moisture on the anthers and increases bursting making more pollen grains available for pollination. Others prefer to collect the pollens when they are ready for pollination after emasculation. A single flower produces thousands of pollen grains, and hence, in principle, should be enough to pollinate several flowers. In practice, a single flower is used to pollinate one or two female flowers. This is simply to make sure that at least a couple of pollen grains land on the stigma of each emasculated flower. For pollination, the corolla should be opened with the finger or forceps and the pollen tapped gently onto the top of the emasculated flower. The pollinated flower should be tagged properly and any nearby flowers on the node removed.

5.5 Breeding Objectives

The breeding objective determines the direction of any breeding programme. Soybean breeding priority areas, which are eventually the breeding objectives, are described below.

Agronomic traits

Agronomic traits come first whenever we talk of any breeding programme. These are traits related to adaptability to the existing climatic and geographical environments and overall performance with respect to seed yield. Major agronomic traits include plant height, lodging resistance, maturity, number of pods per plant, number of seeds per pod, 100-seed weight and, of course, seed yield.

Tall plant height is not a desirable trait, and hence most soybean breeders aim to develop a dwarf and compact plant type. This reduces lodging and makes cultural and harvesting operations easier. A number of studies have shown a positive correlation between plant height and lodging (Lee *et al.*, 1996a, 1996b; Panthee *et al.*, 2007). Lodging not only impedes harvesting of crops, but also reduces the yield. There is also a negative correlation between lodging and seed yield (Table 5.1). It is clear that lodging reduces the exposure of plants to sunlight, and hence reduces total photosynthesis. Some reports have shown that even if the plants start filling grains before lodging, the 100-seed weight is low after lodging. It is obvious that because of reduced accumulation of photosynthates, the 100-seed weight will be less. Furthermore, lodged plants are attacked by insects and diseases. Especially in disease development, the micro-environment becomes conducive

Table 5.1. Genetic and phenotypic correlation coefficients among agronomic traits in an F_6-derived soybean population developed from N87-984-16 × TN93-99 (reprinted with permission from Panthee *et al.*, 2007).

Trait	Days to flower	Seed-filling period	Maturity	Lodging	Plant height	Yield
Days to flower	–	-0.62^{c}	0.17^{b}	0.43^{c}	0.35^{c}	-0.30^{c}
Seed-filling period	-0.37^{c}	–	0.58^{c}	0.09^{ns}	0.08^{ns}	0.42^{c}
Maturity	0.07^{ns}	0.52^{c}	–	0.17^{b}	0.35^{c}	0.17^{b}
Lodging	0.20^{b}	0.13^{ns}	0.03^{ns}	–	0.60^{c}	-0.35^{c}
Plant height	0.16^{b}	0.06^{ns}	0.49^{c}	0.58^{c}	–	-0.29^{c}
Yield	0.20^{b}	0.29^{c}	0.06^{ns}	-0.94^{c}	-0.11^{a}	–

ns, non-significant.
Values below and above the diagonal are genetic and phenotypic correlations, respectively.
^{abc}Significant at the 0.05, 0.01 and 0.001 probability levels, respectively.

because of high humidity created around the lodged plants. Furthermore, irrigation, the application of pesticide sprays and harvesting are difficult in a lodged soybean field.

Breeding to manipulate plant maturity is related to cropping systems and growing regions. In a country such as the USA, where there are several distinct climatic zones and only one crop per year is grown, breeding for a particular climatic zone, called maturity groups, is quite common. Longer-maturity cultivars are grown in the south, whereas shorter-maturity cultivars are grown in the north. However, considering the global situation, where multiple crops are often grown in a year, short-maturity cultivars are favoured over long ones. Generally, there is a positive correlation between maturity period and seed yield, but less yield from early-maturity cultivars is compensated for by another crop. Therefore, breeding for maturity period is of great importance in areas with multiple-crop growing systems.

Seed yield is the number one objective throughout the world irrespective of cropping system or any other factors. While improvements in other traits might be the breeding objective in a breeding programme, yield cannot be undermined. This is because farmers are paid on the basis of seed yield and marketing is based on seed weight. Therefore, yield is at the centre of all breeding programmes. Improvements in other traits are either directly related to the yield or there is an indirect relationship with seed yield. Heritability estimates of yield and yield-contributing traits have an important role in selection because selection response is the function of heritability and selection intensity. In other words, selection intensity can be low to select a desirable genotype for a trait with high heritability estimates. Unfortunately, several studies have shown that yield has a low heritability estimate (Burton, 1987). Heritability estimates for yield and other related traits are given in Table 5.2.

Table 5.2. Heritability estimates of some of the agronomic traits in an F_6-derived soybean population (reprinted with permission from Panthee *et al.*, 2007).

Trait	Heritability estimate
Days to flower	0.47
Seed-filling period	0.65
Maturity	0.21
Lodging	0.43
Height	0.63
Yield	0.12

Seed composition

Soybean seed is a major source of oil (20%) and protein (40%). In addition to these two components, it is rich in isoflavone, phytate, sugar and other nutritional components. Soybean breeding for oil and protein has been a major objective of several breeding programmes throughout the world, and is still valid. There are rarely any soybean breeding programmes without this objective. However, at present there is a trend towards breeding soybean for quality traits, including oil quality (Pantalone *et al.*, 2004) and protein quality (Panthee *et al.*, 2006a, 2006c).

Oil quality consists of fatty acid composition. There are five predominant fatty acids in soybean oil: palmitic (C16:0), stearic (C18:0), oleic (C18:1), linoleic (C18:2) and linolenic (C18:3) acids (Wilson, 2004). Depending upon the use of the soybean oil, different concentrations of a particular fatty acid are desirable. The higher the number of carbon bonds, the greater the level of unsaturation, indicating that the oil is more reactive. It is known from past research that saturated and polyunsaturated fatty acids are not desirable for human consumption because they become rancid in a short time. Increases in oleic acid and decreases in linolenic acid make the oil better for human consumption. Increases in saturated fatty acids may improve applications of soybean oil such as in cosmetics. The fatty acid concentration can be manipulated in a breeding programme if the genetics of the trait are well known and desired sources of germplasm are available. It is known that fatty acid composition is a quantitative trait (Diers and Shoemaker, 1992; Wilson *et al.*, 2002). Realizing this fact, researchers have identified the quantitative trait loci (QTL) associated with various fatty acids (Spencer *et al.*, 2003; Panthee *et al.*, 2006b). This has provided important information for the manipulation of fatty acid profiles in soybean through marker-assisted selection (MAS). Tremendous progress has been made in identifying and manipulating fatty acid composition by breeding, as reported in several studies (Johnson *et al.*, 2001; Alt *et al.*, 2002; Wilson *et al.*, 2002; Hyten *et al.*, 2004).

Breeding for protein quality consists of improving amino acid composition (Kwanyuen *et al.*, 1998). A major function of proteins in nutrition is to supply adequate amounts of required amino acids (Friedman and Brandon, 2001). Amino acids are the principal building blocks of enzymes and other proteins. Twenty different amino acids are required for the growth and development of human and animal bodies. These amino acids are classified into two groups: essential and non-essential. Non-essential amino acids are readily available or can be synthesized by animals, hence they need not be supplied from external sources. Essential amino acids cannot be synthesized by animals, but play a crucial role in metabolic processes. The essential amino acids are lysine, histidine, leucine, isoleucine, valine, methionine, threonine, tryptophan and phenylalanine (D'Mello, 2003), although some nutritionists do not include histidine as an essential amino acid. Differences in classifying the amino acids as essential or non-essential are based on the type of animal and its nutritional requirements. For example, humans can produce ten of the 20 amino acids, whereas swine can produce only nine. Other amino acids must be supplied in the feed. Failure to obtain an adequate quantity of even a single essential amino acid leads to degradation of the body's proteins to obtain the deficient amino acid. Unlike fat and starch, the body does not store excess amino acids for later use. Therefore, the amino acids must be obtained from food every day. As mentioned before, soybean is rich in protein, but it does not contain a balanced composition of amino acids. Mainly, it is deficient in the sulphur-containing amino acids methionine and cysteine. Recently, soybean breeders have started to address this issue and breeding for protein quality has been a breeding objective (Panthee *et al.*, 2006c). As a result of these efforts, three breeding lines have been released (Panthee and Pantalone, 2006). These lines are being used in other breeding programmes to improve the protein quality.

Abiotic stress tolerance

Breeding for stress tolerance is an objective in soybean breeding in various parts of the world, but it is location-specific because of the magnitude and nature of the problem. The major forms of stress reported in soybean are for aluminium (Bianchi Hall *et al.*, 2000), drought (Egli *et al.*, 1984) and salinity (Hong and Pak, 1999). Investigation of the genetics behind stress tolerance and breeding soybean for one or more of these stress tolerances has been reported from various parts of the world (Qin *et al.*, 2000). Researchers have found QTL associated with stress tolerance (Specht *et al.*, 2001), but they have yet to be used in breeding programmes for MAS. Furthermore, as in other species, it has been very challenging to develop stress-tolerant varieties of soybean by conventional as well as molecular approaches.

Herbicide tolerance

The development of herbicide-tolerant soybean was, in the past, a top-priority breeding objective. Several efforts were made to achieve this objective through conventional breeding (Fehr, 1987), particularly in industrialized countries where agriculture is fully mechanized and there is heavy use of herbicides. However, there was no progress until the late 1990s. With the identification of the Roundup Ready gene and its successful incorporation into soybean through genetic transformation, the development of herbicide-tolerant soybean became a routine process (Owen, 2000). This completely changed the breeding objective of soybean for herbicide tolerance to focus on Roundup tolerance. In the USA, >75% of the soybean area is under Roundup Ready soybean production. This soybean is not affected by the Roundup herbicide, thus keeping the soybean field free from weeds. This gene has now also been transferred into a number of soybean cultivars of different genetic backgrounds through back-crossing.

Disease resistance

A number of fungal, bacterial and viral diseases are found in soybean. Major diseases of soybean include charcoal rot, *Fusarium* wilt, *Rhizoctonia* root rot, sudden-death syndrome, anthracnose, brown stem rot, phomopsis, *Sclerotonia* stem rot, stem canker, soybean cyst nematode (SCN) and Asian soybean rust, to name just a few of a long list (Grau *et al.*, 2004; Tulin and Lacy, 2004). Significant yield loss is caused by these diseases. It has been reported that SCN or Asian soybean rust can completely wipe out a crop under severe conditions. Considering the magnitude of the problem, the development of resistant soybean varieties has been the breeding objective of a number of breeding programmes. Some of the diseases are epidemic in nature, while others are endemic with more severe problems in one particular location than others. Therefore, which disease gets more priority in a soybean breeding programme is the matter of the magnitude of the problem. However, the development of disease resistance in soybean varieties is a major objective in soybean breeding for two reasons: cost of production and environmental protection. The use of a resistant variety helps to reduce production costs by minimizing the use of fungicides. Reduced fungicide use in turn helps to protect the environment.

Resistance can be introgressed from wild relatives or exotic germplasms. For example, Concibido *et al.* (1996) mapped a major partial-resistance locus on linkage group 'G' near restriction fragment length polymorphism marker C006V in plant introduction (PI) 209332, which was effective against three SCN race isolates tested. On the basis of this finding they were able to perform MAS. Soybean rust resistance has been identified in *G. tomentella* and introduced into cultivated species (Schoen *et al.*, 1992; Singh *et al.*, 1998). There is a long list of such developments, but resistance breeding continues as new pathogen races evolve and the host–pathogen response continues to change.

Insect resistance

The development of insect-resistant soybean varieties would reduce pesticide use in controlling insects, but little success has been achieved. The army worm (*Spodoptera exigua*), leaf beetles (*Cerotoma trifurcata*), soybean looper (*Pseudoplusia includens*), aphids (*Aphis glycines*) and grasshoppers (*Melanoplus* species) feed on soybean leaf, while the stink bug (*Acrosternum hilare)*, corn earworm (*Heliothis zea*) and soybean stem borer (*Dectes texanus texanus*) feed on pods and stems (Boethel, 2004). The development of insect resistance in soybean is not a priority for most breeding programmes, probably because the magnitude of loss caused by insects is not as severe as that by disease. In addition, there are difficulties in combining high yield with insect resistance. Although breeding efforts have resulted in insect-resistant cultivars, the yield potential of these cultivars is generally lower than that of conventional cultivars under conditions of light insect pressure. For example, an extensive evaluation of germplasm from the US Department of Agriculture collection in the late 1960s identified three Japanese PIs resistant to a number of insects in soybean. These PIs (namely PI 171451, PI 227687 and PI 229358) were resistant to bean beetle (*Epilachna varivestis*), soybean looper (*Pseudoplusia includens*), velvet bean caterpillar (*Anticarsia gemmatalis*), cabbage looper (*Trichoplusia ni*) and corn earworm (*Heliothis zea*). However, because of the linkage drag, the resistance gene could not be transferred into a variety with high yielding ability (Boerma and Walker, 2005). Insect-resistance genes have been transferred into soybean through genetic transformation on an experimental basis (Mazier *et al.*, 1997; Dang and Wei, 2007; Homrich *et al.*, 2008). However, it is not as popular as herbicide tolerance in soybean and in other crops such as cotton (*Gossypium* species), probably because field resistance is not effective.

Functional foods

Because of increased health awareness in the public, there is a tremendous demand for functional foods that contain enhanced levels of phytochemicals that are beneficial for human health. In soybean, such phytochemicals are isoflavones, fatty acids, amino acids, phytic acids, phytoestrogens, glucosides and saponin glycosides. Isoflavones have been reported to have several health benefits such as in breast cancer, prostate cancer and cardiovascular diseases (Menon *et al.*, 1998; Ji *et al.*, 1999). There are three major groups of isoflavones in soybean – genistein, daidzein and glycitein – that have positive health effects. It has been reported that there is a genetic variation for this important phytochemical in soybean (Ding *et al.*, 1995), and breeding objectives aim to increase the concentration of desirable isoflavones. To this end, the molecular breeding approach has also been applied by mapping the QTL associated with isoflavones and eventually adopting MAS (Primomo *et al.*, 2005, 2006). Furthermore, similar work in detecting QTL associated with phytoestrogen (Kassem *et al.*, 2004) has prompted the adoption of the MAS approach in improving this phytochemical in soybean seed. Other phytochemicals have yet to be characterized in detail and the health benefits realized. In the future, improvement in functional foods is likely to become one of the top objectives of soybean breeding programmes.

5.6 Breeding Procedures

Genetic variation is the foundation for the development of a new variety. This variation is available either in natural collections of germplasms (described above) or in artificial populations developed through hybridization. The characterization of germplasms and making a selection may limit options to include multiple traits in a single genotype; hence, selection in a population through hybridization is more common in the development of soybean varieties (Fehr, 1987).

Hybridization and advancement of the generation

Hybridization is performed between superior parents selected on the basis of breeding objectives. These important steps (the selection of parents and hybridization) in breeding programmes have already been described above. In a successful cross, a small pod can be seen after a week. The success rate of hybridization varies from 10% to 75% in soybean, depending upon the experience of the breeder (Fehr, 1987). Generally, hybridization is performed during the normal soybean growing season in the field, although some private companies hybridize in greenhouses throughout the year. After hybridization, the next step is generation advancement to produce the inbred lines. Soybean is a self-pollinated plant. Simply growing different generations will result in selfing to advance the generation to produce inbred lines. Recombinant inbred lines are grown at shuttle breeding stations to minimize the time required to achieve homozygosity in the population. In a large country such as the USA, shuttle breeding can be performed by growing soybean in the south of the country (e.g. Florida) during the winter and in the north during the summer. In smaller countries, the same effect can be achieved by collaboration between countries with appropriate environments. Furthermore, three generations can be grown near the equator, where the photoperiod remains constant throughout the year, or where about 12–14 h of light are available. After growing for five generations a soybean line will be at an almost homozygous state, after which selection can be made in homozygous lines.

Selection methods

A soybean breeder must consider two major points before performing any selections. These are the level of homozygosity and the management of a population to achieve that level of homozygosity (Fehr, 1987). In principle, the later the generation selected, the better the additive genetic effect, which is already fixed in the population. This means that there is no dominance effect in the performance of lines (Kearsey and Pooni, 1996) and hence no further segregation. There are two forms of selection: individual plant or line. In an early generation (F_2), individual plant selection is carried out on the basis of visual performance. Pod setting, number of seeds per pod, plant height, disease infection, leaf pubescence colour and plant type are traits a soybean breeder can scan at a glance and select in the F_2 population. Later-generation selection is based on yield trial data. In either case, the selection response is based on the heritability of the trait and the number of plants or lines selected, called selection intensity. To obtain a desirable level of selection response, number of selects can be adjusted on the basis of the heritability of the trait. Plants selected at the F_2 generation could be homozygous or heterozygous, which simply cannot be distinguished on the basis of the visual phenotype. Therefore, lines developed from these selects may vary significantly in later generations. In contrast to this, later generations are already in a homozygous state, and hence the performance of lines developed from such selects is less likely to change on further testing or yield trials. In other words, selection on the basis of a possible dominance effect is likely to segregate, whereas that with an additive effect is fixed in the population. The following methods are used to advance the generation and achieve homozygosity before making any selections in soybean.

Single seed descent

When advancing a generation to produce inbred lines, it is difficult to handle a large number of lines or plants in a population. At the same time, it is detrimental to narrow the genetic base before starting any selection. As a compromise, soybean breeders prefer to plant at least a single seed from as many individuals of the F_2 population as possible to maintain the available genetic base all the way up to F_5 or F_6, at which the entire population is at a working homozygous stage (Brim, 1966). In practice, rather than taking a single seed from an individual plant, breeders harvest a single pod from a single plant and use it for planting the next season, called the 'modified single seed descent' method. After threshing, seeds are mixed well and divided into two equal parts, with one part planted the next season and the other kept in reserve. By doing this the population size is kept constant, still maintaining the wide genetic base. Individual plants are selected on the basis of visual performance and individual lines are developed from the selected plants. These lines are compared for their performance in a single row at the beginning and in four-row plots at later stages.

Pedigree selection

In this method, selection starts at an early stage of F_2 by making a visual selection. Plant-to-row is planted at F_3 and selection between and within rows is continued until the F_5 or F_6 stage. This means that superior plants within a row are selected and grown in plant-to-row next season, and selection is also made between rows. The performance of a single row or plant can be traced all the way back to F_2. This is the best method of selection, but involves a lot of manpower and resources in record-keeping and systematic planting. However, selection is limited to the normal growing season and cannot take place either in the greenhouse or winter nursery (Fehr, 1987).

This is the main disadvantage of this method, and the long time taken to develop a new variety makes it unpopular with soybean breeders.

Bulk selection

This is a very simple method of selection. Individual plants are selected on the basis of visual evaluation and seeds are bulked every year. When the entire population is at a homozygous or working homozygous stage, plant-to-row is planted and rows are selected. The major condition of this method is that the growing conditions should be favourable to enhance the performance of the population for which selection is going to be made. For example, if variety is going to be developed for relatively poor soil, the inbreeding population generation should be advanced in a similar environment. This is undesirable for soybean breeders, as it may be difficult to find similar conditions in which to advance the generation in greenhouses or winter nurseries.

Recurrent selection

Desirable allele frequency is increased by selecting individuals from a population of hybrids produced from a cross of selected individuals. In soybean, the cycle of making a selection followed by inter-crossing among the selected individuals is continued until a desirable phenotype is achieved.

Backcross selection

In this method, F_1 plants produced from a cross of selected parents are crossed again with one of the parents; this process is continued until a desirable gene is transferred into the genetic background of interest. Generally, disease resistance genes, quality or any colour-conferring genes are transferred through this method. The F_1 is crossed back with the parent with a desirable genetic background.

Testing at early stages

Early-generation testing as a breeding procedure for self-pollinated crops consists of testing heterogeneous families, followed by the selection of homozygous lines from superior families (St. Martin and Geraldi, 2002). Traditionally, this includes test-cross evaluation of partially inbred plants in outcrossing species and recurrent selection procedures. This may start at the F_1 or F_2 stage and continue until the F_3 or F_4 stage. When the concept is applied in development of homozygous cultivars in a self-pollinated species such as soybean, the selection of homozygous lines from superior heterogeneous families permits the breeder to exploit the genotypic variance provided by inbreeding (Kearsey and Pooni, 1996) and to develop cultivars of suitable uniformity. Thus, the procedure has two phases: selection among heterogeneous families and selection of homozygous potential cultivars from superior families. This is not a good method of selection since the performance of better or selected lines could be because of heterozygous conditions, and hence could be unwanted. However, a large number of lines makes it less likely that useful materials will be lost.

St. Martin and Geraldi (2002) assessed the effectiveness of F_1-, F_2- and F_3-derived soybean families and compared them with unselected lines of the same generation to determine genetic gain. The three early generations produced similar genetic gains in seed yield, averaging approximately 4% genetic gain. The selection of F_1-derived families for yield increased plant height and lodging, but the other two selection procedures were satisfactory in this respect. To maximize genetic gain for yield while avoiding undesirable changes in lodging in an early-generation testing programme, they found F_2-derived families to be an appropriate early generation with which to maximize genetic gain (St. Martin and Geraldi, 2002). This finding needs to be evaluated in multiple populations.

Testing at later stages

In contrast to early-generation testing, lines can be selected and tested from homozygous populations. Homozygosity is achieved by the single seed descent or pedigree method, as described above. In this method, lines are evaluated on the basis of yield performance. Since most of the lines are already at the homozygous stage, the performance of the selected lines is less likely to be altered because of additive gene effects (Kearsey and Pooni, 1996). It should be noted that additive gene effects have higher heritability estimates, meaning the performance of lines is likely to remain constant generation after generation once the line or cultivar is released. Therefore, while it may be costly to conduct late-generation testing, it is far preferable to early-generation testing. Kearsey and Pooni (1996) have clearly shown that there will be more additive genetic gain with late generations.

Multi-location trials

Once superior lines have been identified, the performance of the lines is tested at several locations to estimate the effects of genotype, environment and their interactions (G × E interaction) (Primomo *et al.*, 2002). Depending on the adaptability of the lines, the performance may vary significantly at different locations. A non-significant interaction between genotype and environment indicates a similar performance of the lines across locations, which is also regarded as the lines having high plasticity. Superior performance at a given location has a high G × E interaction, indicating that the line may be suitable for a certain location or a limited number of locations. Therefore, it is important to determine the G × E interaction of selected lines before releasing a soybean variety.

Release of variety or germplasm

On the basis of the performance of lines across locations and growing environments and in comparison to the standard check, a genotype may be released as a cultivar or germplasm. Generally, the performance of lines is tested over multiple years. A proposal is prepared on the basis of yield and other relevant data and submitted by the soybean breeder for the release of a cultivar or germplasm to the variety and germplasm release committee. There may be some variation in terms of the name or composition of this committee in different countries, but its main purpose is to judge whether the newly proposed material is worth releasing as a new variety for commercial production. The performance of all proposed genotypes is evaluated by the variety and germplasm release committee of a country with respect to a previously released variety. If found superior, it is released for commercial production or as a breeding line to be used as a parent. The name of the new variety may be proposed by the breeder and approved by the committee. Released cultivars are grown as a new variety and germplasms are used as a parent in further breeding programmes.

5.7 Seed Production

Seed production and distribution is a vital part of any breeding programme. All of the technology developed through breeding efforts is packaged into the seed, which is delivered to the end users: the farmers. Addressing the problems of farmers and making the soybean industry beneficial is the goal of the breeding programme. To make sure that the end-product of the breeding efforts – the seed – has the targeted traits, and to supply it in a required quantity, there are three different categories of seed production.

Breeder's seed

The soybean breeder produces a small quantity of the newly released cultivar or germplasm for further multiplication under his/her own supervision. At least 1 kg of breeder's seed is provided to the seed-producing agency for foundation seed production (see below). The breeder maintains the seed of released germplasm for distribution to other breeding programmes and genetic research. The purity of breeder's seed is 100%. Since the breeder has to provide this category of seeds for foundation seed production, he/she must produce this category of seeds every year. To distinguish it from other categories of seeds, breeder's seed is tagged with a 'white tag'.

Foundation seed

A seed-producing agency or public research farm produces foundation seeds using breeder's seed as source seeds. A seed-certification agency closely supervises foundation seed production in the field and its quality in the laboratory. It is distinguished from other categories of seeds using the same colour as breeder's seed (i.e. 'white tag'). In some countries, 'registered seed' is produced from foundation seed under the close inspection of a certifying agency; a 'purple tag' is used to distinguish this category.

Certified seeds

Certified seed is produced using foundation (or registered) seed in a large scale for commercial soybean production. It is tagged with a 'blue tag' to distinguish it from other categories of seeds. This tag also indicates that the seed has been certified by a certification agency to ensure that there is standard varietal purity and standard germination and that it is free from weed seeds.

5.8 Future Prospects

Because of the significant economic contribution of soybean in major soybean-producing countries such as the USA, Brazil, Argentina, China, India and others, there are robust public soybean breeding programmes. However, private breeding programmes are stronger in those countries because of soybean's industrial importance. In developed and industrialized countries, private breeding programmes are taking over the public programmes. This indicates that public breeding programmes are becoming a lower priority for universities and governments. However, important aspects of research and regular breeding programmes need to be maintained for the future. They should not move to the private programme entirely, since some components that do not look promising economically still need to be maintained due to their potential importance in the future.

In this chapter, mostly conventional breeding techniques have been discussed. However, molecular breeding is becoming popular in several soybean breeding programmes. This needs to be integrated into breeding programmes to enhance the efficiency of variety development and increase the precision of gene introgression. In fact, several private companies have already started using molecular breeding approaches to screen thousands of early-generation plants and lines. This has been helpful in advancing useful materials, and has hence saved a lot of resources. The integration of molecular and conventional breeding can enhance the efficiency of selection and produce an output to address problems of importance and market demand. Future soybean breeding programmes should follow this path.

References

Alt, B.J., Fehr, W.R. and Welke, G.A. (2002) Selection for large seed and high protein in two- and three-parent soybean populations. *Crop Science* 42, 1876–1881.

Bianchi Hall, C.M., Carter Jr, T.E., Bailey, M.A., Mian, M.A.R., Rufty, T.W., Ashley, D.A., Boerma, H.R., Arellano, C., Hussey, R.S. and Parrott, W.A. (2000) Aluminum tolerance associated with quantitative trait loci derived from soybean PI 416937 in hydroponics. *Crop Science* 40, 538–545.

Bodanese-Zanettini, M.H., Lauxen, M.S., Richter, S.N.C., Cavalli-Molina, S., Lange, C.E., Wang, P.J. and Hu, C.Y. (1996) Wide hybridization between Brazilian soybean cultivars and wild perennial relatives. *Theoretical and Applied Genetics* 93, 703–709.

Boerma, H.R. and Walker, D.R. (2005) Discovery and utilization of QTLs for insect resistance in soybean. *Genetica* 123, 181–189.

Boethel, D.J. (2004) Integrated management of soybean insects. In: Boerma, H.R. and Specht, J.E. (eds) *Soybeans: Improvement, Production, and Uses*, 3rd edn. Agronomy Monograph 16. American Society of Agronomy, Crop Science Society of America, and Soil Science Society of America, Madison, WI, USA, pp. 833–882.

Brim, C.A. (1966) A modified pedigree method of selection in soybeans. *Crop Science* 6, 220.

Burton, J.W. (1987) Quantitative genetics: Results relevant to soybean breeding. In: Wilcox, J.R. (ed.) *Soybeans: Improvement, Production, and Uses*, 2nd edn. Agronomy Monograph 16. American Society of Agronomy, Crop Science Society of America, and Soil Science Society of America, Madison, WI, USA, pp. 211–247.

Carlson, J.B. and Lersten, N.R. (2004) Reproductive morphology. In: Boerma, H.R. and Specht, J.E. (eds) *Soybeans: Improvement, Production, and Uses*, 3rd edn. Agronomy Monograph 16. American Society of Agronomy, Crop Science Society of America, and Soil Science Society of America, Madison, WI, USA, pp. 59–95.

Concibido, V.C., Denny, R.L., Lange, D.A., Orf, J.H. and Young, N.D. (1996) RFLP mapping and marker-assisted selection of soybean cyst nematode resistance in PI 209332. *Crop Science* 36, 1643–1650.

Concibido, V.C., La Vallee, B., McLaird, P., Pineda, N., Meyer, J., Hummel, L., Yang, J., Wu, K. and Delannay, X. (2003) Introgression of a quantitative trait locus for yield from *Glycine soja* into commercial soybean cultivars. *Theoretical and Applied Genetics* 106, 575–582.

Dang, W. and Wei, Z.M. (2007) An optimized Agrobacterium-mediated transformation for soybean for expression of binary insect resistance genes. *Plant Science* 173, 381–389.

Diers, B.W. and Shoemaker, R.C. (1992) Restriction fragment length polymorphism analysis of soybean fatty acid content. *Journal of the American Oil Chemists' Society* 69, 1242–1244.

Ding, A.L., Sun, J.M., Chang, R.H. and Dong, H.R. (1995) The preliminary analysis of isoflavone content in Chinese soybean cultivars. *Soybean Genetics Newsletter* 22, 24–31.

D'Mello, J.P.F. (2003) Amino acids as multifunctional molecules. In: D'Mello, J.P.F. (ed.) *Amino Acids in Animal Nutrition*, 2nd edn. CABI Publishing, Wallingford, Oxon, UK, pp. 1–14.

Egli, D.B., Orf, J.H. and Pfeiffer, T.W. (1984) Genotypic variation for duration of seedfill in soybean. *Crop Science* 24, 587–592.

Fehr, W.R. (1987) Breeding methods for cultivar development. In: Wilcox, J.R. (ed.) *Soybeans: Improvement, Production, and Uses*, 2nd edn. Agronomy Monograph 16. American Society of Agronomy, Crop Science Society of America, and Soil Science Society of America, Madison, WI, USA, pp. 249–293.

Friedman, M. and Brandon, D.L. (2001) Nutritional and health benefits of soy proteins. *Journal of Agricultural and Food Chemistry* 49, 1069–1086.

Gardner, M.E., Hymowitz, T., Xu, S.J. and Hartman, G.L. (2001) Physical map location of the *Rps1-k* allele in soybean. *Crop Science* 41, 1435–1438.

Goldman, I.L., Carter Jr, T.E. and Patterson, R.P. (1989) Differential genotypic response to drought stress and subsoil aluminum in soybean. *Crop Science* 29, 330–334.

Grau, C.R., Dorrance, A.E., Bond, J.E. and Russin, J.S. (2004) Fungal diseases. In: Boerma, H.R. and Specht, J.E. (eds) *Soybeans: Improvement, Production, and Uses*, 3rd edn. Agronomy Monograph 16. American Society of Agronomy, Crop Science Society of America, and Soil Science Society of America, Madison, WI, USA, pp. 679–764.

Hegstad, J.M., Nickell, C.D. and Vodkin, L.O. (1998) Identifying resistance to *Phytophthora sojae* in selected soybean accessions using RFLP techniques. *Crop Science* 38, 50–55.

Homrich, M.S., Passaglia, L.M.P., Pereira, J.F., Bertagnolli, P.F., Pasquali, G., Zaidi, M.A., Altosaar, I. and Bodanese-Zanettini, M.H. (2008) Resistance to *Anticarsia gemmatalis* Hubner (Lepidoptera, Noctuidae) in transgenic soybean (*Glycine max* (L.) Merrill, Fabales, Fabaceae) cultivar IAS5 expressing a modified Cry1Ac endotoxin. *Genetics and Molecular Biology* 31, 522–531.

Hong, Y. and Pak, Y. (1999) Identification of chiro-inositol-containing phospholipids and changes in their metabolism upon salt stress in soybean seedlings. *Phytochemistry Oxford* 51, 861–866.

Hymowitz, T. (2004) Speciation and cytogenetics. In: Boerma, H.R. and Specht, J.E. (eds) *Soybeans: Improvement, Production, and Uses*, 3rd edn. Agronomy Monograph 16. American Society of Agronomy, Crop Science Society of America, and Soil Science Society of America, Madison, WI, USA, 97–136.

Hyten, D.L., Pantalone, V.R., Saxton, A.M., Schmidt, M.E. and Sams, C.E. (2004) Molecular mapping and identification of soybean fatty acid modifier quantitative trait loci. *Journal of the American Oil Chemists' Society* 81, 1115–1118.

Ji, S., Willis, G.M., Frank, G.R., Cornelius, S.G. and Spurlock, M.E. (1999) Soybean isoflavones, genistein and genistin, inhibit rat myoblast proliferation, fusion and myotube protein synthesis. *Journal of Nutrition* 129, 1291–1297.

Johnson, S.L., Fehr, W.R., Welke, G.A. and Cianzio, S.R. (2001) Genetic variability for seed size of two- and three-parent soybean populations. *Crop Science* 41, 1029–1033.

Kassem, M.A., Meksem, K., Iqbal, M.J., Njiti, V.N., Banz, W.J., Winters, T.A., Wood, A. and Lightfoot, D.A. (2004) Definition of soybean genomic regions that control seed phytoestrogen amounts. *Journal of Biomedicine and Biotechnology* 2004, 52–60.

Kearsey, M.J. and Pooni, H.S. (1996) *The Genetical Analysis of Quantitative Traits*. Chapman and Hall, New York, USA.

Kwanyuen, P., Wilson, R.F. and Burton, J.W. (1998) Soybean protein quality. In: Koseoglu, S.S., Rhee, K.C. and Wilson, R.F. (eds) *Oilseeds and Edible Oils Processing*, Vol. I. AOCS Press, Champaign, IL, USA, pp. 284–289.

Lee, S.H., Bailey, M.A., Mian, M.A.R., Carter Jr, T.E., Ashley, D.A., Hussey, R.S., Parrott, W.A. and Boerma, H.R. (1996a) Molecular markers associated with soybean plant height, lodging, and maturity across locations. *Crop Science* 36, 720–735.

Lee, S.H., Bailey, M.A., Mian, M.A.R., Shipe, E.R., Ashley, D.A., Parrott, W.A., Hussey, R.S. and Boerma, H.R. (1996b) Identification of quantitative trait loci for plant height, lodging, and maturity in a soybean population segregating for growth habit. *Theoretical and Applied Genetics* 92, 516–523.

LeRoy, A.R., Fehr, W.R., and Cianzio, S.R. (1991) Introgression of genes for small seed size from *Glycine soja* into *G. max*. *Crop Science* 31, 693–697.

Mazier, M., Chaufaux, J., Sanchis, V., Lereclus, D., Giband, M. and Tourneur, J. (1997) The cryic gene from *Bacillus thuringiensis* provides protection against *Spodoptera littoralis* in young transgenic plants. *Plant Science* 127, 179–190.

Menon, L.G., Kuttan, R., Nair, M.G., Chang, Y.C. and Kuttan, G. (1998) Effect of isoflavones genistein and daidzein in the inhibition of lung metastasis in mice induced by B16F-10 melanoma cells. *Nutrition Cancer* 30, 74–77.

Orf, J.H., Diers, B.W. and Boerma, H.R. (2004) Genetic improvement: Conventional and molecular-based strategies. In: Boerma, H.R. and Specht, J.E. (eds) *Soybeans: Improvement, Production, and Uses*, 3rd edn. Agronomy Monograph 16. American Society of Agronomy, Crop Science Society of America, and Soil Science Society of America, Madison, WI, USA, pp. 417–450.

Owen, M.D.K. (2000) Current use of transgenic herbicide-resistant soybean and corn in the USA. *Crop Protection* 19, 765–771.

Pantalone, V., Walker, D., Dewey, R. and Rajcan, I. (2004) DNA marker-assisted selection for improvement of soybean oil concentration and quality. In: Wilson, R., Stalker, H. and Brummer, E. (eds) *Legume Crop Genomics*. AOCS Press, Champaign, IL, USA, pp. 283–311.

Panthee, D.R. and Pantalone, V.R. (2006) Registration of soybean germplasm lines TN03-350 and TN04-5321 with improved protein concentration and quality. *Crop Science* 46, 2328–2329.

Panthee, D.R., Pantalone, V.R., Sams, C.E., Saxton, A.M., West, D.R., Orf, J.H. and Killam, A.S. (2006a) Quantitative trait loci controlling sulfur containing amino acids, methionine and cysteine, in soybean seeds. *Theoretical and Applied Genetics* 112, 546–553.

Panthee, D.R., Pantalone, V.R. and Saxton, A.M. (2006b) Modifier QTL for fatty acid composition in soybean oil. *Euphytica* 152, 67–73.

Panthee, D.R., Pantalone, V.R., Saxton, A.M., West, D.R. and Sams, C.E. (2006c) Genomic regions associated with amino acid composition in soybean. *Molecular Breeding* 17, 79–89.

Panthee, D.R., Pantalone, V.R., Saxton, A.M., West, D.R. and Sams, C.E. (2007) Quantitative trait loci for agronomic traits in soybean. *Plant Breeding* 126, 51–57.

Patzoldt, M.E., Tyagi, R.K., Hymowitz, T., Miles, M.R., Hartman, G.L. and Frederick, R.D. (2007) Soybean rust resistance derived from *Glycine tomentella* in amphiploid hybrid lines. *Crop Science* 47, 158–161.

Primomo, V.S., Falk, D.E., Ablett, G.R., Tanner, J.W. and Rajcan, I. (2002) Genotype × environment interactions, stability, and agronomic performance of soybean with altered fatty acid profiles. *Crop Science* 42, 37–44.

Primomo, V.S., Poysa, V., Ablett, G.R., Jackson, C.J., Gijzen, M. and Rajcan, I. (2005) Mapping QTL for individual and total isoflavone content in soybean seeds. *Crop Science* 45, 2454–2464.

Primomo, V.S., Poysa, V., Ablett, G.R., Jackson, C.J., Gijzen, M. and Rajcan, I. (2006) Mapping QTL for individual and total isoflavone content in soybean seeds. *Crop Science* 46, 495–496 (erratum).

Qin, Z., Rui, Y., Zhiang, H., Min, Z. and Xin, D. (2000) Cloning and characterization of a molecular marker associated with salt tolerance from soybean cultivars. *Soybean Genetics Newsletter* 27, 4.

Rebetzke, G.J., Pantalone, V.R., Burton, J.W. and Carter Jr, T.E. (1997) Genotypic variation for fatty acid content in selected *Glycine max* × *Glycine soja* populations. *Crop Science* 37, 1636–1640.

Schoen, D.J., Burdon, J.J. and Brown, A.H.D. (1992) Resistance of *Glycine tomentella* to soybean leaf rust *Phakopsora pachyrhizi* in relation to ploidy level and geographic distribution. *Theoretical and Applied Genetics* 83, 827–832.

Sebolt, A.M., Shoemaker, R.C. and Diers, B.W. (2000) Analysis of a quantitative trait locus allele from wild soybean that increases seed protein concentration in soybean. *Crop Science* 40, 1438–1444.

Singh, R.J., Kollipara, K.P. and Hymowitz, T. (1993) Backcross (Bc2-Bc4)-derived fertile plants from *Glycine max* and *Glycine tomentella* intersubgeneric hybrids. *Crop Science* 33, 1002–1007.

Singh, R.J., Kollipara, K.P. and Hymowitz, T. (1998) Monosomic alien addition lines derived from *Glycine max* (L.) Merr. and *G. tomentella* Hayata: Production, characterization, and breeding behavior. *Crop Science* 38, 1483–1489.

Soytech Inc. (2007) *Oilseed Statistics*. Soytech Inc., Bar Harbor, ME, USA.

Specht, J.E., Chase, K., Macrander, M., Graef, G.L., Chung, J., Markwell, J.P., Germann, M., Orf, J.H. and Lark, K.G. (2001) Soybean response to water: A QTL analysis of drought tolerance. *Crop Science* 41, 493–509.

Spencer, M.M., Pantalone, V.R., Meyer, E.J., Landau Ellis, D. and Hyten Jr, D.L. (2003) Mapping the *Fas* locus controlling stearic acid content in soybean. *Theoretical and Applied Genetics* 106, 615–619.

St. Martin, S.K. and Geraldi, I.O. (2002) Comparison of three procedures for early generation testing of soybean. *Crop Science* 42, 705–709.

Stokstad, E. (2004) Plant pathologists gear up for battle with dread fungus. *Science* 306, 1672–1673.

Tulin, S.A. and Lacy, G.H. (2004) Viral, bacterial and phytoplasmal diseases of soybean. In: Boerma, H.R. and Specht, J.E. (eds) *Soybeans: Improvement, Production, and Uses*, 3rd edn. Agronomy Monograph 16. American Society of Agronomy, Crop Science Society of America, and Soil Science Society of America, Madison, WI, USA, pp. 765–820.

Wang, D., Arelli, P.R., Shoemaker, R.C. and Diers, B.W. (2001) Loci underlying resistance to Race 3 of soybean cyst nematode in *Glycine soja* plant introduction 468916. *Theoretical and Applied Genetics* 103, 561–566.

Weng, C., Yu, K., Anderson, T.R. and Poysa, V. (2001) Mapping genes conferring resistance to *Phytophthora* root rot of soybean, *Rps1a* and *Rps7*. *Journal of Heredity* 92, 442–446.

Weng, W.M., Gao, X.S., Zhuang, N.L., Xu, M.L. and Xue, Z.T. (1995) The glycinin A3B4 mRNA from wild soybean *Glycine soja* Sieb. *Plant Physiology* 107, 665–666.

Wilson, R. (2004) Seed composition. In: Boerma, H.R. and Specht, J.E. (eds) *Soybeans: Improvement, Production, and Uses*, 3rd edn. Agronomy Monograph 16. American Society of Agronomy, Crop Science Society of America, and Soil Science Society of America, Madison, WI, USA, pp. 621–678.

Wilson, R., Burton, J., Pantalone, V. and Dewey, R. (2002) New gene combinations governing saturated and unsaturated fatty acid composition in soybean. In: Kuo, T. and Gardner, H. (eds) *Lipid Biotechnology*. Marcel Dekker, Inc., Basel, New York, USA, pp. 95–113.

Witcombe, J.R. and Virk, D.S. (2001) Number of crosses and population size for participatory and classical plant breeding. *Euphytica* 122, 451–462.

Zakharova, E.S., Epishin, S.M. and Vinetski, Y.P. (1989) An attempt to elucidate the origin of cultivated soybean via comparison of nucleotide sequences encoding glycinin B4 polypeptide of cultivated soybean, *Glycine max*, and its presumed wild progenitor, *Glycine soja*. *Theoretical and Applied Genetics* 78, 852–856.

6 Soybean Yield Physiology: Principles and Processes of Yield Production

Dennis B. Egli
Department of Plant and Soil Sciences, University of Kentucky, Lexington, Kentucky, USA

6.1 Introduction

Soybean (*Glycine max* (L.) Merrill) is a legume with typical C_3 photosynthesis (Shibles *et al.*, 1987); consequently, photosynthesis is inherently limited by competition between CO_2 and O_2 for the active site on Rubisco (ribulose-1,5-bisphosphate carboxylase oxygenase) (Ogren, 1984). Soybean photosynthesis responds to CO_2 concentrations above ambient (Egli *et al.*, 1970; Acock *et al.*, 1985) and reaches a maximum at between 20°C and 30°C (Shibles *et al.*, 1987). As a legume, most of its nitrogen comes from N_2 fixation in the nodules, although nodulation and nitrogen fixation can be suppressed by NO_3 in the soil solution (Harper, 1987; Sinclair, 2004). Soybean produces a seed that contains roughly 380 g kg^{-1} protein and 200 g kg^{-1} oil, a combination that is approached by only a few crops (Egli, 1998; Wilson, 2004). The combination of C_3 photosynthesis and a seed high in protein and oil limits soybean yield (Sinclair, 2004). For example, the yield of soybean is, on the average, roughly one third that of maize (*Zea mays* L.), a high-yielding C_4 crop that produces a seed high in starch and is often grown in the same environment as soybean (Specht *et al.*, 1999; Egli, 2008a).

Yield – the weight of seeds from a unit area – is ultimately determined by photosynthesis; plants accumulate dry matter primarily through fixation of carbon by the photosynthetic enzymes in the leaves. Yield, therefore, will be determined, in large part, by the photosynthetic capacity of the plant community integrated over time. The ability of the seeds to accumulate dry matter during seed filling is also an important part of the yield production process and it is controlled, in part, by the characteristics of the seed (Egli, 1998, 2006).

Between the production of assimilate by photosynthesis and the accumulation of dry matter by the seeds lie a multitude of complex interlocking physiological processes and cycles that play crucial roles in yield production.

Many of these processes are well understood at the enzyme, process and organelle level (see, for example, reviews by Shibles *et al.*, 1987; Harper, 1987; Wilson, 1987; Egli and Crafts-Brandner, 1996). It has proven difficult, however, to use this information to understand the production of yield, which is primarily a phenomenon of the whole plant and the plant community (Thronley, 1980; Trewavas, 1986). This chapter focuses on processes operating at the organ, plant and plant community level.

Murata's three phases of yield production (Murata, 1969) capture the essence of the overall process very nicely. The three phases are as follows:

- Phase I: formation of organs for nutrient absorption and photosynthesis (vegetative growth).
- Phase II: formation of flower organs and the 'yield container' (flowering and pod set).
- Phase III: production, accumulation and translocation of 'yield contents' (seed filling).

Phases I and II partially coincide in soybean with vegetative growth (node and leaf production), continuing until nearly the end of phase II (Beaver *et al.*, 1985; Egli *et al.*, 1985a). Phases II and III relate directly to the yield components, seeds per unit area and weight per seed or seed size, commonly used to describe yield (Egli, 1998). Fruits and seeds per unit area are determined during phase II and seed size is fixed during phase III.

Murata's phases clearly define the sequential nature of the yield production process – first the plant grows vegetatively and produces the photosynthetic machinery, followed by flowering and production of fruits and seeds and finally the production of the yield 'contents' during seed filling. They also focus attention on the time component of yield production, an important and often neglected aspect of the process.

The objective of this chapter is to discuss the yield production process in soybean, highlighting the processes and characteristics of the plant that play important roles in determining yield. The focus will be on the whole plant and plant community to develop a general framework to use when analyzing the effect of the environment or plant characteristics on yield. Understanding the complex interactions of plant growth and the environment integrated over time is difficult at best, but facing it with a clear concept of how yield is produced will help. In fact, the yield production process at the community level is not as complex as is often thought and simple models at this level can greatly aid our understanding, which, in turn, leads to better management decisions and more efficient and productive cropping systems.

6.2 Vegetative Growth (Phase I)

Yield production begins with vegetative growth (Murata, 1969) when the 'formation of organs for nutrient absorption and photosynthesis' provides the machinery to produce yield. The rate of canopy photosynthesis is determined by the inherent photosynthetic capacity of the leaves, environmental conditions (temperature, solar radiation, CO_2 concentration, nutrient and water availability) and the proportion of the incident solar radiation absorbed by the plant canopy. The presence of weeds, diseases and insects may also affect canopy photosynthesis either directly (e.g. reducing leaf area, shading the plants) or indirectly by, for example, causing water stress by increasing community water use. The supply of solar radiation available for photosynthesis has two components: the incident radiation determined by location (latitude and elevation), time of year and atmospheric conditions; and the proportion of the incident radiation intercepted and absorbed by the leaves.

Leaf area and interception of solar radiation

Radiation interception is closely associated with leaf growth and leaf area during the initial stages of vegetative growth (Shibles and Weber, 1965; Taylor *et al.*, 1982; Wells, 1991). Leaf area, commonly described by the leaf area index (LAI), the ratio of the leaf area to the ground area (Watson, 1947), increases steadily after seedling emergence and eventually reaches a maximum at or before growth stage R5 (Wells, 1991; Board and Harville, 1996; Setiyono *et al.*, 2008). Node production (Egli *et al.*, 1985a) and vegetative mass (Egli and Leggett, 1973) reach their maximum near the beginning of seed filling or growth stage R5, suggesting that R5 approximates the end of the vegetative growth phase. This pattern holds for cultivars with indeterminate and determinate growth habits (Egli and Leggett, 1973; Egli *et al.*, 1985a), although much of the vegetative growth after growth stage R1 on determinate types occurs on branches (Egli *et al.*, 1985a).

There is a linear relationship between LAI and radiation interception until radiation interception is maximized, after which further increases in LAI do not increase interception (Shibles and Weber, 1965; Heilman *et al.*, 1977; Taylor *et al.*, 1982). The critical LAI is often between 2.0 and 3.5 (Shibles and Weber, 1965; Heilman *et al.*, 1977; Taylor *et al.*, 1982), although higher values (5.0–6.8) have been reported in wide rows (Taylor *et al.*, 1982; Wells, 1991). Maximum interception usually occurs sooner in narrow rows and at higher plant populations (Wells, 1991). Taylor *et al.* (1982) found that 0.25 m rows required an LAI of 3.0 for maximum radiation interception compared with an LAI of 4.5 in 1.0 m rows.

The length of the vegetative growth phase (emergence to growth stage R5) and, therefore, the maximum vegetative mass and LAI, is directly related to cultivar maturity (Egli, 1993, 1994; Edwards and Purcell, 2005). For example, in Kentucky, maturity group (MG) IV cultivars planted in mid-May produced 44% more vegetative mass than MG I cultivars in 23 more days (Table 6.1). Zeiher *et al.* (1982) obtained 80% more vegetative mass from MG V cultivars than from MG II cultivars in a 34-day longer vegetative phase.

Table 6.1. Effect of cultivar maturity and planting date on vegetative growth characteristics of soybean (average of 2 years) grown at Lexington, Kentucky, USA (38°N) (adapted from Egli and Bruening, 2000).

	Nodes (no. per m²)		Vegetative growth phase[c] (days)		Maximum vegetative mass[d] (g per m²)	
Maturity Group[a]	Early[b]	Late[b]	Early	Late	Early	Late
I	827	662	65	54	566	513
II	850	638	73	60	535	533
III	1071	755	82	64	821	633
IV	1578	798	88	68	815	656
LSD (0.05)	163				88	

LSD, least significant difference.
[a]One cultivar in each maturity group.
[b]Average planting dates: early, May 18; late, June 25.
[c]Days from planting to growth stage R5.
[d]Above-ground vegetative mass at growth stage R5.

The initial rate of vegetative development and, presumably, LAI accumulation is not affected by cultivar maturity (Zeiher *et al.*, 1982), so early and late cultivars could reach the critical LAI (LAI producing 95% radiation interception) at roughly the same number of days after seedling emergence. Delayed flowering of later-maturing cultivars, however, provides more time to reach the critical LAI before the beginning of reproductive growth. Late-maturing cultivars may, therefore, be more tolerant of short periods of stress early in vegetative growth or of wide-row culture. Early cultivars may not reach the critical LAI before growth stage R1 unless they are grown in narrow rows (Board and Harville, 1994). The tendency for soybean to be more responsive to narrow rows in the Midwest than in the south of the USA (Johnson, 1987) may simply reflect the earlier flowering and shorter vegetative growth periods of the Midwestern types. A longer period before flowering may have been an advantage in wide-row production systems, but it is not as important in modern systems that can easily accommodate very narrow rows.

Canopy photosynthesis

Canopy photosynthesis (CO_2 fixation expressed on a ground area basis) (Wells, 1991) and crop growth rate (an indirect estimate of canopy photosynthesis) (Shibles and Weber, 1965; Buttery, 1969; Board and Harville, 1994) are directly related to LAI and radiation interception during the initial stages of vegetative growth. Canopy photosynthesis, therefore, increases rapidly during early vegetative growth (Larson *et al.*, 1981; Christy and Porter, 1982; Acock *et al.*, 1985) and reaches its maximum level when radiation interception is complete. There is no evidence for an optimum

LAI in soybean (Shibles and Weber, 1965); consequently, there is no change in canopy photosynthesis (or crop growth rate) as LAI increases beyond the level required for maximum radiation interception. Canopy photosynthesis, crop growth rate and ultimately yield will be reduced if the canopy does not reach maximum interception by the beginning of reproductive growth (Lee *et al.*, 2008). Reaching maximum interception before flowering does not contribute directly to yield, but it may aid weed control (Buhler and Hartzler, 2004). Jiang and Egli (1995) found that stress that reduced vegetative growth before growth stage R1 had no effect on yield if there was enough LAI to ensure maximum radiation interception by the beginning of flowering.

The rate of evapotranspiration follows the increase in LAI and radiation interception, especially when a dry soil surface limits soil evaporation (Heatherly and Elmore, 2004). Consequently, the rapid early development of leaf area associated with narrow rows and high populations may increase evapotranspiration (Heatherly and Elmore, 2004).

Radiation-use efficiency (RUE; dry matter produced per unit of intercepted solar radiation) is often used to evaluate the productivity of crop communities since it represents an estimate of the efficiency with which the community converts solar radiation into dry matter (Sinclair and Muchow, 1999). It seems to provide a simple characterization of productivity, but estimates are often quite variable since they require determination of dry matter accumulation and radiation interception (Sinclair and Muchow, 1999), making it difficult to detect small differences between treatments or cultivars.

RUE is sensitive to any variation in photosynthesis, including those caused by environmental conditions (e.g. temperature, water stress, nutrient availability) and plant species. Since soybean is a C_3 legume that produces leaves with high protein levels and seeds with a high energy content, the maximum RUE (i.e. measured under non-stress conditions) is less than in many other crops. Sinclair and Muchow (1999) concluded after an extensive review that the average RUE for soybean is 1.02 g MJ^{-1} (based on total solar radiation), which is lower than estimates for maize (1.6–1.7 g MJ^{-1}) and wheat (*Triticum* species) (1.4–1.7 g MJ^{-1}).

6.3 Flowering and Pod Set (Phase II)

The appearance of the first flower (growth stage R1) marks the beginning of reproductive growth and the beginning of Murata's (1969) phase II – the formation of flower organs and the 'yield container'. Phase II ends shortly after the beginning of growth stage R6, at which time all fruits that will survive to maturity are established and there will be no changes in the number of fruits during the rest of reproductive growth (Fig. 6.1) (Board and Tan, 1995; Egli, 1997). The potential size of the yield container is determined by the number of seeds that are set (a function of fruit number and seeds per fruit) and the potential size of the seed (Egli, 1998).

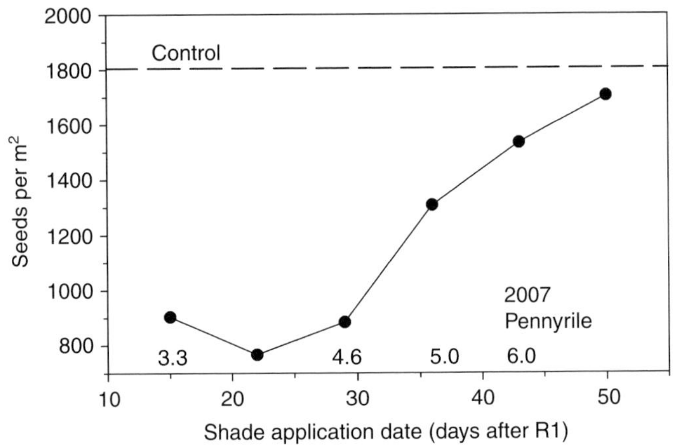

Fig. 6.1. Effect of shade on seed production in the field. Shade cloths that reduce the incident solar radiation by 80% were placed over the plants at six dates after growth stage R1 and left in place until maturity. The dashed line represents the seed number on control (unshaded) plants. There was no significant difference ($P = 0.05$) between seed number on the control and shaded plants at the last shade treatment. Reproductive growth stages are shown on the x-axis. Plants were irrigated to minimize water stress (Egli, 2007, unpublished data).

Flowering profiles

The soybean plant has a long flowering period that typically can be as short as 20 days or as long as 40 days, although periods of up to 90 days have been documented (Hansen and Shibles, 1978; Gai *et al.*, 1984; Dybing, 1994; Zheng *et al.*, 2002). The total length of the period is somewhat misleading because >70% of the flowers are produced in less than half of the total period (Hansen and Shibles, 1978; Gai *et al.*, 1984; Nakamoto *et al.*, 2001). Fruit production (the appearance of fruits ≥10 mm in length; Egli and Bruening, 2002a) follows the same pattern as flower production, with 70–80% of the fruits produced in just 12 days of a 30- to 40-day fruit production period (Fig. 6.2).

Fruit production always continues past growth stage R5 for several cultivars in field and greenhouse experiments, but is usually complete by growth stage R6 or shortly thereafter (Egli and Bruening, 2006a). Cultivars with determinate growth habits have shorter fruit production periods than those with indeterminate growth habits. The length of the period varies among years and it is shorter in delayed plantings, but it is not affected by CO_2 enrichment or low plant density, changes that increase the productivity of the plant (Saitoh *et al.*, 1998; Nakamoto *et al.*, 2001; Egli and Bruening, 2006a).

Determination of fruit number

Linking fruit and seed number to canopy photosynthesis during phase II provides a simple, straightforward mechanism that is supported by experimental

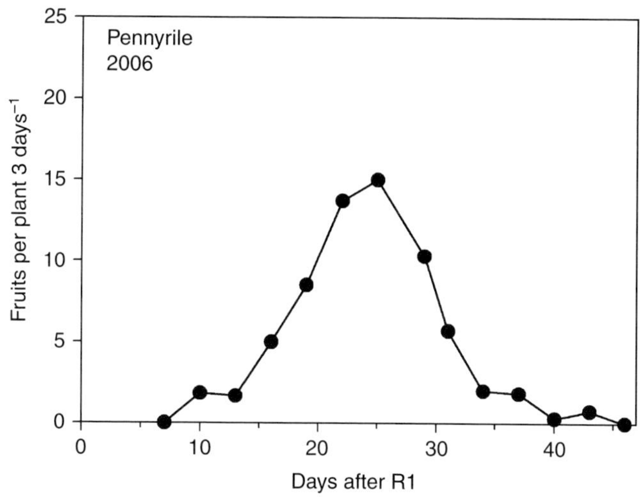

Fig. 6.2. Fruit production on the soybean cultivar Pennyrile (maturity group IV, indeterminate growth habit) growing in the field. Fruits that were ≥10 mm in length were counted and marked at 3-day intervals, as described by Egli and Bruening (2002a). Growth stage R5 occurred approximately 26 days after growth stage R1 (Egli and Rucker, 2006, unpublished data).

results and explains environmental variation in the size of the yield container. The treatments that affect photosynthesis during phase II always result in a corresponding change in fruit and seed number. Increasing photosynthesis with CO_2 enrichment or extra light increases fruit and seed number (Hardman and Brun, 1971; Schou *et al.*, 1978) while shade, water stress and defoliation reduce fruit and seed number (Shaw and Laing, 1966; Egli and Zhen-wen, 1991; Board and Tan, 1995; Andrade and Ferrerio, 1996). Modifying photosynthesis during only part of phase II usually affects fruit and seed number, but the effect is always less than when the treatment is applied during the entire period (Shaw and Laing, 1966; Schou *et al.*, 1978; Jiang and Egli, 1995).

Some researchers have suggested that the rate of photosynthesis may be affected by the size of the reproductive sink. Reducing sink size often reduces photosynthesis (Mondal *et al.*, 1978; Goldschmidt and Huber, 1992), but increasing sink size above its normal level with extra light or high CO_2 levels during flowering and pod set did not increase yield, suggesting that photosynthesis was not stimulated (Hardman and Brun, 1971; Ackerson *et al.*, 1984). Higher levels of single leaf or canopy photosynthesis during the early stages of seed filling (Dornhoff and Shibles, 1970; Ghorashy *et al.*, 1971) have been interpreted as a response to a larger sink (Shibles *et al.*, 1987). Such increases, however, do not always occur (Christy and Porter, 1982). Direct evaluation of this hypothesis with an isolated-node system found no effect of sink size on photosynthesis during seed filling (Egli and Bruening, 2003).

Relating the rate of photosynthesis to reproductive sink size is completely untenable with the argument that fruit number and sink size are determined by the assimilate supply (Egli, 1998). The latter mechanism is well supported by extensive experimentation in many crops and provides a fundamental explanation for many crucial relationships that are important in the production of yield; there seems to be no reason at this time to accept the alternative hypothesis that photosynthesis and ultimately yield is determined by sink size.

Competition for assimilate between vegetative and reproductive sinks could reduce fruit and seed numbers (Egli, 1998). Vegetative growth may have continued past the end of phase II and limited flower and seed set in older cultivars (Gay *et al.*, 1980), but, in modern cultivars, vegetative growth stops at growth stage R5 (Egli *et al.*, 1985a), just before the end of flowering and pod set. Most of the fruits that survive to maturity, however, are produced before growth stage R5 (Egli and Bruening, 2006a), but it is not yet clear if this competition limits fruit number.

The soybean plant has two mechanisms by which fruit and seed number are adjusted to match the assimilate supply. First, flower production responds to environmental conditions and varies among cultivars; second, not all flowers produce fruits and not all fruits survive until maturity. Variation in flowers per plant or per unit area probably plays a major role in matching fruit and seed number to the general level of productivity of the environment (Fig. 6.3), while abortion probably makes a larger contribution when the community experiences a large shift in environmental conditions (i.e. short-term stress) during phase II (Jiang and Egli, 1993).

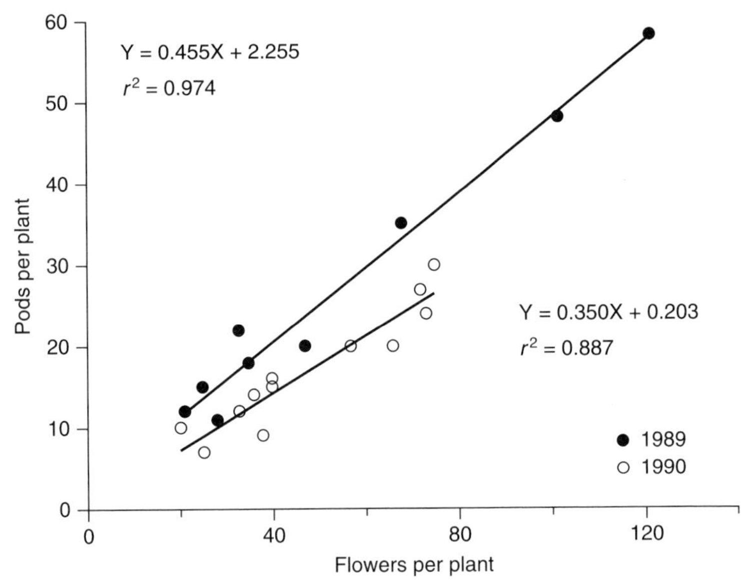

Fig. 6.3. Relationship between pods per plant and flowers per plant. Data from two field experiments using cultivars from maturity groups 00–V. Shade cloths that reduce the incident solar radiation by 30% and 63% were placed over the plants at growth stage R1 and left in place until maturity to create differences in plant growth. All plants were irrigated to minimize water stress (reprinted with permission from Jiang and Egli, 1993).

Flower production

The number of flowers is directly related to the number of nodes (Egli, 2005), although there is evidence that environmental conditions can influence the number of flowers per node (Jiang and Egli, 1993). The number of nodes per plant is influenced by environmental conditions during vegetative growth and cultivar maturity (Table 6.1). Late-maturing cultivars generally have more nodes as a result of their longer vegetative growth period (Jiang and Egli, 1993), while delayed planting shortens the vegetative growth period and reduces node number (Egli and Bruening, 2000). In field experiments, manipulating the photoperiod has lengthened phase II and increased nodes, fruits and yield (Kantolic and Slafer, 2001, 2005).

Abortion and abscission

Abortion and abscission of flowers and fruits is always high, with estimates ranging from 36% to 81% (van Schaik and Probst, 1958; Hansen and Shibles, 1978; Jiang and Egli, 1993). In fact, high levels of abortion can occur in high-yield environments (50% of the flowers and fruits have been seen to abort with a yield of nearly 4000 kg ha^{-1}) (Egli, 1993; Jiang and Egli, 1993). Abortion is rarely caused by failure of the pollination process (Abernathy *et al.*, 1977), but it can be influenced by manipulation of the assimilate supply (Mann and Jaworski, 1970; Heitholt *et al.*, 1986; Jiang and Egli, 1993; Miceli *et al.*, 1995).

Abortion and abscission occur at several stages of reproductive development with flowers (Kato *et al.*, 1955; van Schaik and Probst, 1958; Huff and Dybing, 1980; Heitholt *et al.*, 1986), immature pods (Hansen and Shibles, 1978; Heitholt *et al.*, 1986) and immature seeds (Duthion and Pigeaire, 1991; Westgate and Peterson, 1993) all participating in the process. The first flowers at a node (Huff and Dybing, 1980; Heitholt *et al.*, 1986; Nakamoto *et al.*, 2001) or on whole plants (Brevedan *et al.*, 1978; Hansen and Shibles, 1978; Yoshida *et al.*, 1983; Egli and Bruening, 2006a) have lower rates of abortion than late-developing flowers.

Reproductive failure is very much a part of the early stages of fruit development. In fact, fruits that reach their maximum length and seeds that are past the cell-division phase of growth rarely abort (Duthion and Pigeaire, 1991; Westgate and Peterson, 1993; Egli and Bruening, 2006a). A major role for the time of fruit development suggests that abortion may result from competition for assimilate between rapidly growing fruits from early flowers and small fruits from later developing flowers (Bruening and Egli, 1999; Egli and Bruening, 2002a). This competition may be an entirely intranodal phenomenon with little competition between nodes (Egli and Bruening, 2006b).

The capacity to adjust flower number and to abort flowers and developing fruits allows soybean to match the size of the yield container to the productive capacity of the plant in most environments; not all crops (e.g. maize, sunflower) (Vega *et al.*, 2001) share this capacity. The yield container, however, responds only to environmental conditions during phase II; consequently, a change in the environment after phase II can result in an incorrectly sized container requiring a change in seed size to match the assimilate supply.

Seed characteristics

The characteristics of the developing seed also play an important role in determining fruit and seed number. There is an inverse relationship between genetic variation in individual seed growth rate (rate of dry matter accumulation) and fruit and seed number; cultivars with high seed growth rates produce fewer seeds, given the same assimilate supply, than cultivars with low seed growth rates (Egli, 1993, 1998, 2006). The adjustment to seed growth rate seems to be a matter of balancing the total assimilate needs of the seeds (seed number × assimilate requirement per seed) with the assimilate supply (Egli, 1998, 2006). There is some evidence that genetic variation in seed growth rate is also inversely related to flower number (Jiang and Egli, 1993). Genetic differences in seed growth rate are usually related to genetic differences in seed size (Egli, 1998, 2006). Thus, selection for large seeds usually results in a reduction in seed number and no effect on yield (classic yield component compensation). Genetic variation in seed size is common and substantial in soybean (Hartwig, 1973), even among commercial cultivars, leading to cultivar variation in fruit number that is not related to the availability of assimilate, the productivity of the environment or yield (Egli, 1998).

Phase II is a critical phase in the production of yield because this is when the size of the yield container (number of fruits and seeds and potential seed size) is determined and that size sets the upper limit on yield. The processes regulating size are very efficient, but seed size often varies because the size of the yield container is not always perfectly matched to canopy photosynthesis during seed filling.

6.4 Seed Filling (Phase III)

Seed filling (Murata's phase III) marks the beginning of the accumulation of yield. All previous activities are simply preliminary events – essential, but just preparation for the 'production, accumulation and translocation of yield contents'. At the beginning of seed filling there is no yield present (depending somewhat upon the definition of the beginning of seed filling), but all systems are in place to produce yield. Surprisingly, the actual production of yield occurs in only a relatively small proportion of the total growth cycle of the crop. The seed-filling period in soybean is usually 30–40 days long (using growth stage R5–R7) (Egli, 2004), which represents ≤40% of the total growth cycle for most cultivars grown in their area of adaptation (Egli, 1994, 2004); so, more than half of the growth cycle is spent on preliminary events. The proportion spent filling seeds tends to decrease in later-maturing cultivars with longer total growth cycles because the seed-fill duration does not increase proportionately (Egli, 1994).

The seed-filling period starts when the seeds begin to accumulate dry matter and ends at physiological maturity (maximum seed dry weight;

TeKrony *et al.*, 1979). While it is easy to define the seed-filling period, it is more difficult to measure it. Plant growth stages (R5–R7; Fehr and Caviness, 1977) are often used for non-destructive plant and community estimates (Egli *et al.*, 1984; Agudelo *et al.*, 1986). Complete seed growth curves may be used to estimate the time from 5% to 95% (or 10–90%) of maximum seed weight (see, for example, Johnson and Tanner, 1972, with maize). The effective filling period (= final seed size / seed growth rate) (Daynard *et al.*, 1971) provides an estimate that avoids the difficulty of modelling the lag phases at the beginning and end of seed growth (Egli, 1998). All methods provide acceptable estimates, but they cannot be compared. The growth stage method produces longer estimates for cultivars with determinate growth habits than for cultivars with indeterminate growth habits, because growth stage R5 occurs earlier in the seed-filling period in determinate types (Agudelo *et al.*, 1986; Pfieffer and Egli, 1988). The advantage of the growth stage method is that it does not require destructive sampling as do the effective filling period and growth curve methods (Egli, 2004).

Seeds cannot grow without a supply of assimilate so the photosynthetic productivity of the plant community during phase III is important. The total seed growth rate (g per m^2 day^{-1}) is directly affected by the assimilate supply (Table 6.2) (Egli, 1999) through effects on seed number and individual seed growth rate (Egli and Bruening, 2001). There are two sources of assimilate: current photosynthesis and remobilization of stored carbohydrates. The potential contribution from stored carbohydrates (starch) is apparently relatively small in soybean (<15% of total seed mass without adjusting for respiration losses in one series of experiments) (Egli, 1997); much less than that reported for wheat (model estimates suggest a contribution of 20–50%) (Gent, 1994) and sunflower (*Helianthus annuus* L., 22–27%) (Hall *et al.*, 1989).

Table 6.2. Effect of assimilate supply on seed growth rate, average of 1993–1995 (adapted from Egli, 1997, 1999).

		Yield (g per m^2)		Total seed growth rate[c] (g per m^2 day^{-1})	
Cultivar	Total growth cycle (days)[a]	Control	Shade[b]	Control	Shade
Early (MG I)					
Kasota	95	359	262	14.0	9.2
Hardin	92	345	261	11.8	7.8
Late (MG V)					
Essex	139	362	286	10.3	8.2
Hutcheson	143	381	291	11.5	8.4
LSD (0.05)		27			

LSD, least significant difference; MG, maturity group.
[a]Days from planting to physiological maturity (growth stage R7).
[b]Shaded (solar radiation reduced by 63%) from just after the beginning of growth stage R6 until maturity. Irrigated to minimize water stress.
[c]Total seed growth rate = yield divided by seed-fill duration.

Seed growth rate

The individual seed growth rate decreases as the temperature drops below 22–23°C (Egli and Wardlaw, 1980; Egli, 1998). Since the total seed growth rate (g per m^2 day^{-1}) is simply the individual seed growth rate (mg $seed^{-1}$ day^{-1}) multiplied by the number of seeds per unit area, temperature or other environmental effects on individual seed growth rate should be reflected in the total seed growth rate. The individual seed growth rate and, therefore, the total seed growth rate is relatively tolerant of water stress (Meckel *et al.*, 1984; Westgate *et al.*, 1989), although stress levels that cause large reductions in photosynthesis would probably eventually reduce it. Seed growth is not very sensitive to the nitrogen supply (Egli *et al.*, 1985b; Hayati *et al.*, 1996). In fact, seeds in an *in vitro* culture system required only 18 mM nitrogen in the media to maintain normal rates of dry matter accumulation, while 135 mM was required to maintain the normal seed nitrogen concentration. Genetic increases in seed protein concentration had no effect on the individual seed growth rate (Egli and Bruening, 2007).

Some aspects of individual seed growth are controlled by the seed (e.g. genetic variation in seed growth rate) and some by the mother plant through the assimilate supply (Egli, 1998, 2006). Thus, the capabilities of the vegetative plant, the characteristics and capabilities of the seeds (the sink), and the environment, all play a role in determining the total seed growth rate (g per m^2 day^{-1}).

Leaf senescence

The productivity of the soybean canopy begins to decline shortly after the beginning of seed filling (Larson *et al.*, 1981; Wells *et al.*, 1982; Acock *et al.*, 1985; Boerma and Ashley, 1988) as senescence destroys the photosynthetic machinery and nitrogen is exported from the leaf (Crafts-Brandner and Egli, 1987). The decline in photosynthesis represents an obvious enigma; the plant has progressed through the preliminary phases of the yield production process (phases I and II) and now, when the main event begins, the photosynthetic capacity of the plant is slowly destroyed. This does not seem to be, on the surface, a rational strategy for high yield, but it results in an efficient use of nitrogen and is a strategy that is followed by soybean and all grain crops.

The nitrogen exported from the senescing leaf is translocated to the developing seed (Morris and Weaver, 1983) where it can account for up to 100% of the seed nitrogen at maturity (Egli *et al.*, 1978; McBlain and Hume, 1981; Zeiher *et al.*, 1982; Egli *et al.*, 1983). The remobilization of nitrogen is probably a result of senescence (Hayati *et al.*, 1995, 1996) not a cause and, therefore, models based on a hypothesized seed nitrogen demand (Sinclair and de Wit, 1975, 1976; Frederick and Hesketh, 1994) probably do not provide an accurate depiction of the interaction of seed growth and leaf senescence in soybean.

Seed-fill duration

The duration of seed fill is under genetic control, but it is also influenced by environmental conditions. Numerous researchers have demonstrated significant genotypic variation in seed-fill duration (Hanway and Weber, 1971; Gay *et al.*, 1980; Egli *et al.*, 1984). Seed-fill duration can be modified by direct selection (Metz *et al.*, 1984, 1985; Salado-Navarro *et al.*, 1985; Smith and Nelson, 1987; Pfieffer and Egli, 1988) with estimates of heritability ranging from –0.20 to 1.02. Plant breeders have also inadvertently lengthened the seed-fill duration when selecting for yield (McBlain and Hume, 1981; Boerma and Ashley, 1988; Shiraiwa and Hashikawa, 1995; Kumudini *et al.*, 2001).

Seed-fill duration increases as temperature drops below 30°C in many crops (Egli, 2004), but Egli and Wardlaw (1980) found little difference between 20°C and 30°C in soybean. Boote *et al.* (1996) found that they could improve the predictability of the CROPGRO soybean simulation model in cool environments by making the seed-fill duration less sensitive to temperature than earlier stages of reproductive growth.

Water stress shortens the seed-filling period (Meckel *et al.*, 1984; de Souza *et al.*, 1997) by accelerating leaf senescence in soybean (de Souza *et al.*, 1997). This acceleration is not reversed when stressed soybean plants are returned to well-watered conditions after a single application of 3–5 days of stress early in seed filling (Brevedan and Egli, 2003), suggesting that relatively short periods of stress during seed filling may have a greater than expected effect on yield. Nitrogen stress also accelerates leaf senescence and shortens the seed-filling period (Boon-Long *et al.*, 1983; Hayati *et al.*, 1995), but supplying high levels of nitrogen to the plant does not prevent senescence and nitrogen redistribution (Egli *et al.*, 1978).

Seeds produced by late-developing flowers often have shorter seed-filling periods than seeds from early-developing ones (Gbikpi and Crookston, 1981; Spaeth and Sinclair, 1984; Egli *et al.*, 1987a). The seeds start growing at different times, but they reach physiological maturity at nearly the same time (Spaeth and Sinclair, 1984)

Seed-fill duration is regulated by the plant through the supply of assimilate to the developing seed and by the characteristics of the seed (Egli, 1998). Seeds cannot grow without assimilate, so when senescence is complete and the photosynthetic apparatus is destroyed, the seeds must stop growing. On the other hand, seeds can mature normally when there are green leaves on the plant and assimilate is still available (Egli, 1998). Limits placed on the ability of the developing seed to increase in volume by pod or seed structures may trigger seed maturation when ample assimilate is still available (Egli, 1990, 1998).

Physiological maturity marks the end of phase III and the end of the yield production process; there will be no more production, accumulation and translocation of 'yield contents' after physiological maturity. Seed moisture concentration at physiological maturity is approximately 550 g kg^{-1} (wet weight basis), so the seeds will not be ready for harvest until they have dried to a harvestable level (TeKrony *et al.*, 1979). Environmental conditions during this period can increase harvest losses and reduce the realized yield, but these concerns have nothing to do with the processes involved in the production of yield.

6.5 Determination of Yield

Maximum yield requires contributions from all of Murata's three phases: the photosynthetic capacity of the canopy, determined by LAI and environmental conditions must be at a maximum level; the reproductive sink (the yield container) must be large enough to accommodate all the available assimilate; and the seeds must have the ability to synthesize and accumulate storage materials (the yield contents). Yield limitations can occur during any phase of this system, but in typical field environments, they are much more likely during phases II or III.

Seed number

Most of the environmental variation in yield is associated with variations in seed number, because this is determined first in the yield production sequence (Egli, 1998) and, therefore, represents the first opportunity for the plant to adjust the size of the yield container to the productivity of the environment. Since seed number is related to canopy photosynthesis during flower and pod set (phase II), the large yield container needed for high yield requires maximum photosynthesis during this period. Seed number and yield will be reduced if photosynthesis is limited by environmental conditions or radiation interception. There is no evidence that photosynthesis prior to the beginning of flowering and pod set (i.e. prior to growth stage R1) has any direct effect on seed number (Jiang and Egli, 1995) or yield (Lee *et al.*, 2008) if the leaf area is large enough to maximize radiation interception by the beginning of phase II (growth stage R1). The pre-flowering environment could, however, reduce LAI to the point that it limits radiation interception during phase II, thereby indirectly affecting fruit and seed number, and yield.

It is not yet completely clear if photosynthesis must be at a maximum level throughout phase II to produce maximum fruit numbers. Reducing photosynthesis during part of the flowering and pod set period (growth stage R1–R5/R6) always reduces seed number (Shaw and Laing, 1966; Schou *et al.*, 1978; Jiang and Egli, 1995). In greenhouse experiments, seed numbers have been unable to recover from early shade treatments when the shade was removed midway through the flowering and pod set period (Egli and Bruening, 2005) because the increase in photosynthesis did not lengthen the flowering period or stimulate flower production. Relatively short periods of shade (covering 12–30% of phase II), however, had only limited effects on seed number (a significant reduction in only one comparison out of eight in a 2-year study with two cultivars) (Egli, unpublished data, 2007). These results suggest that the soybean plant may be able to tolerate short periods

(7–10 days) of reduced photosynthesis without experiencing reductions in fruit and seed number. This insensitivity may be related to the relatively long periods (up to 12 days) of assimilate depravation required to trigger fruit abortion (Egli and Bruening, 2006b). Although canopy photosynthesis during phase II determines the size of the yield container, the temporal relationship between these two variables is not yet clearly understood; it is, however, important in the constantly fluctuating environment in the field.

Seed size – the second yield component – may or may not be related to yield, depending upon the source of the variation. Genetic variation in seed size is usually unrelated to yield when it is associated with variation in individual seed growth rate; instead, increases in individual seed growth rate and seed size results in a reduction in seed number (Egli and Zhen-wen, 1991) and yield does not change – a classic example of yield component compensation (Egli, 1998). Genetic variation in seed size that is related to the duration of seed fill is often related to yield. Such variation exists (Swank *et al.*, 1987), but is much less common than differences associated with individual seed growth rate.

Variation in seed size that is a result of variation in environmental conditions during seed filling is directly related to yield (Table 6.3). Variation in assimilate availability can affect seed growth rate and increase or decrease seed size and yield (Egli *et al.*, 1985b; Egli, 1997). Water stress during seed filling also plays an important role in determining seed size by accelerating leaf senescence, shortening the seed-filling period and reducing seed size (de Souza *et al.*, 1997; Brevedan and Egli, 2003; Egli and Bruening, 2004).

Separation of phase II and III in time creates the opportunity for interactions between number and size. All environmental variation in yield would be due to seed number in an environment that is perfectly constant during phase II and III and seed size would not change. The environment in the real world, however, is constantly changing and seed size must compensate when the environment becomes more or less favourable after fruit and seed number are fixed. Seed size and yield will decrease if environmental conditions

Table 6.3. Effect of reducing photosynthesis during seed filling on soybean yield and yield components (adapted from Egli, 1997).

	Yield (g per m^2)		Seed no. (no. per m^2)		Seed size ($mg\ seed^{-1}$)	
Cultivar[a]	Control	Shade[b]	Control	Shade[b]	Control	Shade[b]
Early	352	262**	2280	2065*	154	127**
Late	372	288**	2540	2205*	146	132**

[a]Average of two cultivars from maturity group I (early) or maturity group V (late).
[b]A shade cloth that reduced incident radiation by 63% was placed over the plots at growth stage R5 (beginning seed fill).
*,**Significantly different from the control within a maturity group at $P = 0.05$ and 0.01, respectively, based on LSD.

during seed filling deteriorate (Table 6.3) (Egli *et al.*, 1978); if the environment improves, size and yield will increase (Egli *et al.*, 1985b).

The ability of the plant to respond to improvements in the environment during seed filling depends upon the capacity of the seed to increase its size (i.e. upon its potential seed size). Seeds harvested from a single plant or a plant community exhibit a substantial range in size, with the largest seeds frequently 60% larger than the mean size (Egli *et al.*, 1987b). Part of this variation is related to the time of flower development, with the largest seeds usually found in pods from flowers that developed early in phase II (Egli *et al.*, 1987b). Fruit removal treatments (80% removal) after the target pods reach full size do not completely eliminate variations in seed size, suggesting that part of the variation may be related to variations in pod size (Egli *et al.*, 1987b). Seed size increases after de-podding (75–80% pod removal) have been found to range from 30% to 114% in greenhouse and field experiments with three cultivars (Egli *et al.*, 1985b, 1989). Borras *et al.* (2004), in a thorough summary of the literature (including some of the data cited here), reached much the same conclusion, reporting increases of 12–114%. Apparently, potential size is usually much larger than the actual size (as much as twice as large), indicating that soybean is capable of significant adjustments in seed size when needed. Of course, there is no practical limit to how much a seed can decrease in size in an unfavourable environment.

Source-sink limitations

The sequential determination of yield components in soybean, with seed number coming first, indicates that soybean yield is source-limited. Canopy photosynthesis determines the size of the yield container, so photosynthesis (the source) must be the primary factor limiting yield. Yield may be sink-limited if photosynthesis increases during phase III and the yield container cannot accommodate all of the extra assimilate (Borras *et al.*, 2004). Sink limitations are rare because: (i) most seeds are apparently substantially smaller than their potential size, indicating that they could increase in size if needed; (ii) the tendency for solar radiation during seed filling to be less than during phase II results in excess capacity in the yield container (Egli, 1999; Egli and Bruening, 2001; Borras *et al.*, 2004); and (iii) radical improvements in environmental conditions after phase II are relatively rare given the typical persistence in environmental conditions in the field.

Partitioning

Increases in crop yield, particularly by cereals, are often attributed to changes in partitioning (Hay and Porter, 2006). This implies that assimilate that once ended up in vegetative plant parts is now shifted to the seed, where it contributes to yield with no change in total biomass. From this viewpoint, partitioning appears to be a simple concept – assimilate is 'divided' among several competing sinks – but, in reality, it is far more complicated. Partitioning is often measured as the end result – the dry matter in various plant parts – with little consideration given to the mechanisms responsible for the final result. The harvest index (seed mass / total biomass at maturity = seed mass / [vegetative mass + reproductive mass]) (Donald, 1968) is such a measure and is commonly used as an indicator of partitioning at maturity in many crops. It is not as popular for soybean as it is for wheat and other cereals, probably because leaf and petiole abscission during seed filling makes it difficult to estimate vegetative biomass at maturity (a limitation not shared by the cereals or maize). Schapaugh and Wilcox (1980), however, reported that cultivar differences in the harvest index (including abscised leaves and petioles) are maintained when the harvest index is based on the standing crop at maturity. Standing crop estimates, however, are biased upward as a result of the missing vegetative biomass. Estimates of the harvest index for soybean (based on standing crop) cluster around 0.50 with a range from 0.35 to 0.65 (Salado-Navarro *et al.*, 1993; Ball *et al.*, 2000; Kumudini *et al.*, 2001; Pedersen and Lauer, 2004), which is in the range reported for modern wheat and maize cultivars (Hay and Porter, 2006).

A shift in partitioning from vegetative to reproductive sinks during phase II and III would increase yield, but, as discussed earlier, vegetative growth in modern cultivars stops before the end of phase II, so the potential increase may be limited. Increasing yield by lengthening the duration of seed fill will also increase the harvest index (yield will increase with no change in vegetative biomass) (Egli, 2004), but this represents only an apparent change in partitioning since there is no transfer of assimilate from vegetative to reproductive plant sinks. An increase in the harvest index is not always associated with higher yield; it can be a result of decreasing vegetative biomass with no change in yield, as often happens when comparing cultivars with a range in maturities (Zeiher *et al.*, 1982; Egli, 1998).

Much remains to be learned about partitioning, but relying on simple indices describing the results of the partitioning process (e.g. the harvest index) adds little to our understanding of the process and can be misleading. When discussing the harvest index, Charles-Edwards (1982) concluded that 'It seems more logical and the problems of increasing grain yields more tractable, to look directly at the phenological, physiological and environmental determinants of grain yield.'

Partitioning in time

Time is an important attribute of the yield production process for two reasons. First, because yield (or total biomass) is always determined by a rate of growth expressed over a specific time interval; and second, because the potential productivity at any location is partially determined by the time available for crop growth (de Wit, 1967). The basic resource for crop growth is the solar radiation available when temperatures are suitable for growth, so time is an important determinant of the energy a crop has available to support dry matter accumulation. How the crop uses time, therefore, becomes an important part of the yield production process.

The total growth cycle of soybean cultivars grown in the USA varies from just over 100 to 150 days from planting to 95% brown pods (Egli, 1998). Cultivars from MG 00–V have been found to vary from 80 to 134 days (planting to growth stage R7) when grown under irrigation in Lexington, Kentucky (38°N) (Egli, 1994). Most of this variation was associated with differences in the vegetative growth phase (Egli, 1993, 1994), but the length of the vegetative phase was not closely associated with yield (Fig. 6.4) (Egli, 1993; Egli and Bruening, 2000). Increasing leaf area (and vegetative mass) does not increase canopy photosynthesis or the crop growth rate after radiation interception reaches a maximum (Shibles and Weber, 1965), so there is no yield benefit from the higher LAI.

The length of the flowering and pod set period (phase II) is less variable than vegetative growth, but has been seen to exhibit a 13-day increase (65%) as the total growth cycle increased from 80 to 134 days (Egli, 1994). It is not yet clear if a longer phase II consistently leads to more fruits and seeds and higher yields as some data suggest (Egli and Bruening, 2000; Kantolic and Slafer, 2001, 2005), especially when there is evidence that synchronous flowering also increases fruit set (Egli and Bruening, 2002b).

Seed-fill duration (phase III) (estimated by the effective-filling period) does not change after the total growth duration exceeds 100 days, but is often shorter for durations <100 days (Zeiher *et al.*, 1982; Egli, 1994). Late-maturing cultivars do not necessarily provide a longer seed-filling period and, therefore, from this viewpoint, do not provide an inherent yield advantage.

Soybean, like other grain crops, is not very efficient at utilizing time to produce yield (Egli, 1998). As growth duration increases beyond a minimum,

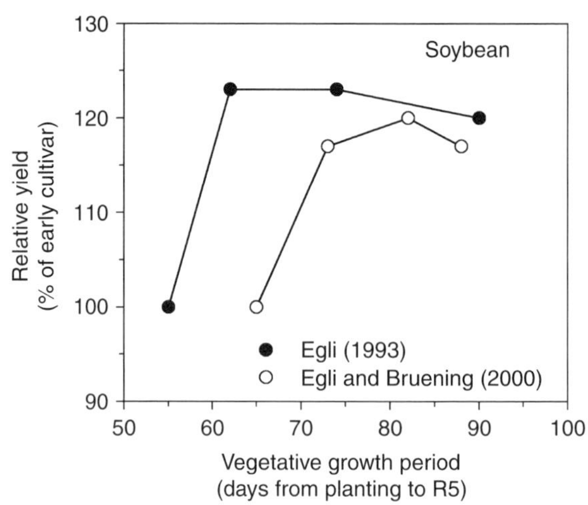

Fig. 6.4. Relationship between the length of the vegetative growth period (phase I) and yield of field-grown soybean cultivars from maturity groups 00–V. The plants were irrigated to minimize water stress (adapted from Egli, 1993; Egli and Bruening, 2000).

yield does not necessarily continue to increase. Cultivars with longer growth cycles can be used to more closely match the available time, but they simply grow more vegetative material and there is no obvious mechanism available to transfer that extra vegetative mass into yield. This phenomenon explains the curvilinear relationship between maximum vegetative mass and yield (Board and Modali, 2005), which results in a negative relationship between yield and harvest index (Egli, 1998; Board and Modali, 2005).

This inefficient use of time, however, provides one benefit; it is possible to achieve near-maximum yield with a relatively short growth cycle (Egli, 1993, 1997). Short-season cultivars would require less irrigation water, which would conserve water when supplies are short and reduce production costs without sacrificing yield. Short cycles may create opportunities to shift critical growth stages to environments that are more favourable for yield. For example, in the southern USA, the early soybean production system consistently produces high yields with early-maturing cultivars planted early so that they mature before drought conditions develop (Heatherly, 1999). Finally, short-season cultivars increase opportunities for multiple cropping, which is one way to overcome that the inherent inefficiency of grain crops in utilizing the time and solar radiation available in long growing seasons.

6.6 Future Yield Growth

Increases in yield have helped increase soybean production since soybean became an important grain crop (Fig. 6.5; Egli, 2008a). Higher yields are a result of improvements in the plant via plant breeding and improvements in the plant's environment via crop management (Egli, 2008a). The global population is expected to increase by roughly 20% (an additional 14 billion people) in the next 20 years (UN, 2006), which will require continued increases in crop yield to maintain per capita food supplies.

There must be a limit to yield; one ultimately determined by the solar radiation at the surface of the earth and the ability of the plant to convert solar energy into plant tissues. Defining this limit is difficult, but estimates provide a perspective from which to consider future yield growth.

Evans (1993) defined yield potential as 'the yield of a cultivar when grown in environments to which it is adapted, with nutrients and water non-limiting and with pests, diseases, weeds, lodging and other stresses effectively controlled'. Yield potential is determined only by the environment (principally temperature and solar radiation, since water and nutrients are not limiting and stresses are effectively controlled) and the plant's ability to convert solar energy into harvestable plant tissues. Yield potential may be limited by the plant or the plant's environment; consequently, it varies by cultivar and environment. Yield potential is higher than attainable yields (yields obtained by skilful use of the best available technology) and actual yields or the average yield of an area, district, state or country (Loomis and Conner, 1992).

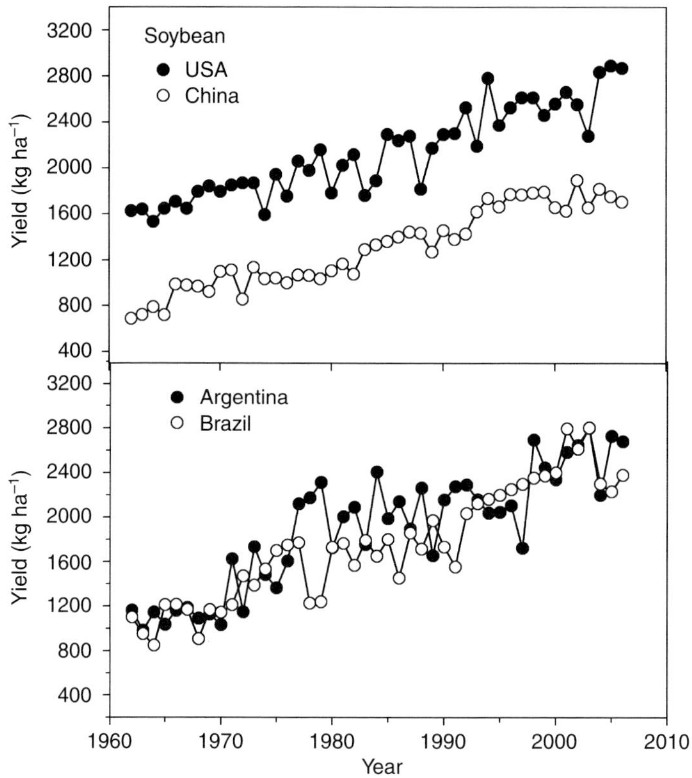

Fig. 6.5. Average soybean yields from 1962 to 2006 (data from FAO, 2008).

Yield potential is a straightforward concept, but it is difficult to estimate because it requires measuring yield in a stress-free environment. It is nearly impossible to be sure that any experiment is completely stress free. Record yields, winning yields in yield contests and yields from crop simulation models are sometimes used as estimates of yield potential (Evans, 1993; Specht *et al.*, 1999). Record soybean yields summarized by Evans (1993) range from 5600 to 7400 kg ha^{-1} (produced in 1966–1977). In an extensive summary of contest-winning yields from the USA (Iowa, Missouri and Nebraska, 1966–1998), Specht *et al.* (1999) found that most winning yields were >4500 kg ha^{-1}, with only one >6725 kg ha^{-1}. A record yield in north China of nearly 6000 kg ha^{-1} has been reported by Liu *et al.* (2008). Specht *et al.* (1999) concluded that yield potential is approximately 8000 kg ha^{-1}, but a recent contest-winning yield reached 10,414 kg ha^{-1} (155 bushels acre^{-1}) (Lamp, 2007).

It is difficult to evaluate the validity of these records and contest-winning yields. Do they provide true estimates of yield potential or are they simply a result of contest 'fever' and the desire to win? Evaluation would be easier if yield data were supported by other measures of growth (e.g. duration of growth phases, rate of dry matter accumulation, photosynthesis), but usually only yield data are available. If the measurement of yield, however, is accurate (a question rarely addressed by anyone measuring yield), all methods provide a minimum estimate of yield potential (i.e. it could be higher than the estimate, but not lower).

Yield potential is often used to estimate how much actual yields can increase – in other words, how long until actual yield equals potential yield, or how large is the exploitable yield gap (the difference between actual or attainable yield and yield potential) (Cassman *et al.*, 2003). If yield potential is static in time, an estimate of the potential provides a direct estimate of when yield will stop increasing. If, however, yield potential increases with time, changes in the exploitable yield gap depend upon the relative growth rates of the two entities and predictions of the future are far less clear. Yield potential should increase with time as genetic manipulation of the plant improves its ability to utilize solar energy for growth. Arguments that yield potential is stable over time for maize (Duvick and Cassman, 1999) have been based on the premise that genetic yield improvement is primarily a matter of increases in stress tolerance (Tollenaar and Wu, 1999). There is some evidence that genetic improvement of soybean has improved stress tolerance (Boyer *et al.*, 1980), but many other improvements in the plant, such as improved partitioning (Gay *et al.*, 1980) and longer seed-fill duration (Gay *et al.*, 1980; Boerma and Ashley, 1988; Kumudini *et al.*, 2001), have surely contributed to an increase in yield potential.

All of the estimates of potential yield cited here (generally >5000 kg ha^{-1}) are substantially higher than actual yields in the USA and other major soybean-producing countries (Fig. 6.5). Country yields represent the average of environments with a wide range of production potential; county yields represent a smaller area and, therefore, are more variable, with some being well above and some well below the mean for the state or country. The highest county yields in high-yield US environments (Iowa, Illinois, Nebraska-irrigated) are rarely >4000 kg ha^{-1} (USDA-NASS, 2008), substantially below current estimates of yield potential. These comparisons suggest that there is a substantial exploitable yield gap, leaving room for improvement without an increase in yield potential. It is not at all clear what technology (whether new or simply the judicious application of present technology) is needed to narrow the gap between actual and potential yield in modern soybean production systems. This question must be answered before the practical and economic aspects of such an effort can be evaluated.

Recent examination of soybean yield trends in the USA found no convincing evidence that yield growth rates at the county or state levels are decreasing or that yield is reaching a plateau (Egli, 2008a, 2008b). There were, however, permanent yield plateaus in high-stress, low-yield environments (non-irrigated areas of Arkansas and Nebraska and counties in Kentucky with >60% of the crop grown as a second crop after wheat), where there was no significant increase in yield between 1972 and 2003 (Egli, 2008b). Clearly, the changes that have driven yields upward in more productive environments are completely ineffective under stress.

No one can say with any certainty what will happen to soybean yield in the near future. It seems reasonable to assume that, given the availability of adequate genetic variation (Fehr, 1999; St. Martin, 1999) and the growing contribution from biotechnology and molecular breeding approaches, yields will continue to increase. It seems likely that genetic improvement will make a larger contribution to yield gains in the future than it has in the past as benefits of improved crop management practices experience diminishing returns (i.e. each improvement of the crop's environment makes the next improvement more difficult) (Egli, 2008a).

References

Abernathy, R.H., Palmer, R.G., Shibles, R. and Anderson, J.C. (1977) Histological observations on abscising and retained soybean flowers. *Canadian Journal of Plant Science* 57, 713–716.

Ackerson, R.C., Havelka, V.D. and Boyle, M.G. (1984) CO_2-enrichment effects on soybean physiology. II. Effects of stage-specific CO_2 exposure. *Crop Science* 24, 1150–1154.

Acock, B., Reddy, V.R., Hodges, H.F., Baker, D.N. and McKinion, J.M. (1985) Photosynthetic response of soybean canopies to full-season carbon dioxide enrichment. *Agronomy Journal* 77, 942–947.

Agudelo, O., Green, D.E. and Shibles, R. (1986) Methods of estimating developmental periods in soybean. *Crop Science* 26, 1226–1230.

Andrade, F.H. and Ferreiro, M.A. (1996) Reproductive growth of maize, sunflower and soybean at different source levels during grain filling. *Field Crops Research* 48, 155–165.

Ball, R.A., Purcell, L.C. and Vories, G.D. (2000) Short season soybean yield compensation in response to population and water regime. *Crop Science* 40, 1070–1078.

Beaver, J.S., Cooper, R.L. and Martin, R.J. (1985) Dry matter accumulation and seed yield of determinate and indeterminate soybeans. *Agronomy Journal* 77, 675–679.

Board, J.E. and Harville, B.G. (1994) A criterion for acceptance of narrow-row culture in soybean. *Agronomy Journal* 86, 1103–1106.

Board, J.E. and Harville, B.G. (1996) Growth dynamics during the vegetative period affects yield of narrow-row, late-planted soybean. *Agronomy Journal* 88, 567–572.

Board, J.E. and Modali, H. (2005) Dry matter accumulation predictors for optimal yield in soybean. *Crop Science* 45, 1790–1799.

Board, J.E. and Tan, Q. (1995) Assimilatory capacity effects on soybean yield components and pod number. *Crop Science* 35, 846–851.

Boerma, H.R. and Ashley, D.A. (1988) Canopy photosynthesis and seed-fill duration in recently developed soybean cultivars and selected plant introductions. *Crop Science* 28, 137–140.

Boon-Long, P., Egli, D.B. and Leggett, J.E. (1983) Leaf N and photosynthesis during reproductive growth in soybeans. *Crop Science* 23, 617–620.

Boote, K.J., Jones, J.W. and Pickering, N.B. (1996) Potential uses and limitations of crop models. *Agronomy Journal* 88, 704–716.

Borras, L., Slafer, G. and Otegui, M.A. (2004) Seed dry weight response to source-sink manipulations in wheat, maize and soybean: A quantitative reappraisal. *Field Crops Research* 86, 131–146.

Boyer, J.S., Johnson, R.R. and Saupe, S.G. (1980) Afternoon water deficits and grain yields in old and new soybean cultivars. *Agronomy Journal* 72, 981–986.

Brevedan, R.E. and Egli, D.B. (2003) Short periods of water stress during seed filling, leaf senescence and yield of soybean. *Crop Science* 43, 2083–2088.

Brevedan, R.E., Egli, D.B. and Leggett, J.E. (1978) Influence of N nutrition on flower and pod abortion and yield of soybeans. *Agronomy Journal* 70, 81–84.

Bruening, W.P. and Egli, D.B. (1999) Relationship between photosynthesis and seed number at phloem isolated nodes in soybean. *Crop Science* 39, 1769–1775.

Buhler, D.G. and Hartzler, R.G. (2004) Weed biology and management. In: Boerma, H.R. and Specht, J.E. (eds) *Soybeans: Improvement, Production, and Uses*, 3rd edn. Agronomy Monograph 16. American Society of Agronomy, Crop Science Society of America, and Soil Science Society of America, Madison, WI, USA, pp. 883–918.

Buttery, B.R. (1969) Analysis of the growth of soybean as affected by plant populations and fertilizer. *Canadian Journal of Plant Science* 49, 675–684.

Cassman, K.G., Dobermann, A., Walters, D.T., and Yang, Y. (2003) Meeting cereal demand while protecting natural resources and improving environmental quality. *Annual Review of Environmental Resources* 28, 15–58.

Charles-Edwards, D.A. (1982) *Physiological Determinants of Crop Growth*. Academic Press, North Ryde, NSW, Australia.

Christy, A.L. and Porter, C.A. (1982) Canopy photosynthesis and yield in soybean. In: Govindjee (ed.) *Photosynthesis, Development, Carbon Metabolism and Plant Productivity*. Vol. II. Academic Press, New York, USA, pp. 499–511.

Crafts-Brandner, S.J. and Egli, D.B. (1987) Sink removal and leaf senescence in soybean: Cultivar effects. *Plant Physiology* 85, 662–666.

Daynard, T.B., Tanner, J.W. and Duncan, W.G. (1971) Duration of the grain filling period and its relationship to grain yield in corn, *Zea mays* L. *Crop Science* 11, 45–48.

de Souza, P.I., Egli, D.B. and Bruening, W.P. (1997) Water stress during seed filling and leaf senescence in soybean. *Agronomy Journal* 89, 807–812.

de Wit, C.T. (1967) Photosynthesis: Its relationship to overpopulation. In: San Pierto, A., Green, F.A. and Army, T.J. (eds) *Harvesting the Sun*. Academic Press, New York, USA, pp. 315–320.

Donald, C.M. (1968) In search of yield. *Journal of Australian Institute of Agricultural Science* 28, 171–178.

Dornhoff, G.M. and Shibles, R.M. (1970) Varietal differences in net photosynthesis of soybean leaves. *Crop Science* 10, 42–45.

Duthion, C. and Pigeaire, A. (1991) Seed lengths corresponding to the final stage in seed abortion in three grain legumes. *Crop Science* 31, 1597–1583.

Duvick, D.L. and Cassman, K.G. (1999) Post-green revolution trends in yield potential of temperate maize in the North-Central United States. *Crop Science* 37, 1622–1630.

Dybing, C.D. (1994) Soybean flower production as related to plant growth and seed yield. *Crop Science* 34, 489–497.

Edwards, J.T. and Purcell, L.C. (2005) Soybean yield and biomass responses to increasing plant population among diverse maturity groups. I. Agronomic characteristics. *Crop Science* 45, 1770–1777.

Egli, D.B. (1990) Seed water relations and the regulation of the duration of seed growth in soybean. *Journal of Experimental Botany* 41, 243–248.

Egli, D.B. (1993) Cultivar maturity and potential yield of soybean. *Field Crops Research* 32, 147–158.

Egli, D.B. (1994) Cultivar maturity and reproductive growth duration in soybean. *Journal of Agronomy and Crop Science* 173, 249–254.

Egli, D.B. (1997) Cultivar maturity and responses of soybean to shade stress during seed filling. *Field Crops Research* 52, 1–8.

Egli, D.B. (1998) *Seed Biology and the Yield of Grain Crops*. CAB International, Wallingford, UK.
Egli, D.B. (1999) Variation in leaf starch and sink limitations during seed filling in soybean. *Crop Science* 39, 1361–1368.
Egli, D.B. (2004) Seed-fill duration and yield of grain crops. *Advances in Agronomy* 83, 243–279.
Egli, D.B. (2005) Flowering, pod set and reproductive success in soya bean. *Journal of Agronomy and Crop Science* 191, 283–291.
Egli, D.B. (2006) The role of the seed in the determination of yield of grain crops. *Australian Journal of Agricultural Research* 57, 1237–1247.
Egli, D.B. (2008a) Comparison of corn and soybean yields in the United States: Historical trends and future prospects. *Agronomy Journal* 100, S-79–S-88.
Egli, D.B. (2008b) Soybean yield trends from 1972 to 2003 in mid-western USA. *Field Crops Research* 106, 53–59.
Egli, D.B. and Bruening, W.P. (2000) Potential of early-maturing soybean cultivars in late plantings. *Agronomy Journal* 92, 532–537.
Egli, D.B. and Bruening, W.P. (2001) Source-sink relationships, seed sucrose levels and seed growth rates in soybean. *Annals of Botany* 88, 235–242.
Egli, D.B. and Bruening, W.P. (2002a) Flowering and fruit set dynamics during synchronous flowering at phloem-isolated nodes in soybean. *Field Crops Research* 79, 9–19.
Egli, D.B. and Bruening, W.P. (2002b) Synchronous flowering and fruit set at phloem isolated nodes in soybean. *Crop Science* 42, 1535–1540.
Egli, D.B. and Bruening, W.P. (2003) Increasing sink size does not increase photosynthesis during seed filling in soybean. *European Journal of Agronomy* 19, 289–298.
Egli, D.B. and Bruening, W.P. (2004) Water stress, photosynthesis, seed sucrose levels and seed growth in soybean. *Journal of Agricultural Science* 142, 1–8.
Egli, D.B. and Bruening, W.P. (2005) Shade and temporal distribution of pod production and pod set in soybean. *Crop Science* 45, 1764–1769.
Egli, D.B. and Bruening, W.P. (2006a) Temporal profiles of pod production and pod set in soybean. *European Journal of Agronomy* 24, 11–18.
Egli, D.B. and Bruening, W.P. (2006b) Fruit development and reproductive survival in soybean: Position and age effects. *Field Crops Research* 98, 195–202.
Egli, D.B. and Bruening, W.P. (2007) Accumulation of nitrogen and dry matter by soybean seeds with genetic differences in protein concentration. *Crop Science* 47, 359–366.
Egli, D.B. and Crafts-Brandner, S.J. (1996) Soybean. In: Zamski, E. and Schaffer, A.A. (eds) *Photoassimilate Distribution in Plants and Crops*. Marcel Dekker, Inc., New York, USA, pp. 595–623.
Egli, D.B. and Leggett, J.E. (1973) Dry matter accumulation patterns in determinate and indeterminate soybean. *Crop Science* 13, 220–222.
Egli, D.B. and Wardlaw, I.F. (1980) Temperature response of seed growth characteristics of soybean. *Agronomy Journal* 72, 560–564.
Egli, D.B. and Zhen-wen, Y. (1991) Crop growth rate and seed number per unit area in soybean. *Crop Science* 31, 439–442.
Egli, D.B., Pendleton, J.W. and Peters, D.B. (1970) Photosynthetic rate of three soybean communities as related to carbon dioxide levels and solar radiation. *Agronomy Journal* 62, 411–414.
Egli, D.B., Leggett, J.E. and Duncan, W.G. (1978) Influence of N stress on leaf senescence and N redistribution in soybean. *Agronomy Journal* 74, 375–379.
Egli, D.B., Meckel, L., Phillips, R.E., Radcliffe, D. and Leggett, J.E. (1983) Moisture stress and N redistribution in soybean. *Agronomy Journal* 75, 1027–1031.

Egli, D.B., Orf, J.H. and Pfeiffer, T.W. (1984) Genotypic variation for duration of seed fill in soybean. *Crop Science* 24, 587–592.
Egli, D.B., Guffy, R.D. and Leggett, J.E. (1985a) Partitioning of assimilate between vegetative and reproductive growth in soybean. *Agronomy Journal* 77, 917–922.
Egli, D.B., Guffy, R.D., Meckel, L.W. and Leggett, J.E. (1985b) The effect of source-sink alterations on soybean seed growth. *Annals of Botany* 55, 395–402.
Egli, D.B., Wiralaga, R.A., Bustamam, T., Zhen-wen, Yu, and TeKrony, D.M. (1987a) Time of flowering opening and seed mass in soybean. *Agronomy Journal* 79, 697–700.
Egli, D.B., Wiralaga, R.A. and Ramseur, E.L. (1987b) Variation in seed size in soybean. *Agronomy Journal* 79, 463–467.
Egli, D.B., Ramseur, E.L., Zhen-wen, Yu, and Sullivan, C.H. (1989) Source-sink alterations affect the number of cells in soybean cotyledons. *Crop Science* 29, 732–735.
Evans, L.T. (1993) *Crop Evolution, Adaptation and Yield*. Cambridge University Press, Cambridge, UK.
FAO (2008) Food and Agriculture Organization of the United Nations. Available at: http://faostat.fao.org/site/291/ (accessed 11 May 2009).
Fehr, W.R. (1999) The importance and future prospects for genetic diversity for yield improvement. In: Kauffman, H.R. (ed.) *Proceedings of the 6th World Soybean Research Conference*, Urbana-Champaign, IL, USA, pp. 186–187.
Fehr, W.R. and Caviness, C.E. (1977) *Stages of Soybean Development*. Iowa State University, Special Report 80, Ames, IA, USA.
Frederick, J.R. and Hesketh, J.D. (1994) Genetic improvement in soybean: Physiological attributes. In: Slafer, G.A. (ed.) *Genetic Improvement of Field Crops*. Marcel Dekker Inc., New York, USA, pp. 237–286.
Gai, J., Palmer, R.G. and Fehr, W.R. (1984) Bloom and pod set in determinate and indeterminate soybeans grown in China. *Agronomy Journal* 76, 979–984.
Gay, S., Egli, D.B. and Reicosky, D.A. (1980) Physiological aspects of yield improvement in soybeans. *Agronomy Journal* 72, 387–391.
Gbikpi, P.J. and Crookston, R.K. (1981) Effect of flowering date on accumulation of dry matter and protein in soybean seeds. *Crop Science* 21, 652–655.
Gent, M.P.N. (1994) Photosynthate reserves during grain filling in winter wheat. *Agronomy Journal* 86, 159–167.
Ghorashy, S.R., Pendleton, J.W., Peters, D.B., Boyer, J.S. and Beuerlein, J.E. (1971) Internal water stress and apparent photosynthesis with soybeans differing in pubescence. *Agronomy Journal* 63, 674–676.
Goldschmidt, E.E. and Huber, S.C. (1992) Regulation of photosynthesis by end-product accumulation in leaves of plants storing starch, sucrose, and hexose sugars. *Plant Physiology* 99, 1443–1448.
Hall, A.J., Conner, D.J. and Whitfield, D.M. (1989) Contribution of pre-anthesis assimilate to grain filling in irrigated and water stressed sunflower crops. I. Estimates using labeled carbon. *Field Crops Research* 20, 95–112.
Hansen, W.R. and Shibles, R. (1978) Seasonal log of the flowering and podding activity of field-grown soybeans. *Agronomy Journal* 70, 47–50.
Hanway, J.J. and Weber, C.R. (1971) Dry matter accumulation in eight soybean (*Glycine max* L. Merrill) varieties. *Agronomy Journal* 63, 227–230.
Hardman, L.L. and Brun, W.A. (1971) Affects of atmospheric carbon dioxide enrichment at different development stages on growth and yield components of soybeans. *Crop Science* 11, 886–888.
Harper, J.E. (1987) Nitrogen metabolism. In: Wilcox, J.R. (ed.) *Soybeans: Improvement, Production, and Uses*, 2nd edn. Agronomy Monograph 16. American Society of

Agronomy, Crop Science Society of America, and Soil Science Society of America, Madison, WI, USA, pp. 497–533.

Hartwig, E.E. (1973) Varietal development. In: Caldwell, B.E. (ed.) *Soybeans: Improvement, Production, and Uses*. Agronomy Monograph 16. American Society of Agronomy, Madison, WI, USA, pp. 187–210.

Hay, R.K.M. and Porter, J.R. (2006) *The Physiology of Crop Yield*, 2nd edn. Blackwell Publishing Ltd, Oxford, UK.

Hayati, R., Egli, D.B. and Crafts-Brandner, S.J. (1995) Carbon and nitrogen supply during seed filling and leaf senescence in soybean. *Crop Science* 35, 1063–1069.

Hayati, R., Egli, D.B. and Crafts-Brandner, S.J. (1996) Independence of nitrogen supply and seed growth in soybean: Studies using an *in vitro* culture system. *Journal of Experimental Botany* 47, 33–40.

Heatherly, L.G. (1999) Early soybean production system (ESPS). In: Heatherly, L.G. and Hodges, H.F. (eds) *Soybean Production in the Midsouth*. CRC Press, Boca Raton, FL, USA, pp. 103–118.

Heatherly, L.G. and Elmore, R.W. (2004) Managing inputs for peak production. In: Boerma, H.R. and Specht, J.E. (eds) *Soybeans: Improvement, Production, and Uses*, 3rd edn. Agronomy Monograph 16. American Society of Agronomy, Crop Science Society of America, and Soil Science Society of America, Madison, WI, USA, pp. 451–536.

Heilman, J.L., Kanemasu, E.T. and Paulsen, G.M. (1977) Estimating dry-matter accumulation in soybean. *Canadian Journal of Botany* 55, 2196–2201.

Heitholt, J.J., Egli, D.B. and Leggett, J.E. (1986) Characteristics of reproductive abortion in soybean. *Crop Science* 26, 589–595.

Huff, A. and Dybing, C.D. (1980) Factors affecting shedding of flowers in soybean (*Glycine max* (L.) Merrill). *Journal of Experimental Botany* 31, 751–762.

Jiang, H. and Egli, D.B. (1993) Shade induced changes in flower and pod number and fruit abscission in soybean. *Agronomy Journal* 85, 221–225.

Jiang, H. and Egli, D.B. (1995) Soybean seed number and crop growth rate during flowering. *Agronomy Journal* 87, 264–267.

Johnson, D.R. and Tanner, J.W. (1972) Comparisons of corn (*Zea mays* L.) inbreds and hybrids grown at equal leaf area index, light penetration, and population. *Crop Science* 12, 482–485.

Johnson, R.R. (1987) Crop management. In: Wilcox, J.R. (ed.) *Soybeans: Improvement, Production, and Uses*, 2nd edn. Agronomy Monograph 16. American Society of Agronomy, Crop Science Society of America, and Soil Science Society of America, Madison, WI, USA, pp. 355–390.

Kantolic, A.G. and Slafer, G.A. (2001) Photoperiod sensitivity after flowering and seed number determination in indeterminate soybean cultivars. *Field Crops Research* 72, 109–118.

Kantolic, A.C. and Slafer, G.A. (2005) Reproductive development and yield components in indeterminate soybean as affected by post-flowering photoperiod. *Field Crops Research* 93, 212–222.

Kato, I., Sakaguchi, S. and Naito, Y. (1955) Anatomical observations of fallen buds, flowers, and pods of soya-bean, *Glycine max* M. *Tokai-Kinki National. Agriculture Experiment Station Bulletin* 2, 159–168.

Kumudini, S., Hume, D.J. and Chu, G. (2001) Genetic improvement in short season soybeans: I. Dry matter accumulation, partitioning, and leaf area duration. *Crop Science* 41, 391–398.

Lamp, J. (2007) Kip Cullers sets soybean record at 155 bu./acre. *Corn and Soybean Digest*, Available at: http://cornandsoybeandigest.com (accessed 30 October 2008).

Larson, E.M., Hesketh, J.D., Woodley, J.T. and Peters, D.B. (1981) Seasonal variations in apparent photosynthesis among plant stands of different soybean cultivars. *Photosynthesis Research* 2, 3–20.

Lee, C.D., Egli, D.B. and TeKrony, D.M. (2008) Soybean response to plant population at early and late planting dates in the Mid-South. *Agronomy Journal* 100, 1–6.

Liu, X., Jin, J., Wang, G. and Herbert, S.J. (2008) Soybean yield physiology and development of high-yield practices in Northeast China. *Field Crops Research* 105, 157–171.

Loomis, R.S. and Conner, D.J. (1992) *Crop Ecology: Productivity and Management in Agricultural Systems*. Cambridge University Press, Cambridge, UK.

Mann, J.D. and Jaworski, E.G. (1970) Comparison of stress which may limit soybean yields. *Crop Science* 10, 620–624.

McBlain, B.A. and Hume, D.J. (1981) Reproductive abortion, yield components, and nitrogen content in three early soybean cultivars. *Canadian Journal of Plant Science* 61, 499–505.

Meckel, L., Egli, D.B., Phillips, R.E., Radcliffe, D. and Leggett, J.E. (1984) Effect of moisture stress on seed growth in soybeans. *Agronomy Journal* 75, 1027–1031.

Metz, G.L., Green, D.E. and Shibles, R.M. (1984) Relationship between soybean yield in narrow rows and leaflet, canopy and developmental characters. *Crop Science* 24, 457–462.

Metz, G.L., Green, D.E. and Shibles, R.M. (1985) Reproductive duration and date of maturity in populations of three wide soybean crosses. *Crop Science* 25, 171–176.

Miceli, F., Crafts-Brandner, S.J., and Egli, D.B. (1995) Physical restriction of pod growth alters development of soybean plants. *Crop Science* 35, 1080–1085.

Mondal, M.H., Brun, W.A. and Brenner, M.L. (1978) Affects of sink removal on photosynthesis and senescence in leaves of soybean (*Glycine max* L.) plants. *Plant Physiology* 61, 394–397.

Morris, D.R. and Weaver, R.W. (1983) Mobilization of ^{15}N from soybean leaves as influenced by Rhizobial strains. *Crop Science* 23, 1111–1114.

Murata, Y. (1969) Physiological responses to nitrogen in plants. In: Eastin, J.D., Haskins, F.A., Sullivan, C.Y. and Van Bavel, C.H.M. (eds) *Physiological Aspects of Crop Yield*. ASA-CSSA, Madison, WI, USA, pp. 235–259.

Nakamoto, H., Zheng, S.H., Furuya, T., Tanaka, K., Yamazaki, A. and Fukuyama, M. (2001) Effects of long-term exposure to atmospheric carbon dioxide enrichment on flowering and podding in soybean. *Journal of Faculty of Agriculture, Kyushu University* 46, 23–29.

Ogren, W.L. (1984) Photorespiration: Pathways, regulation and modification. *Annual Review of Plant Physiology* 35, 415–442.

Pedersen, P. and Lauer, J.G. (2004) Response of soybean yield components to management system and planting date. *Agronomy Journal* 96, 1372–1381.

Pfeiffer, T.W. and Egli, D.B. (1988) Heritability of seed-filling period estimates in soybean. *Crop Science* 28, 921–925.

Saitoh, K., Isobe, S. and Kuroda, T. (1998) Significance of flower differentiation and development in the process of determining yield: Relation between the number of pods and flowers. *Japanese Journal of Crop Science* 67, 70–78.

Salado-Navarro, L.R., Sinclair, T.R. and Hinson, K. (1985) Comparisons among effective filling period, reproductive period duration, and R5–R7 in determinate soybeans. *Crop Science* 25, 1050–1054.

Salado-Navarro, L.R., Sinclair, T.R. and Hinson, K. (1993) Changes in yield and seed growth traits in soybean cultivars released in the southern USA from 1945 to 1983. *Crop Science* 33, 1204–1209.

Schapaugh, W.T. and Wilcox, J.R. (1980) Relationships between harvest indices and other plant characteristics in soybean. *Crop Science* 20, 529–533.
Schou, J.B., Jeffers, D.L. and Streeter, J.G. (1978) Effects of reflectors, black boards, or shades applied at different stages of plant development on yield of soybeans. *Crop Science* 18, 29–34.
Setiyono, T.D., Weiss, A., Specht, J.E., Cassman, K.G. and Dobermann, A. (2008) Leaf area index simulation in soybean grown under near-optimal conditions. *Field Crops Research* 108, 82–92.
Shaw, R.H. and Laing, D.R. (1966) Moisture stress and plant response. In: Pierre, W.H., Kirkham, D., Pesek, J. and. Shaw R. (eds) *Plant Environment and Efficient Water Use*. ASA, Madison, WI, USA, pp. 73–94.
Shibles, R.M. and Weber, C.R. (1965) Leaf area, solar radiation interception and dry matter production by soybean. *Crop Science* 5, 575–577.
Shibles, R.M., Secor, J. and Ford, D.M. (1987) Carbon assimilation and metabolism. In: Wilcox, J.R. (ed.) *Soybeans: Improvement, Production, and Uses,* 2nd edn. Agronomy Monograph 16. American Society of Agronomy, Crop Science Society of America, and Soil Science Society of America, Madison, WI, USA, pp. 535–588.
Shiraiwa, T. and Hashikawa, U. (1995) Accumulation and partitioning of nitrogen during seed filling in old and modern soybean cultivars in relation to seed production. *Japanese Journal of Crop Science* 64, 754–759.
Sinclair, T.R. (2004) Improved carbon and nitrogen assimilation for increased yield. In: Boerma, H.R. and Specht, J.E. (eds) *Soybeans: Improvement, Production, and Uses,* 3rd edn. Agronomy Monograph 16. American Society of Agronomy, Crop Science Society of America, and Soil Science Society of America, Madison, WI, USA, pp. 537–568.
Sinclair, T.R. and de Wit, C.T. (1975) Comparative analysis of photosynthetic and nitrogen requirements in the production of seeds by various crops. *Science* 18, 565–567.
Sinclair, T.R. and de Wit, C.T. (1976) Analysis of the carbon and nitrogen limitations to soybean yield. *Agronomy Journal* 68, 319–324.
Sinclair, T.R. and Muchow, R.C. (1999) Radiation use efficiency. *Advances in Agronomy* 65, 215–265.
Smith, J.R. and Nelson, R.L. (1987) Predicting yield from early generation estimates of reproductive growth periods in soybean. *Crop Science* 27, 471–474.
Spaeth, S.C. and Sinclair, T.R. (1984) Soybean seed growth. I. Timing of growth of individual seeds. *Agronomy Journal* 76, 123–127.
Specht, J.E., Hume, D.J. and Kumudini, S. (1999) Soybean yield potential – a genetic and physiological perspective. *Crop Science* 39, 1560–1570.
St. Martin, S. (1999) Strategies for reaching future yield goals. In: Kauffman, H.R. (ed.) *Proceedings of the 6th World Soybean Research Conference*, Urbana-Champaign, IL, USA, p. 132.
Swank, J.C., Egli, D.B. and Pfieffer, T.W. (1987) Seed growth characteristics of soybean genotypes differing in duration of seed fill. *Crop Science* 27, 85–89.
Taylor, H.M., Mason, W.K., Bennie, A.T.P. and Rowse, H.R. (1982) Response of soybeans to two row spacings and two soil water levels. I. An analysis of biomass accumulation, canopy development, solar radiation interception and components of seed yield. *Field Crops Research* 5, 1–14.
TeKrony, D.M., Egli, D.B., Balles, J., Pfieffer, T. and Fellows, R.J. (1979) Physiological maturity in soybean. *Agronomy Journal* 71, 771–775.
Thronley, J.H.M. (1980) Research strategy in the plant sciences. *Plant, Cell and Environment* 3, 233–236.

Tollinaar, M. and Wu, J. (1999) Yield improvement in temperate maize is attributable to greater stress tolerance. *Crop Science* 39, 1597–1604.

Trewavas, A. (1986) Understanding the control of plant development and the role of growth substances. *Australian Journal of Plant Physiology* 13, 447–457.

UN (2006) World population prospects: The 2006 version. United Nations, New York, USA. Available at: http://esa.un.org/unpp (accessed 4 November 2008).

USDA-NASS (2008) National Agricultural Statistics Service. USDA-NASS, Washington DC. Available at: www.nass.usda.gov/index.asp (accessed 4 November 2008).

van Schaik, P.H. and Probst, A.H. (1958) The inheritance of inflorescence type, peduncle length, flowers per node and percent flower shedding in soybeans. *Agronomy Journal* 59, 98–102.

Vega, C.R.C., Andrade, F.H., Sadras, V.O., Urhart, S.A. and Valentinuz, O.R. (2001) Seed number is a function of growth: A comparative study in soybean, sunflower, and maize. *Crop Science* 41, 748–754.

Watson, D.J. (1947) Comparative studies on the growth of field crops: I. Variation in net assimilation rate and leaf area between species and varieties, and within and among years. *Annals of Botany* 11, 41–76.

Wells, R. (1991) Soybean growth response to plant density: Relationship among canopy photosynthesis, leaf area and light interception. *Crop Science* 31, 755–761.

Wells, R.L., Schulze, L.L., Ashley, D.A., Boerma, H.R. and Brown, R.H. (1982) Cultivar differences in canopy apparent photosynthesis and their relationship to seed yield in soybeans. *Crop Science* 22, 886–890.

Westgate, M.E. and Peterson, C.M. (1993) Flower and pod development in water-deficient soybeans (*Glycine max* L. Merr.). *Journal of Experimental Botany* 44, 109–117.

Westgate, M.E., Schussler, J.R., Reicosky, D.C. and Brenner, M.L. (1989) Effects of water deficits on seed development in soybean. II. Conservation of seed growth rate. *Plant Physiology* 91, 980–985.

Wilson, R.F. (1987) Seed metabolism. In: Wilcox, J.R. (ed.) *Soybeans: Improvement, Production, and Uses*, 2nd edn. Agronomy Monograph 16. American Society of Agronomy, Crop Science Society of America, and Soil Science Society of America, Madison, WI, USA, pp. 643–686.

Wilson, R.F. (2004) Seed composition. In: Boerma, H.R. and Specht, J.E. (eds) *Soybeans: Improvement, Production, and Uses*, 3rd edn. Agronomy Monograph 16. American Society of Agronomy, Crop Science Society of America, and Soil Science Society of America, Madison, WI, USA, pp. 621–677.

Yoshida, K., Nomura, F. and Gotoh, K. (1983) Significance of intra-plant flowering date in soybean seed production. 2. Number of flowers, podding efficiency, nodal distribution of pods and yield components among different flowering dates. *Japanese Journal of Crop Science* 52, 567–573.

Zeiher, C., Egli, D.B., Leggett, J.E. and Reicosky, D.A. (1982) Cultivar differences in N redistribution in soybeans. *Agronomy Journal* 74, 375–379.

Zheng, S.H., Nakamoto, H., Yoshikawa, K., Furuya, T. and Fukuyama, M. (2002) Influences of high night temperatures on flowering and pod setting in soybean. *Plant Production Science* 5, 215–218.

7 Agro-techniques for Soybean Production

Guriqbal Singh, Hari Ram and Navneet Aggarwal
Department of Plant Breeding and Genetics, Punjab Agricultural University, Ludhiana, Punjab, India

7.1 Introduction

Different varieties of a crop are developed, which may vary in maturity period, growth habit, seed size and so on. Not only different crops, but also different varieties of a crop may need specific agro-techniques for realizing high yields. Agronomic practices such as tillage, time of sowing, method of sowing, depth of sowing, plant population, plant geometry, seed priming, mulching, intercropping, nutrient management, water management and weed management may influence the productivity of a crop through effects on germination, emergence, crop growth and development, phenology, disease and insect pest infestation. These practices are known as agro-techniques. Agro-techniques should be used in such a way as to not only produce high crop yields, but also to reduce the costs of production by utilizing resources and inputs judiciously and taking care of the environment. This chapter discusses some of the important agro-techniques for raising a successful crop of soybean (*Glycine max* (L.) Merrill).

7.2 Tillage/Seedbed Preparation

Tillage is physical manipulation of the soil. It is done to create conditions conducive for good germination and plant growth, control weeds, mix fertilizers and manures into the soil, incorporate the straw of a previous crop or a green manure into the soil and so on. For a good seedbed preparation for soybean, two or three cultivations, harrowings or ploughings are generally sufficient for most soils. Tillage intensity as well as type of tillage, however, may vary with the presence or absence of residue from the previous crop, weeds and the soil type.

Pedersen and Lauer (2003) reported similar soybean yields with no-tillage and conventional tillage systems. The seed yield may be similar with disk-chisel tillage, strip tillage and no-tillage systems (Perez-Bidegain *et al.*, 2007). However, soil texture is an important factor influencing tillage response. On a heavy clay soil, broadcasting seed onto a no-till soil followed by disking may result in lower plant stand than broadcasting onto a tilled surface (Popp *et al.*, 2000). On sandy soils, no-till and conventional tillage provide similar yields, whereas on silt loam and clay soils, yields are generally higher with conventional tillage than with no-tillage (Hairston *et al.*, 1990). In general, in no-tillage systems, a seed yield either similar to or higher than that obtained with conventional tillage may be the result of better yield attributes such as seed mass, seed number per m^2 and pod number per m^2; these have been reported to be 15%, 9% and 9% greater, respectively, in a no-tillage compared to a conventional tillage system (Pedersen and Lauer, 2004). Furthermore, preference for a tillage system may vary depending on the production system; conventional tillage may be preferred for full-season systems and a conservation tillage system for double-cropping systems (Popp *et al.*, 2000). Irrigation facilities may also influence the performance of soybean grown under different tillage systems. Under irrigated conditions soybean gives similar yields under conservation and conventional tillage systems, whereas under non-irrigated conditions yields are slightly higher with a conventional than conservation tillage system (Parsch *et al.*, 2001).

If weeds are not a serious problem then soybean may be sown without any seedbed preparation (PAU, 2009). However, when soybean and the succeeding crop are sown with a no-till drill for 3–4 years, yields tend to decrease during the later years (Table 7.1). This is mainly due to problems with perennial weeds despite chemical weed control, and to some extent to reduced uptake of nutrients, as reflected by the presence of higher amounts of nutrients in the soil (Table 7.1). No-till sowing saves energy, reduces

Table 7.1. Seed yield of soybean and available macronutrient status of soil as influenced by tillage management practices in soybean-based (soybean–wheat [*Triticum aestivum*], soybean–field pea [*Pisum sativum*] and soybean–lentil [*Lens culinaris*]) cropping systems (adapted from Prakash *et al.*, 2004).

Tillage management practice	Soybean seed yield (kg ha^{-1})					Available macronutrient status (0–15 cm) of soil after a 4-year cropping system (kg ha^{-1})		
	1999	2000	2001	2002	Mean	Nitrogen	Phosphorus	Potassium
No-till	1863	2396	2903	296	1865	336.6	23.3	102.2
Minimum	1886	2535	3535	1272	2307	322.6	24.7	93.6
Conventional	2096	2667	3340	1341	2361	305.5	20.3	87.0
CD (P = 0.05)	NS	NS	274	170	246	7.1	0.7	2.1

CD, critical difference; NS, not significant.

production costs, improves soil physical and chemical properties, helps with timely sowing to utilize residual soil moisture for proper germination, checks environmental pollution due to lesser use of diesel and ensures timely sowing on a large area, which results in higher yields. Therefore, the area under no-till sowing of various crops is increasing in various countries, including the USA (Parsch *et al.*, 2001). However, no-till sowing practice should be followed only for a short period (e.g. for one or two seasons) to avoid any decline in yields due to problems with weeds, insect pests, diseases or any changes in the physical properties of the soil.

7.3 Time of Sowing

Time of sowing is a non-monetary input that influences the productivity of soybean to a great extent. Both too early and too late sowings result in drastic reductions in yields. Soil and air temperatures of 13–16°C are necessary for germination and seedling growth of soybean, but further increases in temperature up to about 32°C are better (Christmas, 2008). The optimum time of sowing is determined on the basis of various factors such as weather parameters (e.g. minimum and maximum temperatures, photoperiod, relative humidity, rainfall) during the crop growing season, maturity duration of the genotype, soil type, moisture availability at sowing and so on.

The optimum time for soybean sowing may vary at different locations due to different climatic conditions. In India, mid-June to the first week of July is the optimum sowing time for the North Hill and North Plain zones, whereas mid-June to mid-July is optimum for the North-Eastern and Central zones and mid-June to end of July is optimum for the Southern zone (Chauhan and Joshi, 2005). In Wisconsin, USA, crop sown in early May has been found to produce a higher yield than that sown in late May (Table 7.2), with the early planting producing a higher seed number, pod number and harvest index than the late planting (Pedersen and Lauer, 2004). Cultivars may show differential responses to sowing time. For example, in Arlington, USA, cultivar CX 232 yielded 7% higher when sown in early May

Table 7.2. Influence of date of sowing on seed yield of soybean at Arlington and Hancock, Wisconsin, USA (adapted from Pedersen and Lauer, 2003).

		Seed yield (kg ha^{-1})			
Location	Sowing date	1997	1998	1999	2000
Arlington	Early May (3–6 May)	3330	4590	4300	3750
	Late May (23–27 May)	3470	4430	4010	3490
	LSD (0.05)	130	80	110	210
Hancock	Early May (8–13 May)	3630	5250	4580	3790
	Late May (26 May–3 June)	3510	5120	2930	3330
	LSD (0.05)	NS	NS	820	360

LSD, least significant difference; NS, not significant.

(4370 kg ha^{-1}) than in late May (Pedersen and Lauer, 2003), whereas no sowing-date effect was observed for two other cultivars. In another study, the yield was decreased from 3000 to 2900 to 2800 kg ha^{-1} when sowing was delayed from an early (6–21 May) to intermediate (20–27 May) to late (4–11 June) planting date, respectively (Perez-Bidegain *et al.*, 2007).

In Lincoln, Nebraska, USA, delayed sowing after 1 May led to a significant seed yield decline of 17 kg ha^{-1} day^{-1} in 2003 and 43 kg ha^{-1} day^{-1} in 2004 (Bastidas *et al.*, 2008). In Hyderabad, Andhra Pradesh, India, a delay in planting in the winter season caused a progressive decline in soybean yield from 2573 to 2396, 2193 and 1975 kg ha^{-1} when sowing was delayed from 15 October to 4 November, 24 November and 14 December, respectively (Murthy *et al.*, 2001). In the summer season, the crop yielded 1425, 1538, 1295 and 1141 kg ha^{-1} when sown on 5 January, 25 January, 14 February and 6 March, respectively. In Ludhiana, Punjab, India, 25 May, 10 June and 25 June sowings yielded 1685, 2210 and 1909 kg ha^{-1} (Singh *et al.*, 2000). Lower yields from 25 May and 25 June sowing dates were attributed due to very high and very low dry matter accumulation, respectively, on the two sowing dates. In another study conducted at Ludhiana, crop sown on 24 May, 8 June, 24 June and 8 July yielded 1414, 1419, 1363 and 723 kg ha^{-1} (Singh and Jolly, 2004b). Other researchers have also reported lower yields of soybean with delayed planting (Oplinger and Philbrook, 1992; Egli and Bruening, 2000; De Bruin and Pedersen, 2008a).

A timely sown crop generally results in higher yields than late-sown crop, unless there is a specific problem such as drought, waterlogging, high incidence of insect pests and disease or lodging. Higher yields in the timely sown crop may be due to better plant growth and yield attributes, longer maturity duration and higher agroclimatic indices such as growing-degree days, heliothermal units and photothermal units (Table 7.3). Lower yields in late-planted crops could be due to a variety of reasons, including shifting of the reproductive phase into less favourable environment (shorter days, lower temperatures and insolation), less availability of soil moisture and a shorter growth period.

Table 7.3. Agroclimatic indices at physiological maturity, total dry matter and seed yield of soybean cultivar PK 416 sown on different dates at Ludhiana, Punjab, India (adapted from Hundal *et al.*, 2003).

Sowing date	Accumulated growing-degree days (°C day)	Accumulated heliothermal units (°C day h)	Accumulated photothermal units (°C day h)	Total dry matter (kg ha^{-1})	Seed yield (kg ha^{-1})
6 June 1997	2448	19001	32201	7348	1632
23 June 1997	2301	18390	29913	6116	1462
7 June 1999	2669	21085	35843	6163	1723
21 June 1999	2544	20097	33888	5862	1528

Table 7.4. Effect of planting date and maturity group on vegetative growth and reproductive growth characteristics, yield and yield components of soybean in Lexington, Kentucky, USA (average across years) (adapted from Egli and Bruening, 2000).

		Maturity group			
Character	Sowing date	I	II	III	IV
Node number (no. per m^2)	Early (mid May)	827	850	1071	1578
	Late (late June)	662	638	755	798
Above-ground vegetative mass (g per m^2)	Early (mid May)	566	535	821	815
	Late (late June)	513	533	633	656
Length of flowering and pod set (R1–R5) (days)	Early (mid May)	26	32	37	37
	Late (late June)	21	23	25	26
Seed-filling period (R5–R7) (days)	Early (mid May)	31	31	31	37
	Late (late June)	31	33	35	35
Yield (g per m^2)	Early (mid May)	343	403	410	400
	Late (late June)	291	312	313	300
Seeds per m^2	Early (mid May)	2141	2164	2575	2391
	Late (late June)	1881	1604	1940	1850
Seed size (mg seed^{-1})	Early (mid May)	161	186	159	171
	Late (late June)	155	195	162	162

Early flowering and a shorter vegetative growth phase in late-planted soybean results from the combined effect of photoperiod and temperature. In late-planted crop, due to early flowering, plants are shorter and have fewer nodes, resulting in fewer seeds per unit area, lower seed size and ultimately lower seed yields (Table 7.4). Delayed planting shortens the flowering and pod set period, but not the seed-filling period. Under late planting, soybean yield may be increased to some extent by the use of a high plant population and narrow rows.

Planting date not only influences the seed yield, but also the quality of soybean oil. The quality of soybean oil can be improved by reducing palmitic acid (16:0) and linolenic acid (18:3). Early planting (24–29 May) has been found to decrease linolenic acid, while late planting (22–28 June) decreased palmitic acid levels in modified fatty acid breeding lines of soybean (Ray *et al.*, 2008).

7.4 Method of Sowing

Soybean is sown in rows on a flat bed either in a well-prepared field or as a no-till crop. It is also sown on raised beds. In some areas it is grown as a sole or mixed crop, while in others intercropping with cereals, oilseeds, grain legumes and fibre crops is also practised.

The crop should be sown using a drill as this ensures the desired spacing (between as well as within rows) and depth, thus resulting in the proper plant stand and consequently higher yields than a crop sown by the

Table 7.5. Seed yield (kg ha^{-1}) of soybean as influenced by planting methods under two pre-plant tillage methods at two locations in the USA (adapted from Popp *et al.*, 2000).

	Conventional tillage		Conservation tillage	
Location	Broadcast sowing	Conventional drill sowing	Broadcast sowing	No-till drill sowing
Keiser (full-season)	3270	3300	3060	3340
Pine Tree (full-season)	1410	1490	1390	1440
Pine Tree (double-crop)	1430	1510	1380	1660

broadcast method. Drill sowing is more important in conservation tillage systems than with conventional tillage (Table 7.5).

7.5 Depth of Sowing

The crop should be sown at the proper depth to ensure optimum germination and emergence. If the crop is sowed very deeply, seedlings will not emerge as the food stored in the cotyledons will be exhausted before the coleoptile emerges. Conversely, in the case of shallow sowing the surface soil becomes dry very quickly, particularly in areas where the air temperature is very high at sowing, leaving very little moisture for seeds to imbibe. Sowing of soybean at a 2.5–5.0 cm depth is considered optimum (Pedersen and Lauer, 2003; Chauhan and Joshi, 2005; PAU, 2009). Christmas (2008) advocated sowing of soybean to 2.5–3.7 cm depth only, as deeper sowings are expected to reduce emergence.

7.6 Plant Population and Planting Geometry

An optimum plant population is a prerequisite for realizing high seed yields. If the plant population is below the optimum mark, high seed yields cannot be obtained with any measures. The optimum plant population may be ensured by using an adequate quantity of good-quality seed. Furthermore, seed treatment against seed-borne diseases prior to sowing helps to check plant loss due to diseases.

The optimum seed rate of soybean varies with the seed size, plant type and the maturity period of the genotype. Generally, 62.5–75.0 kg seed ha^{-1} is considered optimum (PAU, 2009). Chauhan and Joshi (2005) summarized the information on soybean production technology for different agroclimatic zones of India and reported optimum plant populations of 0.4 million plants ha^{-1} for the North Hill and North Plain zones and 0.4 to 0.6 million plants ha^{-1} for the Central, Southern and North-Eastern zones. Furthermore, row and plant spacings can be maintained at 45 × 5 cm in the North Hill zone, 45–60 × 5–8 cm in the North Plain zone and

30–45 × 5 cm in the Central, Southern and North-Eastern Zones. In Pune, Maharashtra, India, a planting geometry of 30 × 10 cm (0.333 million plants ha^{-1}) has been found to provide the highest oil and seed yields of soybean, followed by 45 × 5 cm (0.444 million plants ha^{-1}) (Table 7.6). With constant row spacings but increasing plant spacings, plant height tends to decrease while branches per plant, pods per plant, seeds per plant and 100-seed weight increase.

Soybean is generally sown in rows, and the spacings between rows could vary depending upon the growth habit, sowing time, soil type and so on. In the early soybean production system, which is practised in some parts of the USA, yields are higher with a narrow row spacing of 23 cm (Holshouser and Whittaker, 2002) or ≤40 cm (Bowers *et al.*, 2000). Many other studies have also shown that soybean planted at narrow row spacings provides higher yields than that planted at wider row spacings, such as 38 versus 76 cm (De Bruin and Pedersen, 2008c), 19 versus 57 cm (Andrade *et al.*, 2002), 19 versus 38 cm (Kratochvil *et al.*, 2004) and 23 versus 46 cm (Holshouser and Whittaker, 2002). Soybean in narrow rows exhibits higher yields than that in wide rows in general and in late-planting dates in particular. Increased seed yields in response to closer rows could be due to an improvement in light interception during the critical period for seed set (Andrade *et al.*, 2002) or late pod fill (i.e. stages R6–R7) (Bennie *et al.*, 1982), increased leaf area index (Holshouser and Whittaker, 2002) and higher photosynthesis (Bennie *et al.*, 1982).

Row spacing may influence both main-stem and branch seed yields. In one study (Norsworthy and Shipe, 2005), when soybean was grown in narrow (19 cm) and wide (97 cm) rows at the recommended seeding rates, main-stem yields accounted for 45% and 69% of the total seed yield in wide and narrow rows, respectively, whereas branch seed yields accounted for 55% and 31%. Therefore, only genotypes with more branching should be selected for wider rows, whereas less-branching genotypes should be preferred for narrow rows.

Table 7.6. Effect of planting geometry and plant density on growth, yield attributes and yield of soybean in Pune, Maharashtra, India (adapted from Halvankar *et al.*, 1999).

Planting geometry (cm)	Plant density (million plants ha^{-1})	Plant height (cm)	Pods per plant	Branches per plant	Seeds per plant	100-seed weight (g)	Oil yield (kg ha^{-1})	Seed yield (kg ha^{-1})
30 × 5	0.666	58.9	27.7	3.0	47.1	12.2	571	3239
30 × 10	0.333	51.4	47.0	5.0	83.4	12.5	623	3529
30 × 15	0.222	48.9	69.5	6.6	122.8	12.8	574	3234
45 × 5	0.444	54.1	36.6	4.0	67.3	12.4	599	3391
45 × 10	0.222	49.3	61.8	6.3	116.8	12.5	543	3048
45 × 15	0.148	46.3	85.1	6.8	154.6	13.0	522	2932
CD (P = 0.05)	–	1.8	2.9	0.3	5.0	0.3	42	237

CD, critical difference.

As the late-planted crop has shorter plants with lower plant biomass, the plant population may be increased to realize higher seed yields from late-planted crops. In the mid-south of the USA, the optimum plant population has been reported to range from 108,000 to 232,000 plants ha^{-1} for May-sown crop compared with 238,000–282,000 plants ha^{-1} for June-sown crop (Lee *et al.*, 2008). Row spacings may also influence the required plant population. Plant populations of 194,000–290,800 plants ha^{-1} and 157,300–211,800 plants ha^{-1} are required for locations where soybean is sown at 38 and 76 cm row spacings, respectively (De Bruin and Pedersen, 2008b).

Soybean yields do not continue to increase at high plant population densities due to decreased radiation-use efficiency (Purcell *et al.*, 2002). In the case of glyphosate-resistant genotypes of soybean, similar seed yields with 20% reduced seeding rates (345,800 seeds ha^{-1} for full season and 444,600 seeds ha^{-1} for double crop production) than standard have been obtained (Kratochvil *et al.*, 2004), indicating that seed rate may be reduced with an additional profit to the range of $14.30–27.72 ha^{-1}. In an earlier study, glyphosate-resistant soybean at 370,000 seeds ha^{-1} and 620,000 seeds ha^{-1} produced similar yields when sown in narrow rows without moisture stress (Norsworthy and Frederick, 2002). Since high plant populations involve high seeding rates and consequently high production costs, it is not the seed yield but the net returns that matter most to farmers.

7.7 Straw Mulching

In some areas, rainfall may occur after sowing and before soybean emergence. Rainfall followed by high temperatures results in crust formation, which reduces the plant stand to unacceptable levels. In such situations the crust needs to be broken mechanically or straw mulch applied as soon as field conditions permit walking into the field. Straw of wheat (*Triticum aestivum*), rice (*Oryza sativa*) or of any crop should be applied at 3–6 t ha^{-1} to alleviate crust effects on emergence (Mehta and Prihar, 1973; Singh and Jolly, 2008). In areas where rainfall is expected during the sowing period, straw mulch may be applied after sowing as a matter of routine. After emergence of the crop, the straw may be removed as far as obtaining normal emergence is concerned or retained if its other beneficial effects, such as moisture retention or weed smothering, are desired.

Straw mulch lowers the maximum and increases the minimum soil temperature in the seed zone. It also increases the soil moisture content, which enhances the rate as well as final count of seedling emergence in soybean (Mehta and Prihar, 1973; Chaudhri and Das, 1978). In Japan, plant residue mulch has been found to increase the minimum soil temperature by 3°C during the early stage of soybean growth, which resulted in improved growth and pod yield (Kitoh and Yoshida, 1996).

7.8 Seed Priming

Prior to sowing, soaking seeds in ordinary water or in a chemical/nutrient solution for a specific duration helps in obtaining higher and faster emergence. Seed priming is of paramount importance in conditions of sub-optimal soil moisture at the time of sowing. Soybean germination is improved by soaking seeds in water or a 20% gibberellic acid solution (Nalawadi *et al.*, 1973).

The duration of seed priming is critical for obtaining good emergence. Soaking seed for >2h may prevent the germination of some soybean cultivars completely (Wadud and Kosar, 1997) while having a beneficial effect in others. Soaking soybean seeds in water for 1–8h may cause water-uptake injury to imbibing seeds and consequently reduce germination (Woodstock and Taylorson, 1981). Such types of injury, however, may be avoided by reducing the initial rate of water uptake osmotically by soaking seeds in 30% polyethlylene glycol. The germination of aged soybean seeds has also been improved by priming seeds with polyethlylene glycol solution (Park *et al.*, 1999). Seed priming, however, is not commonly used for raising soybean in various parts of the world.

7.9 Intercropping/Mixed Cropping

Intercropping and mixed cropping are age-old practices being followed by farmers. In mixed cropping, seeds of two or more crops are sown by the broadcast method; in intercropping, seeds of two or more crops are sown in rows in a specific pattern. Intercropping is more common than mixed cropping due to the ease of controlling weeds, spraying chemicals and harvesting. In earlier times, intercropping was followed to ensure the success of at least one crop in the event of failure of the other(s) due to reasons mostly related to climate or climate-induced. In the present-day agriculture, however, intercropping is followed to achieve greater productivity and net returns per unit area per unit of time.

Soybean is grown in the intercropping system with a number of crops including maize (*Zea mays*), sorghum (*Sorghum bicolor*), black gram (*Vigna mungo*), sugarcane (*Saccharum officinarum*), pigeon pea (*Cajanus cajan*), pearl millet (*Pennisetum typhoides*), groundnut (*Arachis hypogaea*) and cotton (*Gossypium* species). The sowing pattern (row ratio) is selected in such a way that the productivity of the main crop is either not adversely affected or is affected to the least extent when compared with sole cropping.

In intercropping systems, time and space are best utilized, resulting in higher total crop productivity as well as net income to farmers. Furthermore, intercropping may help in checking disease and insect pests and in smothering weeds. To avoid or lessen competition among crops, care should be taken to select only those crops or crop varieties that are compatible. Plant height, leaf area and crop duration are some of the important aspects that need to be considered when planning intercropping. Crop management practices such as irrigation, fertilizer application and chemical weed control should not be harmful to the companion crop.

7.10 Nutrient Management

Soybean seeds are rich in protein as well as oil, which is probably why the nutrient requirements of soybean are generally higher than those of other grain legumes. The crop needs to be fertilized as per soil test basis. In general, soybean requires the application of 5–10 t farmyard manure (FYM), 20 kg nitrogen, 80 kg P_2O_5, 20 kg K_2O and 20 kg sulphur ha^{-1} (Chauhan and Joshi, 2005). The application of 125% of the recommended dose of fertilizers in soybean–wheat cropping systems increases the seed yield of both crops significantly over the recommended dose of nutrients (recommended dose: 20:60:20 and 120:60:40 kg ha^{-1} of nitrogen:phosphorus:potassium [NPK] to soybean and wheat, respectively) (Jain *et al.*, 2005). The nutrients may be supplied through any of the commonly available fertilizers that contain them. However, the availability of a fertilizer and the price determine its use by farmers. When providing a crop with phosphorus, it is better to use single superphosphate as it contains not only phosphorus but also sulphur. The entire fertilizer dose is applied at the time of sowing, except in some cases where nitrogen is top-dressed or applied as foliar spray at the reproductive phase.

Soil or foliar application of nitrogen to soybean at the reproductive phase is a subject of debate as it has been found to be beneficial in some (Salvagiotti *et al.*, 2008) but not in other studies (Barker and Sawyer, 2005). The in-season application of 84 kg N ha^{-1} at R2 or between the R4 and R5 growth stages either through broadcast or subsurface banding did not influence soybean yields significantly (Schmitt *et al.*, 2001). Similarly, the application of nitrogen to the soil up to 168 kg ha^{-1} at either the R3 or R5 growth stage did not increase soybean seed yield (Freeborn *et al.*, 2001).

Nutrient application should be based on the cropping system rather than a single crop basis, as nutrients applied to the previous crop may not be fully utilized by the crop and a considerable amount may be left for use by the succeeding crop. In addition, the climatic conditions also warrant using fertilizers on a cropping system basis. In the soybean–wheat cropping system, if the recommended dose of phosphorus has been applied to wheat then the phosphorus dose to be applied to soybean may be reduced by 25% (from the otherwise recommended soybean dose) (PAU, 2009). This is possible in some areas such as in northern India, where moisture due to monsoon rains coupled with high temperatures make the phosphorus applied to a previous wheat crop during the winter season available to the soybean crop.

The application of 30 kg N ha^{-1} to soybean and 120 kg N ha^{-1} to wheat has been found to substantially increase the productivity of both soybean and wheat over lower doses of nitrogen application to soybean (Ramesh and Reddy, 2004). The application of phosphorus to rainy-season crops (soybean and others) leaves residual phosphorus for succeeding winter-season crops (Nimje, 2003). FYM has an important role in agriculture in meeting the nutrient demands of crops. However, there are only a few examples of FYM use in soybean. Similar is true for vermicompost. These bulky organic manures are perhaps used more in high-nutrient-demanding crops such as cereals.

Biofertilizers such as *Bradyrhizobium*, phosphate-solubilizing bacteria and plant growth-promoting rhizobacteria have a unique role in soybean production. The use of biofertilizers, especially *Bradyrhizobium*, has been found to greatly increase the productivity of soybean (Alves *et al.*, 2003; Albareda *et al.*, 2009). When biofertilizers are used, the dose of chemical fertilizers may be reduced to obtain similar yield levels as with the use of chemical fertilizers only. However, the application of recommended levels of NPK (32 kg N, 34.4 kg P and 33.6 kg K ha^{-1}) with FYM (5 t ha^{-1}) and biofertilizers has been found to increase yield attributes, protein and oil content and seed yield of soybean over the sole application of recommended levels of NPK (Singh and Rai, 2004). The integrated use of 1 t FYM ha^{-1}, 26.4 kg P ha^{-1} and biofertilizers (*Bradyrhizobium japonicum* and *Pseudomonas* species) provides higher yields over no or lower doses of nutrients (Gautam *et al.*, 2003). The combined use of chemical fertilizers and biofertilizers, therefore, ensures less use of chemical fertilizers, thereby resulting in lower production costs and consequently higher net returns for farmers, as well as reduced environmental pollution.

There are few examples of the use of micronutrients in soybean. This could be due to the lack of response to micronutrients in improving soybean yields. Freeborn *et al.* (2001) observed no significant increase in seed yield with foliar-applied boron up to 0.56 kg ha^{-1} at the R3 or R5 stage. However, a significant response to soil- (at V2 stage) or foliar-applied (at R2 stage) boron up to 1.5 kg ha^{-1} has been reported (Cirak *et al.*, 2006). Soils with deficient levels of micronutrients are expected to respond to the application of one or more micronutrients in obtaining high seed yields of soybean.

7.11 Water Management

Where possible, soybean should be sown in proper moisture conditions by applying pre-sowing irrigation for obtaining good emergence, early vigour of the crop and efficient use of nutrients. Afterwards the crop may need 1–5 irrigations, depending upon rainfall, soil type and maturity group.

If the crop has to be grown as a rainfed crop, efforts should be made to use every drop of rain water as judiciously as possible. The crop should be sown as soon as possible after receiving the first rainfall during or near the optimum sowing period. Water harvesting should be encouraged and the same water may be used for irrigating the crop at critical growth stages. Wherever possible the crop should be irrigated to meet its water requirements. Irrigation may be applied using different methods such as flood irrigation, furrow irrigation and sprinkler irrigation. Irrigation increases soybean yields significantly compared to no irrigation (Heatherly and Spurlock, 2000). However, the furrow and flood irrigation methods have been found to have similar effects in influencing the seed yield of soybean in maturity groups IV and V. Under compelling circumstances, saline water may be used for irrigation either in rotation or mixed with fresh water.

Flowering and podding are the most critical stages. Based on the availability of water, irrigation should be scheduled in such a way that the crop does not suffer water stress during these stages.

7.12 Weed Management

Weeds compete with crop plants for moisture, nutrients and light and thereby reduce crop yields. The reduction in crop yield depends upon the quantum of weed flora, weed species present and the duration of crop–weed competition. Weeds may cause losses in soybean yield of 35–83% (Yadav *et al.*, 1999; Chandel and Saxena, 2001; Kewat and Pandey, 2001; Vyas and Jain, 2003; Singh *et al.*, 2004; Billore *et al.*, 2007; Singh, 2007; Singh and Jolly, 2009). The initial 45-day period is considered to be the critical crop–weed competition period in soybean. Weed removal during this period is therefore a must for realizing high seed yields. Weed plants that emerge late in the crop cycle produce less seeds than early-emerged weed plants (Clay *et al.*, 2005) as the soybean canopy can reduce photosynthetically active radiation penetration from 50% to 100% at R3 or later growth stages. Late-emerged weeds may not reduce soybean yields by much, but they may remain green to interfere with harvest and can produce seeds, which can be a problem in the future. They therefore also need removal.

The major weed flora that infest soybean crops include *Caesulia axillaries* Roxb., *Cyperus rotundus* (L.) Link., *Digitaria sanguinalis* L., *Echinochloa colona* L., *Commelina benghalensis* L., *Acalypha indica* L., *Anotic monthuloni* Hook., *Trianthema portulacastrum* L., *Digera arvensis* L., *Phyllanthus niruri* L., *Dactyloctenium aegypticum* Beauv., *Cynodon dactylon* (L.) Pers., *Amaranthus spinosis* L., *Chenopodium album* L., *Setaria faberi* Herrm., *Amaranthus retroflexus* L., *Ambrosia artemisiifolia* L., *Leucas aspera* Spreng., *Euphorbia hirta* L. and *Abutilon theophrasti* L.

Weeds must be controlled at the appropriate time using suitable methods if high seed yields of soybean are to be realized. Weeds may be controlled using cultural practices, employing mechanical means or spraying herbicides. However, an integrated weed management approach is considered to be the best in present-day intensive agriculture. In soybean, weeds may be effectively controlled by one or two hand-weedings or hoeings, which are generally performed 30 and 45 days after sowing. However, manual weed control is cumbersome and costly. Furthermore, as soybean is grown during the rainy season in various parts of the world, timely weed control is not possible due to frequent rains. The non-availability of labour for weed removal, particularly at the critical period of crop–weed competition, further demands other less labour-intensive strategies for effective weed control.

Higher plant populations and narrow row spacing may help in checking weed growth and consequently increasing soybean seed yields. The use of a 150 kg ha^{-1} seed rate has been reported to reduce the weed population and dry weight of weeds significantly and increase the soybean seed yield when compared to 100 and 125 kg ha^{-1} seed rates (Yadav *et al.*, 1999), possibly due to suppressed weed emergence and establishment. Weed control and soybean yields are higher in narrow rows than wide rows (Mickelson and Renner, 1997; Nelson and Renner, 1998). *In situ* mulching with weeds 30 days after sowing has also been found to be effective in keeping weeds under control and providing high seed yields of soybean (Singh, 2005, 2007).

Soil solarization (a method of harvesting solar energy during the hottest period by covering the soil surface with transparent polyethylene sheet when the soil has high moisture content) is an important method for weed seed aging (Singh *et al.*, 2004). In Madhya Pradesh, India, a mean maximum soil temperature of 56.4°C at the surface and 53.6°C at 5 cm, 44.3°C at 10 cm and 39.4°C at 15 cm soil depths has been reported with transparent polyethylene mulching; this was hotter than non-solarized plots by 10.2°C, 9.4°C, 5.1°C and 3.4°C, respectively. Soil solarization for 5 weeks helped to reduce the emergence of many weed species, with a resulting marked increase in soybean crop growth, yield attributes and seed yield (Singh *et al.*, 2004).

Various pre-plant, pre-emergence (PE) and post-emergence (POE) herbicides have been found to be effective for weed control in soybean (Table 7.7). Some promising herbicides for effective weed control in soybean include pre-plant incorporation of cloransulam (Reddy, 2000), PE application of cloransulam (Reddy, 2000) and imazaquin (Reddy, 2000) and POE spraying of imazamox (Nelson and Renner, 1998) and imazethapyr (Nelson and Renner, 1998).

The optimum dose of a herbicide may vary depending upon the soil texture, climatic conditions, weed flora, stage of the crop, stage of weeds and so on. At a higher dose, a herbicide may have some phytotoxicity on the soybean crop, as reported in the case of metribuzin (Kewat and Pandey, 2001). Some herbicides may have adverse effects on nodulation and nitrogen fixation. Therefore, only herbicides that are not only effective for controlling weeds but are also safe for soybean–rhizobia symbiosis should be used.

In glyphosate-resistant soybean, the application of glyphosate at 840 g a.e. ha^{-1} at the V5 growth stage has been reported to have no effect on vegetative growth, reproductive development or seed yield (Nelson and Renner, 2001). In glyphosate-resistant soybean, a single application of glyphosate can prevent yield loss in narrow-row (18 cm) sown crop, whereas in wide-row (76 cm) sown crop, late-emerging weeds may warrant a second application (Mulugeta and Boerboom, 2000).

A single herbicide may not control all weed species very effectively. However, tank mixtures of some herbicides such as cloransulam-methyl and diphenyl ether increase the spectrum of weed control and consequently the soybean seed yield over the application of these herbicides alone (Pline *et al.*, 2002). The integration of a herbicide with hand weeding – for example,

Table 7.7. Some promising herbicides for controlling weeds in soybean.

Herbicide	Dose	Time of application	Seed yield of soybean ($kg\ ha^{-1}$) with different treatments: Herbicide treatment	Two-hand weedings	Weedy check	Reference
Alachlor	2 kg ha^{-1}	Pre-emergence	1526	1710	1041	Vyas and Jain (2003)
Alachlor	2 kg ha^{-1}	Pre-emergence	1194	1282	377	Chandel and Saxena (2001)
Alachlor	2 kg ha^{-1}	Pre-emergence	3585	3318	2666	Singh (2005)
Alachlor	2 kg ha^{-1}	Pre-emergence	1530	1678	698	Singh and Jolly (2004a)
Alachlor	2 kg ha^{-1}	Pre-emergence	3216	3196	1910	Singh (2007)
Alachlor (10%) granules	2 kg ha^{-1}	Pre-emergence	1291	1543	782	Yadav *et al.* (1999)
Anilofos	1.75 kg ha^{-1}	Pre-emergence	1330	1282	377	Chandel and Saxena (2001)
Clomazone	1 kg ha^{-1}	Pre-emergence	1654	1858	955	Pandya *et al.* (2005)
Clomazone	1 kg ha^{-1}	Pre-emergence	2962	3196	1910	Singh (2007)
Fluchloralin	1 kg ha^{-1}	Pre-plant incorporation	1946	–	1217	Billore *et al.* (2007)
Flumioxazin	60 g ha^{-1}	Pre-plant incorporation	1922	–	1217	Billore *et al.* (2007)
Flumioxazin	45 g ha^{-1}	Pre-emergence	2125	–	1217	Billore *et al.* (2007)
Imazamox + imazethapyr 5%	75 g ha^{-1}	Post-emergence	1622	1710	1041	Vyas and Jain (2003)
Imazethapyr	100 g ha^{-1}	Post-emergence	1642	1282	377	Chandel and Saxena (2001)
Metolachlor	1 kg ha^{-1}	Pre-emergence	1296	–	964	Singh *et al.* (2004)
Metolachlor (5%) granules	2 kg ha^{-1}	Pre-emergence	1339	1543	782	Yadav *et al.* (1999)
Metribuzin	0.50 kg ha^{-1}	Pre-emergence	2342	2457	1217	Kewat and Pandey (2001)
Metribuzin	0.75 kg ha^{-1}	Pre-emergence	2160	2457	1217	Kewat and Pandey (2001)
Pendimethalin	1.5 kg ha^{-1}	Pre-emergence	2085	2457	1217	Kewat and Pandey (2001)
Propaquizafop	50 g ha^{-1}	Post-emergence	1038	1282	377	Chandel and Saxena (2001)
Quizalofop ethyl	50 g ha^{-1}	Post-emergence	1552	1710	1041	Vyas and Jain (2003)
S-metolachlor	750 g ha^{-1}	Pre-emergence	1524	1710	1041	Vyas and Jain (2003)
S-metolachlor	750 g ha^{-1}	Pre-emergence	1907	1678	698	Singh and Jolly (2004a)

hand weeding plus clomazone 1 kg ha^{-1} (PE) (Pandya *et al.*, 2005; Singh, 2005, 2007), pendimethalin 0.45 kg ha^{-1} (PE) (Singh, 2005), pendimethalin 1 kg ha^{-1} (PE) (Singh, 2007) or fenoxaprop-p-ethyl 175 g ha^{-1} (POE) (Pandya *et al.*, 2005) – or the integrated use of PE and POE herbicides (Reddy, 2000; Barnes and Oliver, 2005) provides effective weed control and higher soybean yields than the sole use of a PE herbicide.

7.13 Conclusions

Various agro-techniques need to be followed to realize high soybean seed yields. However, the lack of mechanization is one of the most serious problems for the successful cultivation of soybean in many parts of the world. The mechanization of various operations such as sowing, weeding and harvesting/threshing is the need of the day. Happy Seeder sowing of some crops under such situations has been tested and found satisfactory (Sidhu *et al.*, 2007), and this now needs to be studied in soybean. Similarly, the use of combine harvesters/threshers for threshing the soybean crop with enough precision to avoid adversely affecting seed quality and subsequently germination should be studied and popularized among farmers.

There is a need to lower the production costs of raising a successful crop of soybean. Strategies need to be worked out to use the inputs judiciously and more efficiently. Fertilizers may have some residual effects. Therefore, farmers should be advised to apply fertilizers on a cropping system basis. Furthermore, the quality of rhizobia should be ensured as rhizobia inoculations often fail to enhance soybean productivity due to poor quality, which could result at the production level itself or during storage or transportation.

References

Albareda, M., Rodríguez-Navarro, D.N. and Temprano, F.J. (2009) Soybean inoculation: Dose, N fertilizer supplementation and rhizobia persistence in soil. *Field Crops Research* 113, 352–356.

Alves, B.J.R., Boddey, R.M. and Urquiaga, S. (2003) The success of BNF in soybean in Brazil. *Plant and Soil* 252, 1–9.

Andrade, F.H., Calviño, P., Cirilo, A. and Barbieri, P. (2002) Yield responses to narrow rows depend on increased radiation interception. *Agronomy Journal* 94, 975–980.

Barker, D.W. and Sawyer, J.E. (2005) Nitrogen application to soybean at early reproductive development. *Agronomy Journal* 97, 615–619.

Barnes, J.W. and Oliver, L.R. (2005) Preemergence weed control in soybean with cloransulam. *Weed Technology* 18, 1077–1090.

Bastidas, A.M., Setiyono, T.D., Dobermann, A., Cassman, K.G., Elmore, R.W., Graef, G.L. and Specht, J.E. (2008) Soybean sowing date: The vegetative, reproductive and agronomic impacts. *Crop Science* 48, 727–740.

Bennie, A.T.P., Mason, W.K. and Taylor, H.M. (1982) Responses of soybeans to two row spacings and two soil water levels, 3: Concentration, accumulation and translocation of 12 elements. *Field Crops Research* 5, 31–43.

Billore, S.D., Vyas, A.K., Joshi, O.P., Pandya, N. and Khan, I.R. (2007) Bio-efficacy of flumioxazin in soybean (*Glycine max*). *Indian Journal of Agricultural Sciences* 77, 642–644.

Bowers, G.R., Rabb, J.L., Ashlock, L.O. and Santini, J.B. (2000) Row spacing in the early soybean production system. *Agronomy Journal* 92, 524–531.

Chandel, A.S. and Saxena, S.C. (2001) Effect of some new post-emergence herbicides on weed parameters and seed yield of soybean (*Glycine max*). *Indian Journal of Agronomy* 46, 332–338.

Chaudhri, K.G. and Das, D.K. (1978) Emergence of soybean, cotton and maize seedlings as influenced by crust formation in salt-affected soils. *Journal of the Indian Society of Soil Science* 26, 106–111.

Chauhan, G.S. and Joshi, O.P. (2005) Soybean (*Glycine max*) – the 21st century crop. *Indian Journal of Agricultural Sciences* 75, 461–469.

Christmas, E.P. (2008) Plant populations and seeding rates for soybeans. Agronomy Guide, Purdue University, Cooperative Extension Service, West Lafayette, Indiana.

Cirak, C., Odabas, M.S., Kevseroglu, K., Karaca, E. and Gulumser, A. (2006) Response of soybean (*Glycine max*) to soil and foliar applied boron at different rates. *Indian Journal of Agricultural Sciences* 76, 603–606.

Clay, S.A., Kleinjan, J., Clay, D.E., Forcella, F. and Batchelor, W. (2005) Growth and fecundity of several weed species in corn and soybean. *Agronomy Journal* 97, 294–302.

De Bruin, J.L. and Pedersen, P. (2008a) Soybean cultivar and planting date response to soil fumigation. *Agronomy Journal* 100, 965–970.

De Bruin, J.L. and Pedersen, P. (2008b) Soybean seed yield response to planting date and seeding rate in the Upper Midwest. *Agronomy Journal* 100, 696–703.

De Bruin, J.L. and Pedersen, P. (2008c) Effect of row spacing and seeding rate on soybean yield. *Agronomy Journal* 100, 704–710.

Egli, D.B. and Bruening, W.P. (2000) Potential of early-maturing soybean cultivars in late plantings. *Agronomy Journal* 92, 532–537.

Freeborn, J.R., Holshouser, D.L., Alley, M.M., Powell, N.L. and Orcutt, D.M. (2001) Soybean yield response to reproductive stage soil-applied nitrogen and foliar-applied boron. *Agronomy Journal* 93, 1200–1209.

Gautam, P., Agnihotri, A.K. and Pant, L.M. (2003) Effect of phosphorus rate and *Pseudomonas* species in combination with *Bradyrhizobium japonicum* and farmyard manure on seed yield and yield attributes of soybean (*Glycine max*). *Indian Journal of Agricultural Sciences* 73, 426–428.

Hairston, J.E., Jones, W.F., McConnaughey, P.K., Marshall, L.K. and Gill, K.B. (1990) Tillage and fertilizer management effects on soybean growth and yield on three Mississippi soils. *Journal of Production Agriculture* 3, 317–323.

Halvankar, G.B., Varghese, P., Taware, S.P. and Raut, V.M. (1999) Influence of planting geometry and variety on seed yield and related parameters in soybean (*Glycine max*). *Indian Journal of Agronomy* 44, 601–604.

Heatherly, L.G. and Spurlock, S.R. (2000) Furrow and flood irrigation of early-planted, early-maturing soybean rotated with rice. *Agronomy Journal* 92, 785–791.

Holshouser, D.L. and Whittaker, J.P. (2002) Plant population and row-spacing effects on early soybean production systems in the Mid-Atlantic USA. *Agronomy Journal* 94, 603–611.

Hundal, S.S., Singh, H., Kaur, P. and Dhaliwal, L.K. (2003) Agroclimatic models for growth and yield of soybean (*Glycine max*). *Indian Journal of Agricultural Sciences* 73, 668–670.

Jain, V., Jain, V., Vishwakarma, S.K. and Sharma, R.S. (2005) Maximization of productivity of soybean (*Glycine max*)–wheat (*Triticum aestivum*) system in Kymore Plateau and Satpura Hill Zone of Madhya Pradesh. *Indian Journal of Agronomy* 50, 19–21.

Kewat, M.L. and Pandey, J. (2001) Effect of pre-emergence herbicides on weed control in soybean (*Glycine max*). *Indian Journal of Agronomy* 46, 327–331.

Kitoh, M. and Yoshida, S. (1996) Mulching effect of plant residues on soybean growth and physiochemical properties of soil. *Japanese Journal of Soil Science and Plant Nutrition* 67, 40–48.

Kratochvil, R.J., Pearce, J.T. and Harrison Jr, M.R. (2004) Row-spacing and seeding rate effects on glyphosate-resistant soybean for Mid-Atlantic production systems. *Agronomy Journal* 96, 1029–1038.

Lee, C.D., Egli, D.B. and TeKrony, D.M. (2008) Soybean response to plant population at early and late planting dates in the Mid-South. *Agronomy Journal* 100, 971–976.

Mehta, A.P. and Prihar, S.S. (1973) Seedling emergence in soybean and cotton as affected by seed-bed characteristics and surface mulches. *Indian Journal of Agricultural Sciences* 43, 45–49.

Mickelson, J.A. and Renner, K.A. (1997) Weed control using reduced rates of post-emergence herbicides in narrow and wide row soybean. *Journal of Production Agriculture* 10, 431–437.

Mulugeta, D. and Boerboom, C.M. (2000) Critical time of weed removal in glyphosate-resistant *Glycine max*. *Weed Science* 48, 35–42.

Murthy, V.R., Satyanarayana, V. and Mohammad, S. (2001) Impact of seeding time on yield components and yield of soybean (*Glycine max*) genotypes in Southern Talengana zone of Andhra Pradesh. *Indian Journal of Agricultural Sciences* 46, 314–318.

Nalawadi, U.G., Raj, P. and Krishnamurthy, K. (1973) Improvement in the seed germination of soybean varieties by pre-soaking treatments. *Indian Journal of Agricultural Sciences* 43, 546–550.

Nelson, K.A. and Renner, K.A. (1998) Weed control in wide- and narrow-row soybean (*Glycine max*) with imazamox, imazethapyr and CGA-277476 plus quizalofop. *Weed Technology* 12, 137–144.

Nelson, K.A. and Renner, K.A. (2001) Soybean growth and development as affected by glyphosate and postemergence herbicide tank mixtures. *Agronomy Journal* 93, 428–434.

Nimje, P.M. (2003) Effect of phosphorus fertilization on soybean (*Glycine max*)-based cropping sequences under rainfed conditions. *Indian Journal of Agricultural Sciences* 73, 191–193.

Norsworthy, J.K. and Frederick, J.R. (2002) Reduced seeding rate for glyphosate-resistant, drilled soybean on the Southeastern Coastal Plain. *Agronomy Journal* 94, 1282–1288.

Norsworthy, J.K. and Shipe, E.R. (2005) Effect of row spacings and soybean genotype on mainstem and branch yield. *Agronomy Journal* 97, 919–923.

Oplinger, E.S. and Philbrook, B.D. (1992) Soybean planting date, row width and seeding rate response in three tillage systems. *Journal of Production Agriculture* 5, 94–99.

Pandya, N., Chouhan, G.S. and Nepalia, V. (2005) Influence of integrated weed management on yield and economic viability of soybean (*Glycine max*) varieties grown at different crop geometries. *Indian Journal of Agricultural Sciences* 75, 510–512.

Park, E., Choi, Y.S., Jeong, J.Y. and Lee, S.S. (1999) Effect of priming on germination of aged soybean seeds. *Korean Journal of Crop Science* 44, 74–77.

Parsch, L.D., Keisling, T.C., Sauer, P.A., Oliver, L.R. and Crabtree, N.S. (2001) Economic analysis of conservation and conventional tillage cropping systems on clayey soil in eastern Arkansas. *Agronomy Journal* 93, 1296–1304.

PAU (2009) *Package of Practices for Kharif Crops of Punjab*. Punjab Agricultural University, Ludhiana, India.

Pedersen, P. and Lauer, J.G. (2003) Soybean agronomic response to management system in the Upper Midwest. *Agronomy Journal* 95, 1146–1151.

Pedersen, P. and Lauer, J.G. (2004) Response of soybean yield components to management system and planting date. *Agronomy Journal* 96, 1372–1381.

Perez-Bidegain, M., Cruse, R.M. and Ciha, A. (2007) Tillage system by planting date interaction effects on corn and soybean yield. *Agronomy Journal* 99, 630–636.

Pline, W.A., Wilcut, J.W. and Edmisten, K.L. (2002) Postemergence weed control in soybean (*Glycine max*) with cloransulam-methyl and diphenyl ether tank-mixtures. *Weed Technology* 16, 737–742.

Popp, M.P., Oliver, L.R., Dillon, C.R., Keisling, T.C. and Manning, P.M. (2000) Evaluation of seedbed preparation, planting method and herbicide alternatives for dryland soybean production. *Agronomy Journal* 92, 1149–1155.

Prakash, V., Bhattacharaya, R. and Srivastava, A.K. (2004) Effect of tillage management on yield and soil properties under soybean (*Glycine max*)-based cropping system in mid-hills of north-western Himalayas. *Indian Journal of Agricultural Sciences* 74, 573–577.

Purcell, L.C., Ball, R.A., Reaper, J.D. and Vories, E.D. (2002) Radiation use efficiency and biomass production in soybean at different plant population densities. *Crop Science* 42, 172–177.

Ramesh, P. and Reddy, K.S. (2004) Productivity and nutrient balance of soybean (*Glycine max*)-based cropping systems as influenced by nitrogen levels in deep Vertisols of central India. *Indian Journal of Agronomy* 49, 140–142.

Ray, C.L., Shipe, E.R. and Bridges, W.C. (2008) Planting date influence on soybean agronomic traits and seed composition in modified fatty acid breeding lines. *Crop Science* 48, 181–188.

Reddy, K.N. (2000) Weed control in soybean (*Glycine max*) with cloransulam and diclosulam. *Weed Technology* 14, 293–297.

Salvagiotti, F., Cassman, K.G., Specht, J.E., Walters, D.T., Weiss, A. and Dobermann, A. (2008) Nitrogen uptake, fixation and response to fertilizer N in soybeans: A review. *Field Crops Research* 108, 1–13.

Schmitt, M.A., Lamb, J.A., Randall, G.W., Orf, J.H. and Rehm, G.W. (2001) In-season fertilizer nitrogen applications for soybean in Minnesota. *Agronomy Journal* 93, 983–988.

Sidhu, H.S., Manpreet-Singh, Humphreys, E., Yadvinder-Singh, Balwinder-Singh, Dhillon, S.S., Blackwell, J., Bector, V., Malkeet-Singh and Sarbjeet-Singh (2007) The Happy Seeder enables direct drilling of wheat into rice stubble. *Australian Journal of Experimental Agriculture* 47, 844–854.

Singh, G. (2005) Effect of chemical and mechanical methods on weed management, growth and grain yield of soybean [*Glycine max* (L.) Merrill]. *Indian Journal of Weed Science* 37, 131–132.

Singh, G. (2007) Integrated weed management in soybean (*Glycine max*). *Indian Journal of Agricultural Sciences* 77, 675–676.
Singh, G. and Jolly, R.S. (2004a) Effect of herbicides on the weed infestation and grain yield of soybean (*Glycine max*). *Acta Agronomica Hungarica* 52, 199–203.
Singh, G. and Jolly, R.S. (2004b) Effect of time of sowing on the productivity of soybean genotypes. In: Moscardi, F., Hoffmann-Campo, C.B., Saraiva, O.F., Galerani, P.R., Krzyzanowski, F.C. and Carrão-Panizzi, M.C. (eds) *Abstracts of Contributed Papers and Posters of VII World Soybean Research Conference and VI International Soybean Processing and Utilization Conference*. Foz do Iguassu, PR, Brazil, p. 189.
Singh, G. and Jolly, R.S. (2008) Improving emergence of soybean (*Glycine max*) with straw mulch and other practices under normal and crusted soil conditions. *Acta Agronomica Hungarica* 56, 275–283.
Singh, G. and Jolly, R.S. (2009) Integrated weed management in soybean (*Glycine max*). In: Qiu, L., Guan, R., Jin, J., Song, Q., Guo, S., Li, W., Wang, Y., Han, T., Liu, X., Yu, D., Jiang, L. and Peng, D. (eds) *Developing a Global Soy Blueprint for a Safe Secure and Sustainable Supply. Abstracts for oral presentations and posters of the World Soybean Research Conference VIII*, Beijing, China, 10–15 August 2009. The Chinese Academy of Agricultural Sciences, Beijing, China, p. 34.
Singh, R. and Rai, R.K. (2004) Yield attributes, yield and quality of soybean (*Glycine max*) as influenced by nutrient management. *Indian Journal of Agronomy* 49, 271–274.
Singh, S., Singh, K. and Kler, D.S. (2000) Influence of planting time and plant geometry/density on periodic dry matter accumulation and seed yield of soybean. *Crop Research* 20, 76–80.
Singh, V.P., Dixit, A., Mishra, J.S. and Yaduraju, N.T. (2004) Effect of period of soil solarization and weed-control measures on weed growth and productivity of soybean (*Glycine max*). *Indian Journal of Agricultural Sciences* 74, 324–328.
Vyas, M.D. and Jain, A.K. (2003) Effect of pre- and post-emergence herbicides on weed control and productivity of soybean (*Glycine max*). *Indian Journal of Agronomy* 48, 309–311.
Wadud, S. and Kosar, R. (1997) Studies on the germinating of soybean. *Sarhad Journal of Agriculture* 13, 533–540.
Woodstock, L.W. and Taylorson, R.B. (1981) Soaking injury and its reversal with polyethylene glycol in relation to respiratory metabolism in high and low vigour soybean seeds. *Physiologia Plantarum* 53, 263–268.
Yadav, R.P., Shrivastava, U.K. and Dwivedi, S.C. (1999) Effect of seed rate and weed-control methods on yield of soybean (*Glycine max*) under rainfed condition. *Indian Journal of Agronomy* 44, 795–799.

8 Nutrient Management in Soybean

A. Subba Rao and K. Sammi Reddy
Indian Institute of Soil Science, Nabi Bagh, Bhopal, Madhya Pradesh, India

8.1 Introduction

Soybean (*Glycine max* (L.) Merrill), the 'golden bean', is an important crop in the world in terms of its use in human food and cattle feed. Soybean seeds contain about 40% protein and 21% oil on a dry weight basis. Soybean protein contains many essential amino acids, for both human and animals, mainly lysine, tryptophan, methionine and cystine. Lecithin, extracted from soybean oil, is used for everything from pharmaceuticals to protective coatings. It is a natural emulsifier and lubricant. Soybean is the best and the cheapest source of protein for human beings and animals. Soybean oil finds its way into products such as margarine, salad dressings and cooking oils. The soybean is the best natural source of dietary fibre.

Four major soybean producers – the USA, Brazil, China and Argentina – account for 90–95% of the global soybean production. In 2007, the productivity of soybean varied from 82 kg ha^{-1} in Tajikistan to 1235 kg ha^{-1} in India, 3535 kg ha^{-1} in Turkey and 7368 kg ha^{-1} in Georgia (FAO, 2009). Productivity of 3.3–3.6 t ha^{-1} in Turkey, Egypt and Italy was higher compared to productivity of 2.8–2.9 t ha^{-1} in three major soybean-producing countries (the USA, Brazil and Argentina).

Even though the area under soybean in India has increased rapidly from 0.03 million ha in 1970 to 2.6 million ha in 1990 and 5.7 million ha in 2000, its productivity is very low. Increased soybean production in India is mainly driven by an increased area sown to the crop. In 2008, the soybean production was 10.9 million t. However, the average productivity in India has hovered around 1 t ha^{-1}, which is very low compared to the potential yield of 2.5 t ha^{-1}.

The major reasons for low productivity of soybean are: (i) erratic rainfall distribution, late arrival and early withdrawal of monsoons; (ii) prolonged waterlogging in soybean fields due to heavy rains, particularly during the early stages of plant growth; (iii) insect pests and diseases; (iv) emerging multinutrient deficiencies (e.g. of nitrogen, phosphorus, sulphur, zinc, iron, boron) with the application of only nitrogen and phosphorus to major crops by farmers, and that often at lower rates than recommended; (v) sulphur deficiency due to farmers' preference for diammonium phosphate (DAP) as source of phosphorus rather than single superphosphate (SSP); and (vi) low use of rhizobial inoculants.

The productivity of soybean in major countries has already reached a plateau. Therefore, it is necessary to improve the productivity of soybean in second-rung countries and sustain higher productivity in major soybean-producing countries to increase the overall production of soybean. This can be achieved through efficient nutrient management techniques.

8.2 Soil and Climate Requirements

Soybean can be grown on a wide variety of well-drained soils, but thrives best on clay loam soils. The crop is better adapted for production on clay than either corn or cotton. It is also suited for production on muck. Soybean prefers a slightly acid soil (pH 6.0–6.5) (McLean and Brown, 1984). However, it grows quite well on calcareous clay soils (pH 7.5) if the free lime level is not too high. In India, 80% of soybean is grown on medium to deep black soils with a pH from 7.5 to 8.2. Soybean is rated as a moderately salt-tolerant crop and the reported salinity threshold is about 5 dS m^{-1} (Maas, 1986). Although soybean is classified as the warm-season crop, its cultivation now extends from the tropics to 52°N. The major commercial production of soybean is between 25°N and 45°N latitude and at altitudes of <1000 m (Fageria *et al.*, 1991). Soybean is a temperature-sensitive crop and is usually grown in environments with temperatures between 10°C and 40°C during the growing season (Whigham, 1983). It is a short-day plant, but cultivars may differ with respect to the maximum dark period required to induce flowering.

8.3 Nutrient Requirements

The mineral elemental composition of soybean plants can vary considerably according to soil fertility and is affected by disequilibrium between nutrients in the soil. Under optimal conditions, however, the plants show a fairly uniform composition regardless of region. Carbon, hydrogen and oxygen from the air make up 90% of dry matter production. However, these cannot be assimilated unless the other major and minor elements are present in the soil in sufficient quantity. In decreasing order of importance, these essential elements are nitrogen, potassium, calcium, magnesium, phosphorus and sulphur. Per hectare, a soybean crop yielding 2.5 t seed removes about 125 kg nitrogen, 23 kg phosphorus, 101 kg potassium, 22 kg sulphur, 35 kg calcium, 19 kg magnesium, 192 g zinc, 866 g iron, 208 g manganese and 74 g copper from the soil (Pasricha and Tandon, 1989; Tandon, 1989).

The nutrient requirements of a crop vary according to soil and climatic conditions, cultivar, yield level, cropping system and management practices. Soybean can fix atmospheric nitrogen if the proper *Bradyrhizobium* bacteria are present in the soil or if the seed is properly inoculated. The plants start to fix substantial amounts of atmospheric nitrogen approximately 4 weeks after germination. Most estimates show that soybean derives between 25% and 75% of its nitrogen from fixation (Deibert *et al.*, 1979).

Deficient, sufficient and high concentrations of nutrients for upper fully developed trifoliate of soybean prior to pod set have been compiled by Fageria *et al.* (1991) from Small and Ohlrogge (1973) and Rosolem (1980). The respective concentrations are 40, 45–55 and 56–70 g kg^{-1} for nitrogen, 1.5, 2.6–5 and 6–8 g kg^{-1} for phosphorus, 12.5, 17–25 and 26–28 g kg^{-1} for potassium, 2.0, 3.6–20 and 21–30 g kg^{-1} for calcium and 1.0, 2.6–10 and 11–15 g kg^{-1} for magnesium. Similarly deficient, sufficient and high concentrations, respectively, for micronutrients are 30, 51–350 and 351–500 mg kg^{-1} for iron, 14, 21–100 and 101–250 mg kg^{-1} for manganese, 10, 21–50 and 51–75 mg kg^{-1} for zinc, 10, 21–55 and 56–80 mg kg^{-1} for boron, 4, 10–30 and 31–50 mg kg^{-1} for copper and 0.4, 1–5 and 6–10 mg kg^{-1} for molybdenum. These data provide some guidelines for understanding the mineral requirements of the soybean crop.

Hanway and Weber (1971) studied the relative uptake of nitrogen, phosphorus and potassium by indeterminate soybean under field conditions. The total accumulation of nitrogen, phosphorus and potassium in the plants followed a pattern similar to that of dry matter accumulation. Accumulation was slow early in the growth stage but then became rapid, and nutrients accumulated at constant daily rates at between 54 and 100 days after emergence. Approximately 80% of the total accumulation of these nutrients occurred during the 46-day period from 54 to 100 days after emergence. The order of concentration of macronutrients in stems and leaves, with the highest first, is nitrogen, potassium, calcium, phosphorus, magnesium and sulphur, while the order of concentration in the seeds is nitrogen, potassium, phosphorus, sulphur, calcium and magnesium. When the mean nutrient uptake is compared with the mean nutrient removal in seeds, 80% of absorbed phosphorus, 78% of absorbed nitrogen and only 53% of absorbed potassium is removed with the seeds. The remaining phosphorus, nitrogen and potassium are returned to the soil as stems and leaves and are recycled when the soil organic matter is mineralized. Although micronutrients are required in smaller quantities than macronutrients, they are essential in soybean nutrition and removal in seeds as a percentage of nutrients in tops is also high. The order, in descending order of importance, is molybdenum, zinc, copper, chlorine, manganese, boron and iron.

Nitrogen is required in the greatest quantity of all plant nutrients absorbed from soil. It is present in all amino acids, which are the building blocks of protein, nucleic acids and chlorophyll (Jones *et al.*, 1991). Soybean plants can use nitrogen released by mineralization, residual soil nitrogen, fertilizer nitrogen or atmospheric nitrogen, which is converted into a usable form in root nodules through a symbiotic relationship between *Bradyrhizobium japonicum* bacteria and the soybean plant. While the soil is the primary source of nitrogen for many crops, soybean obtains 65–85% of its needs through the symbiotic nitrogen fixation process. A high rate of nitrogen fertilizer suppresses nitrogen fixation and most specialists recommend either no fertilizer nitrogen or a modest application of 30–50 kg ha^{-1} either at sowing or just before flowering. Some researchers have noted a favourable effect of nitrogen applied at the time of sowing on nitrogen fixation, root nodule weight and activity (Eaglesham *et al.*, 1983).

Phosphorus, although required in far lower quantities than either nitrogen or potassium, is critical to rapid growth and proper development. The most noteworthy functions of phosphorus in plants include energy storage and transfer, membrane function and genetic transfer (Marschner, 1995). Phosphorus is critical for root and plant canopy development and seed production. The phosphorus content of harvested soybean seed is approximately 0.50–0.58%. Therefore, the maintenance of available soil phosphorus requires at least this range to be returned to the soil each year. A minimal tissue phosphorus level at flowering of ≥0.31% has been suggested by Bell *et al.* (1995). A range of 0.26–0.50% for the most recently matured leaves prior to pod set is considered sufficient (Jones *et al.*, 1991). Sammi Reddy *et al.* (1997) found 0.25–0.26% phosphorus in the third fully opened trifoliate leaf at 50% flowering stage to be the optimum concentration in the most commonly grown soybean variety (JS 335) in India.

The crop takes up about 125 kg ha^{-1} potassium. Potassium is particularly important in plant physiology – it is very mobile and is involved in the transport of assimilates, the activation of many enzymes, the water economy of the plant and photosynthesis. It favours the formation of nodules and hence nitrogen fixation. It improves disease and stress tolerance. It has a large effect on yield, increasing grain weight and protein content, although it slightly decreases the oil content.

Soybean requires a large amount of calcium, taking up 50–90 kg ha^{-1}, but only 20% of this is removed in the grain. Calcium has a beneficial effect on nodulation, either directly or through improvement of soil pH; it is difficult to distinguish between its direct and indirect effects. The necessity of calcium for plant growth can easily be demonstrated by interrupting calcium supply to the roots. The growth rate is immediately reduced, and after some days the root tips become brown and gradually die (Mengel and Kirkby, 1987). Magnesium is the central element in the chlorophyll molecule and is a co-factor in the activation of many enzymes. It has multiple functions. It is needed for photosynthesis and CO_2 assimilation is restricted when the calcium content falls too low. It plays an important part in plant symbiotic nitrogen fixation.

Sulphur is needed for the synthesis of certain amino acids and thus in the formation of proteins. It is involved in the formation of chlorophyll. Soybean plants use nearly as much sulphur as phosphorus or magnesium and the removal by seeds may be 27–66% of that absorbed by stems and leaves. The sulphur concentration of the third fully opened trifoliate leaf of soybean collected at 50% flowering has been found to range from 0.15% to

0.47% and showed positive relationship with Bray's percentage yield. Using scatter plots, a concentration of 0.34% sulphur has been established as the critical concentration (A.N. Ganeshamurthy, Bangalore, 2008, personal communication).

Micronutrients are absorbed in smaller quantities by the soybean plant than are nitrogen, phosphorus, potassium and, sometimes, calcium, magnesium and sulphur. Their role is equally as important, however, and deficiencies of micronutrients lead to severely depressed growth and yield. Zinc activates several enzymes and is involved in nitrogen metabolism in the plant and the formation of protein. Iron is an essential constituent of chlorophyll and necessary for respiration and photosynthesis. Manganese has important roles in the metabolic processes, such as activating enzymes, chlorophyll synthesis, photosynthesis and nitrate reduction. Copper plays an important role in chloroplast functioning and improving photosynthesis. Its deficiency can reduce growth and yield by reducing the rate of photosynthesis. Molybdenum is required for the activity of two important enzymes, nitrate reductase and nitrogenase, which are essential for nitrate reduction and atmospheric nitrogen fixation. It is also needed for the proper functioning of root nodules and nitrogen assimilation; deficiency in the field looks similar to a nitrogen deficiency. Boron is required in meristematic activity, and hence in the growth of shoot tips and roots and of the floral organs.

Cobalt is the other beneficial element required in addition to micronutrients. It is essential for efficient nitrogen fixation. Cobalt has beneficial effects on nodule number and weight and on plant nitrogen content when it is supplemented through soil or foliar application.

8.4 Symptoms of Nutrient Deficiencies and Toxicities

Plants that are deficient in nitrogen fade in colour and the leaves become pale green with a yellowish tinge. The leaves may later become distinctly yellow over their entire surface (Borkert and Sfredo, 1994). This symptom usually appears first on the bottom leaves, but spreads quickly to upper parts. The symptoms appear last in the younger leaves since nitrogen, being highly mobile in the plant, is translocated from older tissues to the younger leaves. The plant's growth rate is reduced.

Phosphorus-deficiency symptoms are not well defined. The main symptoms are a retarded growth rate and spindly plants with small, dark-green or bluish-green leaflets. Because of the high mobility of phosphorus in the plant, under deficient conditions the younger leaves deplete phosphorus from the older leaves, causing symptoms to appear on the older leaves first (Borkert and Sfredo, 1994).

In potassium-deficient plants, the leaf margins turn yellow first, with the yellowing spreading inwards. The centre and base of the leaf remain green. In serious cases the leaf margins die (Fauconnier, 1986). As a result of the mobility of potassium, these symptoms generally appear first on the more mature leaves. Extremely potassium-deficient soybean produces wrinkled, small, misshapen seeds and maturity is delayed.

Reduced growths of meristematic tissue at the stem, leaf and root tip are the major characteristics of calcium deficiency. Due to the immobility of calcium in the plant, deficiency symptoms generally first appear on the younger leaves and in the growing apexes (Borkert and Sfredo, 1994).

Magnesium deficiency in soybean appears as interveinal yellowing, at first on the older leaves. Yellowing is apparent first in the basal leaves and as the deficiency becomes more acute it eventually reaches the younger leaves. Magnesium-deficiency symptoms can be difficult to recognize, as they can be confused with those of potassium, iron or manganese deficiency (Fauconnier, 1986).

Young trifoliate leaves of sulphur-deficient soybean plants first develop paling, then turn severely chlorotic. Chlorosis starts from the leaf margins and spreads towards the midrib. The margins and tips of young leaves become necrotic and rolled. Severe deficiency leads to premature defoliation and flowering and fruiting are reduced (Singh, 1999).

Chlorophyll production is severely reduced in iron-deficient plants. Its deficiency always begins in the younger leaves. In the initial stages of symptom development, the areas between soybean leaf veins turn yellow. As it becomes more severe, the veins also begin to turn yellow and finally the whole leaf turns almost white. Brown necrotic spots may occur near leaf edges (Borkert and Sfredo, 1994).

Zinc-deficient leaves and shoots are smaller than normal. Interveinal yellowing or browning is particularly seen on the lower leaves and this can lead to the death of the leaf tissue. Molybdenum-deficiency symptoms are similar to those of nitrogen deficiency because molybdenum is essential for nitrogen fixation and nitrate reduction. Deficiency of copper generally causes necrosis of the tips of young leaves. This proceeds along the margins of the leaves, resulting in a withered appearance. Soybean growth is reduced and the colour of the plant changes to greyish-green, bluish-green or olive green. High amounts of copper in the nutrient medium are toxic to growth for most plant species. This appears to affect, in part, the ability of copper to displace other cations, particularly iron, from physiologically important sites. Chlorosis is, therefore, a commonly observed symptom of copper toxicity, superficially resembling iron deficiency (Borkert and Sfredo, 1994).

Abnormal or retarded growth of the apical growing points is the major characteristic of boron deficiency. The youngest leaves are wrinkled, often thickened and of a darkish-blue-green colour. Irregular chlorosis between the intercostal veins may occur. The leaves and stems become brittle, indicating a disturbance in transpiration. As the deficiency progresses, internode elongation slows down, the terminal growing point dies and flower formation is restricted or inhibited. Soybean is sensitive to high boron in soils. Boron is toxic to soybean at levels only slightly above those required for normal growth. Toxicity is, however, more usually observed in arid and semi-arid regions, where boron levels are frequently high. The boron status of irrigation water is particularly important in these regions. The toxic effects of boron may result in leaf-tip yellowing followed by progressive necrosis. This begins at the tip and margins and finally spreads between the lateral veins towards the midrib. The leaves take on a scorched appearance and drop prematurely (Borkert and Sfredo, 1994).

Aluminium is important in many acidic soils because of its toxic effects on plants. The foliar symptoms of aluminium toxicity resemble those of phosphorus deficiency – there is an overall stunting, leaves are small and dark green, leaf tips yellow and die and maturity is delayed. Aluminium toxicity in soybean appears as an induced calcium deficiency or reduced calcium transport within the plant, causing curling and rolling young leaves and a collapse of growing points or petioles. Aluminium-injured soybean roots are characteristically stubby and brittle. Root tips and lateral roots become thick and may turn brown. The root system as a whole appears coralloid, with many stubbly lateral roots but little fine branching.

8.5 Nutritional Constraints of Soils Supporting Soybean-based Cropping Systems

The USA, Brazil, China and Argentina are the major soybean-producing countries. India ranks third after Argentina and Brazil to have registered a phenomenal growth in the production of soybean. In Brazil, soybean is mostly grown on oxisols, alfisols, ultisols and entisols. Out of these soybean-growing areas in Brazil, fertile alfisols constitute only 1.7% of the area (Borkert and Sfredo, 1994). Oxisols and ultisols together constitute 64% of the area. Entisols, mainly quartz sands of low fertility, are the third most predominant soils supporting soybean in Brazil. Oxisols are low in organic matter and deficient in calcium, magnesium and potassium. In some sandy areas, zinc and manganese deficiencies may occur. Deficiency of molybdenum is found in some oxisols. Both oxisols and ultisols are low in organic matter and nitrogen deficiency is a general constraint for crop production. Entisols derived from quartz materials are consistently deficient in potassium.

In India, soybean is mostly grown on vertisols and associated soils in central parts of the country. Vertisols, in general, are low to medium in organic matter content. Intensive cultivation with double cropping coupled with imbalanced fertilizer application by farmers (only nitrogen and phosphorus, and that at lower rates than recommended) has led to the emergence of multinutrient deficiencies in the soybean-growing areas of central India. A survey conducted in central India has shown that more than half of the soil samples are low (<0.5%) in organic carbon (Table 8.1). The total nitrogen of all the fields tested was low (<0.07% in the top 0–15 cm of soil). Almost all fields were low in potentially mineralizable nitrogen (<94 mg kg^{-1} soil). More than half of the fields were low in available phosphorus (<4.9 mg kg^{-1} soil). The available potassium concentration of the soils analysed was generally high (>56 mg kg^{-1} soil). About 46% of soil samples were found to be deficient in available sulphur. Overall, 58% of soil samples were

Table 8.1. Available nutrient status (mean ± standard deviation) in soils of farmers' fields in Rajgarh and Vidisha districts of Madhya Pradesh, India (N = 500) (reprinted with permission from Sammi Reddy *et al.*, 2007).

Parameter	Mean nutrient status	Soil samples deficient (%)
Soil organic carbon (%)	0.51 ± 0.06	50
Total nitrogen (%)	0.053 ± 0.007	100
Potentially available nitrogen (mg kg^{-1} soil)	94 ± 15.2	97
Available phosphorus (mg kg^{-1} soil)	7.4 ± 2.7	52
Available potassium (mg kg^{-1} soil)	163.8 ± 35.7	0
Available sulphur (mg kg^{-1} soil)	10.9 ± 5.0	46
Available zinc (mg kg^{-1} soil)	0.50 ± 0.26	58
Available iron (mg kg^{-1} soil)	4.87 ± 0.96	40
Available manganese (mg kg^{-1} soil)	15.3 ± 5.1	0
Available copper (mg kg^{-1} soil)	2.67 ± 0.74	0

low in available zinc. About 40% of soil samples were found to be deficient in available iron. All of the soils analysed contained sufficient levels of available manganese and copper.

Other specific nutritional constraints found in soybean-growing areas of the world are discussed briefly below:

- Nitrogen deficiency in soybean is rarely seen except when inoculation is omitted on soil not previously cropped with soybean or when the efficiency of nitrogen fixation is constrained by soil acidity or molybdenum deficiency. The virtual absence of nodules in such a case results in pale green or yellow foliage. Anything that hinders nodulation – nematodes, *Fusarium* or, rarely, calcium deficiency – can cause nitrogen deficiency.
- Phosphorus deficiency may occur in almost all acidic tropical soils with low pH and high phosphorus-fixation capacity. In addition to the above, a limited supply of phosphorus reduces the number as well as the efficiency of nitrogen-fixing bacteria. High levels of phosphorus can induce zinc deficiency since high uptake rates of phosphorus are associated with the reduced uptake and translocation of zinc, iron and copper.
- Calcium deficiency is observed on soybean growth on acidic tropical soils. However, the symptoms probably result from a combination of calcium deficiency and aluminium and manganese toxicity (Borkert and Sfredo, 1994). Magnesium deficiency is most likely to occur on sandy, acidic tropical soils low in organic matter, but the application of dolomitic limestone prevents deficiencies of both elements.
- Sulphur deficiencies are most likely on coarse-textured acidic tropical soils that are low in organic matter. Manganese deficiency has been observed in soils high in iron and/or aluminium and in sandy oxisols that have been overlimed (Borkert and Sfredo, 1994). Additions of lime and phosphorus may reduce zinc availability and cause deficiency. Zinc deficiency is most common in regions of limited rainfall where the

surface soil has been partly or entirely removed by erosion or land levelling.

- Iron deficiency or 'iron chlorosis' in soybean is most common on soils with an alkaline pH (>7.0) containing calcium carbonate (Varco, 1999). Thus, the term 'lime-induced chlorosis' is sometimes used to describe this condition. With increasing pH and the presence of calcium carbonate, the availability of iron is drastically reduced. Soybean varieties differ in their susceptibility to iron deficiency, and iron absorption may be genetically controlled. It appears that tolerant varieties increase the availability of iron in the root zone through root secretion of various compounds (Brown and Jolley, 1989).
- Generally, manganese availability decreases as soil pH increases. Manganese deficiencies can be induced on calcareous soils when chelated iron is applied to soybean, especially during early growth when temperatures are cool (Moraghan, 1985).
- Zinc deficiency can be a problem on high pH or alkaline soils that are low in organic matter. Problems with zinc deficiency can also arise where the tropical soil depth is reduced by erosion or land-forming operations since most available zinc is associated with organic matter concentrated in the surface soil. Excessive available phosphorus can also induce zinc deficiency.
- Lability of molybdenum decreases with decreasing soil pH or increasing acidity. This is one of the primary factors involved in determining pH or liming requirements for soybean.

Effect of liming

By nature, soils become more acidic with time and degree of weathering. Minerals within the soil that buffer acidity can eventually become depleted. Problems arise for soybean when the soil pH declines to <6.0. The availability of some essential nutrients decreases, while the availability of other elements increases to the point of toxicity.

Tropical and temperate acid soils should undoubtedly be limed to produce high soybean yields. In India, Kalia *et al.* (1984) obtained a significant increase in soybean grain and plant residue yields with liming as compared to without liming during two cropping seasons. Liming acid soils decreases toxic levels of aluminium and manganese, increases the availability of nutrients such as phosphorus, calcium, magnesium and molybdenum, and increases microbial activity, most notably nitrogen-fixing bacteria. Thus, the soybean response to liming can be due to numerous factors.

Liming acid soils is very important for improving nodulation as there is no nodulation below pH 5. Liming also helps in improving phosphate utilization by soybean in leached tropical and subtropical soils through the elimination of phosphorus fixation by iron and aluminium, and eliminating aluminium and manganese toxicity. Liming also interacts with potassium and boron; several crops have shown a limited response to applied potassium or boron in the absence of lime, but spectacular responses after liming. A large number of trials in Brazil have shown spectacular effects of liming at 3 to >10 t ha^{-1} and the effects persisted over several years (Fauconnier, 1986). The choice of liming material is often dictated by proximity to supplies and rates of application should be calculated from the neutralizing value of the material available.

In developing countries such as India, farmers are reluctant to apply large quantities of liming materials for reclamation due to economic reasons. Hence, Rattan (2007) suggested ameliorating the acid soil with minimum quantities of lime with the application of all other macro- and micronutrients at recommended rates in a balanced way so that farmers can achieve reasonably good yields of soybean on acid soils. In studies conducted in seven states of India, the application of lime at 200–300 kg ha^{-1} improved crop yields by 16–48%. Application of half of the recommended rate of nitrogen, phosphorus and potassium (NPK) with lime was at par with or superior to the full dose of NPK without lime (Sharma and Sarkar, 2005).

8.6 Efficient Use of Applied Nutrients and Biological Systems

Nitrogen

Nitrogen nutrition of leguminous crops presents a complex and somewhat paradoxical problem to scientists. Nitrogen fertilization of these crops is uncommon and the comparatively yield gains are small. Unpredictability in soybean yield response to fertilizer nitrogen is likely related to many factors. Some of these include nitrogen fertilization repressing nitrogen fixation, variability in soil nitrogen-supplying capacity, soil water availability and environmental conditions in general. Supplementation of the symbiotically fixed nitrogen by nitrogen fertilizers has been considered as a means of increasing plant growth. Research has shown that the process of symbiotic nitrogen fixation does not occur efficiently if large quantities of nitrogen fertilizers are applied to legumes. It has, however, been observed that applications of low levels of starter nitrogen may circumvent the 'nitrogen hunger' stage of seedling growth to improve early root and leaf development, which, in turn, stimulates subsequent nodule formation and dinitrogen fixation. Bacterial invasion of the root cortex and initial nodule development may require about 20–25 days. Singh *et al.* (1998) recommended the application of moderate rates of 35 kg N ha^{-1} as a starter dose for soybean on nitrogen-deficient soils.

Phosphorus

Soybean response to applied phosphorus depends on soil acidity, soil organic matter level and clay content. Clay content affects the interpretation of soil test values obtained by extraction, and values for clay soils will likely be very different from those for sandy soils. Therefore, phosphorus fertilizer recommendations will depend on soil texture. In Indian black soils with a clay content of around ≥40%, soybean has been found to respond significantly when the Olsen's phosphorus in the soil is <7.3 mg P kg^{-1} soil (Subba Rao and Ganeshamurthy, 1994). Field studies conducted in this area for assessing the response of soybean to phosphorus application revealed that soybean yields increased with the application of 26 kg P ha^{-1}. The higher rates of phosphorus at 39 and 52 kg ha^{-1} were on the yield plateau region and did not further enhance yields (Subba Rao *et al.*, 1997). In Brazil, for very low, low, medium and high phosphorus soils, the recommendations are 35–43, 26–43, 17–30 and 13–21 kg P ha^{-1}, respectively (Fauconnier, 1986). In Mediterranean or semi-arid regions, soils are not leached and are more often alkaline than acid. At a rainfall of <700 mm, soybean can be grown under irrigation. Soluble phosphorus fertilizers should be used on these soils. In the absence of soil analysis to specify the rate more closely, the recommendation is for 26–43 kg P ha^{-1}. Sammi Reddy *et al.* (1997) developed critical limits of soil solution (external) phosphorus concentration for soybean grown in central India. A soil solution concentration of 0.10 mg P l^{-1} in solution was found to be critical for soybean in vertisols. In temperate regions, soils have usually been under cultivation for a long time and many farmers are fully conversant with the behaviour of their soil in relation to phosphorus. Soluble fertilizers such as SSP, triple superphosphate, monoammonium phosphate and DAP are suitable for soybean on neutral or nearly neutral soils. Phosphorus should be applied at 32–43 kg ha^{-1} (Fauconnier, 1986).

Considerable importance is attached to better application methods for increasing the efficiency of phosphorus. Much research has been conducted to compare drilling or placement of phosphorus to surface broadcast and to assess the feasibility of drilling with seed. For higher phosphorus-use efficiency in grain legumes, the results are overwhelmingly in favour of drilling/placement of phosphorus below the soil surface and into the root zone. In general, it is advised to place phosphate fertilizer a few centimetres to the side of and below the seed in order to improve phosphorus fertilizer efficiency. This is applicable in non-intensive cropping at low yield levels, but, in order to obtain a vigorous root system resistant to drought, it is preferable to raise the phosphorus content of the whole plough layer by broadcast application before cultivating. Field trials conducted in Brazil on very low phosphorus soil (<1 ppm by Mehlich's method) showed that the initial broadcast heavy dressings (86 kg P ha^{-1} was broadcast in the first year, then 22 kg P ha^{-1} was placed) were more effective than placement of the smaller dressing (22 kg P ha^{-1} per annum placed). The residual value of the initial dressing was of the order of 50% in the second year and 30% in the third year, reducing thereafter towards zero in subsequent years (Fauconnier, 1986).

Traditional phosphorus fertilizers (triple superphosphate, SSP, monoammonium phosphate and DAP) have a high phosphorus content and high water solubility. In the last decade, however, other phosphorus fertilizers have become available, including partially acidulated phosphate rock and calcined phosphate rock, known in Brazil as *termofosfato*. A finely ground rock phosphate heated to about 1000°C, termofosfato is a good option for the corrective fertilization of acidic tropical soils because of its low cost compared to the highly soluble phosphates. Other phosphatic fertilizers that are soluble in 2% citric acid (e.g. slag, dicalcium phosphate and thermophosphates) or even finely ground rock phosphate are often as effective as soluble phosphates on these acidic soils, but there is variability in the efficiency of rock phosphate according to source. SSP is especially effective because of its sulphur content.

Continuous application of phosphatic fertilizers to soybean may lead to a built-up of available phosphorus in the soil, which reduces the subsequent rate of phosphorus application. Several studies have been conducted to evaluate the response of soybean and other crops to residual phosphorus in the soil (Bolland and Barrow, 1991; Subba Rao and Ganeshamurthy, 1994). Subba Rao *et al.* (1996) studied the residual effects of applied phosphorus to either soybean or wheat in a 3-year experiment on vertisol soils under soybean–wheat rotation. The results revealed that the application of 39 kg P ha^{-1} to soybean had a significant effect on the yields of two subsequent crops (wheat and soybean), whereas the same amount of phosphorus applied to wheat had a significant effect on only one subsequent crop (soybean). From this, the researchers developed a residual phosphorus management system in which the application of 39 kg P ha^{-1} either to soybean or wheat in a soybean–wheat rotation produced a statistically similar yield as the application of 26 kg P ha^{-1} to both soybean and wheat crops. This system saves about 13 kg P ha^{-1} $year^{-1}$ during a soybean–wheat rotation (Sammi Reddy *et al.*, 2003).

The response of soybean to phosphorus application depends upon the level of available phosphorus in the soil. It is necessary to apply maintenance dose of phosphorus that equals the amount of phosphorus removed by the previous crop. Soil phosphate maintenance fertilization in soybean–wheat rotation on a vertisol (available phosphorus 5.84 mg kg^{-1}) has shown that the application of phosphorus at a rate equivalent to phosphorus removal by each crop, through 5 t farmyard manure (FYM) plus 8 kg fertilizer P ha^{-1} or 10 t FYM ha^{-1} to soybean and 10 kg fertilizer P ha^{-1} to wheat is good enough to obtain the target of 2 t soybean and 4 t wheat yields ha^{-1} and helps to maintain phosphorus fertility at near the initial level (Table 8.2) (Reddy, 2007b).

Potassium

Response to potassium fertilization depends greatly on soil-available potassium and mineralogy, as well as on other factors limiting crop growth such as rainfall and the availability of other essential plant nutrients. Experiments in both tropical and temperate climates have shown that the crop responds well to potassium, both recently applied and as residues in the soil.

In tropical and subtropical regions, the dominant clay mineral in lateritic soils is kaolinite, hence soils are low in total cations as a result of leaching.

Table 8.2. Soil test maintenance phosphorus requirement and its relationship with crop yield and phosphorus removal (uptake under different phosphorus supply strategies (reprinted with permission from Reddy, 2007b).

PSS[a]	STMPR of soybean–wheat rotation (kg ha^{-1} $year^{-1}$)[b]	Yield levels of rotational crops at STMPR (mg ha^{-1})		Total annual phosphorus removal at STMPR (kg ha^{-1} $year^{-1}$)	STMPR to phosphorus removal ratio
		Soybean	Wheat		
PSS-I	36.1 (22.2 + 13.9)	1.91	4.10	25.2	1.4
PSS-II	26.3 (16.2 + 10.1)	1.86	4.06	23.4	1.1
PSS-III	24.1 (14.8 + 9.3)	1.90	4.01	23.7	1.0

PSS, phosphorus supply strategy; STMPR, soil test maintenance phosphorus requirement.
[a]PSS-I, PSS-II and PSS-III imply phosphorus supply through inorganic (fertilizer), organic (farmyard manure) and integrated (fertilizer + farmyard manure) sources, respectively, to soybean. Phosphorus supply to wheat was solely through fertilizer under all strategies.
[b]Figures in parentheses indicate the phosphorus rates for component crops of annual soybean–wheat rotation obtained by splitting STMPR in the same ratio of 1.6:1 as was used in the treatments for soybean and wheat.

The cation exchange capacity (CEC) varies from 1 to 7 me 100 g^{-1} soil. Potassium retention is very poor. Non-exchangeable reserves are also low and it is not possible to improve the potassium status of these soils, as can be done in temperate regions (Fauconnier, 1986). Following several years' fertilization with phosphorus alone, crops suffer from potassium deficiency and this reduces the efficiency of phosphate fertilizer. It is, therefore, important to restore equilibrium by applying potash (Rosolem, 1980). Similarly, long-term experiments conducted in India with the soybean–wheat system on silty clay loam soil have also shown that 21 years' continuous application of only recommended rates of nitrogen and phosphorus without potassium resulted in a significant decrease in the available potassium content of soil, which was even lower than in unfertilized plots (Mahapatra *et al.*, 2007).

Recommended rates of potassium for soybean vary according to the soil potassium content (Fauconnier, 1986). The recommended rate in Peru varies from 80 to 120 kg ha^{-1} for low-potassium soils to 20–40 kg ha^{-1} for highly fertile soils. In the subhumid tropics (e.g. India), black or red soils containing montmorillonite are not so severely leached. Although the soils are more or less clayey, their CEC is 15–60 me 100 g^{-1} and potassium saturation is very variable and often low at 1% to 3.7% (Brar *et al.*, 1986). Long-term experiments and soil investigations are needed to decide upon the rate of potassium-manuring required. In India, a nominal rate of 20–30 kg ha^{-1} is recommended on soils high in available potassium status to arrest potassium mining under intensive cropping. The soils of semi-arid or Mediterranean regions are not heavily leached and have a high CEC. A maintenance rate of 50–83 kg K ha^{-1} is required. In temperate regions, a regular application of 58–125 kg K ha^{-1} is recommended whether the soil is thought to be high or low in potassium and according to expected yield (Fauconnier, 1986).

While phosphate can sometimes be placed close to the seed without causing any trouble, the same does not apply to potash applied as potassium chloride, which adversely affects young plants. For this reason, some experiments on soils have shown no response where a positive response would have been expected. It is best to apply potash before cultivation so that the material is incorporated into the whole plough layer. In some tropical soils with low CEC, potassium is easily leached. By applying potassium fertilizer in the furrow instead of broadcasting it, leaching is reduced but the hazard of seed damage is increased.

At present, potassium chloride is the form of fertilizer most widely used, but with the wider occurrence of sulphur deficiency it is advisable to use sulphate of potash especially on all light-textured soils that are low in organic matter. If the soils are saline, sulphate of potash should be used rather than the muriate.

Potassium fertilization is also based on calibrated soil tests. On acidic tropical soils in Brazil, the critical point above which there is no yield response to potassium is a soil test value of between 50 and 80 mg kg^{-1}. As in the case of phosphorus, this critical point will depend on the texture of the soil. A broad description of critical levels and responses to potassium fertilization on highly weathered soils can be found in Vilela and Ritchey (1985).

Sulphur

Several factors lead to a decline in the sulphur status of soil over time. Plant-available sulphur is derived primarily from the decomposition of plant residues and soil organic matter. Sulphur deficiency is most common in soils that are inherently low in sulphur, have a sandy texture and are low in organic matter and soils that are prone to high leaching. Historically, when ordinary superphosphate – which contains 12% sulphur – was in common usage, sulphur was inadvertently applied to many soils. Kamprath and Jones (1986) summarized nitrogen fertility research in the southern USA and found that a response to fertilization occurred for only two of nine sites. A positive yield response occurred on soils with available sulphur levels of ≤4 ppm, while non-responsive soil had 8 ppm available sulphur and an accumulation of subsoil sulphur within 8 inches of the surface. Employing statistical and graphical methods, 11.2 mg S kg^{-1} soil was established as the critical limit in central India for black soils to isolate sulphur-deficient soils from the rest (Subba Rao and Ganeshamurthy, 1994). In tropical and subtropical soils, the continuous use of sulphur-free, high-analysis fertilizers in soybean-growing areas has led to sulphur deficiency. About 46% of soils in central India, where soybean is grown, are deficient in sulphur.

Sulphur extracted by 0.15% calcium chloride provides a better index of plant-available sulphur in some soils. Based on several studies, soils testing at <10 mg S kg^{-1} soil have been considered to be sulphur deficient. Depending upon the soil test value of sulphur, the optimum rate of application of sulphur to oilseed crops, including soybean, has been found to vary from

Table 8.3. Fertilizer sulphur recommendations based on the available-sulphur status of soils (reprinted with permission from Singh, 1999).

Available sulphur in soil (mg kg^{-1})	Sulphur fertility class	Amount of sulphur to be applied (kg ha^{-1})
<5	Very low	60
6–10	Low	45
11–15	Medium	30
16–20	High	15
>20	Very high	0

15 to 60 kg S ha^{-1} (Table 8.3) (Singh, 1999). Ganeshamurthy *et al.* (1994) conducted several trials on farmers' fields in black soil areas of central India to evaluate the soybean response to sulphur application. The response of soybean to applied sulphur (40 kg ha^{-1}) ranged from 50 to 575 kg ha^{-1} (i.e. the difference between the seed yield obtained with NPK and NPKS treatments), with an average of 277 kg ha^{-1} in sulphur-deficient soils.

Since crops have a greater sulphur requirement at early growth stages, its application may be made prior to sowing or bud initiation or flowering, preferably under moist conditions to ensure high availability for better crop yields. Rathore *et al.* (1995) evaluated the efficiency of different sulphur sources in improving the soybean yield on sulphur-deficient black soils of central India. Ammonium sulphate and SSP were found to be equally good sources of sulphur and were superior to gypsum.

As with phosphorus, the continuous application of sulphur fertilizers to different crop rotations leads to a build-up of soil sulphur status. Therefore, it is essential to utilize the residual sulphur left over in the soils during the next crop for higher efficiency. Singh and Saha (1997) reported the direct and residual effects of different rates of sulphur application on soybean yield in soybean–wheat rotation on black soils in the Madhya Pradesh and Maharashtra states of central India. They found that the application of 20 kg S ha^{-1} to both soybean and wheat or the application of 40 kg S ha^{-1} to either soybean or wheat was found to be adequate for the sulphur requirements of a soybean–wheat rotation. In a 3-year field investigation on a typical vertisol soil under soybean–wheat rotation, Ganeshamurthy and Takkar (1997) found that sulphur applied at 60 kg ha^{-1} to soybean showed residual effects in two succeeding crops (wheat and soybean), while the same applied to wheat showed residual effect in only one succeeding crop (soybean). In this system, therefore, sulphur applied to soybean was more efficiently utilized by the succeeding crops as compared to that applied to wheat.

Calcium and magnesium

Soybean calcium and magnesium needs can be met if dolomitic limestone is used to increase soil pH to ameliorate acid soils.

Micronutrients

Zinc and sometimes manganese, molybdenum or copper are most frequently required for soybean. Zinc, manganese and copper deficiencies have been effectively treated with the sulphate forms of these elements. Care must be taken to achieve uniform distribution; even spreading prevents adverse effects. Iron chlorosis or iron deficiency on high-pH soils is by no means rare. This can be prevented by dusting 100–160 g ha^{-1} iron chelate (DTPA or EDDHA) onto the soil along the rows. Spraying the foliage in early growth with iron sulphate, nitrate or chelate can also give good results (Fauconnier, 1986). In India, field studies have shown that the application of 5 kg Zn ha^{-1} as zinc sulphate to soybean was found adequate to meet the zinc requirement of both soybean and wheat crops in soybean–wheat rotation on black soils of central India (Sammi Reddy *et al.*, 2007). If deficiency is severe, the rate of zinc application may be increased to 10 kg ha^{-1}, which increases the yield benefits significantly. Zinc chelates are added to soils at 10–15 kg ha^{-1} to correct zinc deficiency, but chelates are very expensive compared to zinc sulphate. A basal application of zinc through broadcast is preferred. If soil application is missed, a spray of 0.5% zinc sulphate solution neutralized with 0.25% lime or 0.20% zinc chelate solution two to three times in 300–500 l ha^{-1} at 7- to 10-day intervals between 20 and 45 days after germination also corrects the zinc deficiency. In total, about 3–4 kg zinc chelate or 7–10 kg zinc sulphate ha^{-1} is required for 3–4 foliar sprays (M.V. Singh, Bhopal, 2008, personal communication).

On boron-deficient black soils of central India, an application of 4 kg borax ha^{-1} has been found to be effective for correcting boron deficiency (Acharya *et al.*, 2003). Boronated superphosphate was found to be the best source for soybean–wheat systems as it also supplies phosphorus and sulphur (Shinde *et al.*, 1990).

8.7 Balanced and Integrated Nutrient Management in Soybean-based Cropping Systems

In an era of multiple nutrient deficiencies, a single nutrient approach can lower fertilizer-use efficiency. Balanced nutrition implies that there are no deficiencies, excesses, antagonisms or negative interactions. All deficient nutrients must be at an optimum rate by themselves and in relation to each other, enabling positive interactions to enhance yields. Field trials conducted in different villages of central India on black soils deficient in nitrogen, phosphorus, sulphur and zinc have shown that balanced fertilization (BF) through the application of NPKSZn at recommended rates (25 kg N, 60 kg P_2O_5, 20 kg K_2O, 20 kg S and 5 kg Zn ha^{-1}) produced a higher soybean seed yield by 30–35% over farmers' usual practice (12.5 kg N and 30 kg P_2O_5 ha^{-1}) (Fig. 8.1). Omitting the application of phosphorus (NKSZn treatment) and sulphur (NPKZn treatment) had resulted in a 15–19% reduction in

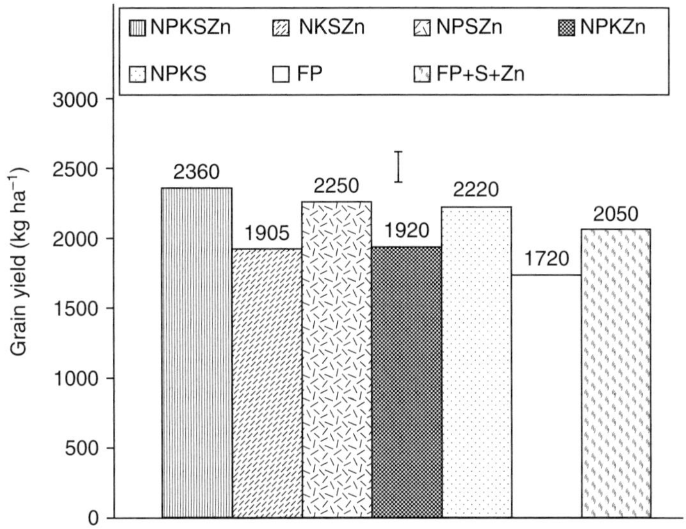

Fig. 8.1. Effect of balanced fertilization on soybean seed yield (I indicates the LSD at $P = 0.05$) (reprinted with permission from Subba Rao *et al.*, 2006).
FP, farmers' practice.

soybean seed yield as compared to NPKSZn treatment. Similarly, the soybean seed yield was reduced significantly when zinc was not applied. The application of sulphur (20 kg ha^{-1}) and zinc (5 kg ha^{-1}) along with farmers' practice produced 19% more soybean yield over farmers' practice alone (Subba Rao *et al.*, 2006).

The interactive advantage of the combined use of all possible sources of nutrients and their scientific management in integrated nutrient management (INM) has proved superior to the use of each component alone for optimum growth, yield and quality of different crops and cropping systems in specific agro-ecological situations. The basic concept underlying the principle of INM is to maintain or adjust plant nutrient supply to achieve a given level of crop production by optimizing the benefits from all possible sources of plant nutrients. The basic objectives of INM are to reduce the inorganic fertilizer requirement or use, restore organic matter in soil, enhance nutrient-use efficiency by synergetic interactions and maintain soil quality in terms of physical, chemical and biological properties. Bulky organic manures may not be able to supply adequate amounts of nutrients; nevertheless, their role is important in meeting the above objectives. Organic manures are known to decrease phosphorus adsorption/fixation and enhance phosphorus availability in phosphorus-fixing soils (Reddy *et al.*, 1999a). Organic anions formed during the decomposition of organic inputs can compete with phosphorus for the same sorption sites and thereby increase phosphorus availability in the soil (Iyamuremye *et al.*, 1996) and improve utilization by crops. Reddy *et al.* (1999b) observed

Table 8.4. Apparent phosphorus recovery (APR) of fertilizer phosphorus in soybean and wheat (mean of 5 years) (reprinted with permission by Elsevier from Reddy *et al.*, 1999b).

	Rate of fertilizer phosphorus (kg P ha^{-1})							
	0	11	22	44	0	11	22	44
Manure dose (t ha^{-1})[a,b]	APR by soybean (%)				APR by wheat (%)			
0 (0)	–	36.6	30.0	19.0	–	35.1	27.8	17.7
4 (5.6)	–	44.1	31.8	19.1	–	33.4	32.8	19.2
8 (11.2)	–	46.5	35.2	22.7	–	35.9	33.9	23.1
16 (22.4)	–	47.0	35.8	23.2	–	40.3	38.5	26.9
Mean	–	43.6	33.2	21.0	–	36.2	33.2	21.6

[a]Manure with fertilizer phosphorus applied to soybean, whereas wheat received only fertilizer phosphorus.
[b]Figures in parentheses are phosphorus applied through manure.

higher apparent phosphorus recovery by a soybean–wheat system on vertisol with a combination of fertilizer phosphorus and manure (Table 8.4). Organic manures play a direct role in supplying macro- and micronutrients and an indirect role by improving the physical, chemical and biological properties of soils. These manures, besides supplying nutrients to the current crop, very often have a substantial residual effect on succeeding crops in the system. The integrated use of chemical fertilizers and FYM has had a marked influence on improving the physical, chemical and biological quality of soil.

In the recent past, several INM strategies have been developed for soybean-based cropping systems in India and other countries. On-farm trials conducted on farmers' fields in central India have revealed that the INM module, comprising '50% NPKS + 5 t FYM ha^{-1} + *Rhizobium* to soybean and 75% NPKS + phosphate-solubilizing bacteria (PSB) to wheat' increased the soybean seed yield by 46% and the wheat grain yield by 24% over farmers' practice (Table 8.5) (Sammi Reddy *et al.*, 2007).

Conservation tillage and double-cropping practices may alter fertilizer nitrogen requirements due to a lowered soil nitrogen supply. Hairston *et al.* (1987), working on an Okolona silty clay soil in Mississippi, found that 25 lb N $acre^{-1}$ (28 kg N ha^{-1}) increased soybean yield only when planted no-till into wheat stubble and demonstrated a lack of benefit when straw was removed, burned or incorporated. In addition, net returns were increased the most in response to nitrogen fertilization for the no-till treatment. In central India, wheat residue incorporation/surface retention with or without FYM/poultry manure resulted in higher crop yields and led to an improvement in organic carbon and nutrient availability of soil under a soybean–wheat system as compared to residue burning. The value to cost

Table 8.5. Economics of integrated nutrient management (INM) in soybean (average of years 2005 and 2006) (reprinted with permission from Sammi Reddy *et al.*, 2007).

Details of the nutrient management option	Mean grain yield (kg ha^{-1})	Gross income (Rs ha^{-1})	Total cost[a] (Rs ha^{-1})	Net returns (Rs ha^{-1})	VCR
100% NPKSZn to soybean and 100% NPKS to wheat (BF)	1,953	21,483	10,919	10,564	0.97
50% NPKS + 5t FYM ha^{-1} to soybean and 75% NPKS to wheat (INM 1)	2,051	22,561	10,520	12,041	1.14
50% NPKS + 5t FYM ha^{-1} + *Rhizobium* to soybean and 75% NPKS + PSB to wheat (INM 2)	2,182	24,002	10,545	13,457	1.28
Farmers' practice	1,727	18,997	9,693	9,304	0.96

BF, balanced fertilization through inorganic fertilizers only; PSB, phosphate-solubilizing bacteria; VCR, value to cost ratio.
100% NPKSZn: soybean – 25 kg N, 60 kg P_2O_5, 20 kg K_2O, 20 kg S and 5 kg Zn ha^{-1}; wheat – 120 kg N, 60 kg P_2O_5, 20 kg K_2O and 20 kg S.
Farmers' practice: soybean – 12.5 kg N and 30 kg P_2O_5 ha^{-1}; wheat – 80 kg N and 50 kg P_2O_5 ha^{-1}.
[a]Total cost includes the cost of fertilizers, manure and all other costs of cultivation.

ratio (VCR) was 14.7 for residue incorporation and 3.1 for residue retention (Reddy, 2007a).

8.8 Soil-test-based Fertilizer Recommendations in Soybean

Liebig's law of minimum states that the growth of plants is limited by the plant nutrient element present in the smallest amount, all others being in adequate quantities. From this, it follows that a given amount of a soil nutrient is sufficient for any one yield of a given percentage nutrient composition. Ramamoorthy *et al.* (1967) established the theoretical basis and experimental proof for the fact that Liebig's law of minimum operates equally well for nitrogen, phosphorus and potassium. This forms the basis for fertilizer application for targeted yields, first advocated by Truog (1960). Among the various methods of fertilizer recommendation, the one based on yield-targeting is unique in the sense that this method not only indicates a soil-test-based fertilizer dose, but also gives the level of yield the farmer can hope to achieve if good agronomic practices are followed in raising the crop. The essential basic data required for formulating fertilizer recommendations for a targeted yield are: (i) the nutrient requirement in kg q^{-1} of produce, grain or other economic produce; (ii) the percentage contribution from the soil-available nutrients; and (iii) the percentage contribution from the applied fertilizer nutrients (Ramamoorthy *et al.*, 1967).

The above mentioned three parameters are calculated as follows (Subba Rao and Srivastava, 2001):

- Nutrient requirement of nitrogen, phosphorus and potassium for grain production (NR):

$$NR\ (kg\ of\ nutrient\ q^{-1}\ of\ grain) = \frac{Total\ uptake\ of\ nutrient\ (kg)}{Grain\ yield\ (q)}$$

- Percentage contribution of nutrient from soil (%CS):

$$\%CS = \frac{Total\ uptake\ in\ control\ plot\ (kg\ ha^{-1})}{Soil\ test\ value\ of\ nutrient\ in\ control\ plot\ (kg\ ha^{-1})} \times 100$$

- Percentage contribution of nutrient from fertilizer (%CF):

$$Contribution\ from\ fertilizer\ in\ treated\ plots\ (CF) = Total\ uptake\ of\ nutrients - (Soil\ test\ values\ of\ nutrients\ in\ fertilizer\ treated\ plots \times CS)$$

$$\%CF = \frac{CF \times 100}{Fertilizer\ dose\ (kg\ ha^{-1})}$$

- Calculation of fertilizer dose (FD):

  The above basic data are transformed into workable adjustment equation as follows:

$$FD = \left(\frac{NR}{\%CF} \times 100 \times T\right) - \left(\frac{\%CS}{\%CF} \times Soil\ test\ value\right)$$

$$FD = (a\ consant \times T) - (b\ consant \times soil\ test\ value\ in\ kg\ ha^{-1})$$

Where T is the yield target (q ha^{-1}).

Ramamoorthy *et al.* (1967) refined the procedure of fertilizer prescription initially given by Truog (1960) and it was later extended to different crops in different soils (Randhawa and Velayutham, 1982). The targeted yield concept strikes a balance between 'fertilizing the crop' and 'fertilizing the soil'. The procedure provides a scientific basis for BF and balance between applied nutrients and soil-available nutrients. In the targeted yield approach, it is assumed that there is a linear relationship between grain yield and nutrient uptake by the crop; as for obtaining a particular yield, a definite amount of nutrients is taken up by the plant. Once this requirement is known for a given yield level, the fertilizer needed can be estimated taking into consideration the contribution from soil-available nutrients.

Conditions for the successful use of the targeted yield approach are as follows:

- The targeted yield approach should be used for similar soils occurring in a particular agro-ecoregion.
- Targets chosen should not be unduly high or low and should be within the range of experimental yield obtained.

Table 8.6. Basic data and fertilizer adjustment equations for soybean (reprinted with permission from Puri and Gorantiwar, 2001).

Nutrient	Basic data NR (kg q^{-1})	CS (%)	CF (%)	Fertilizer adjustment equation	Relative yield kg^{-1} nutrient
Nitrogen				F N = 5.91 T – 0.48 SN	
P_2O_5	1.33	45.4	25.6	F P_2O_5 = 5.2 T – 4.1 SP	7.00
K_2O	4.54	21.3	116	F K_2O = 3.9 T – 0.22 SK	

CF, contribution of nutrient from fertilizer; CS, contribution of nutrient from soil; F, fertilizer; NR, nutrient requirement; T, target yield; SK, soil test potassium (ammonium-acetate-extractable potassium); SN, available nitrogen in soil (kg ha^{-1}); SP, soil test (Olsen's) phosphorus.

Table 8.7. Fertilizer recommendations for soybean at varying soil test values at specific yield targets (reprinted with permission from Puri and Gorantiwar, 2001).

Available nutrient status in soil (kg ha^{-1})		Fertilizer doses (kg ha^{-1}) for yield target of 2.5 t ha^{-1}	
Olsen's phosphorus	Ammonium acetate extractable potassium	P_2O_5	K_2O
5	200	109	53
10	250	89	42
15	300	68	31
20	350	48	20
25	400	27	9
30	450	–	–
35	500	–	–
40	550	–	–

- Adjustment equations must be used within the experimental range of soil test values and cannot be extrapolated.
- Good and recommended agronomic practices need to be followed while raising crops.
- Other micro- and secondary nutrients should not be yield limiting.
- For leguminous crops such as soybean, a minimum dose of nitrogen (20–30 kg ha^{-1}) may be applied.

Puri and Gorantiwar (2001) developed fertilizer adjustment equations for soybean grown in the monsoon season on medium black soils of central India. The basic data and fertilizer adjustment equations for a targeted yield are given in Table 8.6. Fertilizer nitrogen, phosphorus and potassium doses computed from the fertilizer adjustment equations are shown in Table 8.7.

Table 8.8. Seed yield of soybean under farmers' practice and soil test crop response (STCR)-based fertilizer dose (reprinted with permission from Srivastava *et al.*, 2001).

Site	Treatment[a]	Fertilizer dose (kg ha^{-1})			Grain yield
		N	P_2O_5	K_2O	(kg ha^{-1})
Site 1	Farmers' practice	9	23	0	1080
	STCR (1680 kg ha^{-1})	0	57	9	1650
Site 2	Farmers' practice	32	23	0	2030
	STCR (1680 kg ha^{-1})	44	0	19	2260
Site 3	Farmers' practice	34	12	0	1760
	STCR (2000 kg ha^{-1})	42	65	0	2270
Site 4	Farmers' practice	45	82	12	1770
	STCR (2000 kg ha^{-1})	17	37	38	2170
Site 5	Farmers' practice	80	100	36	1550
	STCR (2400 kg ha^{-1})	1	0	0	2390
Site 6	Farmers' practice	23	80	29	2100
	STCR (2400 kg ha^{-1})	34	13	11	2310

[a]Figures in parentheses are the target yield of soybean.

The advantage of application of fertilizer nutrients based on the target yield approach has been demonstrated to the farmers of different villages (Srivastava *et al.*, 2001). The soil-test-based fertilizer dose computed from the above equations produced significantly higher yields over the farmers' practice at all sites (Table 8.8).

8.9 Integrated Fertilizer–Water Management under Dryland Conditions

Soybean, by virtue of its tap-root system, is able to utilize soil moisture from deeper layers of the soil and stand under rainfed conditions and use the scarce moisture as efficiently as possible. Rooting is dense between 5 and 35 cm depth and some roots can reach a depth of 2 m when structure, moisture and fertility of the subsoil are favourable. Soybean water requirements vary with temperature, but on an average a yield of 3.5 t ha^{-1} needs 600 mm water between 1st July and 20th September (Fauconnier, 1986).

Plant nutrient uptake is dependent on diffusion, mass flow and root interception in the soil. Soil moisture influences the solubility of nutrients, their rate of movement towards roots and the rate of root growth. The availability of nutrients to the plant, therefore, is extremely dependent on soil moisture content.

Soybean grown under rainfed condition in the rainy season is generally not irrigated. Dry spells occur frequently in many areas and may be the main cause of soybean yield instability. The damage caused to a soybean crop by a dry spell depends on the intensity of the atmospheric evaporative demand, duration of rain absence, stage of crop development and soil 'state'. Water deficit (i.e. drying of the soil) is reflected in progressive closure of the stomata with a reduction of both transpiration and photosynthesis.

Appropriate soil and crop management can increase soybean yield stability in areas where dry spells occur. One or two more irrigations may be required during the pod-filling stage in cases of early withdrawal of monsoons to avoid yield reductions.

8.10 Economics of Fertilizer Use

The farmers are generally not applying the recommended doses of fertilizers to soybean because of high prices, inadequate supply and risks involved in soybean's cultivation under rainfed situations. However, the efficient use and management of fertilizer is of great value. The common method of determining the profitability potential is a VCR that represents the value of extra crop produced per unit of money invested in fertilizer. A VCR >1 mean a net profit while <1 means a net loss. The higher the VCR, the more attractive the use of fertilizers. Generally, a VCR of 2–2.5 is considered to be attractive for a farmer to adopt fertilizer use.

The economics of different nutrient management options for a soybean–wheat system under farmers' field conditions in representative villages of central India have been calculated (Sammi Reddy *et al.*, 2007). The mean economic analysis of 2 years (2005–2006 and 2006–2007) across 16 field trials in soybean revealed that the INM 2 (50% NPKS + 5 t FYM ha^{-1} + *Rhizobium* to soybean and 75% NPKS + PSB to wheat) produced 44% higher net returns (Rs 13,457 ha^{-1}) to the farmer over farmers' practice (Table 8.5). The VCR of this INM option was also higher as compared to farmer's practice (0.96:1) and BF through only inorganic fertilizers.

Most of the economic analyses of fertilizer usage in different crops have been restricted to a single crop or season. But the economics of different nutrient management options in different crops or cropping systems should be analysed on the basis of the whole cropping system rather than in a single crop because of the many reasons, including: (i) most fertilizers, particularly phosphorus and sulphur, have significant residual value; (ii) the application of micronutrients such as zinc in one crop meets the requirement of the following crop in a cropping system; and (iii) integration of organic manures with inorganic fertilizers reduces the fertilizer rates of subsequent crops due to their residual value. When soybean and wheat crops are considered together as a soybean–wheat system (Table 8.9), an INM option involving fertilizers, manure and biofertilizers (INM 2) has been found to yield the highest net returns (Rs 61,840 ha^{-1}) –24% higher than farmers' practice (Rs 49,178 ha^{-1}). The VCR of this treatment in the soybean–wheat system was also the highest (2.86:1) among all the treatments.

Table 8.9. Economics of INM in soybean–wheat system (reprinted with permission from Sammi Reddy *et al.*, 2007).

Details of the INM modules	Mean grain yield (kg ha^{-1}) Soybean	Wheat	Gross income (Rs ha^{-1})	Total cost (Rs ha^{-1})	Net returns (Rs ha^{-1})	VCR
100% NPKSZn to soybean and 100% NPKS to wheat (BF)	1,953	5,085	83,648	22,668	60,979	2.69
50% NPKS + 5t FYM ha^{-1} to soybean and 75% NPKS to wheat (INM 1)	2,051	4,863	81,841	21,567	60,274	2.79
50% NPKS + 5t FYM ha^{-1} + *Rhizobium* to soybean and 75% NPKS + PSB to wheat (INM 2)	2,182	4,869	83,461	21,621	61,840	2.86
Farmers' practice (FP)	1,727	4,168	69,613	20,435	49,178	2.41

BF, balanced fertilization through inorganic fertilizers only; PSB, phosphate-solubilizing bacteria; VCR, value to cost ratio.
100% NPKSZn: soybean – 25 kg N, 60 kg P_2O_5, 20 kg K_2O, 20 kg S and 5 kg Zn ha^{-1}; wheat – 120 kg N, 60 kg P_2O_5, 20 kg K_2O and 20 kg S.
Farmers' practice: soybean – 12.5 kg N and 30 kg P_2O_5 ha^{-1}; wheat – 80 kg N and 50 kg P_2O_5 ha^{-1}.

Despite better economics of improved nutrient management options in soybean-based cropping systems under field conditions, the adoption rate of these technologies by farmers is very limited, particularly in developing countries such as India. A survey conducted with 100 farmers in the Madhya Pradesh state of central India indicated that about 52% of farmers were applying fertilizers (nitrogen and phosphorus only) to soybean, whereas 80% of farmers were applying fertilizers to wheat (Sammi Reddy *et al.*, 2005). Farmers said that they could not predict the forthcoming risks in soybean production, particularly the incidence of pests, intensity of weed growth and possibility of continuous rains that may not allow farmers to institute timely pest and weed management practices. About 100 field demonstrations were conducted in ten villages of Madhya Pradesh with BF, INM and farmers' practice of nutrient management (Fig. 8.2) (Subba Rao *et al.*, 2008).

In these demonstrations, scientists helped the farmers in imposing the nutrient management options at the time of soybean sowing and then stepped back to allow the farmers to take up other pest and weed management practices as per their schedule. Very interesting results were obtained across the 100 fields. About 50% of the farmers managed their fields well and obtained 2.5–3.7 t ha^{-1} soybean yield with BF and INM and about 1.0–2.5 t ha^{-1} with farmers' practice. The rest of the farmers could not conduct timely weed and pest control measures and achieved only 1.0–2.0 t ha^{-1}

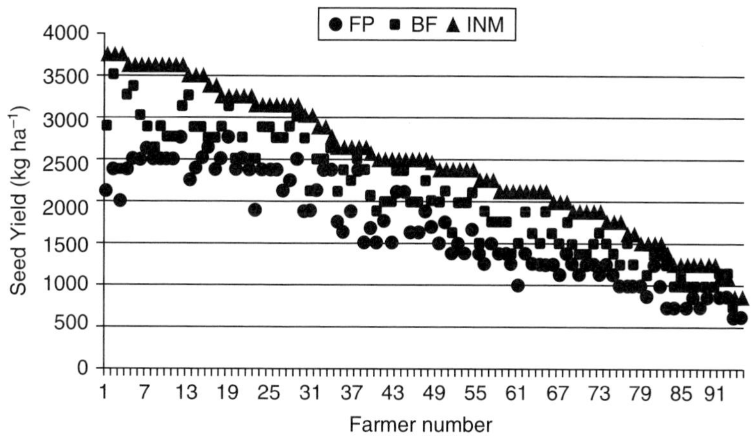

Fig. 8.2. Effect of farmers' practice (FP), balanced fertilization (BF) and integrated nutrient management (INM 2) on soybean seed yield (rates of manure and fertilizer application are as given in Table 8.5) (reprinted with permission from Subba Rao *et al.*, 2008).

yield with BF and INM. If all the farmers could conduct timely weed and pest management practices, soybean yields could easily reach 3.0–3.5 t ha^{-1} with BF and INM. From this example, it can be concluded that to encourage farmers to adopt improved nutrient management options in soybean, it is necessary to remove all of the risks involved in the soybean production, such as pests, weeds and waterlogging. Therefore, improved nutrient management technologies should be recommended and popularized among farmers along with other pest, weed and water management options as a package of practices.

8.11 Summary and Conclusions

Soybean occupies a very significant place in global oilseed production as well as being the predominant source of protein in the vegetarian diet of people. The productivity of soybean in many countries is around 2 t ha^{-1}, with a gap of 1 t ha^{-1} between potential and observed yield. Emerging multinutrient deficiencies (e.g. of nitrogen, phosphorus, sulphur, zinc, iron, boron) in soils, coupled with the application of only nitrogen and phosphorus to major crops by farmers (and that at lower rates than recommended), is one of the major reasons for lower productivity. Sulphur deficiency is also widespread in soybean-growing areas because of the farmers' preference for DAP as a source of phosphorus rather than sulphur-containing SSP.

Research in the recent past has conclusively shown that soybean productivity can be increased and sustained through optimum nutrient management. The application of nitrogen at 35 kg ha^{-1} as a starter is recommended on nitrogen-deficient soils. Optimum rates of application of phosphorus and potassium depend upon their availability in the soil. If soils contain higher available phosphorus, maintenance fertilization is recommended. Repeat applications of phosphorus and sulphur lead to their accumulation in the soil, which may reduce the rates of application in subsequent crops. Therefore, these fertilizers should be applied on the basis of soybean-based cropping systems as a unit, rather than as a single crop.

Among the micronutrients, zinc deficiency is most widespread. An application of 5 kg Zn ha^{-1} as zinc sulphate to soybean has been found to adequately meet the zinc requirements of both soybean and wheat crops in soybean–wheat rotation on black soils of central India. The application of 4 kg borax ha^{-1} has been found to be effective in correcting boron deficiency on black soils of India. Keeping in mind multinutrient deficiencies, the balanced application of all deficient nutrients at the recommended rates may be essential for sustaining higher productivity. Among all of the nutrient management options, the conjoint use of organic manures, biofertilizers and fertilizers in INM has proved efficient and economical in achieving high yields of soybean. The application of fertilizer nutrients based on target yield and soil test value could be helpful in saving on costly fertilizers. Many small and marginal farmers in developing countries feel that the application of fertilizers is risky as there are no guarantees that they will be able to harvest a good soybean crop due to other associated problems (e.g. pests, weeds and waterlogging) in soybean production. Therefore, improved nutrient management technologies should be recommended and popularized among farmers along with other pest, weed and water management options as a package of practices.

References

Acharya, C.L., Subba Rao, A., Biswas, A.K., Sammi Reddy, K., Yadav, R.L., Dwivedi, B.S., Shukla, A.K., Singh, V.K. and Sharma, S.K. (2003) *Methodologies and Package of Practices on Improved Fertilizer Use Efficiency under Various Agro-climatic Regions for Different Crops/Cropping Systems and Soil Conditions.* Indian Institute of Soil Science, Bhopal, India.

Bell, P.F., Hallmark,W.B., Sabbe, W.E. and Dombeck, D.G. (1995) Diagnosing nutrient deficiencies in soybean, using M-DRIS and critical nutrient level procedures. *Agronomy Journal* 87, 859–865.

Bolland, M.D.A. and Barrow, N.J. (1991) The effect of level of application on the residual value of super phosphate on a sandy soil in south-western Australia. *Fertilizer Research* 29, 163–172.

Borkert, C.M. and Sfredo, G.J. (1994) Fertilizing tropical soils for soybean. In: *Tropical Soybean: Improvement and Production.* Publication Division, Food and Agriculture Organization (FAO) of the United Nations, Rome, Italy, pp. 175–200.

Brar, M.S., Subba Rao, A. and Sekhon, G.S. (1986) Potassium status of five Benchmark soil series from Southern India. *Journal of Potassium Research* 2, 51–58.

Brown, J.C. and Jolley, V.D. (1989) Plant metabolic responses to iron-deficiency stress. *Biological Sciences* 39, 546–551.

Deibert, E.J., Jeriego, M.D. and Olson, R.A. (1979) Utilization of ^{15}N fertilizer by nodulating and non-nodulating soybean isolines. *Agronomy Journal* 71, 717–723.

Eaglesham, A.R.J., Hassouna, S. and Seegers, R. (1983) Fertilizer N effects on N_2 fixation by cowpea and soybean. *Agronomy Journal* 75, 61–66.

Fageria, N.K., Baligor, V.C. and Jones, C.A. (1991) *Growth and Mineral Nutrition of Field Crops*. Marcel Dekkar, Inc., New York, USA.

FAO (2009) *FAOSTAT*. Food and Agriculture Organization of the United Nations, Rome, Italy. Available at: http://faostat.fao.org (accessed 3 December 2009).

Fauconnier, D. (1986) *Soya: Fertilizers for Yield and Quality*. IPI Bulletin No. 9, International Potash Institute, Bern, Switzerland.

Ganeshamurthy, A.N. and Takkar, P.N. (1997) Residual management of sulphur applied to soybean or wheat in soybean–wheat cropping system on vertisols. *Australian Journal of Soil Research* 35, 199–208.

Ganeshamurthy, A.N., Subba Rao, A. and Sammi Reddy, K. (1994) Phosphorus and sulphur constraints for soybean, wheat and gram production on some swell-shrink soils. *Agropedology* 4, 99–106.

Hairston, J.E., Sanford, J.O., Pope, D.F. and Horneck. D.A. (1987) Soybean–wheat double cropping: Implications from straw management and supplemental nitrogen. *Agronomy Journal* 79, 281–286.

Hanway, J.J. and Weber, C.R. (1971) Accumulation of N, P and K by soybean (*Glycine max* L. Merrill) plants. *Agronomy Journal* 63, 406–408.

Iyamuremye, F., Dick, R.P. and Baham, J. (1996) Organic amendments and phosphorus dynamics: III. Phosphorus speciation. *Journal of Soil Science* 161, 444–451.

Jones, J.B., Wolf, B., Jr and Mills, H.A. (1991) *Plant Analysis Hand Book*. Micro-Macro Publishing, Athens, Greece.

Kalia, B.D., Awasthi, O.P. and Singh, C.M. (1984) Response of soybean to nitrogen, phosphorus, liming and inculation under mid-hill condition of Kangra Valley. *Indian Journal of Agronomy* 29, 199–204.

Kamprath, E.J. and Jones, U.S. (1986) Plant response to sulfur in the southeastern United States. In: Tabatabai, M.A. (ed.) *Sulfur in Agriculture*. Agronomy Monograph No. 27. ASA, CSSA and SSSA, Madison, WI, USA, pp. 323–343.

Maas, E.V. (1986) Salt tolerance of plants. *Applied Agricultural Research* 1, 12–25.

Mahapatra, P., Singh, R.P., Singh B.P., Mishra, B. and Sarkar, A.K. (2007) Long-term effects of fertilizer, organic manure, and amendments on soil health, crop productivity and sustainability. SAAC (BAU) Technical Bulletin No. 4/2007. Birsa Agricultural University, Ranchi, Jharkhand, India.

Marschner, H. (1995) *Mineral Nutrition of Higher Plants*, 2nd edn. Academic Press, San Diego, CA, USA.

McLean, E.O. and Brown, J.R. (1984) Crop response to lime in the Midwestern United States. In: Adams, F. (ed.) *Soil Acidity and Liming*, 2nd edn. Agronomy Service 12, ASA, CSSA and SSSA, Madison, WI, USA, pp. 167–303.

Mengel, K. and Kirkby, E.H. (1987) *Principles of Plant Nutrition*. International Potash Institute, Bern, Switzerland.

Moraghan, J.T. (1985) Manganese deficiency in soybean as affected by FeEDDHA and low soil temperature. *Soil Science Society of America Journal* 49, 1584–1586.

Pasricha, N.S. and Tandon, H.L.S. (1989) Fertilizer management in oilseed crops. In: Tandon, H.L.S. (ed.) *Fertilizer Recommendations for Oilseed Crops – A Guide Book*. Fertilizer Development and Consultation Organization, New Delhi, India, pp. 27–60.

Puri, G. and Gorantiwar, S.M. (2001) Improved fertilizer recommendations for the crops of Madhya Pradesh. In: Subba Rao, A. and Srivastava, S. (eds) *Soil Test Based Fertilizer Recommendations for Targeted Yields of Crops*. Proceedings of the National Seminar on Soil Testing for Balanced and Integrated Use of Fertilizers and Manures. Indian Institute of Soil Science, Bhopal, India, pp. 144–160.

Ramamoorthy, B., Narasimham, R.L. and Dinesh, R.S. (1967) Fertilizer application for specific yield targets of Sonora 64. *Indian Farming* 17(5), 43–45.

Randhwa, N.S. and Velayutham, M. (1982) Research and development programme for soil testing in India. *Fertiliser News* 27, 35–36.

Rathore, G.S., Khamparia, R.S, Gupta, G.P., Dubay, S.B., Sharma, B.L. and Tomar, V.S. (1995) Research Bulletin on 'Twenty-five years of micronutrient research in soils and crops of Madhya Pradesh (1967–92)'. Jawaharlal Nehru Krishi Vishwa Vidyalaya, Jabalpur, India.

Rattan, R.K. (2007) Proceedings of and recommendations from the symposium on Nutrient Management in Acid Soils. Bulletin of the Indian Society of Soil Science, No. 25, Indian Society of Soil Science, New Delhi, India, pp. 1–3.

Reddy, D.D. (2007a) Technology for managing mechanical harvest borne wheat residue in soybean–wheat system on vertisols. In: Sammi Reddy, K., Srivastava, S. and Sharma, A.K. (eds) *IISS Perspective Plan – Vision 2025*. India Institute of Soil Science (IISS), Bhopal, India, p. 14–15.

Reddy, D.D. (2007b) Technology for soil test maintenance P requirement for soybean–wheat system. In: Sammi Reddy, K., Srivastava, S. and Sharma, A.K. (eds) *IISS Perspective Plan – Vision 2025*. India Institute of Soil Science (IISS), Bhopal, India, p. 11.

Reddy, D.D., Subba Rao, A. and Takkar, P.N. (1999a) Phosphate adsorption behaviour of a Typic Haplustert in relation to long-term use of fertilizer P and manure under soybean–wheat rotation. *Journal of the Indian Society of Soil Science* 47, 425–430.

Reddy, D.D., Subba Rao, A., Sammi Reddy, K. and Takkar, P.N. (1999b) Yield sustainability and phosphorus utilization in soybean–wheat system on vertisols in response to integrated use of manure and fertilizer phosphorus. *Field Crops Research* 62, 180–190.

Rosolem, C.A. (1980) *Mineral Nutrition and Fertilization of Soybean*. Technical Bulletin 6. Potassium and Phosphate Institute, Piracicaba, Brazil.

Sammi Reddy, K., Subba Rao, A. and Takkar, P.N. (1997) Optimum external and internal P requirements of soybean and wheat for maximizing production on a phosphorus deficient Typic Haplustert. *Journal of the Indian Society of Soil Science* 45, 297–301.

Sammi Reddy, K., Subba Rao, A., Singh, S. and Reddy, D.D. (2003) Crop responses to residual phosphorus and economic evaluation in soybean–wheat system on vertisols. *Journal of Maharashtra Agricultural Universities* 28, 12–16.

Sammi Reddy, K., Kumar, N., Sharma, A.K., Acharya, C.L. and Dalal, R.C. (2005) Biophysical and sociological impacts of farmyard manure and its potential role in meeting crop nutrient needs: A farmers' survey in Madhya Pradesh, India. *Australian Journal of Experimental Agriculture* 45, 357–367.

Sammi Reddy, K., Rao, D.L.N., Singh, M., Mohanty, M., Subba Rao, A., Dixit, S.K., Pandey, M. and Singh, G. (2007) Enhancing productivity of soybean–wheat system by balanced and integrated nutrient management on black soils of Madhya Pradesh. ICAR-ACIAR Project Bulletin No. 1, Indian Institute of Soil Science, Bhopal, pp. 1–17.

Sharma, P.S. and Sarkar, A.K. (2005) Managing acid soils for enhancing productivity. Technical Bulletin, Natural Resource Management Division, Indian Council of Agricultural Research, New Delhi, India.

Shinde, D.A., Solankey, B.S., Verma, R.P., Marathe, B.S. and Dixit, P.K. (1990) Response of soybean to boron and phosphorus application in Kamaliakheri (Vertic Ustochrept) soil series as influenced by micro-climate. *Fertilizer News* 35, 31–33.

Singh, M.V. (1999) Sulphur management for oilseed and pulse crops. All India Coordinated Research Project of Micro- and Secondary Nutrients and Pollutant Elements in Soils and Plants. Indian Institute of Soil Science, Bhopal, India.

Singh, M.V. and Saha, J.K. (1997) 27th Annual Report of All India Coordinated Research Project of Micro- and Secondary Nutrients and Pollutant Elements in Soils and Plants for the year 1995–96. Indian Institute of Soil Science, Bhopal, India.

Singh, M., Kundu, S., Tripathi, A.K. and Takkar, P.N. (1998) Nitrogen. In: Tomar, K.P., Rupa, T.R. and Ramana, S. (eds) *A Decade of Progress – 1988–1999*. IISS Bulletin No. 1. Indian Institute of Soil Science (IISS), Bhopal, India, pp. 29–42.

Small, H.G., Jr and Ohlrogge, A.J. (1973) Plant analysis and an aid in fertilizing soybeans and peanuts. In: Walsh, L.M. and Beaton, J.D. (eds) *Soil Testing and Plant Analysis*. Soil Science Society of America, Madison, WI, USA, pp. 315–327.

Srivastava, S., Subba Rao, A. and Biswas, A.K. (2001) Results of participatory diagnosis of constraints and opportunities (PDCO) based trials from the state of Uttar Pradesh. In: Subba Rao, A. and Srivastava, S. (eds) *Development of Farmers' Resource Based Integrated Plant Nutrient Supply Systems: Experience of FAO-ICAR-IFFCO Collaborative Project*. All India Coordinated Research Project for Investigations on Soil Test Crop Response Correlation, Indian Institute of Soil Science, Bhopal, India, pp. 10–49.

Subba Rao, A. and Ganeshamurthy, A.N. (1994) Soybean response to applied phosphorus and sulphur on vertic ustochrepts in relation to available phosphorus and sulphur. *Journal of the Indian Society of Soil Science* 42, 606–610.

Subba Rao, A. and Srivastava, S. (eds) (2001) *Soil Test Based Fertilizer Recommendation for Targeted Yields of Crops*. Proceedings of the National Seminar on Soil Testing for Balanced and Integrated Use of Fertilizers and Manures, Indian Institute of Soil Science, Bhopal, India.

Subba Rao, A., Sammi Reddy, K. and Takkar, P.N. (1996) Residual effects of phosphorus applied to soybean or wheat in soybean–wheat cropping system on a Typic Haplustert. *Journal of Agricultural Sciences, Cambridge* 127, 325–330.

Subba Rao, A., Sammi Reddy, K. and Takkar, P.N. (1997) Evaluation of phosphorus efficiency in soybean and wheat cropping system in relation to fertilizer P on a Typic Haplustert. *Indian Journal of Agricultural Sciences* 67, 518–522.

Subba Rao, A., Sammi Reddy, K., Rao, D.L.N., Singh, M., Mohanty, M. and Dixit, S.K. (2006) Annual progress report of the research project entitled 'Integrated manure nutrient management in soybean/wheat cropping systems on Vertisols in Madhya Pradesh and Queensland'. Indian Institute of Soil Science, Bhopal, Madhya Pradesh, India, pp. 1–39.

Subba Rao, A., Sammi Reddy, K., Rao, D.L.N., Singh, M., Mohanty, M. and Dixit, S.K. (2008) Annual progress report of the research project entitled 'Integrated manure nutrient management in soybean/wheat cropping systems on Vertisols in Madhya Pradesh and Queensland'. Indian Institute of Soil Science, Bhopal, Madhya Pradesh, India, pp. 1–42.

Tandan, H.L.S. (1989) *Secondary and Micronutrient Recommendations for Soils and Crops – A Guide Book*. Fertilizer Development and Consultation Organization, New Delhi, India.

Truog, E. (1960) Fifty years of soil testing. In: *Transactions of 7th International Congress of Soil Science*, Volume III, Commission IV, Paper No. 7, pp. 46–53.

Varco, J.J. (1999) Nutrition and fertility requirements. In: Heatherly, L.C. and Hodges, H.F. (eds) *Soybean Production in the Midsouth*. Boca Raton Press, Florida, USA, pp. 53–70.

Vilela, L. and Ritchey, K.D. (1985) Potassium in intensive cropping system on highly weathered soils. In: Mundson, R.D. (ed.) *Potassium in Agriculture*. American Society of Agronomy, Madison, WI, USA, pp. 1155–1175.

Whigham, D.K. (1983) Soybean. In: *Productivity of Field Crops under Different Environments*. International Rice Research Institute, Los Banos, Philippines, pp. 205–225.

9 Water Management in Soybean

Guriqbal Singh
Department of Plant Breeding and Genetics, Punjab Agricultural University, Ludhiana, Punjab, India

9.1 Introduction

Water is an important input for assured agricultural production. Important sources of water for raising agricultural crops include rainfall, canal water, underground water and moisture in the soil profile. Soybean (*Glycine max* (L.) Merrill) is mainly grown as a rainfed crop in many parts of the world. Erratic distribution of rainfall, both in terms of total amount and its distribution during the growing season, adversely affects soybean production. Waterlogging may also occur in specific situations. However, drought is more common than waterlogging and both affect soybean production to varied degrees.

To obtain high yields the crop needs to be irrigated at the critical stages, using either underground water, canal water or conserved rain water. The response to irrigation may depend upon various factors. The water-use efficiency (WUE) can be influenced by the management of the crop and the soil. WUE can be improved by following two types of agronomic approaches (Singh, 1998): (i) reducing evaporation from the soil surface by mulching, modifying the plant population and spacing, selecting varieties with rapid early growth, early sowing and/or application of fertilizers; and (ii) increasing the water supply to plants by rain-water harvesting, supplemental irrigation, cultivation to improve infiltration and reduced runoff, weed control, multiple or relay cropping and/or selecting varieties with deep root systems.

This chapter discusses the water requirements of soybean, the effects of moisture (drought as well as waterlogging) on nodulation, nitrogen fixation, growth and yield, irrigation scheduling, optimum irrigation schedules, irrigation management under saline conditions, conservation and efficient use of rain water and factors affecting WUE in soybean.

9.2 Water Requirement of Soybean

The water requirement of soybean varies depending upon the growing season, crop cultivar, irrigation method, rainfall and so on. For example, a crop grown during very hot summer months requires higher amounts of water than that grown during milder months. Cultivars of longer duration are expected to need higher amounts of moisture than those of short duration. The efficiency of different irrigation methods varies, which consequently affects the water requirements of soybean. When rainfall is optimum and well distributed, no or very little irrigation may be required for realizing high seed yields.

The simulated long-term annual average net irrigation requirement for soybean has been reported to be 367 mm (Lamm *et al.*, 2007). However, the water requirement (or the amount of applied irrigation water) varies with soil moisture depletion (Al-Assily and Mohamed, 2002).

In some studies, irrigation is applied to crops based on an irrigation water to cumulative pan evaporation (IW:CPE) ratio. Under irrigation regimes of IW:CPE 0.80, 0.60 and 0.40, water consumption in soybean has been reported to be 450–533, 350–438 and 250–393 mm, respectively (Rajendran and Lourduraj, 2000).

In lysimeter studies in Rajasthan, India, during a crop period of 108 days, soybean has recorded 662 mm evapotranspiration (ET), the average ET being 6.12 mm/day (Singh *et al.*, 2001). In another lysimeter study (Singh and Prakash, 2001), out of a total water input of 840 mm, the ET of soybean was found to be 484 mm at 50% depletion of available soil moisture in an 88-day period. Furthermore, ET has significant positive correlation with various meteorological parameters such as pan evaporation, sunshine hours and mean maximum temperature and significant negative correlation with relative humidity. The ET of soybean has been found to be 574–619 mm in Turkey (Dogan *et al.*, 2007a) and 725–800 mm in Lebanon (Karam *et al.*, 2005).

9.3 Effects of Moisture on Soybean

Effects of drought on nodulation, nitrogen fixation, growth and yield

Drought may affect various physiological and morphological aspects of the soybean plant, which in turn affect nodulation, nitrogen fixation, growth and yield. The effect of drought stress on a plant may depend on various factors including the stage, severity and duration of stress. Although significant levels of osmotic adjustment have been reported in soybean (Likoswe and Lawn, 2008), osmotic adjustment is of little benefit for leaf survival.

Water stress may have varied effects on the plant, ranging from visually unnoticeable effects to wilting and death. Export of photoassimilates

Table 9.1. Effect of water stress (applied at the beginning of growth stage R6) and the timing of stress relief on yield and yield components of soybean grown in soil-filled pots (adapted from Brevedan and Egli, 2003).

Treatment	Pods per plant	Seeds per plant	Seed size (mg per seed)	Seed yield (g per plant)
Non-stressed (control)	91	211	233	49.2
Early stress relief (returned to non-stress watering level after 3 days of stress)	88	208	212	44.2
Late stress relief (returned to non-stress watering level after 6 days of stress)	87	200	195	39.2
Continuous stress (pots received 40% of the water needed to bring non-stressed control to pot capacity) until maturity	86	193	156	30.1

from the leaves is affected by water stress (Ohashi *et al.*, 2000). With a soil moisture deficit, the chlorophyll content is lowered (Velu, 1999) and various physiological parameters such as photosynthetic rate, net carbon assimilation efficiency, total area of stomatal apertures per unit leaf area, transpiration rate and WUE are decreased considerably (Ghosh *et al.*, 2006b). Water stress even for short periods (3–13 days) during the seed-filling stage (R6) rapidly reduces the carbon exchange rate (Brevedan and Egli, 2003), thereby resulting in earlier maturity, smaller seed size and lower seed yields (Table 9.1). Pod abortion occurs in drought-stressed soybean (Liu *et al.*, 2004), which may be due to a low availability of photosynthate in leaves and an impaired ability of pods to utilize sucrose.

Nodulation is adversely affected by drought or low moisture stress. Drought affects the number, size and weight of nodules. Furthermore, the pattern of nodule formation on roots (i.e. whether on tap or lateral roots) is also influenced by drought. Nitrogen fixation, in terms of percentage nitrogen fixation and the amount of nitrogen fixed, is influenced greatly in soybean by moisture availability and is decreased with water stress (Mohamed, 1995; Ohashi *et al.*, 2000; Ray *et al.*, 2006). Genotypes of soybean do vary in their ability to tolerate drought in terms of nitrogen fixation (Serraj *et al.*, 1997; Sinclair *et al.*, 2007). Furthermore, drought tolerance in some genotypes (e.g. Jackson) may be due to a large nodule size (King and Purcell, 2001), which helps with better photosynthate and water allocation, relative water content and water supply for ureide export.

In soybean, water stress decreases biomass production (Ohashi *et al.*, 2000; Hajare *et al.*, 2001), seed yield (Purcell *et al.*, 2004), root surface area (Benjamin and Nielsen, 2006), root length, plant height, leaf area, dry weight of all plant organs, seed yield and the number of branches, flowers, pods and seeds (Ghosh *et al.*, 2000a; Noureldin *et al.*, 2002b), with the effects increasing with rises in the intensity and duration of the water deficit. The reduction in biomass production due to moisture stress is the greatest at the pod formation and seed-filling stages (Hajare *et al.*, 2001). When drought stress occurs between the initiation of flowering and seed fill, the total seed yield is decreased. This is mainly because of reductions in branch vegetative growth and consequent reductions in seed number and branch seed yield, rather than due to any effect on main-stem seed yield (Frederick *et al.*, 2001). Drought stress during the reproductive phase decreases pod set, which may be due to decreases in water potential and increases in the abscisic acid content of flowers and pods 3–5 days after anthesis (Liu *et al.*, 2003). The above-ground biomass and seed yield of the soybean have been found to be reduced by 16% and 4%, respectively with deficit irrigation at the R2 stage and by 6% and 28%, respectively, with deficit irrigation at the R5 stage (Karam *et al.*, 2005). Dogan *et al.* (2007b) reported that water stress at the R3, R5 or R6 results in substantial yield reductions compared with full irrigation, with the greatest reduction with water stress at the R6 stage.

Effects of waterlogging on nodulation, nitrogen fixation, growth and yield

Like drought, excessive soil moisture or waterlogging also has adverse effects on soybean plants (Thomas and Sodek, 2005). Waterlogging may be caused by heavy rainfall or over-irrigation and the problem is more common in fine-textured soils. Furthermore, excessive soil moisture is also experienced in paddy fields due to the formation of a hard pan owing to puddling operations. This excessive soil moisture induces growth losses in the succeeding soybean crop, particularly during vegetative growth. The problem is more severe in crops sown on a flat bed. Crops sown on raised beds or ridges do not generally experience the adverse effects of excessive soil moisture (van Cooten and Borrell, 1999; Seong *et al.*, 2000a). However, the height of the raised bed or ridge also matters. In a rice field in Korea, the total dry matter accumulation was found to be severely decreased until the growth stage of R5 when soybean was sown at a 10-cm ridge height as compared to at ridge heights of 30 and 50 cm (Seong *et al.*, 2000a).

In the case of flooding soon after sowing, a rapid in-rush of water into soybean seeds results in physical disruption of seeds, consequently reducing seedling emergence as well as seedling growth (Nakayama *et al.*, 2005). Due to excessive soil moisture under poor drainage, crop growth and seed development in soybean are adversely affected (Rao *et al.*, 1999; Seong *et al.*, 1999) and soybean cultivars do differ in response to excessive soil moisture (Seong *et al.*, 2000b). Furthermore, decreased contents of nitrogen, phosphorus, potassium, calcium, magnesium and copper in soybean leaves have also been reported due to excessive soil water stress (Seong *et al.*, 1999), which may be due to decreased plant growth.

9.4 Irrigation Scheduling

Irrigation depth

Irrigation depth varies depending upon the source of irrigation (water discharge), soil type, stage of crop growth, weather conditions and so on. When water discharge is low, farmers tend to apply light irrigation to cover a greater area per unit of time. In heavy soils, heavy irrigations are avoided as these can cause waterlogging or excessive moisture, leading to an adverse effect on crop growth and yields. A crop sown in suboptimal moisture conditions may require a light irrigation within a few days after sowing to improve germination, especially in light soils. Irrigation depth also varies depending upon the prevailing weather conditions. If there is wind, irrigation should be avoided or only a light irrigation should be applied, otherwise there may be lodging.

Irrigation methods

In the soybean, flood irrigation is a common practice in flat-bed-sown crop. However, where the crop has been sown on raised beds, furrow irrigation is applied. Some water-saving high-tech irrigation techniques such as sprinkler and drip irrigation are either not used or very rarely used in soybean. However, the choice of irrigation method is also determined by factors such as the source of irrigation, surface topography and soil texture.

Furrow- and flood-irrigated soybean crops show better growth, seed yields and net returns than non-irrigated crops, and both irrigation systems are equally good in all of these parameters (Table 9.2). Drip irrigation not only uses less water than sprinkler irrigation but also maintains a higher soil temperature, leading to a higher emergence rate and enhanced seedling growth (Wang *et al.*, 2000). Surface and subsurface drip irrigation systems can be used. However, in the case of subsurface drip irrigation, at lower depth a high moisture content with low soil oxygen concentration may cause hypoxia. Oxygation (aerated irrigation water) can ameliorate hypoxia and increase soybean yields (Bhattarai *et al.*, 2008). Although drip and

Table 9.2. Effect of irrigation method on plant characters, seed yield and net returns of soybean grown following rice at Stoneville, Mississippi, USA (average across years and cultivars from maturity groups IV and V) (adapted from Heatherly and Spurlock, 2000).

Irrigation method	Plant height (cm)	Seed weight (mg per seed)	Seed yield (kg ha^{-1})	Net returns (\$ ha^{-1})
Non-irrigated (control)	77	129	2658	206
Furrow-irrigated	85	148	3720	347
Flood-irrigated	86	148	3702	393

sprinkler irrigation systems are known for their irrigation water economy, these systems are not very popular among farmers for soybean crops due to their initial high costs.

Irrigation indices

Irrigation can be based on parameters such as crop stage, weather conditions and soil moisture status.

In the case of soil moisture status, irrigation is applied when the soil moisture is depleted to a certain level (Al-Assily and Mohamed, 2002). Irrigation may also be applied based on the crop's growth stage (de Costa and Shanmugathasan, 2002). Furthermore, a computer program method, tensiometer or gypsum block may be used for deciding when to irrigate (Thompson *et al.*, 2002). Irrigation may be applied to the crop based on any parameter. However, the main objectives of all of these are that the crop yields are high and irrigation water use is efficient.

Effect of irrigation on soybean

As with other crop plants, soybean requires optimum soil moisture for high seed yields. If there is an adequate availability of moisture in the soil either due to rainfall or otherwise, the crop may not need any irrigation. However, in the absence of adequate soil moisture, irrigation has to be applied to obtain proper crop growth and high crop yields.

Soybean yields increase as irrigation increases (Gercek *et al.*, 2009). Soybean is known to respond significantly to irrigation during flowering and seed formation (Bharambe *et al.*, 2002). Similar yields can be obtained by applying a single irrigation at the R4, R5 or R6 stage (Sweeney *et al.*, 2003), which are, however, about 20% higher than with no irrigation. The beneficial effect of irrigation on seed yields may be due to an increased number of seeds per plant or increased weight per seed. Irrigation at R4 increases seeds per plant, whereas irrigation at R5 or R6 increases weight per seed (Sweeney *et al.*, 2003). In the case of a saturated soil culture (continuously irrigated), soybean root and nodule growth, nitrogenase activity, leaf conductance to water vapour and seed yields are higher than with conventional irrigation (Troedson *et al.*, 1989); the seed yields are possibly higher due to greater photosynthesis, sustained nodules during pod growth and consequently a continuous nitrogen supply to the seeds.

9.5 Optimum Irrigation Schedules

Irrigation intervals affect leaf area index, crop growth rate, plant height, harvest index, total dry matter or seed yield of soybean (Osman *et al.*, 2000; Yazdani *et al.*, 2007). In Egypt, irrigation at 15-day intervals throughout the growing season has been found to be the best for the growth and yield of soybean (Osman *et al.*, 2000). With shorter irrigation intervals the yield may increase; however, WUE and net income may decrease. The irrigation interval should therefore be such that water is used judiciously, resulting in high yields, WUE and net returns.

The crop may need a number of irrigations at different growth stages. Various parameters such as the maximum leaf area index, fraction of incoming radiation intercepted, maximum total biomass, number of pods, pod growth stage and harvest index are increased with the number of stages irrigated (de Costa and Shanmugathasan, 2002). However, for realizing high seed yields the crop should not suffer for want of moisture at the critical growth stage. When there is a limited quantity of water available, irrigation should be applied at the R5 stage only (Kim *et al.*, 1999) as the daily mean transpiration rate, WUE and seed yield are high with irrigation at this stage. Irrigation during the reproductive phase increases seed yield by increasing seeds per plant and it can be applied either at R1, R4 or R6 (Sweeney and Granade, 2002). Irrigation frequencies have a great effect on the growth and yield of soybean. Six irrigations result in greater plant height, pods per plant, seed yield and oil content percentage than five, four, three or two irrigations (Kazi *et al.*, 2002). In a study by Rabbani *et al.* (2004), the highest seed yield was with irrigation at 20, 40 and 60 days after sowing as opposed to 20, 20 and 40 or 20, 40, 60 and 80 days after sowing.

In a soybean–winter season sorghum (*Sorghum bicolor*) cropping system on a clay soil vertisol in Maharashtra, India, soybean showed a good response to irrigation applied at 0.60 IW:CPE (Bharambe *et al.*, 1999). A higher soybean seed yield has been reported when the crop is irrigated at 75 mm CPE than at 100 or 125 mm CPE (Deolankar and Mogal, 1998).

In India, the seed, straw, oil and protein yields of soybean and the oil and protein contents in soybean seeds were higher under a 0.8 IW:CPE ratio than under a 0.4 or 0.6 IW:CPE ratio in Rajasthan (Kumawat *et al.*, 2000) and Tamil Nadu (Rajendran and Lourduraj, 2000; Ramasamy and Sankaran, 2001). Irrigation at a 0.6 IW:CPE ratio results in lower canopy temperature, lower transpiration rate, higher relative leaf water content and stomatal diffusive resistance, thereby leading to higher seed yield (Elamathi and Singh, 2001).

9.6 Irrigation Management under Saline Conditions

Salinity tolerance of soybean

The emergence and growth of soybean are reduced with the application of saline water (Blanco *et al.*, 2007). However, soybean is more tolerant to salinity during emergence than in its initial development, suggesting that under compelling circumstances saline water may be used for pre-sowing irrigation.

After rainfall, sodic soils develop a thick crust that impedes soybean seedling emergence (McKenzie *et al.*, 2002). However, crops sown on raised beds generally experience less problems with crust formation and, therefore, have less problems with poor emergence. Growth is adversely affected to a great extent when soybean is grown under salt-affected soil conditions (Noureldin *et al.*, 2002a).

Crop yield, nitrogen uptake by the crop and the nitrogen contribution of the soil are decreased when soybean is irrigated with saline water (van Hoorn *et al.*, 2001). Soybean genotypes do vary in salinity tolerance (Bahmaniar and Sepanlou, 2008) and there is a need to develop salinity-tolerant genotypes. Alternatively, the crop should be managed under saline conditions in such a way that the adverse effects of salinity are minimized.

Management of poor-quality water

Good-quality irrigation water is often unavailable in adequate quantities, but there may be compelling circumstances where irrigation has to be applied to save the crop from water stress. In such a situation irrigation may be applied by mixing fresh and saline water, and the ratio of these waters can be decided on the salinity level. Another option is to alternately irrigate the crop with saline water and fresh water.

Planting method may also help in managing crops under saline conditions. With such problematic soils, the crop may be planted on raised bed rather than flat sowing. Furthermore, the planting should not be on the top of the bed; rather, it should be on the side of the bed at a slanted position so that, with evaporation, the salts accumulate on the top of the bed and thereby have little effect on the crop.

9.7 Conservation and Efficient Use of Rain Water

Rain water conservation

Rain is an important source of water that should not be wasted. The efficient use of rain water is of great importance, particularly in water-scarce areas. Heavy-intensity rain may only occur for a short period. If this rain water is not conserved in an efficient way, not only is water lost but soil erosion may also occur. Rain water conservation should therefore be a top priority. The cultivation of fields prior to sowing ensures better rain water conservation. Deep ploughing has further beneficial effects on water conservation. Ridges check the speed of water movement. Therefore, fields should be divided into small parts with ridges for the *in situ* conservation of rain water before the soybean crop is sown. A broad-bed and furrow system increases rainfall infiltration into the soil (Singh *et al.*, 1999a).

Rain water should be conserved during the crop growing period. Dividing the field into small parts with strong ridges helps to retain rain water in the field, rather than allowing it to flow to other areas. With this practice, rain water is utilized by the crop and nutrients are saved from being removed with flowing rain water.

Rain water harvesting and recycling

Rain water should not be allowed to go waste. Rain water should be saved in ponds or reservoirs so that this stored water can be used for irrigation at critical periods of crop growth. Rain water harvesting is especially important in those areas where rainfall occurs for a limited period during the year and other sources of irrigation water are not available.

Influence of rainfall on soybean productivity

Adequate and timely rainfall during the crop season is essential for obtaining high soybean yields under rainfed conditions. A timely onset of rains ensures timely planting of soybean. However, rainfall prior to crop emergence results in poor emergence owing to crust formation or oxygen deficiency (Hamada *et al.*, 2007).

Rainfall may influence soybean yields due to the effect of time of rain, pattern of distribution, intensity of rain and so on. Delayed onset of rains, erratic distribution and high-intensity rainfall are expected to result in low yields (Hajare *et al.*, 2003). With delayed onset of rains the planting of soybean is delayed; with erratic distribution the moisture is not supplied to the plants at the right time or in the right quantity; and with high-intensity rain waterlogging may occur. Using regression analysis, it has been calculated that for obtaining high yields of soybean under the rainfed conditions of central India, genotypes should be planted that flower in approximately 37 days, mature in approximately 92 days and have a seed-fill duration of approximately 33 days (Bhatia and Ramesh, 2009).

9.8 Factors Affecting Water Use and Water-use Efficiency in Soybean

Method of sowing

Flood irrigation is a common practice in flat-bed-sown crops, whereas in raised bed or ridge-sown crops irrigation is applied in furrows. Furrow irrigation has a higher WUE than flood irrigation in soybean. Furrow opening after two rows of soybean provide a significantly higher seed yield and WUE compared with flat sowing without furrow opening (Autkar *et al.*, 2006).

Soybean has a higher WUE in a broad-bed and furrow system with irrigation at 0.6 IW:CPE ratio than in a flat-bed system (Bharambe *et al.*, 1999). The width of the raised bed may influence the crop yields. Although less water is used for irrigation with wider raised beds, the central rows in the bed are not able to benefit from the furrow-applied irrigation. On the other hand, with a narrower width of the raised bed, and consequently more frequent furrows, drainage of excess water is easier and fast in the case of heavy rains. Among 6, 9, 12 and 15 m-wide raised beds, the highest seed yields of soybean have been obtained in the 6 m raised bed and the lowest in the 15 m raised bed (Tomar *et al.*, 1999). A broad-bed and furrow system helps in decreasing run-off and increasing infiltration of rainfall (Singh *et al.*, 1999b). In some areas the plant population and crop growth are adversely affected due to water accumulation in the case of flat-bed-sown crops, whereas the ridge and furrow system of planting helps to avoid these problems (Lakpale *et al.*, 2009), which may ultimately result in high seed yields and WUE.

No-till sowing of crops, including soybean, is gaining in popularity among farmers due to various advantages associated with this method of sowing. The water intake in a no-tilled field is lower than that in a tilled field. In a no-tilled field, less irrigation water is applied at each irrigation, especially in early irrigations, than in a conventionally tilled field, although irrigation may be required more frequently. Mechanical impedance of the surface soil is much higher in no-tillage than in conventional tillage fields; because of this, the root length density and root branching index of soybean are higher in the surface soil layer in no-tilled field and, therefore, more dependent on irrigation (Iijima *et al.*, 2007).

Selection of crop variety

Soybean varieties differ in their WUE (Al-Assily and Mohamed, 2002; Hufstetler *et al.*, 2007). Some varieties may be more tolerant to water stress (or drought) than others. Indeterminate cultivars are known to be better able to recover from water stress than determinate types (Villalobos-Rodriquez and Shibles, 1985). Some studies have reported drought tolerance and high WUE in some genotypes of soybean (Noureldin *et al.*, 2002b). Moisture-efficient genotypes have root and shoot characteristics that allow the plant to use moisture in an effective manner.

Growing season

Irrigation demands depend upon the prevailing weather conditions. Seasonal rainfall and the prevailing temperature are important weather parameters that affect the WUE in soybean. Seed yield as well as WUE in soybean is higher during the *kharif* (rainy) season than during the summer season (Elamathi and Singh, 2000). High temperatures during the summer result in greater losses of moisture.

Plant population

The plant population should be the optimum for realizing high seed yields of soybean. Too low or too high plant populations not only result in poor seed yields, but WUE is also lowered. In the case of suboptimal plant populations, water loss is more through evaporation, while with too high plant population there is more water loss through transpiration and there may be a shortage of water during the reproductive phase.

Rooting characteristics

Root length and the pattern of root growth influence WUE. Deep-rooted genotypes extract water from deeper layers; this is not taken up by shallow-rooted genotypes and is 'lost'. Deep-rooted genotypes can also withstand water stress far better than those with shallow roots.

Genotypes may differ in the pattern of root growth. Some genotypes have a more horizontal root growth, whereas others have a more vertical root growth. Vertical root-growth patterns are expected to provide a higher WUE than horizontal root-growth patterns, along with greater tolerance under drought conditions.

Crop duration

Long-duration genotypes stand in the field for a longer period and, therefore, may require a greater number of irrigations than short-duration genotypes. When grown during the same season, long-duration genotypes are expected to require greater amounts of irrigation water than short-duration ones. Terminal water deficits reduce soybean yields more in the case of long-season cultivars than in short-season ones (Muchow and Sinclair, 1986), as short-duration cultivars mature before the occurrence of water stress or experience water stress for a shorter period.

Sowing time

The sowing time greatly influences crop growth, crop yield and WUE. In Indonesia, December-sown soybean has been found to produce almost twice the yield of January-sown crop (van Cooten and Borrell, 1999), as the early sown crop was able not only to match growth with water supply, but also avoided end-of-season drought.

Method of irrigation

The irrigation method used for raising a soybean crop greatly influences WUE. In flood irrigation, water is applied to the whole area of the field and is many cases the surface is not uniform, resulting in a great deal of wastage of irrigation water. In the case of furrow irrigation, water is applied only in the furrows, covering less area than is normally covered under flood irrigation and, therefore, WUE is normally higher with furrow irrigation.

Alternate furrow irrigation can further enhance WUE by using less water. In sprinkler and drip irrigation systems, WUE is further improved as water losses are checked to a large extent. Sprinkler and drip irrigation systems are normally used in fields with uneven surfaces and in areas where there are severe water shortages. With surface and subsurface drip irrigation systems, the latter has higher irrigation WUE due to lower evaporation loss (Bhattarai *et al.*, 2008). However, the initial high costs involved in establishing these systems is a major factor in these systems remaining unpopular among farmers.

Mulch application

Straw mulches check evaporation losses and thereby improve moisture availability for crop plants, ultimately leading to high seed yields. The surface application of straw mulch and straw incorporation may both have beneficial effects on improving WUE. For example, sugarcane (*Saccharum officinarum*) trash incorporation has been found to increase WUE in soybean (Bharambe *et al.*, 2002).

Depending upon availability, the straw of wheat (*Triticum aestivum*), paddy (*Oryza sativa*) or any other crop can be used for mulching. However, the rate of application should be such that it does not have any adverse effect on the emergence of soybean plants when it is applied immediately after sowing. Alternatively, straw mulch may be applied between crop rows after emergence. Straw mulch application is often not be feasible due to the costs involved in its application. However, if the previous crop is harvested with a combine harvester (e.g. wheat) then soybean may be sown with no-tillage (using Happy Seeder) and wheat straw can serve as mulch.

Irrigation at the critical stage

Irrigation at the critical growth stage may save the crop and results in higher WUE over its application at other crop stages. WUE is higher with irrigation only at R5 than at other stages (Kim *et al.*, 1999). During the early stages, there is little crop cover and consequently evaporation losses are high. Reducing irrigations during the vegetative stage helps to improve WUE (Neyshabouri and Hatfield, 1986) by avoiding large evaporation losses.

Weed control

Weeds, aside from competing for other resources, also compete with crop plants for moisture. Some weed species are more vigorous than soybean plants and thus have greater ability to extract water. Therefore, the removal of weeds at appropriate times is a must for checking crop–weed competition for various resources, including moisture, and thereby improves WUE of soybean.

Fertilizer application

An optimally fertilized crop exhibits proper growth and development, high yield and high WUE. The highest ET and WUE in soybean have been reported with the application of 100% of recommended nitrogen, phosphorus and potassium (NPK) levels plus 10 t farmyard manure ha^{-1} compared to with 100% of recommended NPK levels alone or no fertilizer (Hati *et al.*, 2000).

9.9 Conclusions

An optimum supply of moisture is essential for obtaining high seed yields of soybean. Both drought and waterlogging adversely influence soybean growth and yield. Soybean is mainly grown as a rainfed crop in different parts of the world. There is a need to efficiently manage both the crop and rain water to avoid losses due to moisture stress. Rain water conservation practices need to be promoted so that the conserved water can be used for life-saving irrigation.

Soybean generally responds to irrigation. Optimum irrigation schedules have been calculated for obtaining high productivity. However, the number of irrigation applications depends on the actual availability of irrigation water in an area. Water use and WUE are influenced by various factors such as the method of sowing, crop variety, growing season, plant population, crop duration, method of irrigation, mulch application, irrigation application at the critical stage, weed management and fertilizer application. Water is a limited resource and the availability of water is expected to decrease further in the future. Therefore, there is a need to obtain high yields per drop of water using efficient water management techniques.

References

Al-Assily, K.A. and Mohamed, K.A. (2002) Soil moisture deficit effect on some soybean genotypes production in Upper Egypt. *Arab Universities Journal of Agricultural Sciences* 10, 153–163.

Autkar, K.S., Gawande, R.L., Vyas, J.S. and Ghodpage, R.M. (2006) Effect of land configuration on yield and water use efficiency of soybean in vertisols of Vidarbha (M.S.). *Annals of Plant Physiology* 20, 158–159.

Bahmaniar, M.A. and Sepanlou, M.G. (2008) Influence of saline irrigation water and gypsum on leaf nutrient accumulation, protein, and oil seed in soybean cultivars. *Journal of Plant Nutrition* 31, 485–495.

Benjamin, J.G. and Nielsen, D.C. (2006) Water deficit effects on root distribution of soybean, field pea and chickpea. *Field Crops Research* 97, 248–253.

Bharambe, P.R., Oza, S.R., Shelke, D.K. and Jadhav, G.S. (1999) Effect of irrigation on crop productivity and soil-plant water relation in soybean–*rabi* sorghum cropping system. *Journal of the Indian Society of Soil Science* 47, 689–694.

Bharambe, P.R., Tajuddin, A.H., Oza, S.R. and Shelke, D.K. (2002) Effect of irrigation and crop residue management on crop and soil productivity under soybean–sorghum cropping system. *Journal of the Indian Society of Soil Science* 50, 233–236.

Bhatia, V.S. and Ramesh, A. (2009) Matching soybean (*Glycine max*) phenology for optimum yield under rainfed production system of central India. In: Qiu, L., Guan, R., Jin, J., Song, Q., Guo, S., Li, W., Wang, Y., Han, T., Liu, X., Yu, D., Jiang, L. and Peng, D. (eds) *Developing a Global Soy Blueprint for a Safe Secure and Sustainable Supply. Abstracts for oral presentations and posters of the World Soybean Research Conference VIII*, Beijing, China, 10–15 August 2009. The Chinese Academy of Agricultural Sciences, Beijing, China, p. 26.

Bhattarai, S.P., Midmore, D.J. and Pendergast, L. (2008) Yield, water-use efficiencies and root distribution of soybean, chickpea and pumpkin under different surface drip irrigation depths and oxygation treatments in vertisols. *Irrigation Science* 26, 439–450.

Blanco, F.F., Folegatti, M.V., Gheyi, H.R. and Fernandes, P.D. (2007) Emergence and growth of corn and soybean under saline stress. *Scientia Agricola* 64, 451–459.

Brevedan, R.E. and Egli, D.B. (2003) Short periods of water stress during seed filling, leaf senescence, and yield of soybean. *Crop Science* 43, 2083–2088.

de Costa, W.A.J.M. and Shanmugathasan, K.N. (2002) Physiology of yield determination of soybean [*Glycine max* (L.) Merr.] under different irrigation regimes in the sub-humid zone of Sri Lanka. *Field Crops Research* 75, 23–25.

Deolankar, K.P. and Mogal, A.D. (1998) Response of soybean to irrigation and fertilizer. *Journal of Maharashtra Agricultural Universities* 23, 189–190.

Dogan, E., Kirnak, H. and Copur, O. (2007a) Effect of seasonal water stress on soybean and site specific evaluation of CROPGRO-soybean Model under semi-arid climatic conditions. *Agricultural Water Management* 90, 56–62.

Dogan, E., Kirnak, H. and Copur, O. (2007b) Deficit irrigations during soybean reproductive stages and CROPGRO-soybean simulations under semi-arid climatic conditions. *Field Crops Research* 103, 154–159.

Elamathi, S. and Singh, S.D.S. (2000) Effect of irrigation and management practices for water use efficiency of soybean. *Madras Agricultural Journal* 87, 307–310.

Elamathi, S. and Singh, S.D.S. (2001) Effect of irrigation management on physiology and productivity of soybean (*Glycine max*). *Indian Journal of Agronomy* 46, 145–150.

Frederick, J.R., Camp, C.R. and Bauer, P.J. (2001) Drought-stress effects on branch and mainstem seed yield and yield components of determinate soybean. *Crop Science* 41, 759–763.

Gercek, S., Boydak, E., Okant, M. and Dikilitaş, M. (2009) Water pillow irrigation compared to furrow irrigation for soybean production in a semi-arid area. *Agricultural Water Management* 96, 87–92.

Ghosh, A.K., Ishijiki, K., Toyota, M., Kusutani, A. and Asanuma, K. (2000a) Biomass, growth and matter partitioning in soybean plants under long-term moisture deficit. *Japanese Journal of Tropical Agriculture* 41, 20–29.

Ghosh, A.K., Ishijiki, K., Toyota, M., Kusutani, A. and Asanuma, K. (2000b) Water potential, stomatal dimension and leaf gas exchange in soybean plants under long-term moisture deficit. *Japanese Journal of Tropical Agriculture* 44, 30–37.

Hajare, T.N., Mandal, D.K., Prasad, J. and Patil, V.P. (2001) Effect of moisture stress on biomass yield of soybean (*Glycine max*) in Nagpur district, Maharashtra. *Agropedology* 11, 17–22.

Hajare, T.N., Mandal, D.K. and Challa, O. (2003) Influence of rainfall on productivity of soybean–safflower cropping sequence in dry subhumid condition of Viderbha. *Journal of Maharashtra Agricultural Universities* 28, 89–90.

Hamada, Y., Shaku, I., Sawada, Y. and Kojima, H. (2007) Factors causing submergence damage on soybean seedling emergence under no-till culture. *Japanese Journal of Crop Science* 76, 212–218.

Hati, K.M., Mandal, K.G. and Misra, A.K. (2000) Effect of nutrient management on water use and yield of rainfed soybean grown on a cracking clay soil. *Environment and Ecology* 18, 202–206.

Heatherly, L.G. and Spurlock, S.R. (2000) Furrow and flood irrigation of early-planted, early-maturing soybean rotated with rice. *Agronomy Journal* 92, 785–791.

Hufstetler, E.V., Boerma, H.R., Carter, T.E., Jr and Earl, H.J. (2007) Genotypic variation for three physiological traits affecting drought tolerance in soybean. *Crop Science* 47, 25–35.

Iijima, M., Morita, S., Zegada-Lizarazu, W. and Izumi, Y. (2007) No-tillage enhanced the dependence on surface irrigation water in wheat and soybean. *Plant Production Science* 10, 182–188.

Karam, F., Masaad, R., Sfeir, T., Mounzer, O. and Rouphael, Y. (2005) Evapotranspiration of seed yield of field grown soybean under deficit irrigation conditions. *Agricultural Water Management* 75, 226–244.

Kazi, B.R., Oad, F.C., Jamro, G.H., Jamali, L.A. and Lakho, A.A. (2002) Effect of irrigation frequencies on growth and yield of soybean. *Pakistan Journal of Applied Sciences* 2, 661–662.

Kim, W.H., Hong, B.H. and Ryu, Y.H. (1999) Grain yield and water use efficiency as affected by irrigation at different growth stages. *Korean Journal of Crop Science* 44, 330–338.

King, C.A. and Purcell, L.C. (2001) Soybean nodule size and relationship to nitrogen fixation response to water stress. *Crop Science* 41, 1099–1107.

Kumawat, S.M., Dhakar, L.L. and Maliwal, P.L. (2000) Effect of irrigation regimes and nitrogen on yield, oil content and nutrient uptake of soybean (*Glycine max*). *Indian Journal of Agronomy* 45, 361–366.

Lakpale, R., Tuteja, S.S, Shrivastava, G.K. and Swamy, S.L. (2009) Broadbed furrow and ridge and furrow method of sowing of soybean (*Glycine max* L.) for high rainfall areas of Chhattisgarh plains of India. In: Qiu, L., Guan, R., Jin, J., Song, Q., Guo, S., Li, W., Wang, Y., Han, T., Liu, X., Yu, D., Jiang, L. and Peng, D. (eds) *Developing a Global Soy Blueprint for a Safe Secure and Sustainable Supply. Abstracts for oral presentations and posters of the World Soybean Research Conference VIII*, Beijing, China, 10–15 August 2009. The Chinese Academy of Agricultural Sciences, Beijing, China, p. 194.

Lamm, F.R., Stone, L.R. and O'Brien, D.M. (2007) Crop production and economics in Northwest Kansas as related to irrigation capacity. *Applied Engineering in Agriculture* 23, 737–745.

Likoswe, A.A. and Lawn, R.J. (2008) Response to terminal water deficit stress of cowpea, pigeonpea, and soybean in pure stand and in competition. *Australian Journal of Agricultural Research* 59, 27–37.

Liu, F.L., Andersen, M.N. and Jensen, C.R. (2003) Loss of pod set caused by drought stress is associated with water stress and ABA content of reproductive structures in soybean. *Functional Plant Biology* 30, 271–280.

Liu, F., Jensen, C.R. and Andersen, M.N. (2004) Drought stress effect on carbohydrate concentration in soybean leaves and pods during early reproductive development: Its implication in altering pod set. *Field Crops Research* 86, 1–13.

McKenzie, D.C., Bernardi, A.L., Chan, K.Y., Nicol, H.I., Banks, L.W. and Rose, K.L. (2002) Sodicity v. yield decline functions for a vertisol (grey vertisol) under border check and raised bed irrigation. *Australian Journal of Experimental Agriculture* 42, 363–368.

Mohamed, A.R.A.G. (1995) Nitrogen accumulation, seed yield and water use of three grain legume species grown under different water regimes. *Egyptian Journal of Agronomy* 20, 27–45.

Muchow, R.C. and Sinclair, T.R. (1986) Water and nitrogen limitations in soybean grain production II. Field and model analyses. *Field Crops Research* 15, 143–156.

Nakayama, N., Shimada, S., Takahashi, M., Kim, Y.H. and Arihara, J. (2005) Effects of water-absorbing rate of seed on flooding injury in soybean. *Japanese Journal of Crop Science* 74, 325–329.

Neyshabouri, M.R. and Hatfield, J.L. (1986) Soil water deficit effects on semi-determinate and indeterminate soybean growth and yield. *Field Crops Research* 15, 73–84.

Noureldin, N.A., Hassan, M.Z., Hassaan, R.K. and Abdel-Wahab, S.I. (2002a) Performance of some soybean genotypes in sandy soil as influenced by some abiotic stresses. I. Effect on vegetative growth characteristics. *Annals of Agricultural Science, Cairo* 47, 191–207.

Noureldin, N.A., Hassan, M.Z., Hassaan, R.K. and Abdel-Wahab, S.I. (2002b) Performance of some soybean genotypes in sandy soil as influenced by some abiotic stresses. II. Effect on seed yield and some yield attributes. *Annals of Agricultural Science, Cairo* 47, 209–223.

Ohashi, Y., Saneoka, H. and Fujita, K. (2000) Effect of water stress on growth, photosynthesis, and photoassimilate translocation in soybean and tropical pasture legume siratro. *Soil Science and Plant Nutrition* 46, 417–425.

Osman, A.S., Abido, Y.M.Y. and Allam, S.M.M. (2000) Response of soybean to phosphorus and zinc fertilization under irrigation regime. *Annals of Agricultural Science, Cairo* 45, 229–238.

Purcell, L.C., Serraj, R., Sinclair, T.R. and De, A. (2004) Soybean N_2 fixation estimates, ureide concentration, and yield responses to drought. *Crop Science* 44, 484–492.

Rabbani, M.F., Ashrafuzzaman, M., Hoque, A.M. and Karim, M.A. (2004) Responses of soybean genotypes to different levels of irrigation. *Korean Journal of Crop Science* 49, 131–135.

Rajendran, K. and Lourduraj, A.C. (2000) Effect of irrigation regimes, plant population and phosphorus levels on the yield attributes and yield of soybean [*Glycine max* (L.) Merrill]. *Acta Agronomica Hungarica* 48, 263–269.

Ramasamy, M. and Sankaran, N. (2001) Yield and physiological attributes of soybean as influenced by P mobilizers under varying irrigation regimes. *Madras Agricultural Journal* 88, 21–25.

Rao, S.K., Shrivastava, A.N. and Mishra, R. (1999) Effect of excess soil moisture on seed yield, seed size development and quality in commercial varieties of soybean. *Legume Research* 22, 259–262.

Ray, J.D., Heatherly, L.G. and Fritschi, F.B. (2006) Influence of large amounts of nitrogen on nonirrigated and irrigated soybean. *Crop Science* 46, 52–60.

Seong, R.C., Kim, J.G. and Nelson, C.J. (1999) Dry matter accumulation and leaf mineral contents as affected by excessive soil water in soybean. *Korean Journal of Crop Science* 44, 129–133.

Seong, R.C., Park, S.J. and Shim, S.I. (2000a) Effects of ridge height in dry paddy field on growth and seed yield of soybean cultivars. *Korean Journal of Crop Science* 45, 88–92.

Seong, R.C., Sohn, J.Y. and Shim, S.I. (2000b) Responses of soybean cultivars to excessive soil moisture imposed at different growth stages. *Korean Journal of Crop Science* 45, 282–287.

Serraj, R., Bona, S., Purcell, L.C. and Sinclair, T.R (1997) Nitrogen accumulation and nodule activity of field-grown 'Jackson' soybean in response to water deficits. *Field Crops Research* 52, 109–116.

Sinclair, T.R., Purcell, L.C., King, C.A., Sneller, C.H., Chen, P. and Vadez, V. (2007) Drought tolerance and yield increase of soybean resulting from improved symbiotic N_2 fixation. *Field Crops Research* 101, 68–71.

Singh, P., Alagarswamy, G., Hoogenboom, G., Pathak, P., Wani, S.P., and Virmani, S.M. (1999a) Soybean–chickpea rotation on Vertic Inceptisols II. Long-term simulation of water balance and crop yields. *Field Crops Research* 63, 225–236.

Singh, P., Alagarswamy, G., Pathak, P., Wani, S.P., Hoogenboom, G. and Virmani, S.M. (1999b) Soybean–chickpea rotation on vertic inceptisols: I. Effect of soil depth and landform on light interception, water balance and crop yields. *Field Crops Research* 63, 211–224.

Singh, R. (1998) Rain water management: Principles and practices. In: Yadav, R.L., Singh, P., Prasad, R. and Ahlawat, I.P.S. (eds) *Fifty Years of Agronomic Research in India*. Indian Society of Agronomy, New Delhi, India, pp. 141–154.

Singh, R. and Prakash, C. (2001) Consumptive use and water-use-efficiency of soybean using lysimeter. *Indian Journal of Soil Conservation* 29, 86–87.

Singh, R., Prasad, S.N., Prakash, C. and Chauhan, V. (2001) Evapotranspiration and water-use-efficiency of soybean and mustard under lysimeter in south-eastern Rajasthan. *Indian Journal of Soil Conservation* 29, 22–25.

Sweeney, D.W. and Granade, G.V. (2002) Effect of a single irrigation at different reproductive growth stages on soybean planted in early and late June. *Irrigation Science* 21, 69–73.

Sweeney, D.W., Log, J.H. and Kirkham, M.B. (2003) A single irrigation to improve early maturing soybean yield and quality. *Soil Science Society of America Journal* 67, 235–240.

Thomas, A.L. and Sodek, L. (2005) Development of the nodulated soybean plant after flooding of the root system with different sources of nitrogen. *Brazilian Journal of Plant Physiology* 17, 291–297.

Thompson, A.L., Henggeler, J.C., Bockhold, D.C. and Sudduth, K.A. (2002) Comparison of eight irrigation scheduling tools on soybean and cotton. *International Water and Irrigation* 22, 24–30.

Tomar, S.S., Tembe, G.P. and Sharma, S.K. (1999) Productivity and economic analysis of raised sunken bed technology for increasing agricultural production in black soils of central Madhya Pradesh. *JNKVV Research Journal* 33, 25–33.

Troedson, R.J., Lawn, R.J., Byth, D.E. and Wilson, G.L. (1989) Response of field-grown soybean to saturated soil culture. 1. Patterns of biomass and nitrogen accumulation. *Field Crops Research* 21, 171–187.

van Cooten, D.E. and Borrell, A.K. (1999) Enhancing food security in semi-arid eastern Indonesia through permanent raised-bed cropping: A review. *Australian Journal of Experimental Agriculture* 39, 1035–1046.

van Hoorn, J.W., Katerji, N., Hamdy, A. and Mastrorilli, M. (2001) Effect of salinity on yield and nitrogen uptake of four grain legumes and on biological nitrogen contribution from the soil. *Agricultural Water Management* 51, 87–98.

Velu, G. (1999) Biochemical response of soybean to moisture stress and ameliorants application. *Madras Agricultural Journal* 86, 661–663.

Villalobos-Rodriquez, E. and Shibles, R. (1985) Response of determinate and indeterminate tropical soybean cultivars to water stress. *Field Crops Research* 10, 269–281.

Wang, D., Shannon, M.C., Grieve, C.M. and Yates, S.R. (2000) Soil water and temperature regimes in drip and sprinkler irrigation and implications to soybean emergence. *Agricultural Water Management* 43, 15–28.

Yazdani, F., Allahdadi, I. and Akbari, G.A. (2007) Impact of super absorbent polymer on yield and growth analysis of soybean (*Glycine max* L.) under drought stress condition. *Pakistan Journal of Biological Sciences* 10, 4190–4196.

10 Weed Management in Soybean

J.S. Mishra
Directorate of Sorghum Research, Rajendranagar, Hyderabad, Andhra Pradesh, India

10.1 Introduction

Soybean (*Glycine max* (L.) Merrill) is a globally important oilseed crop. It occupies third place among the nine oilseed crops of India. Being a rainy-season crop, it suffers heavily due to weed competition and losses due to weeds have been one of the major limiting factors in soybean production. Approximately 37% of attainable production is endangered by weed competition worldwide, as compared to 11%, 1% and 11% endangered by pathogens, viruses and animal pests, respectively (Table 10.1) (Oerke, 2006). The wide row spacing and slow initial growth rate of soybean provides a congenial atmosphere for an abundant population and profuse growth of weeds. Due to scarcity of labour for weeding operation and the rainy season, which allows weeds to come in several flushes, weeds are not managed efficiently at the critical stage, which leads to severe competitive stress on crop growth and productivity. Effective weed control is, therefore, one of the most important practices for economical soybean production.

10.2 Major Weeds of Soybean

Being a rainy-season crop and having wide row spacing and a slow initial growth rate, soybean is heavily infested with grasses, sedges and broadleaf weeds. Among these, the infestation and competitive stress of grassy weeds is very serious. Due to higher competitive ability and diversity, barnyard grass (*Echinochloa crus-galli*) is a major weed among grasses and reduces yield considerably. The major weeds of soybean are listed in Table 10.2.

Table 10.1. Estimated potential losses due to weeds, animal pests (arthropods, nematodes, rodents, birds, slugs and snails), pathogens (fungi, bacteria) and viruses, and actual losses due to pest groups in soybean worldwide in 2001–2003 (reprinted with permission from Oerke, 2006).

	Crop losses (%)[a]	
Pests	Potential	Actual
Weeds	37.0 (35–40)	7.5 (5–16)
Animal pests	10.7 (4–16)	8.8 (3–16)
Pathogens	11 (7–16)	8.9 (3–16)
Viruses	1.4 (0–2)	1.2 (0–2)
Total	60.0 (49–69)	26.3 (11–49)

[a]Figures in parentheses indicate variation among 19 regions.

Table 10.2. Major weeds of soybean.

Category	Scientific name	Common name	Family
Annual grasses and sedges	*Echinochloa colona* (L.) Link	Jungle rice	Poaceae
	Cyperus esculentus L.	Yellow nutsedge	Cyperaceae
	Cyperus iria L.	Rice flat sedge	Cyperaceae
	Dactyloctenium aegyptium (L.) Willd.	Crowfoot grass	Poaceae
	Digitaria sanguinalis (L.) Scop.	Large crab grass	Poaceae
	Echinochloa crus-galli (L.) Beauv.	Barnyard grass	Poaceae
	Eleusine indica (L.) Gaertn.	Goose grass	Poaceae
	Fimbristylis spp.	Globe fingerush	Cyperaceae
	Panicum maximum Jacq.	Guinea grass	Poaceae
	Scirpus grossus L.	–	Cyperaceae
	Setaria glauca (L.) Beauv.	Yellow foxtail	Poaceae
Broadleaf weeds	*Amaranthus hybridus*	Pigweed	Amaranthaceae
	Amaranthus viridis L.	Pig weed	Amaranthaceae
	Cassia obtusifolia	Senna	Fabaceae
	Celosia argentea L.	Cock's comb	Amaranthaceae
	Chenopodium album L.	Common lambsquarters	Chenopodiaceae
	Cleome viscosa L.	Cleome	Capparidaceae
	Commelina benghalensis L.	Day flower	Commelinaceae
	Commelina communis L.	Common dayflower	Commelinaceae
	Euphorbia geniculata Orteg.	Wild poinsettia	Euphorbiaceae
	Ipomoea hederacea Jacq.	Ivy-leaved morning glory	Convolvulaceae
	Ipomoea purpurea	Morning glory	Convolvulaceae
	Physalis minima L.	Ground cherry	Solanaceae
	Sida spinosa L.	Spiny sida	Malvaceae
	Trianthema portulacastrum L.	Carpet weed	Aizoaceae
	Xanthium strumarium L.	Cocklebur	Compositae
Perennials	*Cyperus rotundus* L.	Purple nutsedge	Cyperaceae
	Cynodon dactylon (L.) Pers.	Bermuda grass	Poaceae
	Saccharum spontaneum L.	Tiger grass	Poaceae
	Sorghum halepense (L.) Pers.	Johnson grass	Poaceae

10.3 Losses due to Weeds in Soybean

Weeds compete with soybean for solar radiation, nutrients and soil moisture when they are limited and the early-season competition is the most critical. The effects of weed competition depend on the growth habits of the weed species, number and density of weeds, stand and cultivar of soybean crop, climatic factors and competition for nutrients, soil moisture and the mutual shading of crop and weeds. The losses comprise: (i) direct yield losses from competition; (ii) indirect losses from reduced crop quality; (iii) increased costs in harvesting, land preparation and similar operations; and (iv) harbouring of insect pests and diseases. Some weeds also produce and release allelochemicals that adversely affect crop plants. In addition to all of this, weeds remove 30–60 kg N ha^{-1}, 8–10 kg P ha^{-1} and 40–100 kg K ha^{-1} from the soil. Weeds in soybean have been found to deplete soil fertility by taking 53.24 kg N ha^{-1} and 9.30 kg P ha^{-1} under unweeded conditions (Chhokar *et al.*, 1997). Weeds particularly affect crop productivity via competition for inorganic nutrients (Oerke, 2006). Yield loss due to weeds ranges from 20% to 85% depending on the crop cultivar, nature and intensity of weeds, spacing, duration of weed infestation and environmental conditions (Tiwari and Kurchania, 1990; Singh and Singh, 1992; Tiwari *et al.*, 1996).

Effect of weed density on soybean yield

A cursory review of a portion of the weed-competition literature leads to the conclusion increasing weed density decreases yield. However, the weed density–crop yield relationship diverges from linear. A few weeds usually do not affect yield; in addition, the maximum effect – total crop loss – obviously cannot be exceeded and usually occurs at less than maximum weed density (Zimdahl, 1980). Weed competition can thus be represented by a schematic sigmoidal relationship. Barrentine (1974) reported that one *Ipomoea purpurea* plant 30 cm^{-1} of row reduced the soybean yield by 52%. *Cassia obtusifolia* depressed the soybean yield by 19–32% or 34–35% by 7.7 weeds per m^2 on sandy soils at two locations. Forty plants of *Amaranthus hybridus* m^{-1} of row cut the soybean yield by 55% and only one plant m^{-1} of row reduced it by 18% (Moolani *et al.*, 1964). In India, Mishra and Singh (2001) found that ivy-leaved morning glory (*Ipomoea hederacea*), even at 1 per m^2, reduced the soybean yield by 44%. The presence of 5 to 80 *Commelina communis* plants per m^2 caused a 10.6–58.4% reduction in seed yield (Mishra *et al.*, 2002). *Euphorbia geniculata*, another major weed of soybean, reduced the seed yield of soybean by 12–30% with increasing densities from 10–120 plants per m^2 (Mishra and Singh, 2003).

10.4 Critical Period of Crop–Weed Competition

One of the major principles of crop–weed competition is that plants established earlier in the soil try to smother other species of plant arriving at later stages. Their leaves can shade younger plants that are still short in stature and their longer roots can find water and plant nutrients that are out of reach for smaller young plants. The critical period of weed control refers to the part of the crop-growing season in which weeds should be removed in order to prevent crop loss due to weed competition (Zimdahl, 1988; Hall *et al.*, 1992). This depends not only on the nature of weed and crop, environmental conditions and weed density, but also on the period for which weeds are associated with the crop. Grasses are usually the most dominant in competition during the early season, whereas sedges and broad-leaf weeds dominate later in the season. The maximum yield reduction due to weed competition occurs during the first 45 days after sowing (DAS); therefore, weed management should be empasized during this period.

10.5 Weed Management Practices

Weed management in soybean requires an integrated approach that utilizes preventive, cultural, mechanical, chemical and biotechnological methods in a mutually supported manner in the crop production system, with due consideration of economic, environmental and sociological consequences (Yaduraju and Mishra, 2004). Various methods of weed management in soybean (see below) have been used with different degrees of success in different agro-ecological zones and production systems.

Weed prevention

Prevention includes methods that inhibit or delay the introduction and establishment of new weeds. The success of a preventive programme varies with species and the amount of effort devoted to control. The prevention of a weed problem is usually easier and less costly than control or eradication. The following measures are suggested to prevent the introduction of weeds into non-inhabited fields:

- Use 'clean' (free of weed seeds) crop seed for planting.
- Use organic manures only after thorough decomposition to kill weed seeds.
- Clean harvesters and tillage implements before moving to non-weed-infested areas.
- Avoid the transportation or use of soil from weed-infested area.
- Remove weeds that are near irrigation ditches, fencerows, right of ways and other non-crop land.
- Prevent reproduction of weeds.
- Use weed-seed screens to filter irrigation water.
- Restrict livestock movement into non-weed infested areas.

Other practices that are used to prevent and avoid potential weed problem at the state, regional and national levels are weed laws, seed laws and quarantines. In the absence of strict quarantine laws a large number of weeds (including some exotic species) may be introduced into any country through the import of food grains. Lately, however, new laws have been enacted and food-grain movements are closely watched at international ports.

Cultural methods

Despite the great progress in agriculture, manual and mechanical methods continue to be important weed management practices in many parts of the world. Cultural methods are used to complement manual and mechanical methods. Cultural practices are manipulated in such a way that they become more favourable for crop growth and less favourable to weeds. They are not only eco-friendly, but also reduce the use of costly herbicides.

Stale seedbed technique

The principle of flushing out germinable weed seeds before cropping forms the basis of the stale or false seedbed technique, in which soil cultivation may take place days or weeks before planting a crop. This depletes the seed bank in the surface layer of the soil and reduces subsequent weed emergence. Where light rains occur for an extended period before the onset of the monsoon, or where irrigation is available, it may be possible to kill several flushes of weed growth before planting. To ensure success, cropping should be delayed until the main flush of emergence has passed. The emerged weed seedlings may then be killed by light cultivation or application of contact non-residual herbicides such as Paraquat. It is vital not to disturb below the top 1–2 cm of soil, otherwise a further flush of weeds may emerge.

Soil solarization

Soil solarization is a hydrothermal process that results in elevating temperatures to levels that are lethal to many soil-borne plant pathogens, insects, nematodes and weeds and causes other physical and biological changes in the soil that are beneficial to crop growth (Katan, 1981; Stapleton and Lopez, 1988). It involves covering the soil with transparent polyethylene films for 2–6 weeks during the hot summer months. It has the potential to raise the maximum soil temperature by 8–12°C over unfilmed control (Yaduraju, 1993). In a long-term trial conducted at the Indian Agricultural Research Institute, New Delhi, India, it was found that solarization gave a 33% and 52% greater yield of soybean over hand weeding and herbicide treatment, respectively (Yaduraju and Ahuja, 1996). The corresponding increases in the succeeding wheat crop were 10% and 25%. Soil solarization for a period of 32 days has been found to improve the growth of soybean and increase the seed yield by 78% (Kumar *et al.*, 1993). Singh *et al.* (2004) observed that soil solarization for 5 weeks during the summer significantly reduced the population and dry matter production of major weeds and increased the seed yield of soybean. Although very efficient, the solarization has not found wider adaptability as the treatment cost is relatively high. However, with repeated use of the same films the costs can be reduced substantially.

Crop rotation

Crop rotation is a fundamental tool in integrated weed management. The major goal of adopting a particular crop rotation is to reduce the number of weed seeds available for germination in the following season. It can play a long-term role in weed management by preventing particular weed species from adapting to the growth cycle of specific crops (Akobundu, 1987). Certain weed species are often associated with particular crops and the population of such weeds usually increases when that crop is grown in the same field continuously for several seasons. This is because some environment or cultural conditions that favour crop production also tend to favour the weeds. Such weed associations with crops may be discouraged by growing in sequence crops that have sharply contrasting growth and cultural requirements. The more diverse the crops are in a rotation, planting time, growth habit and life cycle, the more effective the rotation will be in controlling weeds. The problem weed *Euphorbia geniculata* has been found to infest more in soybean–chickpea rotation than in soybean–wheat rotation (Mishra and Singh, 2003).

Mulching

Mulching is an age-old practice for moderating soil temperature, conserving soil moisture and controlling weed growth. Materials used for mulch include crop residues, straw, leaves, paper, plastic films, gravel and dry soil. Mulching the soil surface can prevent weed seeds germinating or physically suppress seedling emergence, but is not effective against established perennial weeds. The chemical effects of mulches on weed control include allelopathy, toxic microbial products and pH changes in the soil. Rajput (1980) and Rajput and Sastry (1986) reported that soybean yield increased by 29% and 13% under white plastic and straw mulching, respectively, over control. Mulching at 5 t ha^{-1}, however, has been found to effectively suppress weed growth and increase seed yield, but was not economically effective (Singh *et al.*, 1992; Nimje, 1996).

Plant geometry and plant density

Planting density and pattern modify the crop canopy structure and, in turn, influence weed-smothering ability. A good stand of soybeans, which emerge rapidly and shade the middles early, is helpful in reducing weed competition. With narrower row spacing, the soybean canopy closes earlier, which allows very little light to reach the soil surface or weeds beneath the canopy. Narrow row spacing brings variation in microclimate (i.e. light intensity, evaporation and temperature) at the soil surface.

The establishment of a crop with a more uniform and dense plant distribution may result in better use of light, water and nutrients and lead to greater crop competitive ability. Increased shading at the soil surface smothers weed growth. A soybean seed rate of 125 kg ha^{-1} in rows 20 cm apart has been found to be effective in minimizing weed intensity compared to other sowing management options (Jain and Tiwari, 1992). Yadav *et al.* (1999) observed that a higher seed rate (150 kg ha^{-1}) significantly reduced weed incidence and enhanced the soybean yield as compared with lower seed rates of 100 and 125 kg ha^{-1}. Singh and Bajpai (1994) reported that changes in crop geometry under different methods of sowing did not give significant weed control; however, crop sown at 30 cm row spacing smothered weed growth by 15.0% and 14.2% compared with 40 cm row spacing and the broadcast method of sowing, respectively. A reduction in row spacing from 45 to 25 cm increased the weed control efficiency by 21.7% and grain yield by 15.6% (Nimje, 1996).

Competitive cultivars

Cultivars differ in relative growth rate, spreading habit, height, canopy structure and inherent competitive character and accordingly differ in their weed-suppressing ability. A quick-growing and early canopy-producing cultivar would be expected to be a better competitor against weeds than crops lacking these characters. Seed size within a species also influences competition, with more vigorous plants produced from larger seeds (Spifters and Van Den Bergh, 1982). Tiwari *et al.* (1997) observed that different soybean varieties did not influence the population of barnyard grass or the total weed population or biomass. However, greater weed-control efficiency was noted with the variety Durga followed by JS 80-21, JS 72-44 and JS 76-205 compared with JS 75-46. An increased competitive ability of cultivars has been attributed to early emergence, seedling vigour, increased rate of leaf expansion, rapid creation of a dense canopy, increased plant height, early root growth and increased root size. Future breeding and variety testing programmes should take factors of crop competitive ability into consideration.

Mechanical methods

Hand hoeing and manual weeding are still the most common practices performed for weed control in many countries. The success of mechanical weeding depends upon the stage of weeds, crop geometry and climatic conditions. Hand weeding may also be used after mechanical inter-row weeding to deal with any weeds that are left over in crop rows. Hand weeding once at 30 DAS (Singh and Bajpai, 1994) and twice at 15 and 30 DAS (Dubey *et al.*, 1984; Lokras *et al.*, 1985) showed significant reductions in weed density, with marked increases in grain yield. Upadhyay *et al.* (1992) reported that weeding at 10 and 25 DAS is effective for controlling weeds.

Manual weeding with hand tools or inter-row cultivators is conducted between 3 and 6 weeks after sowing (depending upon the physical condition of the soil during the rainy season). It has been observed that when farmers depend solely on mechanical methods, two weedings are a must to provide season-long weed control. If preceded by pre-plant incorporated or pre-emergence or early post-emergence herbicide application, one inter-row cultivation or hand weeding is sufficient to provide season-long weed control.

Chemical methods

Herbicides are one of the most effective tools for weed management in soybean. In the light of modern cropping patterns and the agro-ecological situation, weed management through herbicides may be an effective and economical method. In some countries such as India, with soybean being a rainy-season crop, timely weed management through mechanical methods alone is often risky as the continuous rains prevent the use of this method. Under such a situation, a herbicidal approach provides effective control of weeds. However, knowledge of weed flora of the particular field is essential before using herbicides for effective and economical weed control. The weed problems must be anticipated for pre-plant and pre-emergence herbicides, since weeds may not have emerged at the time of herbicide application. Herbicide selection is also based on the ability of the product to control important weeds without causing any significant injury to the crop. For maximum weed-control efficiency from pre-emergence herbicides, proper soil moisture is needed within a week after application. Lack of moisture during this period often results in poor weed control. Promising herbicides in soybean are listed in Table 10.3.

Herbicide mixtures and their sequential application

Most herbicides control a group of specific weeds (grasses or broadleaved). However, the soybean crop suffers with mixed-weed flora (grasses, broadleaved and sedges). Therefore, for broad-spectrum weed control it is necessary either to use herbicide mixtures or sequential application. Post-emergence herbicides can be used in sequential application with all pre-planting or pre-emergence herbicides depending upon the nature of the weed flora. Balyan *et al.* (1999b) reported that sequential application of pre-emergence linuron (750–1000 g ha^{-1}) and post-emergence fluazifop (500 g ha^{-1}) provided better control of all weeds than their single application. A mixture of fluazifop-p-butyl (0.50 kg ha^{-1}) and sethoxydim (0.25 kg ha^{-1}) provided broad-spectrum weed control and a higher yield of soybean (Singh *et al.*, 1999). A tank mixture of fomesafen and haloxyfop at 200 and 150 g ha^{-1} and chlorimuron and haloxyfop at 6 and 150 g ha^{-1} provided season-long weed control and produced a grain yield of soybean similar to that achieved in weed-free conditions (Balyan and Malik, 2003).

Table 10.3. Promising herbicides for weed management in soybean.

Herbicide	Dose (g ha^{-1})	Time of application	Site of action	Remarks
Fluchloralin	1000	PPI	Microtubule assembly inhibitor	Apply before planting and incorporate into surface soil immediately after application. Controls many annual grasses and some broadleaved weeds.
Trifluralin	1000–1500	PPI	Microtubule assembly inhibitor	Apply before planting and incorporate into surface soil immediately after application. Controls many annual grasses and some broadleaved weeds.
Pendimethalin	1000	Pre-em or PPI	Microtubule assembly inhibitor	Incorporate into surface soil immediately after application when used before planting. Apply within 2 DAS as pre-em. Ensure sufficient soil moisture at the time of application. Controls many annual grasses and some broadleaved weeds.
Alachlor	1500	Pre-em	Shoot and root inhibitor	Controls many annual grasses and some broadleaved weeds and sedges. Ensure sufficient soil moisture, particularly when granules are used.
Oxyfluorfen	150–200	Pre-em	Protoporphyrinogen oxidase inhibitor	Controls a wide range of weeds including grasses, sedges and broadleaved weeds.
Butachlor	1500	Pre-em	Shoot and root inhibitor	Controls many annual grasses and some broadleaved weeds and sedges. Ensure sufficient soil moisture, particularly when granules are used.
Metolachlor	1000–1500	Pre-em	Shoot and root inhibitor	Controls many annual grasses and some broadleaved weeds.
Oxadiazon	750–1000	Pre-em	Protoporphyrinogen oxidase inhibitor	Controls many annual grasses and some broadleaved weeds.
Metribuzin	500–750	Pre-em	Photosystem II, binding site A inhibitor	Controls many annual grasses and broadleaved weeds.
Imazethapyr	75–100	15–20 DAS	Acetolactate synthase inhibitor	Controls many annual broadleaved weeds, nutsedge and some grasses. Always add a non-ionic surfactant (0.25% v/v) with imazethapyr.
Chlorimuron ethyl	10–12	15–20 DAS	Acetolactate synthase inhibitor	Controls many annual broadleaved weeds and some grasses and sedges.

(*Continued*)

Table 10.3. *continued*

Herbicide	Dose (g ha^{-1})	Time of application	Site of action	Remark
Metsulphuron methyl	4–6	15–20 DAS	Acetolactate synthase inhibitor	Controls many annual broadleaved weeds and some grasses and sedges.
Quizalofop-ethyl	40–50	15–20 DAS	Acetyl-coenzyme A carboxylase inhibitor	Excellent control of annual grasses.
Fenoxaprop-p-ethyl	80–100	15–20 DAS	Acetyl-coenzyme A carboxylase inhibitor	Excellent control of annual grasses. Can be used as sequential application with all pre-planting or pre-em herbicides.
Bentazon	750–1000	15–20 DAS	PS II (C) inhibitor	Controls many annual broadleaved weeds and sedges.
Clomazone	1000–1500	Pre-em	Diterpene inhibitor	Controls many annual grasses.
Fluazifop-p-ethyl	250–500	15–20 DAS	Acetyl-coenzyme A carboxylase inhibitor	Excellent control of annual grasses.
Lactofen	150–200	15–20 DAS	Protoporphyrinogen oxidase inhibitor	Controls many annual broadleaved weeds. Always apply a non-ionic surfactant (0.25% v/v) with lactofen.
Sethoxydim	250–500	15–20 DAS	Acetyl-coenzyme A carboxylase inhibitor	Excellent control of annual grasses.
Acetochlor	1000	Pre-em	Shoot and root inhibitor	Controls many annual grasses.
Imazaquin	50–70	Pre-em/PPI	Acetolactate synthase inhibitor	Controls many annual grasses. For maximum grass control, mix with trifluralin, pendimethalin or metolachlor.
Flumioxazin	30–45	Pre-em	Protoporphyrinogen oxidase (PPO) inhibitor	Provides good control of many annual broadleaved weeds.
Paraquat	1000	Pre-em to soybean	Photosystem I electron diverter	Controls annual weeds. Used to control emerged vegetation in no-till soybean. Good coverage is essential for effective control.
Glyphosate	1000	Pre-em to soybean	EPSPS inhibitor	Controls annual weeds. Used to control emerged vegetation in no-till soybean.
Glyphosate	750–1000	Post-em	EPSPS inhibitor	Apply glyphosate over the top of transgenic (glyphosate-resistant) soybean (Roundup Ready). Controls a wide range of grasses and broadleaved weeds.
Glufosinate	500	Post-em	Glutamine synthetase inhibitor	Apply to transgenic (glufosinate-resistant) soybean (Liberty Link). Controls broadleaved weeds. Less effective on grasses and sedges.

DAS, days after sowing; EPSPS, enolpyruvylshikimate-3-phosphate synthase; post-em, post-emergence; PPI, pre-plant incorporation; pre-em, pre-emergence.

Integrated weed management

Considering the diversity of weeds, no single method of weed control, whether manual, mechanical or chemical, can reach the desired level of efficiency under all situations. The most promising single approach to weed control in crops combines manual, cultural and mechanical methods with herbicides. Herbicides are used as supplements at as low a rate as possible. On environmental grounds, emphasis has been given to judicious combinations of cultural and chemical methods of weed control. In the rainy season, early weed removal may not be possible because of continuous rains and the use of pre-emergence herbicides for removing early weed competition and supplementary hoeing or hand weeding for removing later-emerging weeds may form a package of weed-control practices. The integration of low rates of pre-emergence applications of linuron (750–1000 g ha^{-1}), acetochlor and alachlor (each at 1000 g ha^{-1}) with one hand weeding at 40 DAS has been found to provide excellent control of all weeds (Balyan *et al.*, 1999a, 1999b). Sowing at 30 cm row spacing and manual weeding at 30 DAS or application of fluchloralin (1.0 kg ha^{-1}) has been found to control weeds effectively and increased the grain yield of soybean (Singh and Bajpai, 1994). Nimje (1996) observed that pre-plant incorporation of fluchloralin (1.0 kg ha^{-1}) plus interculturing at 40 DAS provided effective control of weeds in soybean. The integration of alachlor (1.25 kg ha^{-1}) at pre-emergence and one hand weeding at 40 DAS under a crop density of 444,000 plants ha^{-1} (30 × 7.5 cm spacing) has been found to be the most effective method of weed control under the irrigation regime of a ratio of 0.60 irrigation water to cumulative pan evaporation for obtaining a higher yield and economic return (Veeramani *et al.*, 2000).

10.6 Effect of Herbicides on Soil Microflora

Being a legume crop, biological nitrogen fixation (BNF) makes soybean self sustaining for nitrogen nutrition. In addition, inoculation of seeds with *Bradyrhizobium japonicum* is known to boost the potential of BNF of the crop (Mahler and Illwolum, 1981; Pandazou *et al.*, 1990). The effective phase of the symbiosis begins just before contact occurs between the bacteria and root hairs (Fisher and Long, 1992) and, during this period, the association is highly sensitive to the soil environment (Keyser *et al.*, 1993). Not only survival of the bacteria, but also the whole sequence of events that culminates in nodulation may be affected by changes in the soil environment. Herbicides in soil, especially those applied pre-planting and pre-emergence, may exert negative influences on the association, depending on the chemical, its dose and its interaction with the soil properties (Bollich *et al.*, 1985). Negative effects of herbicides on nodulation are associated with sandy soils that are low in organic matter content (Bollich *et al.*, 1985; Moraes *et al.*, 1989). Praharaj and Dhingra (1995) observed that the application of pendimethalin 0.50 kg ha^{-1} had no adverse effect on nodulation and nitrogenase activity and did not influence the efficiency of rhizobial inoculants in terms of BNF in soybean. *Rhizobium* inoculation, irrespective of the method of weed control (chemical or manual) enhanced the BNF and fixed an additional 66.1–74.7 kg N ha^{-1} over uninoculated controls.

10.7 Herbicide Residue in Soil and Food Chain

For effective weed control, herbicides must remain in the soil in an active and available form until their purpose has been accomplished. Longer persistence poses a hazard to subsequent land use and is undesirable. In countries such as India, where intensive cropping is practiced, herbicide residues in the soil from one crop may adversely affect the succeeding crop in a cropping sequence. Herbicides, especially those applied as pre-planting, pre-emergence and early post-emergence, may leave residues in the soil and harvested produce, depending on the nature of the chemical, its dose and its interactions with the soil properties (Sondhia, 2005). An ideal soil-applied herbicide should persist long enough to give an acceptable period of weed control, but not so long that soil residues after the crop harvest limit the choice of subsequent crops that can be grown. When planning crop rotations, producers must consider the injury potential to subsequent crops from herbicide residues. In general, the herbicides recommended for soybean do not leave residues that may pose a hazard to subsequent land use, if applied properly (correct dose, time and method of application). Kewat *et al.* (2001) observed that the half-life of pendimethalin at 1.0 and 1.5 kg ha^{-1} was 24 and 26 days, respectively, in sandy loam soils. Its residues reached a non-detectable level after the soybean harvest, ascertaining safety to the succeeding wheat crop.

10.8 Herbicide-resistant Soybean

The development of safe, effective and relatively inexpensive herbicides coupled with advances in application technology have provided many successful weed management options in crop production. Herbicides offer a better alternative to mechanical weeding. The discovery and use of herbicides has revolutionized agriculture in many developed countries. Despite the several advantages, many concerns such as food safety, ground water and atmospheric contamination, increased weed resistance to herbicides, destruction to beneficial organisms and concern about endangered species have also arisen with the indiscriminate use of herbicides. Moreover, many herbicides are required to manage the complex weed flora found in different crops.

Imparting herbicide resistance to normally herbicide-susceptible crops has been the most extensively exploited area of plant biotechnology. During 1996–2007 global adoption rates for transgenic crops were unprecedented (James, 2007), reflecting grower satisfaction with the products. Transgenic crops offer significant benefits ranging from more convenient and flexible crop management, higher productivity and net returns ha^{-1} to a safer environment through the decreased use of conventional pesticides, all of which collectively contribute to more sustainable agriculture. Despite growing controversy, the area under transgenic crops is increasing at a fast rate. The global area of transgenic crops increased from 1.7 million ha in 1996 to 114.3 million ha in 2007 (James, 2007), of which >63% of crops were tolerant to a specific herbicide. Transgenic or herbicide-resistant soybeans are genetically altered to tolerate herbicides that would normally kill or injure conventional or non-transgenic varieties. The first use of herbicide-resistant soybean was in 1994 with the introduction of sulphonylurea-tolerant varieties. Glyphosate-resistant (GR) and glufosinate-resistant soybeans (Roundup Ready and Liberty Link, respectively) are now commercially available. Herbicide-tolerant soybean is the most dominant transgenic crop, followed by genetically modified corn, transgenic cotton and genetically modified canola. GR soybean varieties have been widely adopted for planting by American farmers since their introduction in 1996.

10.9 Weed Management in Herbicide-resistant Soybean

GR soybean became commercially available in the USA in 1996 (Fernandez-Cornejo and McBridge, 2000). Transgenic soybean resistant to glyphosate provides producers with the flexibility to control a broad-spectrum of weeds with minimal concern for crop damage (Askew *et al.*, 1998; Ateh and Harvey 1999). This has resulted in the rapid adoption of GR soybean by producers (Coulter and Nafgiger, 2007). Glyphosate is the world's most popular herbicide. It is a non-selective, broad-spectrum herbicide that is used extensively as pre-plant, post-directed, spot and pre- and post-harvest applications. It is highly effective against the majority of annual and perennial grasses, sedges and broadleaved weeds. Glyphosate is toxicologically and environmentally benign with low toxicity to non-target organisms, low or no ground water movement and limited persistence (Reddy, 2001). Glyphosate inhibits the biosynthesis of aromatic amino acids (phenylalanine, tyrosine and tryptophan), which leads to the arrest of protein production and prevention of secondary product formation (Franz *et al.*, 1997). GR soybean contains a gene introduced from *Agrobacterium* species. Glyphosate inhibits the enzyme 5-enolpyruvylshikimate-3-phosphate synthase (EPSPS) in the shikimic acid pathway.

The effectiveness of glyphosate depends on the rate and time of application relative to the weed growth stage. To obtain effective weed control in GR soybean, glyphosate must be applied after most weeds have emerged. Glyphosate provides no residual control of weeds. Early-season glyphosate application will control weeds that have already emerged, but weeds emerging after application will escape control. Late-season glyphosate application (prior to canopy closure) will control most weeds, but delaying too long may result in some weeds being too large to control, even with a high glyphosate rate. Therefore, a second application of glyphosate is needed to control problematic or late-emerging weeds (Reddy, 2001). The use of a single broad-spectrum herbicide such as glyphosate may eliminate concerns over possible antagonism associated with tank mixing grass and broadleaf herbicides (Reddy and Whiting, 2000).

Chlorimuron ethyl in post-emergence is commonly used for broadleaf weed control in soybean (Claus, 1987). However, it does not effectively control common lambsquarters (*Chenopodium album* L.) (Monks *et al.*, 1993) or prickly sida (*Sida spinosa*) (Vidrine *et al.*, 1993). Sulphonylurea-tolerant soybean attained through seed mutagenesis has increased tolerance for chlorimuron and other sulphonylurea herbicides due to the insensitive acetolactate synthase enzyme (Sebastian *et al.*, 1989). With the introduction of sulphonylurea-tolerant soybean, single or multiple applications of chlorimuron alone or in combination with other sulphonylurea herbicides, as well as the application of certain imidazolinone and sulphonylurea herbicides alone or in combination, can improve season-long weed control and reduce soybean injury (Culpepper *et al.*, 1997; Moshier and Freed, 1997). In India, soybean suffers from heavy infestations of complex weed flora including grasses, broadleaved weeds, sedges and perennial weeds. These weeds emerge in several flushes and it is very difficult to manage them. Considering the several advantages of using herbicide-resistant soybean, it is worth exploring their possible use under Indian conditions (Yaduraju and Mishra, 2004) and in other countries.

10.10 Economics of Weed Management

A large number of recommendations have been made by research workers for effective weed management in soybean. However, the adoption of these recommendations by farmers is largely dependent upon the economic viability of the treatments. Generally, if weeds are to be managed by manual weeding within 3–4 weeks of sowing, about 50–75 person-days ha^{-1} are needed depending upon the nature and intensity of the weed flora, soil types and ago-ecological conditions. As weed growth increases, the labour requirement also goes up. The cost of manual weedings varies depending upon the wages in different regions (Table 10.4). The economics of herbicide use is given in Table 10.5.

Table 10.4. Economics of manual weeding in soybean (adapted from Yaduraju *et al.*, 2003).

	Economics of manual weeding (Rs ha^{-1})		
	Wages (Rs man-day^{-1})		
Man-days ha^{-1}	50	75	100
50	2,500	3,750	5,000
75	3,750	5,625	7,500
100	5,000	7,500	10,000

Table 10.5. Economics of chemical weed control in soybean (adapted from Yaduraju *et al.*, 2003).

Herbicide	Product cost (Rs kg^{-1} or l^{-1})	Dose (g ai ha^{-1})	Cost of the treatment (Rs ha^{-1})
Alachlor	300	1500–2000	900–1200
Fenoxaprop	1500	75–90	1200–1500
Fluchloralin	480	1000–1500	1500–1600
Imazethapyr	1600	80–100	1300–1600
Pendimethalin	510	1000–1250	1500–1800
Trifluralin	450	1000–1500	1000–1200
Mean (approximately)			1200–1500

10.11 Conclusions

Being a rainy-season crop, weed management in soybean is a challenging task because of the emergence of weeds in flushes, unpredictability of rains, non-workable soil conditions and non-availability of timely labour. Considering the diversity of weeds, no single method of weed control, whether manual, mechanical or chemical, is sufficient to provide season-long weed control under all situations. An integrated weed management system, as a part of integrated crop management system, is an effective, economical and eco-friendly approach for weed management in soybean. A combination of pre-emergence herbicides with manual or mechanical weeding is required for effective weed management. The sequential application of pre- and post-emergence herbicides may provide broad-spectrum weed control. GR (Roundup Ready) soybean varieties have found success in the USA and other major soybean-growing countries. Considering the several advantages of herbicide-resistant soybean, it is worthwhile exploring its possible use in other countries also.

Acknowledgements

The author is grateful to Dr N.T. Yaduraju, National Coordinator, NAIP (ICAR), Krishi Anusandhan Bhavan-II, Pusa, New Delhi, India, for providing technical support on this chapter.

References

Akobundu, I.O. (1987) *Weed Science in the Tropics: Principles and Practices*. Wiley, Chichester, UK.

Askew, S.D., Shaw, D.R. and Street, J.E. (1998) Red rice (*Oryza sativa*) control and seedhead reduction with glyphosate. *Weed Technology* 12, 504–506.

Ateh, C.M. and Harvey, R.G. (1999) Annual weed control by glyphosate in glyphosate-resistant soybean (*Glycine max* L.). *Weed Technology* 13, 394–398.

Balyan, R.S. and Malik, R.K. (2003) Effect of post-emergence herbicides for weed management in soybean (*Glycine max* L.). *Indian Journal of Weed Science* 35, 62–66.

Balyan, R.S., Malik, R.S., Panwar, R.S. and Pahwa, S.K. (1999a) Chemical weed control studies in soybean (*Glycine max* L.). *Indian Journal of Weed Science* 31, 234–238.

Balyan, R.S., Pahwa, S.K., Dhankar, R.S. and Malik, R.K. (1999b) Efficacy of linuron and fluazifop-p-butyl against weeds in soybean (*Glycine max* L.). *Indian Journal of Weed Science* 31, 229–233.

Barrentine, W.L. (1974) Common cocklebur competition in soybean. *Weed Science* 22, 600–603.

Bollich, P.K., Dunigan, E.P. and Jadi, A.W.M. (1985) Effects of seven herbicides on N_2 (C_2H_2) fixation by soybeans. *Weed Science* 33, 427–430.

Chhokar, R.S., Balyan, R.S. and Pahuja, S.S. (1997) Nutrient removal by weeds in soybean (*Glycine max*) under integrated weed management. *Indian Journal of Agronomy* 42, 138–141.

Claus, J.S. (1987) Chlorimuron-ethyl (Classic): A new broadleaf post-emergence herbicide in soybean. *Weed Technology* 1, 114–115.

Coulter, J.A. and Nafziger, F.D. (2007) Planting date and glyphosate timing on soybean. *Weed Technology* 21, 359–366.

Culpepper, A.S., York, A.C., Batts, R.B. and Jennings, K.M. (1997) Sicklepod (*Sena obtusifolia*) management in an ALS-modified soybean (*Glycine max* L.). *Weed Technology* 11, 164–170.

Dubey, S.K., Jain, H.C., Koshta, L.D. and Raghu, J.S. (1984) Integrated weed control in soybean. *Indian Journal of Weed Science* 16, 238–243.

Fernandez-Cornejo, J. and McBridge, W.D. (2000) *Genetically Engineered Crops for Pest Management in US Agriculture*. Agricultural Economics Report 786, Washington, DC, USA.

Fisher, R.F. and Long, S.R. (1992) *Rhizobium*-plant signal exchange. *Nature* 357, 655–659.

Franz, J.E., Mao, M.K. and Sikorski, J.A. (1997) *Glyphosate: A Unique Global Herbicide*. American Chemical Society Monograph 189. American Chemical Society, Washington, DC, USA.

Hall, M.R., Swanton, C.J. and Anderson, G.W. (1992) The critical period of weed control in grain corn (*Zea mays*). *Weed Science* 40, 441–447.

Jain, H.C. and Tiwari, J.P. (1992) Influence of herbicides on the growth and yield of soybean (*Glycine max*) under different spacing and seeding rates. *Indian Journal of Agronomy* 37, 86–89.

James, C. (2007) Global status of commercialized biotech/GM crops: 2007. ISAAA Brief No. 37. International Service for the Acquisition of Agri-Biotech Applications, Ithaca, New York, USA.

Katan, J. (1981) Solar heating (solarization) of soil for control of soil-borne pests. *Annual Review of Phytopathology* 19, 211–236.

Kewat, M.L., Pandey, J. and Kulshrestha, G. (2001) Persistence of pendimethalin in soybean (*Glycine max*)–wheat (*Triticum aestivum*) sequence following pre-emergence application to soybean. *Indian Journal of Agronomy* 46, 23–26.

Keyser, H.H., Somasegaran, P. and Boholool, B.B. (1993) *Rhizobium* ecology and technology. In: Metting Junior, B. (ed.) *Soil Microbial Ecology*. Marcel Dekker, New York, USA, pp. 205–226.

Kumar, B., Yaduraju, N.T., Ahuja, K.N. and Prasad, D. (1993) Effect of soil solarization on weeds and nematodes under tropical Indian conditions. *Weed Research* 33, 423–429.

Lokras, V.G., Singh, V.K., Bisen, C.R. and Tiwari, J.P. (1985) Chemical weed control in soybean. *Indian Journal of Weed Science* 17, 45–48.

Mahler, R.C. and Illwolum, G. (1981) The influence of irrigation and *Rhizobium japonicum* strains on yields of soybeans grown in a Lakeland sand. *Agronomy Journal* 73, 647–651.

Mishra, J.S. and Singh, V.P. (2001) Interference of Ivyleaf morningglory (*Ipomoea hederacea* L.) in soybean (*Glycine max* L.). *Indian Journal of Weed Science* 33, 87–88.

Mishra, J.S. and Singh, V.P. (2003) Interference of *Euphorbia geniculata* in soybean–chickpea cropping system. *Indian Journal of Weed Science* 35, 225–227.

Mishra, J.S., Singh, V.P. and Yaduraju, N.T. (2002) Interference of common dayflower (*Commelina communis* L.) in soybean (*Glycine max* L.). *Indian Journal of Weed Science* 34, 295–296.

Monks, C.D., Wilcut, J.W. and Richburg III, J.S. (1993) Broadleaf weed control in soybean (*Glycine max* L.) with chlorimuron plus acifluorfen or thifensulfuron mixtures. *Weed Technology* 7, 317–321.

Moolani, M.K., Knake, E.L. and Slife, F.W. (1964) Competition of smooth pigweed with corn and soybeans. *Weeds* 12, 126–128.

Moraes, R., Lopez, N.F., Silva, E.A.M. da and Rodrigues, J.J. do (1989) Tolerance and nodulations in soybeans (*Glycine max* (L) Merrill) cultivated on two substrata and treated with four doses of metribuzin. *Revista Ceres* 36, 492–506.

Moshier, L.J. and Freed, B.E. (1997) Performance of sulfonylurea tolerant soybean as influenced by weed control systems. *Weed Science Society of America*, abstract 37, p. 91.

Nimje, P.M. (1996) Effect of row spacing, mulching and weed control on weed growth and yield of soybean (*Glycine max*). *Indian Journal of Agronomy* 41, 427–432.

Oerke, E.C. (2006) Crop losses to pests. *Journal of Agricultural Science* 144, 31–43.

Pandazou, J., Beunard, P. and Saint Macary, H. (1990) Inoculation of soybean (*Glycine max* L.) in Congo. *Agronomie Tropicale* 45, 199–204.

Praharaj, C.S. and Dhingra, K.K. (1995) Effect of pendimethalin on rhizobial efficiency in soybean (*Glycine max* L.). *Indian Journal of Weed Science* 27, 201–204.

Rajput, R.P. (1980) Response of soybean to climate and soil environment. PhD. thesis, Indian Agricultural Research Institute, New Delhi, India.

Rajput, R.P. and Sastry, P.S.N. (1986) Effect of different mulches on yield, its attributes and quality of soybean crop under rainfed conditions. *Indian Journal of Plant Sciences* 4, 22–27.

Reddy, K.N. (2001) Glyphosate-resistant soybean as a weed management tool: Opportunities and challenges. *Weed Biology and Management* 1, 193–202.

Reddy, K.N. and Whiting K. (2000) Weed control and economic comparisions of glyfosate-resistant, sulfonylurea-tolerant, and conventional soybean (*Glycine max*) systems. *Weed Technology* 14, 204–211.

Sebastian, S.A., Fader, G.M., Ulrich, J.L., Forney, D.R. and Chaleff, R.L. (1989) Semidominant soybean mutation for resistance to sulfonylurea herbicides. *Crop Science* 29, 1403–1408.

Singh, G. and Singh, D. (1992) Weed control efficacy of oxyfluorfen, acifluorfen, metribuzin and linuron in soybean (*Glycine max* L.). *Indian Journal of Weed Science* 24, 97–99.

Singh, R., Kalia, B.D. and Negi, P.S. (1992) Integrated weed management in soybean. *Indian Journal of Weed Science* 24, 94–96.

Singh, V.K. and Bajpai, R.P. (1994) Influence of crop geometry and weed control method on yield and economics of rainfed soybean (*Glycine max*). *Indian Journal of Agronomy* 39, 671–673.

Singh, V.P., Mishra, J.S. and Bhan, V.M. (1999) Integrated weed management in soybean. (*Glycine max* L.). *Indian Journal of Weed Science* 31, 267–268.

Singh, V.P., Dixit, A., Mishra, J.S. and Yaduraju, N.T. (2004) Effect of period of soil solarization and weed control measures on weed growth and productivity of soybean (*Glycine max*). *Indian Journal of Agricultural Sciences* 74, 324–328.

Sondhia, S. (2005) Phytotoxicity and persistence of metribuzin residues in black soil. *Toxicological and Environmental Chemistry* 87, 387–389.

Spifters, C.J.T. and Van Den Bergh, J.P. (1982) Competition between crop and weeds: A system approach. In: Holzner, W. and Numata, M. (eds) *Biology and Ecology of Weeds*. Dr W. Junk Publishers, The Hague, The Netherlands, pp. 137–148.

Stapleton, J.J. and Lopez, J.G.G. (1988) Mulching of soils with transparent (solarization) and black polyethylene films to increase growth of annual and perennial crops in south-western Mexico. *Tropical Agriculture (Trinidad)* 65, 29–32.

Tiwari, J.P. and Kurchania, S.P. (1990) Survey and management of weeds in soybean ecosystem in Madhya Pradesh. *Indian Journal of Agricultural Sciences* 60, 672–676.

Tiwari, J.P., Kurchania, S.P., Paradkar, N.R. and Bhalla, C.S. (1996) Bioefficacy of chlorimuron ethyl for weed control in soybean (*Glycine max*). *Indian Journal of Agricultural Sciences* 66, 583–588.

Tiwari, J.P., Kurchania, S.P., Paradkar, N.R. and Bhalla, C.S. (1997) Varietal susceptibility and weed control efficiency of fluazifop-p-butyl in soybean (*Glycine max*). *Indian Journal of Agricultural Sciences* 67, 147–149.

Upadhyay, V.B., Sharma, M., Acharya, V.N. and Nakely, N.R. (1992) Cultural and chemical weed control in soybean in regions of MP. In: *Abstracts of Annual Conference of Indian Society of Weed Science*, Haryana Agricultural University, Hisar, India, p. 77.

Veeramani, A., Palchamy, A., Ramasamy, S. and Rangaraj, G. (2000) Integrated weed management in soybean (*Glycine max*) under different moisture regimes and population densities. *Indian Journal of Agronomy* 45, 740–745.

Vidrine, P.R., Reynolds, D.B. and Griffin, J.L. (1993) Weed control in soybean (*Glycine max* L.) with lactofen plus chlorimuron. *Weed Technology* 7, 311–316.

Yadav, R.P., Srivastava, U.K. and Dwivedi, S.C. (1999) Effect of seed rate and weed control methods on yield of soybean (*Glycine max*) under rainfed condition. *Indian Journal of Agronomy* 44, 795–799.

Yaduraju, N.T. (1993) The role of soil-solarization in weed management. In: *Proceedings of International Symposium. Indian Society of Weed Science*, Hisar, India, 18–20 November 1993, Volume I, pp. 343–349.

Yaduraju, N.T. and Ahuja, K.N. (1996) Effect of solarization with or without weed control on weeds and productivity in soybean–wheat system. In: *Proceedings of Second International Weed Control Congress*, Copenhagen, Denmark, pp. 721–727.

Yaduraju, N.T. and Mishra, J.S. (2004) Integrated weed management in soybean (*Glycine max* L.). *JNKVV Research Journal* 38, 1–12.

Yaduraju, N.T., Mishra, J.S. and Dixit, A. (2003) Weed management in soybean. *Agrolook* July–September, pp. 16–21.

Zimdahl, R.L. (1980) *Weed-Crop Competition: A Review*. International Plant Protection Center, Oregon State University, Corovallis, OR, USA.

Zimdahl, R.L. (1988) The concept and application of the critical weed-free period. In: Altieri, M.A. and Eibman, F.M.L. (eds) *Weed Management in Agroecosystems: Ecological Approaches*. CRC Press, Boca Raton, FL, USA, pp. 145–155.

11 Biological Nitrogen Fixation in Soybean

David L. McNeil
Tasmanian Institute of Agricultural Research, Chair of Agricultural Science, University of Tasmania, Hobart, Tasmania, Australia

11.1 Introduction

Nitrogen is a limiting factor in many biological systems and becoming more so as atmospheric CO_2 levels rise (Johnson, 2006), global food demand increases and attempts to mitigate atmospheric CO_2 levels potentially lead to effects on the production of artificial nitrogen fertilizers. In biological systems, plants require their nitrogen in a fixed form (e.g. ammonia, nitrate or organic compounds). This can be gained directly from environmental sources, through the application of artificial fertilizers or through associations with nitrogen-fixing bacteria. While soybean (*Glycine max* (L.) Merrill) can access all three of these options, this chapter looks at its important ability to symbiotically fix nitrogen and how this interacts with other sources of nitrogen. The nature of the process will also be investigated, including attempts to improve on the process and to characterize interactions of nitrogen fixation with other environmental and biological factors. The chapter mainly deals with biological nitrogen fixation (BNF) in soybean, although salient and relevant examples from other grain legumes have been included as these may apply to soybean also.

Soybean nitrogen fixation

To establish the global importance of soybean nitrogen fixation it is necessary to compare it with global estimates for all forms of nitrogen fixation. Burns and Hardy (1975) estimated global (biological and non-biological) fixed nitrogen of about 175 Tg N year^{-1}. Cleveland *et al.* (1999) gave a range of 100–290 Tg N year^{-1}. Human activities (mostly legume cropping and fertilizer production) have been suggested to result in the fixation of approximately an additional 150 Tg N year^{-1} in 1998 (Galloway, 1998). Non-biological inputs come primarily from fertilizer production, estimated to provide around 100 Tg N $year^{-1}$ in 2008, up from 3.5 Tg N $year^{-1}$ in 1950 (Erisman *et al.*, 2008), with small amounts from combustion and lightning (Nesbitt *et al.*, 2000). In addition, there have been increasing amounts of foliar deposition of nitrogen, with Galloway *et al.* (2004) estimating 103 Tg N $year^{-1}$ in 1995. Much of this nitrogen is recycled nitrogen released from fertilizers, factory and other emissions, burning and other causes.

Two recent reviews of nitrogen fixation by soybean (Herridge *et al.*, 2008; Salvagiotti *et al.*, 2008) estimated that soybean fixes 50–60% and 58% of its total nitrogen globally with a near neutral balance on soil nitrogen owing to the large proportion of total nitrogen removed in the seed. This nitrogen may, however, re-enter the environment via sewage or animal manures at other points in the system, with a total amount based on global production (Herridge *et al.*, 2008) of around 16 Tg N $year^{-1}$. Estimates for high-yielding US-grown soybean crops indicate a greater reliance on applied fertilizers (40%) than in Brazilian crops (20%). Experimentally, reliance has varied between zero and almost 100%. In a meta-analysis of experiments to date, total soybean BNF has been estimated at 16.4 Tg N $year^{-1}$ out of a total of 50–70 Tg N $year^{-1}$ from agricultural systems (Herridge *et al.*, 2008). Herridge *et al.* (2008), therefore, concluded that soybean is the single most important source of BNF among all of the crop legumes (77%), consistent with it composing 68% of global crop legume production.

BNF in soybean, as in all legumes, depends on bacterial reduction of atmospheric nitrogen by the enzyme nitrogenase. The reduction occurs in specific symbiotic associations in legume root nodules. This reduction requires large amounts of energy and reductant (at least 16 molecules of ATP per N_2 molecule). Reduction of NO_3^- by soybean also requires energy (or direct use of photosynthetic electrons) at a similar level (Atkins, 1982), which explains why soybean yields can be high in crops with either high fixation or high fertilizer nitrogen inputs. However, nitrogen fixation is a sensitive process, requiring high levels of phosphorus.

The increasing cost and negative environmental impacts (direct nitrogen pollution, nitrogen oxide gas release and CO_2 pollution) of artificial nitrogen fertilizers mean that soybean, as the world's greatest nitrogen-fixing legume crop, has an important place in global crop production. Much of the nitrogen fixed by the soybean crop is available for subsequent crops in the rotation as crop residues break down or through applications of manures from animals grazed on the crop. A globally averaged estimate for the nitrogen content of soybean residues is 59 kg N ha^{-1} (Salvagiotti *et al.*, 2008). However, with 40 kg N ha^{-1} removed from the soil the result is 19 kg N ha^{-1} returned to the soil with −4 kg N ha^{-1} for nitrogen fixation return alone. However, the levels of nitrogen fixed by soybean vary considerably in response to management, environmental and ecological factors.

This chapter looks at the microbiological, plant and cropping-system components of soybean nitrogen fixation.

11.2 Microbiological Components

The infection process

Soybean forms a symbiosis with a range of *Rhizobium* species. These species have generally evolved in primary (China) and secondary (India) centres of origin. Considerable traditional and, more recently, molecular taxonomy (Appunu *et al.*, 2008; Yang and Zhou, 2008) on these species has now been carried out, creating a generally accepted taxonomy published by Willems (2006). Willems classified the species into fast growing (*Sinorhizobium fredii* and *Sinorhizobium xinjiangense*, previously described as promiscuous nodulation), slow growing (*Bradyrhizobium japonicum*, *Bradyrhizobium elkanii* and *Bradyrhizobium liaoningense*) and intermediate growing (*Mesorhizobium tianshanense*).

Host specificity arises from the interaction of both plant and bacterial genes (Smit *et al.*, 1992; Begum *et al.*, 2001). Spaink (2000) has reviewed the bacterial factors associated with this recognition process. These factors include the 'Nod factors, the extracellular polysaccharides, the lipopolysaccharides, the K-antigens, and the cyclic glucans'. Many of these bacterial nodulation/recognition factors are now well characterized at the molecular level (Zehner *et al.*, 2008, for the type III secretion system). However, a detailed discussion of these systems goes beyond the scope of this chapter. Olroyd and Downie (2008) have looked at the steps beyond initial recognition and how these are coordinated in creating legume nodules. Many of the genes and their effects in both the plant and bacteria are now well described and these detailed effects are covered by Olroyd and Downie (2008). The basic steps, which are similar in most legume infections, are described in Table 11.1.

Table 11.1. Steps in the nodule formation process (summarized from Olroyd and Downie, 2008).

Process	Major steps
Bacterial infection	Attachment to root hairs, proliferation of bacteria, lectin binding and Nod factor signalling.
Infection thread formation, bacterial entry and infection thread growth	Root hair curling and intracellular infection, selection for clonal-specific bacteria, growth of thread into the cortex.
Making of a nodule	A nodule primordial occurs at the cortex close to the site of infection in response to nodulins. This then produces transcription factors, starting the plant nodule formation process. Modulation of auxin, cytokinin and other plant hormones occurs. Hormone precursors are possibly produced by bacteria to stimulate nodulation.
Regulation of level of nodulation	*Nark* genes occur in shoots that regulate, via an intermediate hormone, the numbers of nodules. Ethylene-responsive genes may also occur in roots.

Gwata *et al.* (2004) described the inheritance by the plant of promiscuous nodulation genes for soybean, while Carroll *et al.* (1986) described the creation and isolation of non-nodulating soybean mutants. Infection regulation thus depends on both bacterial and plant genes and may also be regulated by environmental conditions.

Once the bacteria have infected the plant they are enclosed within root nodules in a specialized form known as bacteroids. Soybean has a determinate nodule structure with several bacteroids within a single peribacteroid membrane sac. The complex structure of nodules exists to enable the functioning of nitrogenase enzyme, which is highly sensitive to oxygen damage, and the growth and functioning of the rhizobia, which are obligate aerobes. In nodules, internal oxygen is regulated to 3–30 nM and oxygen diffusion is facilitated through intercellular air spaces and an oxygen-binding haemoprotein (leghaemoglobin). The bacteroids are enclosed within a host-derived peribacteroid membrane that regulates flows between the bacteroids and their environment. The plant provides organic acids as a carbon source for the bacteria and the bacteroids export fixed nitrogen in the form of ammonia. The soybean nodules then modify the ammonia to export ureides to the above-ground plant parts in the xylem (McNeil and LaRue, 1984).

Bacterial genetics and control of fixation

Early papers investigated the ability of the bacteria to regulate nitrogen fixation in response to changes in the environment (e.g. nitrate limitation; McNeil, 1982). However, later evidence suggested that it is the plant that dominates the environmental effect on both the extent of nodulation and regulation of fixation (Carroll *et al.*, 1985, 1987). That the bacteria do have considerable influence over the efficiency of fixation has been repeatedly observed for many years (McNeil *et al.*, 1981). In some instances the additional bacterial efficiency can be explained by particular genes such as hydrogen uptake (*Hup+*) genes (Baginsky *et al.*, 2002). Gaseous hydrogen is released as a natural product of nitrogen fixation, and hydrogen uptake (*Hup+*) genes code for the ability to utilize this hydrogen to recover ATP and thereby reduce the energy demand of nitrogen fixation. Efforts have been made to engineer soybean bacteria for improved fixation, for example via the introduction of *Hup+* genes with a mini-transposon (Bascones *et al.*, 2000) or through upregulation of metabolism enzymes such as trehalose-6-phosphate synthase (Suarez *et al.*, 2008). Whether these modifications lead to field improvements remains questionable due to the need to get the bacteria into the nodules and interactions with other characteristics of the symbiosis, such as plant-regulated expression of rhizobial genes within the bacteroids (Brito *et al.*, 2008). A number of articles have indicated that genetic tolerances of the rhizobia to adverse conditions may be reflected in the symbiosis. For example, Elsheikh and Wood (1995) found that salt-tolerant rhizobia fix more nitrogen in soybean grown under saline conditions.

Table 11.2. Requirements for effective inoculation (summarized from Herridge *et al.*, 2002).

Carrier	Rhizobia	Processes
High water-holding Non-toxic Sterilizable Cheap pH-buffering High cation exchange capacity	Effective nodulators Efficient fixers Broad plant genetics and environmental adaptability Soil nitrate tolerant Easily grown *in vitro* Persistent in soils (spread and survive without host) Genetically stable Compatible with agrochemicals	Regulatory quality control, ensuring both genetic integrity and numbers Use-by dates Storage-environment control Carrier testing for contaminants and toxicities Post-production checking

The great difficulty in making use of the additional rhizobial genetic efficiency has always revolved around introducing the efficient bacteria into the nodules of plants in the field. Research has indicated that high populations, improved delivery mechanisms and quality control can improve field responses. Herridge *et al.* (2002) reviewed the available technologies for inoculant delivery in Australia and quoted data for soybean indicating a linear relationship between log of actual rhizobia applied per seed and both nodule score and final yield. Thus, technologies that can provide high numbers of high-quality rhizobia per seed (e.g. quality-assured granular inoculants) may result in benefits in the field, provided field levels are not already very high. Herridge *et al.* (2002) gave a list of requirements for inoculant carriers and bacteria (Table 11.2). They also indicated that quality control is essential. For example, inoculant quality can decline with age, even where the bacteria survive and numbers remain high in bacterial counts.

Numerous studies from recent publications to those over 20 years old (McNeil *et al.*, 1983; Botha *et al.*, 2004) have indicated the difficulties encountered in getting adequate numbers of more effective introduced rhizobia into soybean nodules in fields where natural rhizobia exist. This is in spite of the introduced rhizobia being more efficient nitrogen fixers and being supplied at high populations, either through liquid or granular inoculants. Maintaining introduced rhizobia is made more difficult by the great decreases in field rhizobial populations that may occur in response to adverse field conditions (Raverkar *et al.*, 2005). A meta-analysis of published field trials (D.L. McNeil, Tasmania, 2009, personal observations) indicated that for each 10% field introduction of US Department of Agriculture (USDA) 110 (a *Hup+* strain) into soybean nodules, final grain yield increased by 1%. Thus, a major problem in interpreting yield responses to inoculation lies around whether there has been any incorporation of the inoculant bacteria into the nodules. The relative efficiency with which bacteria infect the nodules can be altered by soil conditions (Hungria *et al.*, 2001) as well as by the genetics of the bacteria and the relative timing of the attachment to the soybean root (Kosslak *et al.*, 1983). However, in situations where no viable infective rhizobia exist in the soil, the application of inoculants can create spectacular increases in fixation where soil nitrogen is low. Persistence of soybean rhizobia in the soil under adverse conditions can also be an issue.

DNA microarray analysis has been commonly used to identify gene transcription in particular environmental situations. The technology can be used in both free-living bacteria and within bacteroids. An array is available for *Bradyrhizobium* USDA 110 (Chang *et al.*, 2007). As an example of microarray use to characterize gene expression, Cytryn *et al.* (2007) found that >600 genes were upregulated in soybean free-living rhizobia under desiccating conditions. In particular, they identified upregulation of trehalose-production genes. The subsequent increased trehalose levels acted to reduce intracellular water activity and protect the rhizobia from desiccation. Other genes were also upregulated in response to desiccation, including transcription regulators, DNA repair factors and general stress-response protein genes. Suarez *et al.* (2008) went further, indicating that in the common bean (*Phaseolus vulgaris*), the upregulation of trehalose pathway genes in the bacteroids reduced plant sensitivity to drought. The mechanism included alterations of plant metabolism, in that mutant strains that overproduced trehalose were found to have upregulation of >7000 genes in the plant nodule. These data indicate that mechanisms exist for using transgenic technologies in rhizobia to improve nitrogen fixation by soybean, provided the modified bacteria can be introduced into nodules in the field.

Bacterial biochemistry in nodules

Lodwig and Poole (2003) published a thorough review of biochemical metabolism within bacteroids. Therefore, this chapter concentrates on giving a broad outline of the process as well as identifying soybean-specific components. The basic process is the conversion of plant-supplied sucrose to organic acids, which are transported to the bacteroids as a carbon source. The bacteroid uses the energy in the carbon source to fix nitrogen with the return of ammonia to the plant, which is then converted to ureides for xylem-based transport around the plant (McNeil and LaRue, 1984). Recent evidence suggests that there is also transport of an amino acid (alanine) from the bacteroids to the plant (Prell and Poole, 2006) in soybeans, which may account for the fact that nitrogen-fixing soybeans do not transport 100% ureides in their xylem (McNeil and LaRue, 1984). Determinate nodules, such as those of soybean, can also produce large stores of poly-β-hydroxybutyrate, which can be later broken down and used by the bacteroids to drive nitrogen fixation (Lodwig and Poole, 2003). This gives soybean an advantage under prolonged shading or other adverse conditions, in that the bacteroids will not shut down rapidly to avoid the possibility of oxygen damage. Consequently, soybean nitrogen fixation may continue longer and recover more quickly from shading. Regulation of the process of nitrogen fixation by bacteroids is heavily under the control of oxygen availability (Carroll *et al.*, 1987), which is modified by the soybean plant in response to environmental conditions due to the deleterious effect of excess oxygen on nitrogenase (Kuzma *et al.*, 1999). An energy balance for soybean nodules has been derived by Rainbird *et al.* (1984).

11.3 Plant Components

Overview of soybean nitrogen metabolism

Soybean derives most of its nitrogen from either the soil or nitrogen fixation, but some may also be derived from foliar deposition. Estimates for total global foliar nitrogen deposition have been given by Galloway *et al.* (2004) as 31.6 Tg N year^{-1} in 1860, 103 Tg N year^{-1} in 1995 and a projected 195 Tg N year^{-1} in 2050. Globally, this may represent as much as 16% of total plant nitrogen uptake (Sparks, 2009). Many researchers (Hanway, 1979; Qiao, 1997; Qiao and Murray, 1997; Hardarson and Atkins, 2003) have demonstrated that this mode of nitrogen uptake is possible in soybean (for applied urea as well as gaseous nitrogen). Foliar-applied nitrogen can increase soybean nitrogen content without adversely affecting fixation, and can affect root uptake as well as being affected by root nitrogen uptake. Sparks (2009) gave a model of leaf-applied nitrogen incorporation where NO and NO_2 become nitrite or nitrate ions in the apoplast. The nitrate is converted via nitrate reductase to nitrite in the cytoplasm and the nitrite is converted first into NH_4^+ and then into glutamine in the chloroplasts. Atmospheric NH_3 is converted to NH_4^+, which is converted by glutamine synthase to glutamine in either the cytoplasm or chloroplast. There is evidence that late application of foliar nitrogen may compensate for loss of nitrogen fixation and increase yields (Hanway, 1979) as carbohydrate reserves are largely redistributed to the seeds and the fixation process is starved (McNeil and LaRue, 1984). Alternatively, stem nitrogen remobilization has been modelled to achieve the same effect (Sinclair *et al.*, 2003). An alternative proposal is that use of slow-release nitrogen fertilizers may stimulate soybean growth without affecting nitrogen fixation and thereby increase yields and nitrogen fixation at the same time (Takahashi *et al.*, 2003).

With respect to fixed or nitrate-derived root uptake, McNeil and LaRue (1984) demonstrated that while initial products and distributions vary, ultimately the plant distribution is the same. This is consistent with a model of nitrogen distribution in soybean (Sinclair *et al.*, 2003) where no difference was taken in the approach for nitrogen derived from fixation or nitrate reduction. Plants that are fully reliant on nitrogen fixation may, as they establish their symbiosis, use a substantial part of the early available nitrogen for symbiosis development, thereby reducing growth of the tops due to lack of nitrogen. Consequently, final yields may be lower. The greatest expression of this is in supernodulating soybean, where significantly increased amounts of cotyledonary nitrogen reserves have been shown to be used in the roots relative to non-mutant types (Hansen *et al.*, 1990). Commercially, therefore, starter nitrogen is a common recommendation for grain legume crops, including soybean, to allow them to develop effective nodulation without a shortage of plant nitrogen.

Soil nitrate is predominantly transported to the shoots as amino acids, with ≤5% as free nitrate in the xylem (McNeil and LaRue, 1984). Fixed nitrogen is predominantly transported to the tops as ureides (approximately 80% in a field trial low in soil nitrogen). Ureides have a very high nitrogen to carbon ratio and hence are an efficient compound with which to circulate nitrogen around the plant. When adequate soil nitrogen exists, ureides constitute <10% of xylary nitrogen; in stressed situations, on the other hand, ureides that have been created and stored by the plant may be loaded via the stem to consist of up to 30% of total (low) xylary nitrogen. However, as indicated before, none of these compositional differences seems to create changes in the total plant nitrogen distribution (McNeil and LaRue, 1984). Ureides are not a major component of soybean functional or storage nitrogen-containing components. Therefore, the creation and breakdown of ureides in fixation is an essential side process within soybean. Tajima *et al.* (2004) gave a detailed description of the biochemical processes occurring in the nodule forming the ureides. The breakdown processes that occur in the leaves for incorporation of ureides into amino acids in soybean are described by Todd *et al.* (2006). They suggested that both urea and ammonia are released in the leaf-based degradation of soybean ureides for use in producing amino acids and other nitrogen-requiring processes.

Evidence has shown that soybean will fix more nitrogen under elevated CO_2 levels (Prior *et al.*, 2006). This effect seems to be enhanced under drought stress (Serraj *et al.*, 1998), where breakdown of ureides in the leaves increases and consequently leads to less suppression of fixation. Elevated CO_2 has also been demonstrated to significantly increase photosynthesis in soybean, again particularly in drought-stressed situations (Prior *et al.*, 2006). However, supernodulating mutants that might be expected to have greater benefits from increased CO_2 and consequent increased photosynthesis due to their high demand for nodule carbohydrate do not seem to benefit more than their non-mutant parents (Bandara *et al.*, 1998). In a meta analysis of elevated Free-Air CO_2 Enrichment (FACE) experiments on crops, Taub *et al.* (2008) showed that soybean has a significant nitrogen percentage reduction under elevated CO_2. However, due to soybean's ability to fix nitrogen, this decrease was much lower than that observed in cereal crops. Thus, the current evidence suggests that total nitrogen incorporation under elevated CO_2 benefits less than overall growth. Interestingly, an evaluation of upregulated genes using microarrays indicated increases in the upregulation of leaf nitrate transport and metabolism genes under elevated CO_2, suggesting more nitrate use rather than increased fixation (Ainsworth *et al.*, 2006).

11.4 Breeding for Increased Nitrogen Fixation

Bacterial breeding for increased microbial fixation

Considerable effort has been expended in attempts to create bacteria that provide enhanced growth of soybean symbioses through enhanced fixation, either across all environments or in particular situations. Efforts have included traditional selection of bacteria through direct responses of the bacteria or responses of the inoculated plant (selected for nitrate tolerance; McNeil, 1982), mutation of bacteria and use of genetic-engineering technologies to produce altered bacteria. Attempts have also been made to improve bacterial field survival to improve their ability to nodulate plants at a later date. There is no doubt that under laboratory conditions and in the field (if they can form a substantial proportion of the nodules), selected strains can increase soybean fixation. As a consequence, numerous *Rhizobium* culture collections exist worldwide (e.g. the USDA Agricultural Research Service, the National *Rhizobium* Germplasm Collection, the Australian National *Rhizobium* Programme collection) to provide bacteria for rhizobial inoculants. Herridge *et al.* (2002) reviewed the Australian inoculants industry and suggested ways that inoculant production can lead to benefits in international settings. The issue remains around getting the applied effective and efficient bacteria into the soybean nodules in a soil environment with high levels of competition, and maintaining their survival at high levels through natural changes in the environment. Equally important is the observation that most of the plant-level responses in terms of fixation and nodulation amount and continuance through the life of the plant are under the control of the plant, not the rhizobia.

Plant breeding for increased nitrogen fixation

Clearly, breeding for enhanced nitrogen fixation by soybean through the selection of varieties has been a long sought-after goal. Genetically based variation in nitrogen fixation has been shown to exist both naturally and through mutation breeding (Herridge and Rose, 2000). Increased fixation and nodulation is readily achieved by super-, hyper- and promiscuous-nodulating plants (Carroll *et al.*, 1985; Nohara *et al.*, 2006). Other than for these characteristics, soybean fixation characteristics tend to be polygenic in inheritance (Nicolas *et al.*, 2002) and seem to carry over into improved production via increased overall soybean performance (Herridge and Rose, 2000). Graham and Vance (2000) reviewed possible molecular technologies that could increase soybean fixation. However, at this stage none has been successfully used.

Numerous publications exist in which naturally high nitrogen-fixing lines of soybean have been identified. Herridge and Danso (1995) listed a range of lines of Korean origin, as well as lines and cultivars identified by others, that maintain nitrogen fixation under high nitrate conditions. Crosses using the high-fixing lines and elite parents were carried out, but yields were low relative to elite lines. Herridge and Rose (2000) reported little success in improving yields from their extensive long-term programme, which incorporated high nitrogen-fixing capability into elite soybean cultivars from the original Korean lines. No high-yielding, well-adapted and high nitrogen-fixation cultivars have originated from the programme. Subsequently, high fixing lines have been identified and inheritance values plus the locations of quantitative trait loci have been identified for high- and low-fixing lines of Brazilian soybeans (Nicolas *et al.*, 2002, 2006), indicating that considerable efforts are continuing in attempts to achieve breeding gains by this method. There is no convincing evidence that either natural or human selection for increased yield has produced high-yielding, well-adapted and elevated nitrogen-fixing lines, suggesting that with indirect selection, benefits to nitrogen fixation do not accrue. A long-term, statistically significant, downward trend in soybean nitrogen fixation has been suggested by van Kessel and Hartley (2000), indicating that indirect selection has occurred against increased nitrogen fixation. This has led some authors to suggest that the primary benefit from high nitrogen-fixing lines will be a rotational saving of nitrogen for subsequent crops, not yield increases (Herridge and Danso, 1995).

In an attempt to overcome the limitations imposed by natural variations in nitrogen-fixation ability and the need to backcross the character into elite lines, Carroll *et al.* (1985) mutated Bragg soybeans and created a range of super- and hypernodulating soybeans. These original mutants, plus newly created ones, have since been extensively studied agronomically, physiologically and genetically (Hansen *et al.*, 1990; Bandara *et al.*, 1998; Jung *et al.*, 2008). They have also been used to create a comprehensive understanding of the long- and short-distance regulation of nodulation, particularly by the *nark* gene, which regulates nodulation via expression in soybean tops (Kinkema *et al.*, 2006). Generally, when tested in field experiments, the supernodulating lines have lower yields than their parental cultivars (Herridge and Rose, 2000), although their nitrogen content and carry-over nitrogen may be greater. More recently, however, claims have been made that supernodulation genotypes (specifically Kanto 100) can have superior yields. Jung *et al.* (2008) and Takahashi *et al.* (2003) summarized experiments conducted with Kanto 100, a supernodulating backcrossed mutant of cultivar Enrei, in which under non-waterlogging and low-yielding conditions, it outperformed the parental high-yielding, low-fixing, well-adapted cultivar. Interestingly, the backcrossed line also had significantly improved yields relative to the original mutated line, suggesting that other effects had masked the yield benefits of the mutation. Under higher-yielding conditions, Kanto 100 fixed more nitrogen but had a non-significantly different grain yield from its parent cultivar. Taken together, these data suggest that equal yields with higher nitrogen fixation leading to better rotational benefits may be possible using some of the supernodulation genes.

The other major plant-breeding option has centred on the promiscuity of nodulation of the soybean cultivars. Two methods have emerged: (i) to breed for highly promiscuous lines that nodulate with rhizobia naturally present in the soil; and (ii) to breed for lines with limited nodulation ability that will select for better-quality soil rhizobia or may be artificially inoculated with matched efficient rhizobia that can overcome their non-nodulation phenotype. The first method has produced a number of successful cultivars for use in areas where bradyrhizobia are not naturally present, but where the soil contains other rhizobia that are capable of nodulating some soybean genotypes (Gwata *et al.*, 2004). This is particularly of use where good systems to deliver effective inoculants do not exist. Herridge and Danso (1995) described efforts to use soybean containing the non-nodulating *rj1* gene to create plants that do not normally nodulate, followed by selecting effective rhizobia with which to inoculate them. Such rhizobia do exist, but were not more efficient and thus the combination was no better than the naturally nodulating parent type. Similarly, the selection of genotypes that have preference for more efficient rhizobia has been undertaken. At present there is no published evidence for the successful use of these approaches.

Genetic modification may be used to breed for enhanced nitrogen fixation in soybean. Over half of the world's soybeans are presently genetically modified (GM) (http://www.gmo-compass.org) so there are no technical issues to genetic modification in soybeans. Bohm *et al.* (2009) indicated that use of GM soybeans and the herbicides associated with them does not appear to decrease nitrogen fixation in Brazil. Similarly, Powell *et al.* (2007) found that colonization by rhizobia is not inhibited in GM soybeans. However, at present there are no clear objectives for a GM breeding programme. Suzuki *et al.* (2008) enhanced nitrogen fixation in *Lotus japonicas* with an antisense β-1,3-glucanase gene. Similar approaches are likely to work in soybean, but the benefits are unlikely to be greater than those achieved by more traditional breeding for increased nitrogen fixation. The creation of supernodulating (or high nitrogen-fixing) soybean is not a difficult issue; the difficulty is in gaining benefit from enhanced fixation. Therefore, GM breeding would need to seek other targets. With the high sensitivity of nitrogen fixation to phosphorus deficiency, waterlogging, drought, salinity and possibly elevated CO_2 (Table 11.3), there may be opportunities to indirectly select for increased fixation by selecting for improvements in these areas with GM technologies. Viktor *et al.* (2003) suggested a molecular approach to improve the use of plant phosphorus, which may be of advantage for nitrogen fixation in low-phosphorus African soils. Verdoy *et al.* (2006) demonstrated that overexpression of proline production in *Medicago* can lead to reduced sensitivity to drought stress. Graham and Vance (2000) suggested several other alternatives. At the present time, however, none of these approaches has been proven to work in soybean either through their direct benefits or through indirect benefits on improved nitrogen fixation.

11.5 Cropping-system Components

Limitations to system biological nitrogen fixation

A wide range of environmental stresses has been shown to reduce nitrogen fixation in the field in soybean. The fixation response may be due simply to

Table 11.3. Summary of environmental effects observed to specifically inhibit soybean nitrogen fixation, as reported by Zahran (1999), Taub *et al.* (2008) or other authors as indicated.

Stress	Level for effect	Type of effect
Salt	210 mM	Nodulation and fixation suppression. Effect more on nodulation than either symbiont.
Nitrate	>20 kg N ha^{-1}	Suppression of nodulation and fixation. Partially overcome by plant genotypes (McNeil and LaRue, 1984; Carroll *et al.*, 1985).
Nitrate	200 kg N ha^{-1}	Rhizobia population in soil unaffected. Symbiosis highly reduced.
Moisture stress	<5.5% in soil	Reduced inoculant strains in nodules, poor survival of inoculants in soil, reduced nodule formation, reduced nitrogen fixation. Partly overcome by elevated CO_2 (Serraj *et al.*, 1998).
High temperature	>35–40°C >41°C	Reduced nodulation and fixation. Partially overcome by plant genotype, but no direct link between plant growth ability and fixation ability.
High temperature	>37°C	Loss of rhizobia from the soil.
Acidity	pH 4.6	Root colonization adversely affected. Nitrogen-fixation more sensitive than plant growth. Relative tolerance to high aluminium ions of plant and fixation varies (Cline and Kaul, 1990).
Cadmium, nickel, copper, zinc, lead	>10 mg/kg soil >1–5 ppm	Nodulation reduced, nodule structure altered more than effect on plant. Nitrogen-fixation reduced (Huang *et al.*, 1974; Vesper and Weidensaul, 1978; Chen *et al.*, 2003).
Nitrite	>10 mM	No effect on bacterial fixation in nodules.
Herbicide	Recommended field rates	Sethoxydim, alachlor, fluazifop butyl and metolachlor had no effect on fixation, Paraquat reduced fixation. Results often variable in the field.
Dark chilling	<15°C	Reduction in fixation by up to 45% (Riekert van Heerden and Krüger, 2004).
Ozone	1.5× ambient	Reduced fixation, effects overcome by elevated CO_2.
CO_2	<500 ppm	Reduced fixation, but less so than total plant growth.
Proline		Bacteria unable to metabolize proline. Poor fixation in drought, which increases plant proline levels.
Waterlogging	10 days	Fixation inhibited at several growth stages (Jung *et al.*, 2008).

the reduction in host plant growth, but often the symbiosis is more sensitive than the plant to environmental stress. For example, Elsheikh and Wood (1995) showed that nodulation and fixation were more sensitive than plant growth to excess salt stress, and this was partly offset by the use of appropriate *Rhizobium* strains. Zahran (1999) reviewed legume symbioses in general in severe and arid climates, and indicated that salt stress, osmotic stress, high soil moisture deficit, high temperature, soil acidity and alkalinity, herbicides, nutrient deficiency and nutrient toxicity have all been shown to inhibit nodulation and nitrogen fixation. A summary of the observed effects is given in Table 11.3. A number of claims have been made for specific rhizobia or plant variety combinations to perform better under the stress. However, Hardarson and Atkins (2003) stated in their review that the single biggest method for improving nitrogen fixation in legumes is to apply rhizobia where they are not available in the soil. In soybean, this has also been achieved via the use of promiscuous nodulating varieties (Gwata *et al.*, 2004). Agronomically, Hardarson and Atkins (2003) suggested that optimizing for crop production will also optimize for nitrogen fixation, except in circumstances where added nitrogen inhibits fixation. In view of the genetic variation for nitrogen inhibition of fixation, they suggested that nitrogen-tolerant varieties may spare more nitrogen for rotation crops.

Modelling of the benefits of nitrogen-tolerant soybean over a 30-year period in an Australian cropping region has suggested that in rotations there may be residual benefits of about 7 kg N ha^{-1} from using nitrogen-tolerant soybean (Herridge *et al.*, 2001). Yield improvement through the use of improved bacteria has been hindered by the ability to reliably incorporate the applied bacteria in the field. Herridge and Rose (2000) reviewed the literature for plant breeding and came to the conclusion that 'Whilst genetically based variation in N_2 fixation traits has now been demonstrated for soybean and other legume species, incorporating such variation into cultivars has had little success. Future programmes may benefit from increased integration into mainstream legume breeding programmes that are focussed on a broad range of traits.'

Soybean rotational benefits

Within the US corn belt, rotation of corn (*Zea mays*) and soybean is preferred to continuous cropping (Wilhelma and Wortmann, 2004) because the rotation produces greater grain yield of both crops (West *et al.*, 1996). Input costs are often less; in particular, less nitrogen fertilizer is needed for the corn–soybean rotation (Katupitiya *et al.*, 1997) compared with continuous corn. A corn–soybean rotation also reduces deep leaching of nitrate nitrogen relative to continuous corn. Reduced stress from pests and diseases may also improve yields in rotations. This chapter concentrates on the nitrogen-saving benefits. In the first instance, it has already been indicated that in Brazil, 80% of the direct nitrogen for soybean comes from nitrogen fixation, with a global input of around 58%, so there are direct benefits for the grower of the soybean crop (Herridge *et al.*, 2008; Salvagiotti *et al.*, 2008).

The magnitude of the benefit has also been indicated to increase via the use of supernodulating and nitrogen-sparing soybean varieties (Herridge *et al.*, 2001). A crop calculator for the nitrogen benefit of soybean has been developed for Canadian situations (Przednowek *et al.*, 2004), which suggested that soybean is considerably inferior to the other legumes used in the study (pea, chickpea and faba bean). It is possible that because of the large amounts of nitrogen removed via soybean seed (150–200 kg N ha^{-1}), soybean may frequently act as an nitrogen sink rather than a nitrogen source (Herridge *et al.*, 2008; Salvagiotti *et al.*, 2008) and any nitrogen benefit is due to nitrogen sparing rather than nitrogen fixation. However, Ennin and Clegg (2001) found evidence for soybean nitrogen fertilizer replacement values in Nebraska of up to 46 kg N ha^{-1} in rotation with maize (although only at high plant populations). This indicates a potential for substantial nitrogen savings in soybean rotation systems. However, most rotation articles indicate that nitrogen benefit only accounts for part of the rotational advantage. The size of the rotational benefit can also vary significantly (Bundy *et al.*, 1993) depending on soil type and rotation used. However, in all circumstances the large removal of nitrogen in the soybean crop means that the benefit is generally small.

11.6 Conclusions

Soybean is one of the most studied of crops with respect to BNF. A wealth of data exists around its nitrogen fixation, as well as around the physiology and genetics of the inoculating rhizobia and the plant nodulation and fixation processes. Regulation of the nodulation and fixation processes is well understood, with a variety of mutant and selected plant and bacterial lines developed to understand the process. In summary, soybean is the world's foremost single-grain legume crop for fixing nitrogen. Mutants with significantly more or less promiscuous nodulation have been developed, as well as highly efficient rhizobial inoculants. Soybean lines range from non-nodulating to supernodulating. At present, however, the best increases in fixation occur where soybean is planted in soils with no soybean-effective rhizobia. In soils with high natural levels of rhizobia it is difficult (although not impossible) to achieve increased yields via higher-quality inoculated rhizobia. Most other benefits in fixation come from general breeding for improved varieties and good crop agronomy. In some instances, the nodulation and fixation processes are more sensitive than plant growth. What is clear is that including soybean in crop rotations where inadequate nitrogen is available can lead to high soybean yields and improved yields of the non-nitrogen-fixing rotation crops. Some nitrogen-tolerant lines may provide greater amounts of nitrogen for rotation crops. In tropical rotations with sugarcane (*Saccharum officinarum*), Hemwong *et al.* (2009) indicated that soybean can have substantial nitrogen benefits for the subsequent crop. Its rate of degradation was slower than that of peanut (*Arachis hypogaea*) residue and nitrogen benefits therefore survived longer under tropical conditions.

However, soybean residues are generally regarded as breaking down more quickly than the low-nitrogen grass crop residues. The generally higher nitrogen to carbon ratio of soybean residues means they do not tie up as much nitrogen in the degradation process (Nafziger, 2002).

One area that has not been extensively investigated is the effect of increasing CO_2 on the rotational benefit of soybean. Prior *et al.* (2006) showed increased soybean growth and fixation with a higher nitrogen content in residues of soybean grown under FACE trials. This suggests that rotational benefits due to nitrogen fixation may be increasing as global CO_2 levels rise.

References

Ainsworth, E.A., Rogers, A., Vodkin, L.O., Walter, A. and Schurr, U. (2006) The effects of elevated CO_2 concentration on soybean gene expression: An analysis of growing and mature leaves. *Plant Physiology* 142, 135–147.

Appunu, C., N'Zoue, A. and Laguerre, G. (2008) Genetic diversity of native *bradyrhizobia* isolated from soybeans (*Glycine max* L.) in different agricultural-ecological-climatic regions of India. *Applied and Environmental Microbiology* 74, 5991–5996.

Atkins, C.A. (1982) Efficiencies and inefficiencies in the legume/*Rhizobium* symbiosis – a review. *Plant and Soil* 82, 273–284.

Baginsky, C., Brito, B., Imperial, J., Palacios, J.M. and Ruiz-Argüeso, T. (2002) Diversity and evolution of hydrogenase systems in rhizobia. *Applied and Environmental Microbiology* 68, 4915–4924.

Bandara, D.C., Nobuyasu, H., Ofosu-Budu, K.G., Ando, T. and Fujita, K. (1998) Effect of CO_2 enrichment on biomass production, photosynthesis, and sink activity in soybean cv. Bragg and its supernodulating mutant nts 1007. *Soil Science and Plant Nutrition* 44, 179–186.

Bascones, E., Imperial, J., Ruiz-Argueso, T. and Palacios, JM. (2000) Generation of new hydrogen-recycling Rhizobiaceae strains by introduction of a novel hup minitransposon. *Applied and Environmental Microbiology* 66, 4292–4299.

Begum, A.A., Leibovitch, L., Migner, P. and Zhang, F. (2001) Specific flavonoids induced *nod* gene expression and pre-activated *nod* genes of *Rhizobium leguminosarum* increased pea (*Pisum sativum* L.) and lentil (*Lens culinaris* L.) nodulation in controlled growth chamber environments. *Journal of Experimental Botany* 52, 1537–1543.

Bohm, G.M.B., Alves, B.J.R., Urquiaga, S., Boddey, R.M., Xavier, G.R., Hax, F. and Rombaldi, C.V. (2009) Glyphosate- and imazethapyr-induced effects on yield, nodule mass and biological nitrogen fixation in field-grown glyphosate-resistant soybean. *Soil Biology and Biochemistry* 41, 420–422.

Botha, W.J., Jaftha, J.B., Bloem, J.F., Habig, J.H. and Law, I.J. (2004) Effect of soil *bradyrhizobia* on the success of soybean inoculant strain CB 1809. *Microbiological Research* 159, 219–231.

Brito, B., Toffanin, A., Prieto, R.I., Imperial, J., Ruiz-Argueso, T. and Palacios, J.M. (2008) Host-dependent expression of *Rhizobium leguminosarum* bv. *viciae* hydrogenase is controlled at transcriptional and post-transcriptional levels in legume nodules. *Molecular Plant-Microbe Interactions* 21, 597–604.

Bundy, L.G., Andraski, T.W. and Wolkowski, R.P. (1993) Nitrogen credits in soybean–corn crop sequences on three soils. *Agronomy Journal* 85, 1061–1067.

Burns, R.C. and Hardy, R.W.F. (1975) *Nitrogen Fixation in Bacteria and Higher Plants*. Springer-Verlag, New York, USA.

Carroll, B.J., McNeil, D.L. and Gresshoff, P.M. (1985) Isolation and properties of novel soybean (*Glycine max*. L. Merr.) mutants that nodulate in the presence of high nitrate concentrations. In: *Proceedings of the National Academy of Sciences of the United States of America* 82, 4162–4166.

Carroll, B.J., McNeil, D.L. and Gresshoff, P.M. (1986) Mutagenesis of soybean (*Glycine max* L. Merr.) and the isolation of non-nodulating mutants. *Plant Science* 47, 109–114.

Carroll, B.J., Hansen, A.P., McNeil, D.L. and Gresshoff, P.M. (1987) Effect of oxygen supply on nitrogenase activity of nitrate and dark stressed soybean [*Glycine max*. (L.) Merr.] plants. *Australian Journal Plant Physiology* 14, 679–687.

Chang, W.S., Franck, W.L., Cytryn, E., Jeong, S., Joshi, T., Emerich, D.W., Sadowsky, M.J., Xu, D. and Stacey, G. (2007) An oligonucleotide microarray resource for transcriptional profiling of *Bradyrhizobium japonicum*. *Molecular Plant-Microbe Interactions* 20, 1298–1307.

Chen, Y.X., He, Y.F., Yang, Y., Yu, Y.L., Zheng, S.J., Tian, G.M., Luo, Y.M. and Wong, M.H. (2003) Effect of cadmium on nodulation and N_2-fixation of soybean in contaminated soils. *Chemosphere* 50, 781–787.

Cleveland, C.C., Townsend, A.R., Schimel, D.S., Fisher, H., Howarth, R.W., Hedin, L.O., Perakis, S.S., Latty, E.F., von Fischer, J.C., Elseroad, A. and Wasson, M.F. (1999) Global patterns of terrestrial biological nitrogen (N_2) fixation in natural ecosystems. *Global Biogeochemical Cycles* 13, 623–645.

Cline, G.R. and Kaul, K. (1990) Soybean-*Bradyrhizobium* combinations under acidity, low P and high Al stresses. *Plant and Soil* 127, 243–249.

Cytryn, E.J., Sangurdekar, D.P., Streeter, J.G., Franck, W.L., Chang, W.S., Stacey, G., Emerich, D.W., Joshi, T., Xu, D. and Sadowsky, M.J. (2007) Transcriptional and physiological responses of *Bradyrhizobium japonicum* to desiccation-induced stress. *Journal of Bacteriology* 189, 6751–6762.

Elsheikh, E.A.E. and Wood, M. (1995) Nodulation and N_2 fixation by soybean inoculated with salt-tolerant rhizobia or salt-sensitive *bradyrhizobia* in saline soil. *Soil Biology and Biochemistry* 27, 657–661.

Ennin, S.A. and Clegg, M.D. (2001) Effect of soybean plant populations in a soybean and maize rotation, *Agronomy Journal* 93, 396–403.

Erisman, J.W., Sutton, M.A., Galloway, J., Klimont, Z. and Winiwarter, W. (2008) How a century of ammonia synthesis changed the world? *Nature Geoscience* 1, 636–639.

Galloway, J.N. (1998) The global nitrogen cycle: Changes and consequences. *Environmental Pollution* 102 (Suppl. 1), 15–24.

Galloway, J.N., Dentener, F.J., Capone, D.G., Boyer, E.W. and Howarth, R.W. (2004) Nitrogen cycles: Past, present, and future. *Biogeochemistry* 70, 153–226.

Graham, P.H. and Vance, C.P. (2000) Nitrogen fixation in perspective: An overview of research and extension needs. *Field Crops Research* 65, 93–106.

Gwata, E.T., Wofford, D.S., Pfahler, P.L. and Boote, K.J. (2004) Genetics of promiscuous nodulation in soybean: Nodule dry weight and leaf color score. *Journal of Heredity* 95, 154–157.

Hansen, A.P., Peoples, M.B., Brown, P.H., Carroll, B.J. and Gresshoff, P.M. (1990) Nitrogen partitioning during early development of supernodulating soybean [*Glycine max* (L.) Merrill] mutants and their wild-type parent. *Journal of Experimental Botany* 41, 1239–1244.

Hanway, J.J. (1979) Foliar fertilization of soybeans during seed-filling. In: Corbin, F.T. (ed.) *Proceedings of World Soybean Research Conference II*, Westview Press, Boulder, Colorado, USA, pp. 409–416.

Hardarson, G. and Atkins, C. (2003) Optimising biological N_2 fixation by legumes in farming systems. *Plant and Soil* 252, 41–54.

Hemwong, S., Toomsan, B., Cadisch, G., Limpinuntana, V., Vityakon, P. and Patanothai, A. (2009) Sugarcane residue management and grain legume crop effects on N dynamics, N losses and growth of sugarcane. *Nutrient Cycling in Agroecosystems* 83, 135–151.

Herridge, D.F. and Danso, S.K.A. (1995) Enhancing crop legume N_2 fixation through selection and breeding. *Plant and Soil* 174, 51–82.

Herridge, D. and Rose, I. (2000) Breeding for enhanced nitrogen fixation in crop legumes. *Field Crops Research* 65, 229–248.

Herridge, D., Turpin, J.E. and Robertson, M.J. (2001) Improving nitrogen fixation of crop legumes through breeding and agronomic management: Analysis with simulation modelling. *Australian Journal of Experimental Agriculture* 41, 391–401.

Herridge, D., Gemell, G. and Hartley, E. (2002) Legume inoculants and quality control. In: Herridge, D. (ed.) *Inoculants and Nitrogen Fixation of Legumes in Vietnam. ACIAR Proceedings 109e*. Australian Centre for International Agricultural Research, Canberra, ACT, Australia, pp. 105–115.

Herridge, D., Peoples, M.B. and Boddey, R.M. (2008) Global inputs of biological nitrogen fixation in agricultural systems. *Plant and Soil* 311, 1–18.

Huang, C.Y., Bazzaz, F.A. and Vanderhoef, L.N. (1974) The inhibition of soybean metabolism by cadmium and lead. *Plant Physiology* 54, 122–124.

Hungria, M., Campol, R.J., Chueirel, L.M.O., Grange, L. and Megías, M. (2001) Symbiotic effectiveness of fast-growing rhizobial strains isolated from soybean nodules in Brazil. *Biology and Fertility of Soils* 33, 387–394.

Johnson, D.W. (2006) Progressive N limitation in forests: Review and implications for long-term responses to elevated CO_2. *Ecology* 87, 64–75.

Jung, G., Matsunami, T., Oki, Y. and Kokubun, M. (2008) Effects of waterlogging on nitrogen fixation and photosynthesis in supernodulating soybean cultivar Kanto 100. *Plant Production Science* 11, 291–297.

Katupitiya, A., Eisenhauer, D.E., Ferguson, R.B., Spalding, R.F., Roeth, F.W. and Bobier, M.W. (1997) Long-term tillage and crop rotation effects on residual nitrate in the crop zone and nitrate accumulation in the intermediate vadose zone. *Transactions of the American Society of Agricultural Engineers* 40, 1321–1327.

Kinkema, M., Scott, P.T. and Gresshoff, P.M. (2006) Legume nodulation: Successful symbiosis through short- and long-distance signalling. *Functional Plant Biology* 33, 707–721.

Kosslak, R.M., Bohlool, B.B., Dowdle, S. and Sadowsky, M.J. (1983) Competition of *Rhizobium japonicum* strains in early stages of soybean nodulation. *Applied and Environmental Microbiology* 46, 870–873.

Kuzma, M.M., Winter, H., Storer, P., Oresnik, I., Atkins, C.A. and Layzell, D.B. (1999) The site of oxygen limitation in soybean nodules. *Plant Physiology* 119, 399–408.

Lodwig, E. and Poole, P. (2003) Metabolism of *Rhizobium* bacteroids. *Critical Reviews in Plant Sciences* 22, 37–78.

McNeil, D.L. (1982) Variations in the ability of *Rhizobium japonicum* strains to nodulate soybeans and maintain fixation in the presence of nitrate. *Applied and Environmental Microbiology* 44, 647–652.

McNeil, D.L. and LaRue, T.A. (1984) Effect of nitrogen source on ureides in soybean. *Plant Physiology* 74, 227–232.

McNeil, D.L., Croft, L. and Sandhu, T.S. (1981) Response of chickpeas to inoculation with *Rhizobium* in Hawaii. *International Chickpea Newsletter* 3, 26–27.

McNeil, D.L., Borton, S., Amara, D. and Vora, M.S. (1983) Use of antibiotic resistant *Rhizobium* mutants for competition studies with *Cajanus cajan*. *International Pigeonpea Newsletter* 2, 71–72.

Nafziger, E.D. (2002) Cover crops and cropping systems. In: Picklesimer, P. (ed.) *Online Illinois Agronomy Handbook*, 23rd edn. University of Illinois, Urbana-Champaign, Illinois, USA, pp. 54–59.

Nesbitt, S.W., Zhang, R. and Orville, R.E. (2000) Seasonal and global NOx production by lightning estimated from the Optical Transient Detector (OTD). *Tellus B*. 52, 1206–1215.

Nicolas, M.F., Arias, C.A.A. and Hungria, M. (2002) Genetics of nodulation and nitrogen fixation in Brazilian soybean cultivars. *Biology and Fertility of Soils* 36, 109–117.

Nicolas, M.F., Hungria, M. and Arias, C.A.A. (2006) Identification of quantitative trait loci controlling nodulation and shoot mass in progenies from two Brazilian soybean cultivars. *Field Crops Research* 95, 355–366.

Nohara, T., Nakayama, N., Nakamura, T., Takahashi, M., Maruyama, S., Arihara, J. and Shimada, S. (2006) Cultivar differences of nitrogen fixation capacity and its contribution to nitrogen accumulation in soybean grown in the field with a high soil nitrate level. *Japanese Journal of Crop Science* 75, 350–359.

Oldroyd, G.E.D. and Downie, J.A. (2008) Coordinating nodule morphogenesis with rhizobial infection in legumes. *Annual Review of Plant Biology* 59, 519–546.

Powell, J.R., Gulden, R.H., Hart, M.M., Campbell, R.G., Levy-Booth, D.J., Dunfield, K.E., Pauls, K.P., Swanton, C.J., Trevors, J.T. and Klironomos, J.N. (2007) Mycorrhizal and rhizobial colonization of genetically modified and conventional soybeans. *Applied and Environmental Microbiology* 73, 4365–4367.

Prell, J. and Poole, P. (2006) Metabolic changes of rhizobia in legume nodules. *Trends in Microbiology* 14, 161–168.

Prior, S.A., Torbert III, H.A., Runion, G.B., Rogers H.H., Jr, Ort, D.R. and Nelson, R.L. (2006) Free-air carbon dioxide enrichment of soybean: Influence of crop variety on residue decomposition. *Journal of Environmental Quality* 35, 1470–1477.

Przednowek, D.W.A., Entz, M.H., Irvine, B., Flaten, D.N. and Thiessen Martens, J.R. (2004) Rotational yield and apparent N benefits of grain legumes in southern Manitoba. *Canadian Journal of Plant Science* 84, 1093–1096.

Qiao, Z. (1997) Effects of atmospheric nitrogen dioxide on uptake and assimilation of ammonium in soybean plants. *Journal of Plant Nutrition* 20, 1183–1190.

Qiao, Z. and Murray, F. (1997) The effects of root nitrogen supplies on the absorption of atmospheric NO_2 by soybean leaves. *New Phytologist* 136, 239–243.

Rainbird, R.M., Hitz, W.D. and Hardy, R.W.F. (1984) Experimental determination of the respiration associated with soybean/*Rhizobium* nitrogenase function, nodule maintenance, and total nodule nitrogen fixation. *Plant Physiology* 75, 49–53.

Raverkar, K.P., Gupta, S.B. and Rao, D.L.N. (2005) Survival of soybean rhizobia in summer and proliferation during monsoon in vertisols of central India. *Journal of the Indian Society of Soil Science* 53, 591–597.

Riekert van Heerden, P.D. and Krüger, G.H.J. (2004) Dark chilling inhibition of photosynthesis and symbiotic nitrogen fixation in soybean during pod filling. *Journal of Plant Physiology* 161, 599–609.

Salvagiotti, F., Cassman, K.G., Specht, J.E., Walters, D.T., Weiss, A. and Dobermann, A. (2008) Nitrogen uptake, fixation and response to fertilizer N in soybeans: A review. *Field Crops Research* 108, 1–13.

Serraj, R., Sinclair, T.R. and Allen, L.H. (1998) Soybean nodulation and N_2 fixation response to drought under carbon dioxide enrichment. *Plant, Cell and Environment* 21, 491–500.

Sinclair, T.R., Farias, J.R., Neumaier, N. and Nepomuceno, A.L. (2003) Modeling nitrogen accumulation and use by soybean. *Field Crops Research* 81, 149–158.

Smit, G., Swart, S., Lugtenberg, B.J.J. and Kijne, J.W. (1992) Molecular mechanisms of attachment of *Rhizobium* bacteria to plant roots. *Molecular Microbiology* 6, 2897–2903.

Spaink, H.P. (2000) Root nodulation and infection factors produced by rhizobial bacteria. *Annual Review of Microbiology* 54, 257–288.

Sparks, J.P. (2009) Ecological ramifications of the direct foliar uptake of nitrogen. *Oecologia* 159, 1–13.

Suarez, R., Wong, A., Ramirez, M., Barraza, A., Orozco, M.D., Cevallos, M.A., Lara, M., Hernandez, G. and Iturriaga, G. (2008) Improvement of drought tolerance and grain yield in common bean by overexpressing trehalose-6-phosphate synthase in rhizobia. *Molecular Plant-Microbe Interactions* 21, 958–966.

Suzuki, A., Yamashita, K., Ishihara, M., Nakahara, K.I., Abe, M., Kucho, K.I., Uchiumi, T., Higashi, S. and Arima, S. (2008) Enhanced symbiotic nitrogen fixation by *Lotus japonicus* containing an antisense beta-1,3-glucanase gene. *Plant Biotechnology* 25, 357–360.

Tajima, S., Nomura, M. and Kouchi, H. (2004) Ureide biosynthesis in legume nodules. *Frontiers of Science* 9, 1374–1381.

Takahashi, M., Arihara, J., Nakayama, N. and Kokubun, M. (2003) Characteristics of growth and yield formation in the improved genotype of supernodulating soybean (*Glycine max* L. Merr.). *Plant Production Science* 6, 112–118.

Taub, D.R., Miller, B. and Allen, H. (2008) Effects of elevated CO_2 on the protein concentration of food crops: A meta-analysis. *Global Change Biology* 14, 565–575.

Todd, C.D., Tipton, P.A., Blevins, D.G., Piedras, P., Pineda, M. and Polacco, J.C. (2006) Update on ureide degradation in legumes. *Journal of Experimental Botany* 57, 5–12.

van Kessel, C. and Hartley, C. (2000) Agricultural management of grain legumes: Has it led to an increase in nitrogen fixation? *Field Crops Research* 65,165–181

Verdoy, D., De la Pena, T.C., Redondo, F.J., Lucas, M.M. and Pueyo, J.J. (2006) Transgenic *Medicago truncatula* plants that accumulate proline display nitrogen-fixing activity with enhanced tolerance to osmotic stress. *Plant, Cell and Environment* 29, 1913–1923.

Vesper, S.J. and Weidensaul, T.C. (1978) Effects of cadmium, nickel, copper, and zinc on nitrogen fixation by soybeans. *Water, Air & Soil Pollution* 9, 413–422.

Viktor, A., Cordero-Otero, R. and Valentine, A. (2003) Bioengineering beans for phosphate-deficient soils in southern Africa. *South African Journal of Science* 99, 509–511.

West, T.D., Griffith, D.R., Steinhardt, G.C., Kladivko, E.J. and Parsons, S.D. (1996) Effect of tillage and rotation on agronomic performance of corn and soybean: Twenty-year study on dark silty clay loam soil. *Journal of Production Agriculture* 9, 241–248.

Wilhelma, W.W. and Wortmann, C.S. (2004) Tillage and rotation interactions for corn and soybean grain yield as affected by precipitation and air temperature. *Agronomy Journal* 96, 425–432.

Willems, A. (2006) The taxonomy of rhizobia: An overview. *Plant and Soil* 287, 3–14.
Yang, J.K. and Zhou, J.C. (2008) Diversity, phylogeny and host specificity of soybean and peanut bradyrhizobia. *Biology and Fertility of Soils* 44, 843–851.
Zahran, H.H. (1999) *Rhizobium*–legume symbiosis and nitrogen fixation under severe conditions and in an arid climate. *Microbiology and Molecular Biology Reviews* 63, 968–989.
Zehner, S., Schober, G., Wenzel, M., Lang, K. and Göttfert, M. (2008) Expression of the *Bradyrhizobium japonicum* Type III secretion system in legume nodules and analysis of the associated *tts* box promoter. *Molecular Plant-Microbe Interactions* 21, 1087–1093.

12 Storage of Soybean

Prabal K. Ghosh[1] and Digvir S. Jayas[2]
[1]*Food Development Centre, Portage la Prairie, Manitoba, Canada;* [2]*Department of Biosystems Engineering, University of Manitoba, Winnipeg, Manitoba, Canada*

12.1 Introduction

Soybean (*Glycine max* (L.) Merrill) is the world's most valuable oilseed crop, and is an excellent source of plant protein (34–40%) and oil (20%). Soybean is also known as 'meat of the field' in the Orient and 'Cinderella crop' in the USA. Soybeans can be categorized into two distinct types: commodity beans, which are generally crushed by domestic processors; and specialty beans, which are typically consumed as whole beans. Soybeans are consumed in different forms (e.g. oil, protein-rich meal, full-fat soy flour, low-fat protein flour, de-fatted flour, de-fatted flakes and grits, protein concentrates and isolates) and in a variety of foods (e.g. snack foods, bakery products, tofu, meat products, sauces and soups, desserts, drinks, oils, salad dressings), as well as in animal feed (full-fat or de-fatted meals). Soybeans may be yellow, green, brown or black. World soybean production was approximately 216 million t in 2007 from a total harvested area of about 95 million ha (FAO, 2009). The USA is the largest producer of soybeans (71 million t), followed by Brazil (58 million t), Argentina (45 million t), China (16 million t), India (9 million t), Paraguay (4 million t) and other countries. Hundreds of different soybean cultivars are grown around the world.

12.2 Harvesting

Soybean is harvested when the seeds are mature and the foliage is dry. Soybean is a highly productive crop, with a normal production cycle of 90–110 days from planting to harvesting. Typically, soybeans are harvested at 13–15% moisture content to avoid field losses due to shattering of over-dried seeds and to reduce drying time to achieve safe storage moisture content. Harvesting at <12% moisture content creates seed cracks and shattering losses. In developed countries, plants are usually harvested and soybeans are separated and cleaned using combine harvesters; in developing countries this is done by hand followed by manual or semi-automated threshing. Seed size depends on the genotype and production environment.

12.3 Storage

After harvesting, soybeans need to be stored until they are processed or consumed. Storage time can vary from a few months to more than a year. It is advisable to remove foreign material, weed seeds and fines from soybeans and to inspect them for mould, insects or insect-damaged seeds before drying and storage for better and longer storage (Parde *et al.*, 2002). Foreign material within the seed bulk can create non-uniformity in airflow during aeration. This results in poorly aerated areas, which may provide sites for mould or insect growth. Air-screen cleaners can be used for removing foreign materials from soybean seeds. Foreign materials in soybeans are defined in the US Standards as all materials that pass through a 3.2 mm round-hole sieve (Spencer, 1976). Typically, dirt and sand particles, weed seeds and other small grains harvested with the crop are considered to be foreign materials. The presence of weed seeds can create small zones of high moisture content within the bulk, which may facilitate insect or mould growth and lead to the development of hot spots (i.e. localized areas of intense biological activity manifested by the production of metabolic heat, moisture and CO_2 by insects or fungi). In developed countries, soybeans are usually stored in cylindrical steel or concrete silos. Steel silos have a conical top and are usually equipped with ventilating fans. They may be of any size and can hold up to 60,000 t. Filling and emptying operations are carried out using conveyors. In developing countries, bag handling and storage are still predominant, although this is transitioning to bulk handling and storage systems.

When the moisture content of harvested soybeans is unsafe for storage, drying is performed. Drying reduces the risk of deterioration due to seed respiration, mould attack and spontaneous heating. The safe storage of soybeans used as seeds depends upon two major parameters: moisture content and drying temperature. The recommended safe storage moisture contents for soybean destined for two different uses are given in Table 12.1. The safe

Table 12.1. Safe storage moisture content of soybeans (reprinted with permission from White, 2001).

Moisture content (%, wet basis)	Market stock	Seed stock
10–11	4 years	1 year
10–12.5	1–3 years	6 months
13–14	6–9 months	Questionable, check germination
14–15	6 months	Questionable, check germination

storage of seed for seed stock requires storage at a lower moisture content than that for processing (market).

Depending on the moisture content, soybeans are classified as dry (up to 14%), tough (14.1–16%), damp (16.1–18%), moist (18.1–20%) and wet (>20%) (White, 2001). The maximum permissible moisture content limits for soybean grades (US) 1, 2, 3 and 4 are 13%, 14%, 16% and 18%, respectively (White, 2001). The drying temperature for soybeans to be used as food and oil should not exceed 49°C and for seed purposes should be limited to 43°C (White, 2001). The maximum safe drying temperature for in-storage drying is 38°C. The relative humidity (RH) of the drying air is another major factor and determines the moisture content to which the crop will dry. This moisture is referred to as the equilibrium moisture content (EMC) of the grain. Soybeans are hygroscopic materials that either lose (desorption) or gain (sorption) moisture from the surrounding air. Table 12.2 shows the relationship between the temperature and the RH of the drying air and the EMC of soybeans. The moisture content of soybean seeds can be reduced by decreasing the RH of a storage environment at a constant temperature. The hysteresis effect (repeated sorption and desorption cycle) is less pronounced in soybeans than in other cereal grains under the same conditions (Acasio, 1997).

Once the seeds are dried and have attained a safe storage moisture content, the dried seeds are cooled immediately with a proper aeration system (fans with capacity ranges between 0.001 and 0.003 $m^3 s^{-1}$ per m^3 (Hurburgh, 2008). Soybeans with 14.0–14.3% moisture content and maintained at 5–8°C can be stored for >2 years without mould damage, whereas those kept at 30°C are susceptible to mould growth in few weeks and severely damaged in 6 months. Fungal growth in the stored beans can be dangerous because it may result in loss of germination, spot heating and the production of mycotoxins. Mycotoxins are poisonous when eaten or inhaled. Table 12.3 shows the storage conditions at which mould growth can be prevalent in soybeans. Mycotoxins usually develop when stored seeds are contaminated with *Aspergillus* and *Penicillium* moulds. Ochratoxin and citrinin are specifically found in seeds contaminated with *Penicillium*.

Table 12.2. Equilibrium moisture content of soybeans (reprinted with permission from Acasio, 1997).

Drying air temperature (°C)	Relative humidity (%)								
	10	20	30	40	50	60	70	80	90
	Equilibrium moisture content (%)								
5	5.2	6.3	6.9	7.7	8.6	10.4	12.9	16.9	22.4
15	4.3	5.7	6.5	7.2	8.1	10.1	12.4	16.1	21.9
25	3.8	5.3	6.1	6.9	7.8	9.7	12.1	15.8	21.3
35	3.5	4.8	5.7	6.4	7.6	9.3	11.7	15.4	20.6
45	2.9	4.0	5.0	6.0	7.1	8.7	11.1	14.9	–
55	2.7	3.6	4.2	5.4	6.5	8.0	10.6	–	–

Table 12.3. Typical storage conditions for fungal growth in soybeans (reprinted with permission from Sauer *et al.*, 1992).

Relative humidity (%)	Equilibrium moisture content (%)	Fungi
65–70	11–12	*Aspergillus halophilicus*
70–75	12–14	*A. restrictus*, *A. glaucus*, *Wallemia sebi*
75–80	14–16	*A. candidus*, *A. ochraceus*, plus the above
80–85	16–19	*A. flavus*, *Penicillium* spp., plus the above
85–90	19–23	Any of the above

Periodical aeration is required to prevent moisture migration and heating of seeds in storage and to maintain a uniform temperature to retain seed quality. Even at safe storage moisture contents, a non-uniform temperature in the bulk storage creates interseed air current (0.06 m min^{-1} from a temperature gradient of 16.7°C), resulting in moisture migration and deterioration of the seeds. Typically, insects develop and reproduce between 27°C and 35°C, but they become inactive or die at temperatures of <16°C. In cold weather, soybeans can be stored at as high as 14% moisture content, but for safe storage during summer or spring, soybeans should not contain more than 11–12% moisture (Berglund and Helms, 2003).

12.4 Drying

Drying seeds prevents microbial growth, slows enzymatic changes and considerably increases the storage life. It also reduces the grain mass and thus facilitates transportation and handling. Soybean drying is usually accomplished by forcing air through the bulk seeds at different temperatures: natural (ambient), near-ambient (ambient plus 1–5°C), low (ambient plus 5–15°C) and high (50–200°C). Soybean drying is carried out with various mechanical driers, such as low-temperature driers, on-the-floor driers, in-bin driers, medium-temperature driers, tray driers, radical flow driers, multi-duct ventilated flow driers, countercurrent open-flame grain driers and solar driers. Continuous or in-bin driers are commonly used. On-the-floor drying systems or radially ventilated bins are well suited for slow drying.

Mathematical models of drying soybeans

Drying involves simultaneous heat and mass transfer. Mathematical modelling of drying is, therefore, important to predict the temperature and moisture distribution within the seed bulk and to locate the points of high temperature and moisture gradients in the bulk. Alam and Shove (1973) simulated ambient air drying of soybeans using the following equation (1) for EMC based on an experimental desorption isotherm (salt solution: temperature 5–55°C and RH 10–95%):

$$1 - RH = e^{-cT} ab^{M^n} \tag{1}$$

where RH is the relative humidity, decimal; T is the temperature, °F; M is the moisture content, % dry basis (db); and a, b, c and n are coefficients where a = 1.617 × 10^{-4}, b = 2.85274; c = 0.18938 and n = 1.48234.

The latent heat of vaporization of seeds is given by the following relationship (equation 2) (Alam and Shove, 1973):

$$\frac{L}{L_w} = 1 + 0.21624e^{-0.06233M} \tag{2}$$

where L is the latent heat of vaporization of soybeans, J kg^{-1}; and L_w is the latent heat of vaporization of water, J kg^{-1}.

Sabbah *et al.* (1979) modified the fixed-bed corn drying model of Bakker-Arkema *et al.* (1974) to provide the following single-kernel drying model (equation 3) for reversed-direction airflow drying of soybean seed in a fixed bed to obtain uniform final moisture content:

$$\frac{M - M_e}{M_o - M_e} = 0.608(1 - \exp(-K_2t))(\exp(-K_1t) + 0.25\exp(-4K_1t)) + \exp(-K_2t) \tag{3}$$

where,

$$K_1 = 4\pi^2 D/d^2 \tag{4}$$

$$D = (0.0493\exp(-0.59/(M_o - M_e)) + 0.0181(M - M_e))\exp(3137.6/T_g + 273) \tag{5}$$

$$d = 0.6279 + 0.1255M \tag{6}$$

$$K_2 = GF \times AF \times GAF \tag{7}$$

$$GF = \frac{(558.99 - 196.19M)S}{\rho_g} \tag{8}$$

$$AF = 1.342 \times 10^{-6} F_a \frac{P_{sa}}{P_a} \frac{(T_a + 273)^{1.54}}{(0.514 + 0.0036T_a)0.67} \tag{9}$$

$$GAF = xRe^{-y} \frac{ERH - RH}{M - M_e} \tag{10}$$

$$Re = \frac{C_a}{S(558.99 - 196.19M)\gamma_a \rho_a} \tag{11}$$

The ERH and M_e were evaluated using the following equation (equation 12):

$$1 - RH = \exp(-0.18938(T + 273)(100M)^{1.48234}) \quad (12)$$

for Re ≤ 50, x = 0.91 and y = 0.51; and for Re ≥ 50, x = 0.61 and y = 0.41.

where M is the moisture content, decimal, db; M_o is the initial moisture content of the bed, decimal, db; M_e is the equilibrium moisture content, decimal, db; t is the time, h; D is the mass diffusivity of the seed, $cm^2\ h^{-1}$; d is the equivalent diameter of the seed, cm; T_g is the seed temperature, °C; GF is the grain factor; S is the sphericity of the seed (typically 1); ρ_g is the kernel density of soybean seed, kg per m^3; AF is the air factor; F_a is the mass flow rate of the air, kg h^{-1} per m^2; P_{sa} is the saturation vapour pressure of the air at T, Pa; P_a is the atmospheric vapour pressure, Pa; T_a is the ambient temperature, °C; GAF is the grain-air factor; Re is Reynolds number; x and y are constants; RH is the relative humidity; ERH is the equilibrium relative humidity; C_a is the specific heat of air, J g^{-1} °C^{-1}; γ_a is the kinematic viscosity of the air, $m^2\ s^{-1}$; ρ_a is the density of air, kg per m^3; and T is the drying air temperature, °C.

Hutchinson and Otten (1983) recommended the following thin layer drying equation (13) for soybeans dried at 32–49°C at RH ranging from 34 to 65%:

$$MR = \exp(-Kt^N) \quad (13)$$

where,

$$K = 0.0333 + 0.0003T \quad (14)$$

$$N = 0.3744 + 0.00916T(RH) \quad (15)$$

where MR is the moisture ratio; T is the drying air temperature, °C; and RH is the relative humidity of the air, %.

Nuh and Brinkworth (1997) developed the following thin-layer soybean drying model (equation 16) based on fundamental physical properties:

$$\frac{dM}{dt} = -\frac{h_m A_s P_{ve} P_v}{M_c R_w T M_e \left(1 + \dfrac{h_m P_{ve} R}{5D\rho_g R_w T M_e}\right)} \left(\frac{M}{P_v} - \frac{M_e}{P_{ve}}\right) \quad (16)$$

where,

$$h_m = \frac{h}{\rho C_g}\left(\frac{Pr}{Sc}\right)^{2/3} \quad (17)$$

where M is the moisture content, decimal, db; h_m is the convective mass transfer coefficient, m s^{-1}; A_s is the total surface area available for heat and mass transfer in a packed bed, m^2; P_{ve} is the partial vapour pressure at equilibrium, N per m^2; P_v is the partial vapour pressure at the free stream, N per m^2; M_e is the equilibrium moisture content, decimal, db; M_c is the mass of dry crop in bed, kg; R_w is the ideal gas constant for water vapour, 461.5J kg^{-1} K^{-1}; T is the temperature of air–vapour mixture, K; R is the radius of seed, m; D is the diffusion coefficient of water and/or vapour in seed, $m^2 s^{-1}$; ρ_g is the kernel density of dry seed, kg per m^3; h is the convective heat transfer coefficient, W per $m^2 K^{-1}$; C_g is the specific heat of seed, J $kg^{-1} K^{-1}$; Pr is the Prandtl number; and Sc is the Schmidt number.

An axisymmetric, two-dimensional, heat and mass transfer drying model for a single soybean kernel was developed by Rafiee *et al.* (2008). They assumed that the seed geometry was ellipsoidal, moisture movement from the inside to the surface was by both liquid and vapour diffusion and moisture removal from the surface of the seed was by evaporation. In addition, they assumed a uniform initial moisture content distribution throughout the seed at the beginning of drying and used a single value of water diffusion coefficient for the entire seed. Rafiee *et al.* (2008) solved their model using an axisymmetric finite element grid. The following are their heat and mass transfer equations, along with initial and boundary conditions:

Mass transfer equation:

$$\frac{\partial M}{\partial t} = D\left(\frac{\partial^2 M}{\partial x^2} + \frac{\partial^2 M}{\partial y^2}\right) \tag{18}$$

Heat transfer equation:

$$C_g \frac{\partial T}{\partial t} = K\left(\frac{\partial^2 T}{\partial x^2} + \frac{\partial^2 T}{\partial y^2}\right) + L\rho_g \frac{\partial M}{\partial t} \tag{19}$$

Initial conditions:

$$M(x, y, t = 0) = M_o \tag{20}$$

$$T(x, y, t = 0) = M_o \tag{21}$$

Boundary conditions:

$$D\left(1_x \frac{\partial M}{\partial x} + 1_y \frac{\partial M}{\partial y}\right)\bigg|_s = h_m (M_s - M_e) \tag{22}$$

$$K\left(1_x \frac{\partial T}{\partial x} + 1_y \frac{\partial T}{\partial y}\right)\bigg|_s = h_t (T_s - T_a) \tag{23}$$

where M is the moisture content, decimal, db; t is the time, s; D is the diffusivity, $m^2 s^{-1}$; C_g is the specific heat of seed, J $kg^{-1} K^{-1}$; T is the temperature, K; K is the thermal conductivity, W $m^{-1} K^{-1}$; L is the latent heat of vaporization of water, J kg^{-1}; 1_x and 1_y are the direction cosines of the outward drawn normal to the boundary; s is the surface; ρ_g is the kernel density, kg per m^3; h_m is the mass transfer coefficient, m s^{-1}; h_t is the heat transfer coefficient, W per $m^2 K^{-1}$; T_s is the surface temperature, K; and T_a is the ambient temperature, K.

Modelling of stress cracks

Shrinkage is an inevitable phenomenon of crop drying during the removal of moisture. In soybeans, shrinkage is associated with the seed coat cracking. Proper measurement of shrinkage is, therefore, an essential consideration during drying. Both radial and tangential or tensile stresses that develop due to shrinkage are responsible for seed coat cracking. Misra and Young (1980) and Misra *et al.* (1981) developed a finite element technique to calculate radial and tangential or tensile stresses in soybeans during drying, considering the soybean as an elastic sphere. For a one-dimensional radial element within a sphere they provided the following relationships (equations 24 and 25) of both stresses, which are based on linear shrinkage and elasticity:

$$\sigma_r = \frac{E}{(1+\mu)(1-2\mu)}[(1-\mu)\in_r + 2\mu\in_t + (1+\mu)0.00266\,\Delta M] \quad (24)$$

$$\sigma_t = \frac{E}{(1+\mu)(1-2\mu)}[\in_t + \mu\in_r + (1+\mu)0.00266\,\Delta M] \quad (25)$$

where,

$$\ln(E) = 20.12 - 0.55(\ln(m) - 0.87)^2 \quad (26)$$

$$L = L_o(1 - 0.266\times10^{-2}\,\Delta M) \quad (27)$$

$$\in_r = \frac{u_j}{r_j - r_i} - \frac{u_i}{r_j - r_i} \quad (28)$$

$$\in_t = \frac{1}{r}\left(\frac{r_j - r}{r_j - r_i}\right)u_i + \frac{1}{r}\left(\frac{r - r_i}{r_j - r_i}\right)u_j \quad (29)$$

where σ_r is the radial stress, Pa; E is the modulus of elasticity, Pa; μ is the Poison's ratio (0.4); $\in_r$ is the radial strain, m/m; $\in_t$ is the tangential strain, m/m; σ_t is the tangential or tensile stress, Pa; M is the change in moisture content, decimal, db; m is the moisture content, %, wet basis (wb); L is the instantaneous linear dimension; L_o is the original linear dimension; u_j is the displacement at node j, m; u_i is the displacement at node i, m; r_j is the radius at node j, m; r_i is the radius at node i, m.

The maximum shear stress (τ_{max}) at any point in a sphere depends on the following equation (30) (Misra *et al.*, 1981):

$$\tau_{max} = 0.5|\sigma_r - \sigma_t| \quad (30)$$

Since the soybean seed coat is viscoelastic, Mensah *et al.* (1984) modelled the ultimate tensile (breakage) stress and relaxation modulus of the soybean seed coat, as described below:

Tensile stress of the seed coat parallel to the hilum:

$$\sigma_{t_{parallel}} = 40.82\exp(0.945M - 0.00324TM) \quad (31)$$

Table 12.4. Coefficients for equation 33 (reprinted with permission from Mensah *et al.*, 1984).

Temperature (°C)	Moisture content (%, db)	Load direction: parallel (MPa)				Load direction: perpendicular (MPa)			
		E_∞	E_1	E_2	E_3	E_∞	E_1	E_2	E_3
25	8.7	252.9	36.6	29	59.8	216.5	18.9	45.9	59.4
	12.5	127.9	31.8	106.9	93.2	160.6	33.7	45.3	46.9
	15.0	97.5	28.9	90.3	66.9	137.1	30.9	62.7	51.3
	18.6	57.6	25.3	84.0	28.5	125.4	29.4	71.5	59.4
35	5.8	228.7	6.3	52.1	26.7	229.5	22.2	14.7	42.4
	7.0	180.4	5.1	50.7	31.8	181.8	20.4	25.9	49.1
	10.2	92.4	11.4	48.9	72.7	154.5	18.5	54.3	53.7
	14.5	76.8	47.8	18.0	71.4	93.2	27.0	81.4	36.3
45	4.9	204.6	6.3	65.5	30.5	227.5	4.8	68.9	25.9
	7.9	107.2	11.7	129.4	54.6	150.1	0.16	101.7	65.5
	10.0	84.4	15.7	78.3	36.9	111.3	9.6	87.3	52.2
	11.2	71.7	25.5	70.8	11.5	90.7	20.5	78.4	36.3

Tensile stress of the seed coat perpendicular to the hilum:

$$\sigma_{t_{perpendicular}} = 47.8\exp(0.59M - 0.00218TM) \quad (32)$$

The above two equations (31 and 32) hold true for drying at 25–55°C and an RH range of 20–77%.

The following three-term Maxwell model (equation 33) can predict the relaxation modulus of the soybean seed coat (E(t), MPa) over a drying air temperature and moisture content range of 25–45°C and 5–19% db, respectively:

$$E(t) = E_\infty + E_1e^{-0.5t} + E_2e^{-0.05t} + E_3e^{-0.005t} \quad (33)$$

where t is the time, min; and the values of E_∞, E_1, E_2 and E_3 are given in Table 12.4.

Other methods of drying soybeans

Microwave dryers can also be used to dry soybeans. Shivhare *et al.* (1993) used a 2.5 kW microwave operating at 2.54 GHz and used three power levels: 0.067, 0.13 and 0.20 W g^{-1}. Both the germination and bulk density of dried soybeans decreased with increasing power levels. Seed-grade soybeans were achieved at <0.13 W g^{-1} power level, at which germination was >95% after drying. They provided the following model for microwave drying of soybeans:

$$MR_s(t) = \frac{M'(t) - M'_s(t)}{M'_o - M'_s(t)} = \frac{6}{\pi^2}\sum_{n=1}^{\infty}\frac{1}{n^2}\exp\left[-\frac{n^2\pi^2D}{R^2}t\right] \quad (34)$$

where,

$$M'_s(t) = M'_e + (M'_o - M'_e)\exp(-\beta t) \quad (35)$$

$$\beta = 20.3 - 0.493(M'_o - M'_e) \tag{36}$$

$$M'_e = 0.4163 - 0.0718\ln[-(T + 100.288)\ln(RH)] \tag{37}$$

$$D = 76.7\exp\left[\frac{-3791}{T}\right] \tag{38}$$

where $MR_s(t)$ is the modified moisture ratio at time t; M′(t) is the moisture content at time t, %, db; M'_e is the equilibrium moisture content, %, db; M'_o is the initial moisture content, %, db; $M'_s(t)$ is the surface moisture content at time t, %, db; n is the integer; D is the diffusion coefficient, $cm^2 h^{-1}$; R is the equivalent radius, cm; t is the time, h; β is the surface drying coefficient, h^{-1}; T is the temperature, K; and RH is the relative humidity, decimal.

Barrozo *et al.* (1998), Felipe and Barrozo (2003) and Lacerda *et al.* (2004) used an almost similar drying model (equation 39) as equation 34 to describe countercurrent and concurrent moving-bed drying of soybeans. These dryers consume less electrical energy and the product is homogenously dried.

$$MR = \frac{M - M_e}{M_o - M_e} = \frac{6}{\pi^2}\sum_{n=1}^{\infty}\frac{1}{n^2}\exp\left[-\frac{n^2\pi^2 D}{R^2}t\right] \tag{39}$$

The parameters associated with equation 39 are as follows:

$$D = 532320\exp(-4273T`) \tag{40}$$

$$T` = \left(\frac{1}{T} - \frac{1}{273}\right) \tag{41}$$

$$M_e = 0.01((-\exp(-0.00672\,T + 3.02))/\ln(RH))^{0.663} \tag{42}$$

$$RH = \frac{9.146(1.608W/(1 + 1.608W))}{\exp(18.3036 - 3816.44/(T + 46.13))/760} \tag{43}$$

where M is the moisture content, decimal, db; T is the seed temperature, K; and W is the absolute air humidity, kg kg^{-1}.

Soybeans dried in a fluidized bed dryer maintain their nutritional value due to inactivation of urease (an indirect measure of trypsin inhibitor) when the drying temperature is maintained within the range of 120–140°C (Osella *et al.*, 1997; Soponronnarit *et al.*, 2001). Soponronnarit *et al.* (2001) reported optimum conditions based on minimum cost, minimum energy consumption and maximum capacity for fluidized bed drying of soybeans as drying temperature 140°C, bed depth 18 cm and air velocity 2.9 m s^{-1}. These conditions resulted in 27% cracking and 1.7% breakage.

Wiriyampaiwong *et al.* (2003) mentioned that a two-dimensional spouted bed dryer can also inactivate the urease enzyme during soybean drying, as long as the operating temperature is 150°C.

12.5 Drying- and Storage-related Properties

Moisture content determination

To determine the moisture content of unground soybeans, a sample size of 15g should be dried at 103 ± 1°C in an air oven for 72h according to the American Society of Agricultural Engineers' Standard S352.2 (ASABE, 2008).

Physical properties

Knowledge of the physical properties of soybean is important when designing equipment for drying, storage and other processing operations. Measurement of seed dimensions, size and shape is essential for designing cleaners and graders. Sreenarayanan *et al.* (1988) reported the dimensions of an Indian variety of soybean (CO-1) at 7.5% wb moisture content as length 6.2mm, width 5.8mm and thickness 4.6mm. Table 12.5 shows the dimensions of three soybean varieties. Kashaninejad *et al.* (2008) explained the relationship between dimensions of four Iranian soybean varieties with a seed moisture content of 8.2–24.1% wb (Table 12.6).

Bulk density, kernel density and porosity directly influence the airflow distribution in grain mass. Many researchers have provided relationships between these properties and moisture content for a specified range of

Table 12.5. Dimensions of soybeans at 10.5% wb moisture content (reprinted with permission from Milani *et al.*, 2000).

	Soybean variety		
Dimension (mm)	Yellow	Cokar	Centennial
Length	6.89	7.39	6.50
Width	6.48	6.50	5.88
Thickness	5.76	5.57	4.67
Sphericity (%)	0.92	0.87	0.87

Table 12.6. Dimensions of four Iranian soybean varieties as a function of moisture content (%, wb) (reprinted with permission of Elsevier from Kashaninejad *et al.*, 2008).

	Soybean variety			
Dimension (mm)	Williams	BP	LWK	Sahar
Length	7.33 + 0.05M	6.72 + 0.05M	7.01 + 0.05M	7.07 + 0.02M
Width	6.71 + 0.006M	6.26 + 0.02M	6.21 + 0.02M	6.14 + 0.009M
Thickness	5.80 + 0.01M	5.34 + 0.02M	5.22 + 0.01M	5.10 + 0.07M
Mean diameter	6.60 + 0.02M	6.07 + 0.03M	6.21 + 0.01M	5.81 + 0.02M
Sphericity (%)	89.4 – 0.27M	90.2 – 0.21M	86.6 – 0.23M	85.6 – 0.06M

M, moisture content; %, wb.

moisture contents (Table 12.7). Kashaninejad *et al.* (2008) reported the range of values of bulk density, kernel density and porosity of four Iranian soybean cultivars over a moisture content range of 8.2–24.1% wb (Table 12.8).

The angle of repose and coefficient of friction are important for calculating bin wall pressure and for the proper design of storage and handling facilities. Sreenarayanan *et al.* (1988) reported that the angle of repose for an Indian soybean cultivar (CO-I) was 25.5° at 7.5% wb seed moisture content. Values of the angle of repose for two Canadian soybean cultivars (Maple Presto and McCall) have been reported to be 28° and 29°, respectively, at a seed moisture content of 8.1% wb (Muir and Sinha, 1988). Kashaninejad *et al.* (2008) reported that the angle of repose increases with increased

Table 12.7. Relationships of some physical properties of soybeans with moisture.

Property	Equation	Moisture content range (M, % db)	Reference
Bulk density (kg per m^3)	$804 - 4.49M$	6.25–11.6	Tunde-Akintunde *et al.* (2005)
	$891 - 13.8M$	6.7–15.3	Polat *et al.* (2006)
	$0.744 - 0.0026M$	9.9–39.6	Manuwa (2007)
	$0.1149M^2 + 10.832M + 758.51$	10.62–27.06	Isik (2007)
Kernel density (kg per m^3)	$1043 + 2.76M$	6.7–15.3	Polat *et al.* (2006)
	$1.59 - 0.013M$	9.9–39.6	Manuwa (2007)
	$- 0.4036M^2 + 22.236M + 893.21$	10.62–27.06	Isik (2007)
Porosity (%)	$0.43 - 0.012M$	6.25–11.6	Tunde-Akintunde *et al.* (2005)
	$55.9 - 0.761M$	6.7–15.3	Polat *et al.* (2006)
	$0.141 + 0.0049M$	9.9–39.6	Manuwa (2007)
	$-0.333M^2 + 2.146M + 20.497$	10.62–27.06	Isik (2007)
1000 kernel mass (g)	$140.8 + 1.064M$	9.9–39.6	Manuwa (2007)
	$168.97 + 3.171M$	10.62–27.06	Isik (2007)
Terminal velocity (m s^{-1})	$12.94 - 0.21M$	6.25–11.6	Tunde-Akintunde *et al.* (2005)
	$5.27 + 0.268M$	6.7–15.3	Polat *et al.* (2006)
	$-0.0023M^2 + 0.1578M + 6.5324$	10.62–27.06	Isik (2007)

M, moisture content, decimal, db.

Table 12.8. Three physical properties of four Iranian soybean varieties (compiled from Kashaninejad *et al.*, 2008).

	Soybean variety			
Property	Williams	BP	LWK	Sahar
Bulk density (kg per m^3)	693–720	693–730	696–740	700–716
Kernel density (kg per m^3)	1186–1250	1178–1223	1171–1209	1149–1203
Porosity (%)	39.29–44.54	38.01–43.33	36.78–42.45	37.68–41.83

moisture for four Iranian soybean varieties over a moisture content range of 8.2–24% wb (LWK 32.30–36.03°, Sahar 32.15–35.17°, BP 31.88–35.24°, Williams 30.26–35.06°).

Tunde-Akintunde *et al.* (2005) explained the relationship between angle of repose (φ) and seed moisture content over a moisture content (M) range of 0.0625–0.116 db:

$$\varphi = 8.40 + 0.39\ M \quad (44)$$

Similarly, for a different soybean variety (TGX1448-2E), Manuwa (2007) explained the relationship between angle of repose and seed moisture content over a moisture content (M) range of 0.099–0.396 db:

$$\varphi = 22.06 + 0.189\ M \quad (45)$$

Table 12.9 provides the coefficient of friction of bulk soybeans against different surfaces as a function of seed moisture content. Kashaninejad *et al.* (2008) provided relationships between coefficient of friction and moisture content of four Iranian soybean cultivars against different surfaces (Table 12.10). The coefficient of friction increased with increased seed moisture content.

Table 12.9. Coefficient of friction of soybeans

Surface	Moisture content (%, db)	Coefficient of friction	Source
Galvanized steel	8.81	0.20–0.27	Muir and Sinha (1988)
	11.7	0.28	Milani *et al.* (2000)
	6.7–15.3	0.018 + 0.0161M	Polat *et al.* (2006)
	9.9–39.6	0.338 + 0.0042M	Manuwa (2007)
	10.62–27.06	0.1598 + 0.0589ln M	Isik (2007)
Plywood	11.7	0.29	Milani *et al.* (2000)
	6.25–11.6	21.67 – 0.37M	Tunde-Akintunde *et al.* (2005)
	6.7–15.3	0.174 + 0.0128M	Polat *et al.* (2006)
	9.9–39.6	0.423 + 0.0042M	Manuwa (2007)
Glass	11.7	0.18	Milani *et al.* (2000)
	6.25–11.6	16.57 – 0.21M	Tunde-Akintunde *et al.* (2005)
	6.7–15.3	0.105 + 0.0167M	Polat *et al.* (2006)
	10.62–27.06	0.105 + 0.052ln M	Isik (2007)
Aluminium	11.7	0.26	Milani *et al.* (2000)
	10.62–27.06	0.252 + 0.0267ln M	Isik (2007)
Concrete	8.81	0.25–0.33 (steel-trowelled) 0.28–0.34 (wood-trowelled)	Muir and Sinha (1988)
Rubber	10.62–27.06	0.2166 + 0.054ln M	Isik (2007)
Stainless steel	10.62–27.06	0.1468 + 0.0607ln M	Isik (2007)

M, moisture content, decimal, db.

Table 12.10. Friction coefficients of four Iranian soybean cultivars (reprinted with permission of Elsevier from Kashaninejad *et al.*, 2008).

Cultivar	Moisture content (%, wb)	Surface	Equation
Williams	8.20–23.60	Concrete	0.35 + 0.009M
		Galvanized steel	0.35 + 0.007M
		Fibreglass	0.40 + 0.004M
		Plywood	0.33 + 0.006M
		Rubber	0.42 + 0.007M
BP	9.30–24.10	Concrete	0.34 + 0.009M
		Galvanized steel	0.38 + 0.004M
		Fibreglass	0.37 + 0.005M
		Plywood	0.41 + 0.004M
		Rubber	0.43 + 0.005M
LWK	9.60–23.72	Concrete	0.30 + 0.01M
		Galvanized steel	0.35 + 0.006M
		Fibreglass	0.31 + 0.008M
		Plywood	0.33 + 0.006M
		Rubber	0.41 + 0.007M
Sahar	9.00–23.90	Concrete	0.35 + 0.007M
		Galvanized steel	0.24 + 0.012M
		Fibreglass	0.25 + 0.01M
		Plywood	0.19 + 0.016M
		Rubber	0.36 + 0.009M

M, moisture content; %, wb.

Thermal properties

Thermal properties should be measured to predict the moisture and temperature changes in soybeans during the process operations. Thermal properties play an important role in designing drying or cooling systems and in predicting temperature distribution when using these systems. Deshpande *et al.* (1996) proposed the following equation for measurement of thermal conductivity and thermal diffusivity as a function of the moisture content of soybeans (cv. JS-7244) from data over a moisture content range of 0.081–0.25 db:

$$K = 0.3139 + 1.40M \quad (46)$$

$$\alpha = 2.9 \times 10^{-4}(1 + 0.2075M) \quad (47)$$

where K is the thermal conductivity, kJ h^{-1} m^{-1} K^{-1}; α is the thermal diffusivity, $m^2 h^{-1}$; and M is the moisture content, decimal, db.

The values of K ranged from 0.4165 to 0.6320 kJ h^{-1} m^{-1} K^{-1} and the values of α ranged from 2.9401×10^{-4} to 3.0684×10^{-4} m^2 h^{-1} for the range of moisture contents studied. Both thermal conductivity and thermal diffusivity increase with increased moisture content (equations 46 and 47).

Alam and Shove (1973) proposed the following equation (48) to describe the specific heat of soybeans (cv. Wayne) for a moisture content range of 0–0.37 db:

$$C_p = 0.39123 + 0.46057M' \quad (48)$$

where C_p is the specific heat, Btu lb^{-1} F^{-1}; and M′ is the moisture content, % db.

Deshpande and Bal (1999) proposed the following empirical model (equation 49) to determine the specific heat of soybeans over a moisture content range of 0.081–0.25 db:

$$C_p = 1.444(1 + 4.06 \times 10^{-2} M) \quad (49)$$

Where C_p is the specific heat, kJ kg^{-1} K^{-1}; and M is the moisture content, decimal, db.

Diffusion coefficient

An essential transport property related to moisture transfer within seeds during drying is moisture diffusivity. Misra and Young (1980) provided the following Arrhenius-type relationship (equation 50) to determine diffusion coefficient of water in soybean seeds:

$$D = 0.046944e^{-(3437/T)} \quad (50)$$

where D is the diffusivity, m^2 h^{-1}; and T is the temperature, K.

Gely and Giner (2007) reported that the diffusion coefficient varies between 1.78×10^{-11} and 7.28×10^{-11} m^2 s^{-1} when soybeans (cv. Nidera A6381) were dried in a thin layer using air at temperatures between 19°C and 75°C.

Activation energy

Kitic and Viollaz (1984) determined activation energies for the diffusion of water in soybeans (cv. Williams) in the range of 28.9–31.0 kJ mol^{-1} during thin-layer drying over a moisture content range of 42–62% db. Singh and Kulshrestha (1987) reported that the activation energy for the diffusion of water in soybeans (cv. Bragg) during sorption was 44.3 kJ mol^{-1}. Li *et al.* (2002) provided the following equation (51) to determine water diffusivity in soybeans during drying and reported an activation energy of 38.2 kJ mol^{-1}:

$$D = 5.003 \times 10^{-5} e^{-(3.825 \times 10^4 / T)} \quad (51)$$

Gely and Giner (2007) reported activation energies between 16.6 and 28.8 kJ mol^{-1} when soybeans (cv. Nidera A6381) were dried in a thin layer using air at temperatures between 19°C and 75°C.

12.6 Quality Changes During Drying and Storage

Breakage

Preservation of soybean quality is the greatest concern for international marketing, especially for major exporting countries such as the USA and Brazil. The major quality-degrading factors in soybeans are moisture, splits, foreign material and damage (Spencer, 1976). A reduced moisture level can sometimes be responsible for a greater number of splits. Since soybeans have two structurally strong halves attached together with relatively weak bonds, low moisture can ease the splitting. Split soybeans are becoming a concern in modern cargo shipments. Splitting mostly occurs at the discharge end due to the impact of soybeans falling from heights as high as 30 m (Spencer, 1976). Splitting soybeans accelerates the growth of mould and changes in composition (Urbanski *et al.*, 1980). Hot-air drying also causes seed coat and cotyledon cracking, which increase breakage during subsequent handling or conveying (Ting *et al.*, 1980; White *et al.*, 1980). Soybeans are susceptible to cracks and breakage when moisture content falls below 11% wb (Paulsen *et al.*, 1981). Germination of soybeans decreases by >10% when the seeds are dropped from equal heights onto a concrete floor compared to a galvanized iron floor, and damaged seeds lead to microbial or insect infestation (White *et al.*, 1976; Parde *et al.*, 2002). Physical damage to soybeans due to impacts during handling has been mentioned elsewhere (Paulsen *et al.*, 1981; Bartsch *et al.*, 1986). This indicates that great care needs to be taken when handling soybeans.

Discolouration due to internal heating

Temperature in the bulk of soybean seeds can be high at times due to either microbial or insect activity. This can cause oxygen-entrapped hot spots causing fire hazards or the oxidation of seed oil. Furthermore, unlike other cereal grains, the temperature of soybeans can exceed 200°C during heating, which results in 30% loss of dry weight. Therefore, frequent checks on the seed temperature are required.

The discolouration of soybean seeds is another indicator of spoilage. Mature and dry soybean seeds often have an olive to dark-brown seed coat and tan to dark-brown cotyledons. Discolouration occurs due to mould growth or microbial spoilage from improper heating. Insect- or mould-infested soybeans become wrinkled, medium to dark brown, produce mustiness and an off-odour and often have a superficial grey mould covering. Rancid soybeans are often deep pink. Moisture in the beans increases and the colour of the beans decreases with increased storage time, temperature and RH (Saio *et al.*, 1980).

Other quality changes

The total free sugar content significantly decreases with storage time (Hou and Chang, 2004). Fungal growth within the seed bulk creates lumping or caking. During storage, seeds can also undergo physical, physiological and chemical changes such as increased free fatty acid content, decreased seed viability, increased moisture and decomposition of phospholipids or denaturation of protein (Urbanski *et al.*, 1980). Free fatty acids increase with storage moisture and temperature as fats in seed are readily broken down by lipases (White *et al.*, 1976; Sangkram and Noomhorm, 2002). Seeds can also be damaged due to changes in growing conditions (e.g. heat, rain, frost).

Bag storage helps in self-ventilating and cooling, but uneven airflow during aeration causes non-uniform seed temperature, which deteriorates seed quality. The bulk storage is easy to aerate and cool and, therefore, the desired quality of the seeds can be maintained. In the USA, the maximum limits for damaged soybeans are ≥1%, 2%, 5% and 10% for grade numbers 1, 2, 3 and 4, respectively (Wang *et al.*, 2002). The Canadian grades are set by the Canadian Grain Commission (CGC, 2008). The US soybean grades and grade requirements according to the US Department of Agriculture (USDA) Grain Inspection, Packers and Stockyards Administration are given in Table 12.11.

Machine vision systems

Machine vision or image processing is a rapidly advancing area of research and implementation in the area of grain storage and handling. The various accessories used in a typical machine vision system include a camera (still or video, analogue or digital) to grab an image (single kernel or bulk); lighting for acquiring images (e.g. incandescent, halogen or fluorescent); a frame-grabbing board for receiving and storing the image (normally comes with the camera unit); a display monitor; and computer hardware and software for calculating measurements (commercial software or indigenous algorithms). The three interdependent steps of image analysis are image generation, processing and interpretation. The camera generates a signal representing the image-alike human view of the object and passes it to the frame grabber, which converts it to a digital form and assigns a number to represent the intensity in each small area of the image, called 'pixels'. The step of image conversion from analogue to digital can be replaced with the

Table 12.11. US soybean grades and grade requirements (reprinted with permission from USDA, 2007; courtesy of the USDA Grain Inspection, Packers and Stockyards Administration).

	Minimum limit	Maximum limit				
		Damaged kernels				
Grade	Test weight (kg hl^{-1})	Heat (part of total) (%)	Total (%)	Foreign material (%)	Splits (%)	Soybeans of other colours (%)
1	44.8	0.2	2.0	1.0	10.0	1.0
2	43.2	0.5	3.0	2.0	20.0	2.0
3	41.6	1.0	5.0	3.0	30.0	5.0
4	39.2	3.0	8.0	5.0	40.0	10.0

use of a digital camera, which grabs the image in digital form. Finally, the image is processed using image-processing software written in various languages. Machine vision is an attractive, powerful and non-destructive computerized method that has potential for multifunctional automated performances in grain storage and handling.

Damaged soybeans are detected quickly and reliably with proper machine vision systems. Gunasekaran *et al.* (1988) developed a fast algorithm capable of detecting 96% of soybean seed coat cracking and 100% of cotyledon cracking within a few seconds. The images were in greyscale format, in which the good soybeans were pure white and cracks created black streaks across the seed. The detection method was dependent upon manually placing one seed under the camera, but the algorithm was not robust enough to detect the cracks from improperly placed seeds. Although manually placing one seed for image processing is tedious and time consuming and cannot be feasible for industry, the process seems to be worth further study for processing bulk samples in a batch or continuous fashion. Currently in the USA, the government, local elevators (grain-handling facilities) and soybean industries visually inspect and differentiate damaged soybeans from sound ones (Wang *et al.*, 2002). Visual inspection is a subjective process that requires proper human judgement. Machine vision can replace the use of chemicals (e.g. tetrazolium or indoxyl acetate) in manually detecting seed coat damage, making the process more objective, chemical-free and uniform (Gunasekaran *et al.*, 1988).

Wigger *et al.* (1988) and Casady *et al.* (1992) used colour image processing to detect healthy and fungal-damaged seeds. Their average classification accuracy to discriminate healthy versus fungal-damaged seeds was 77–91%. A machine vision system was further applied by Ahmad *et al.* (1994) to characterize morphological features (e.g. major axis, minor axis, axis ratio, roundness, total pixel area, thickness) to identify and classify asymptomatic (sound) from symptomatic (damaged) seeds. Mbuvi *et al.* (1989) mentioned that seed volume is greater in asymptomatic seeds than in those infected by *Fusarium* or *Phomopsis*. Ahmad *et al.* (1999) also classified asymptomatic and symptomatic seeds using a red, green and blue (RGB) colour feature-based multivariate decision model. However, colour alone did not classify the damaged kernels. The overall classification accuracy was 88% for asymptomatic and symptomatic seeds and individual classification accuracies were: asymptomatic, 97%; *Alternaria* species, 30%; *Cercospora* species, 83%; *Fusarium* species, 62%; green immature seeds, 91%; *Phomopsis* species, 45%; soybean mosaic potyvirus (black), 81%; and soybean mosaic potyvirus (brown), 87%. Information obtained from this study will be useful in developing an intelligent automated soybean-grading system.

Near-infrared spectroscopy

Near-infrared (NIR) spectroscopy is used to obtain chemical constituents from solid or liquid samples. It is fast and non-destructive and requires little or no sample preparation. Furthermore, it can provide simultaneous determination of multiple components per measurement with a remote sampling capability and, hence, it can provide real-time information from a process stream. An NIR spectrometer collects spectral information from solid samples either in reflectance mode (700–2500 nm) or in transmittance mode (700–1100 nm). Wang *et al.* (2002) classified sound and damaged (weather, frost, sprout, heat and mould) soybeans using single-seed spectral information in reflectance mode from a diode-array NIR spectrometer (DA7000, Perten Instruments, Springfield, IL). For classification purposes, they used multivariate data analysis using partial least squares (PLS) and artificial neural network (ANN) methods to develop models that correlate the spectral response of each sample at individual wavelengths with the chemical concentration of the sample. For PLS models, classification accuracies of sound and damaged seeds were >99% for a two-class model (sound and damaged). For a six-class model (sound and five damaged classifications) they were as follows: sound, 90%; weather-damaged, 61%; frost-damaged, 72%; sprout-damaged, 54%; heat-damaged, 84%; and mould-damaged, 86%. For ANN models, the classification accuracies were: sound, 100%; weather-damaged, 98%; frost-damaged, 97%; sprout-damaged, 64%; heat-damaged, 97%; and mould-damaged, 83%. Shatadal and Tan (2003) developed a four-class (sound, heat-damaged, green-frost-damaged and stink-bug-damaged) ANN model based on colour features: RGB; means of hue, saturation and intensity; excess red (2R-G-B); excess green (2G-R-B); and excess blue (2B-R-G). The classification accuracies were: sound, 99.6%; heat-damaged, 95%; green-frost-damaged, 90%; and stink-bug-damaged, 50.6%.

Wang *et al.* (2004) also used NIR to classify fungal-damaged soybean seeds. The same diode-array spectrometer as mentioned above was used with MVA models, which provided a classification accuracy of >99% for a two-class PLS model (sound and damaged). For a five-class ANN model (sound, *Cercospora-*, *Phomopsis-*, soybean mosaic virus- and downy mildew-damaged), classification accuracies were 100%, 99%, 84%, 94% and 96%, respectively.

12.7 Insect Infestation and Control in Soybeans

Insect infestation

Approximately 55 species of insects have been reported from stored soybeans (Table 12.12). The extent of post-harvest damage and losses caused by insects to grains and their products is difficult to quantify. Infestation occurs unless proper sanitation procedures have been followed and the seeds dried and cooled. Several of the insects infesting farm-stored seeds are destructive. These include the rusty grain beetle, red flour beetle, confused flour beetle, rice weevil and Indianmeal moth. Infested seed bulk may include eggs, egg shells, larvae and cast larval exoskeletons, pupae and pupal cases and cocoons, and mature insects.

Table 12.12. Insects and moths reported from stored soybean (N.D.G. White, Winnipeg, 2008, personal communication).

Acanthoscelides obtectus (bean weevil)	*Leguminivora glycinivorella* (soybean pod borer)
Ahasverus advena (foreign grain beetle)	*Leichenum canaliculatum* (Madagascar beetle)
Alphitobius diaperinus (lesser mealworm)	*Liposcelis bostrychophila* (psocid)
Alphitobius laevigatus (black fungus beetle)	*Liposcelis entomophila*
Anisopteromalus calandrae	*Lophocaterers pusillus*
Araecerus fasciculatus (coffee bean weevil)	*Mezium sulcatum* (spider beetle)
Attagenus unicolor (black carpet beetle)	*Monanus concinnulus*
Cadra cautella (almond moth)	*Necrobia rufipes* (red-legged ham beetle)
Callosobruchus analis (Graham bean weevil)	*Nemapogon granella* (European grain moth)
Callosobruchus chinensis (southern cowpea weevil)	*Oryzaephilus mercator* (merchant grain beetle)
Callosobruchus maculatus (cowpea weevil)	*Palorus ficicola*
Callosobruchus phaseoli (cowpea weevil)	*Paralipsa gularis* (stored nut moth)
Callosobruchus theobromae	*Plodia interpunctella* (Indianmeal moth)
Carcinops pumilio (predacious hister beetle)	*Ptinus japonicus*
Carcinops troglodytes	*Pyralis manihotalis*
Carpophilus ligneus	*Sitophilus oryzae* (rice weevil)
Carpophilus maculates	*Stator pruininus* (pruinose bean weevil)
Carpophilus marginellus	*Stegobium paniceum* (drugstore beetle)
Corcyra cephalonica (rice moth)	*Tenebroides mauritanicus* (cadelle)
Cryptolestes ferrugineus (rusty grain beetle)	*Thorictodes heydeni*
Cryptolestes pusillus (flat grain beetle)	*Tinea ditella*
Dermestes ater (black larder beetle)	*Tribolium anaphe*
Dinarmus basalis	*Tribolium castaneum* (red flour beetle)
Enicmus minutus (fungus beetle)	*Tribolium confusum* (confused flour beetle)
Ephestia kuehniella (Mediterranean flour moth)	*Trogoderma granarium* (La khapra beetle)
Eupelmus vuilleti	*Trogoderma variabile* (warehouse beetle)
Lasioderma serricorne (cigarette beetle)	*Xylocoris afer*

Some insects (e.g. rice weevil) feed largely on the endosperm, whereas others (e.g. rusty grain beetle) consume the germ. Other pests, such as Indianmeal moth, do not feed, but their larvae cause extensive surface damage to stored seeds with their strong mouthparts. Cadelles and flour beetles first eat the germ and then the endosperm. Quality and quantity losses cause downgrading of seeds and their market value. The distribution of insects in bulk seed is largely influenced by temperature, moisture, CO_2, dockage, insect species, insect density and seed type. Temperature has pronounced effects on insect distribution in grain bins. Insects are sometimes killed by high temperature generated by the metabolism of moulds and, therefore, insects prefer to migrate towards near-optimal portions of seed bulks. During spring, summer and autumn, most of the insects in a bin of bulk seeds are in the upper half of the bin, whereas in mid-winter they are likely to be in the lower half. In mid-summer, the insects are usually uniformly distributed in the four quadrants, whereas in winter most will be found in the south quadrant (Cotton and Wilbur, 1982). Some insects are attracted to moist seeds and also to pockets of dockage in a seed mass.

The US Federal Grain Inspection Service designates soybeans to be infested if any of the following conditions are met: more than one live weevil *or* one live weevil plus five or more other live injurious pest insects *or* no live weevils, but ten or more other live pest insects injurious to stored grain per kilogram (Weinzierl and Higgins, 2008). Soybeans are usually infested with bruchids (*Bruchidius* species) in the field and it is often difficult to detect infested seeds at the time of harvest because the infestation level is too low. Bruchids are small beetles, typically <1 cm long. They breed rapidly in the stored seed and are typically detected after about 3 months of storage. By the time they are detected, the infested seeds are usually unmarketable. Bruchids are considered to be a threat in soybean seeds in the entire supply chain (from harvesting to exporting). Therefore, proper control is necessary to manage bruchid growth in stored beans.

Insect detection

Detecting insect infestation in soybeans as early as possible is of crucial importance in preventing damage to stored seeds. Various techniques and procedures that can be used to detect the presence of insects in seeds during storage. The following paragraphs describe these different methods, which can be adopted to detect insect infestation in soybean storage. Interested readers are advised to consult Neethirajan *et al.* (2007) to obtain more details about the detection techniques for stored-product insects in grain.

Visual examination

Visual examination of grain and grain products in bulk or warehouses is a very simple technique. Visual inspection can identify early stages of moth infestations because they make small clusters of seeds or webbing over the surface of seeds or of the storage structure. Such moth larvae are usually restricted to the top portion of the bulk seeds; therefore, they cannot readily be removed by probe traps. Rice weevil larvae feed on the aleurone layer just under the seed coat, which results in a visible pale scar on the outside of the seed. Seeds infested with mature internal-feeding insects become slightly darkened and the kernel is appreciably softened.

Grain probes and probe traps

Grain probes are generally used to estimate the insect population in grain bins. A grain probe is a hollow metal tube that is inserted into the grain to the desired depth, opened and closed to collect the sample and then removed from the grain and emptied. The sample is then sieved to separate the insects from the seeds. This is a rather laborious method, especially if several locations are sampled to obtain a reliable measure of the numbers and species of insects present.

Probe traps are mainly used for the detection of insects in grains. Loschiavo and Atkinson (1967, 1973) developed a trap that is inserted into seed bulks to collect moving insects. The trap consists of a pointed probe (about 25 mm diameter × 220 mm long) containing glass vials (19 × 65 mm). Above the solid-walled tube is a hollow cylindrical section that is 100 mm long, perforated with 2.2 mm diameter holes. The probe is inserted from the top surface of the seeds with a metal rod. The rod is removed to be used to insert other traps. After 4–7 days (Loschiavo, 1985), the trap is pulled out by a rope tied to the trap. The contents of the trap can be spread onto a flat surface to identify and count any insects and mites that have fallen into the trap.

The number of insects collected in a trap depends on insect mobility, temperature, dockage, the size of the trap perforations, insect species and the length of time the trap is in the seeds. Traps frequently identify infestations that are undetectable with normal seed sampling, because the trap may collect insects from a volume of seeds much larger than the probe. For these reasons, trap counts cannot be used to define the size of an infestation (Muir, 1997).

Pitfall trap

Detection, and to some degree quantification, of insect infestation in an agricultural commodity is provided by a pitfall probe trap. Loschiavo (1974) designed a modified pitfall trap for use in grain-carrying boxcars. A prototype model of the pitfall trap was constructed by soldering the metal lid to a 340 ml glass jar, 8 cm deep with a 28.5 cm inner diameter. A hole, 6.5 cm in diameter, was punched in the lid before soldering. A similar-sized hole was cut out of the bottom of the pail and covered with a piece of perforated brass sheeting of the kind used for insertion-type traps. The brass was soldered to the floor of the pail around the periphery of the hole. The jar could be screwed into the lid from below the pail. This trap detects more insects than insertion-type traps immersed in the seed bulk. Reed *et al.* (1991) reported that the early detection of insect populations in seed is best accomplished by pitfall traps.

Berlese funnel

A Berlese funnel consists of a metal funnel screened at the bottom. It can contain a seed sample of about 1 kg. Heat, usually from a 60 W electric light bulb placed above the seed surface, drives insects and mites down through the screened bottom. The funnel is placed over a jar containing about 50 ml water or 70% alcohol to kill and preserve the insects and mites. The liquid can be studied under a microscope to identify and count the insects and mites that were in the grain sample (Muir, 1997).

Feeding-generated ultrasonic sound

Hidden insect infestation can be detected with ultrasonic signals. The feeding activity of insects is monitored using ultrasonic emissions at a particular frequency. Ultrasonic emissions can be produced by feeding, but not by movement, from early in the first instar to the last instar. The number of feeding events is typically proportional to the number of insects per seed. Rice weevils and moths can be detected using this technique.

Electro-acoustic silo-detection device

A portable device has been developed at a stored-products entomology and acarology laboratory in Bordeaux, France. The device is used to monitor the typical signals produced by insect activity. This device accurately monitors the insect infestation in a bin without sampling. Insect presence is detected long before the infestation becomes a risk for long-term storage (Lessard and Andrieu, 1986).

X-ray

Detection of insects using X-rays has been studied by several researchers, analysing images of the different life stages of insects inside the kernels (Cotton and Wilbur, 1982; Pederson, 1992; Karunakaran *et al.*, 2003). At an experimental level, >97% accuracy can be achieved in detecting the larvae and pupae of rice weevils with X-ray imaging (Karunakaran *et al.*, 2003).

Specific gravity

Infested seeds can be separated from sound ones by placing seed samples in a liquid solution with a specific gravity. This allows uninfested seeds to sink, while infested ones float.

Cracking and flotation

Cracking and flotation are official methods of the Association of Official Analytical Chemists (methods 44.041 and 44.042, respectively; AOAC, 1984). Insect materials are separated and floated to the surface of a solution. The floated materials are collected on filter paper and examined microscopically (Cotton and Wilbur, 1982).

Uric acid

Because uric acid is an important constituent of insect excreta, the measurement of uric acid content in stored seeds can correlate to the extent of insect infestation (Subrahmanyan *et al.*, 1955). However, the uric acid could be from an old infestation rather than a current one.

CO_2 measurement

As insects develop in seeds, they respire and produce CO_2 as a by-product of their metabolism (Pederson, 1992). Howe and Oxley (1952) used a simple gas analyser and determined that 1% CO_2 produced in a standard sample over a 24-h period in a sealed container was indicative of approximately 25 larvae in 450 g seed. An infrared CO_2 analyser is more sensitive and quicker for the routine inspection of hidden insect infestations after harvesting (Zisman and Calderon, 1990). Infrared gas analysers can detect 0.15–0.3% CO_2 developed by 1–2 insects in a kilogram of seeds within 48 h. Infrared gas analysers can also be used to determine the intergranular CO_2 content by sampling from one or more points in a seed bulk.

Nuclear magnetic resonance spectroscopy

Nuclear magnetic resonance spectroscopy is a non-destructive method to determine insect infestation in seeds. It gives the images of peaks coming from water and lipids in larvae, which can be correlated with weevil development and seed kernel weight loss (Pederson, 1992).

Ninhydrin-impregnated paper

Dennis and Decker (1962) used ninhydrin-treated papers to determine insect infestation in seeds. Free amino acids in the body fluids of insects react with ninhydrin and produce purple spots on the paper (Pederson, 1992).

Immunoassay

The 'insect-detect' immunoassay has been reported to provide the most accurate measurement of actual insect infestation when compared to three traditional methods – X-ray analysis, cracking and flotation, and the insect fragment test (Brader *et al.*, 2002).

Near-infrared spectroscopy

NIR spectroscopy is a procedure that can rapidly detect and measure the chemical composition of biological materials. When the wavelength of the incident infrared energy corresponds to the frequency of vibration of a specific molecule, this energy is absorbed by the molecule. Optical sensors measure this absorption and the amount can be related to the concentration of a particular constituent. Dowell *et al.* (1998) used NIR spectroscopy (1000–1660 nm) to detect infestation of weevils and moths.

Locomotor test after irradiation

Insect infestation can be detected by a locomotor test. Different species respond to γ-irradiation in different ways and their locomotor activity and/or ability to disperse is highly affected. The locomotor activity of γ-irradiated beetles in stored products is inversely proportional to the dose applied (Ignatowicz *et al.*, 1994). A lethal dose of 0.3–1.0 kGy for radiation disinfestation has been suggested by Ignatowicz *et al.* (1994) to lower the locomotor activity of confused flour beetles.

Stains

Different stains are used to detect weevil infestations (eggs, larvae, pupae or adults) in seeds. Weevils chew a small hole through the seed coat into the endosperm, in which an egg is deposited. As the ovipositor is withdrawn, the female secrets a gelatinous plug that fills the egg channel so that the egg cavity is difficult to detect without a microscope. Various stains have been discovered that will colour the egg plug without staining the seed coat, unless it has been damaged mechanically (Cotton and Wilbur, 1982). Wongo (1990) found that a water-soluble fluorescent dye, berberine sulphate, can stain egg plugs yellow.

Control of insects, mites, moulds in stored seeds

Insects are usually controlled by cooling seeds with aeration: below 20°C near harvest, then below –20°C in the winter (White, 2001). Rapid disinfestation is obtained by fumigation with phosphine gas above 10°C. Pneumatic movement of seeds will kill insects and mites (White *et al.*, 1997; Paliwal *et al.*, 1999) and will distribute pockets of high-moisture grain. Once seeds have been moved into the commercial handling system, the seed movement kills most insects or mites except *Cryptolestes ferrugineus*, which can be detected in 1–6% of railcars entering terminal elevators.

12.8 Good Storage Practices

Good storage practices have been outlined by White (2001) and include the following procedures:

1. Prepare the bin before storing a new seed type: sweep or vacuum the floor and walls; burn or bury sweepings that contain spoiled or infested seeds; seal cracks to keep out flying insects, rain and snow; and spray the walls and floors with a recommended insecticide.
2. Install an aeration system to reduce temperature gradients and moisture condensation.
3. Dry tough or damp seeds soon after harvest as they are more likely to heat and become infested with insects and mites than dry seeds; then cool after drying.
4. Examine stored seeds every 2 weeks for signs of heating or infestation; either check temperatures, CO_2 levels and insect activity by traps or probe and screen samples.
5. Move heated or infested seeds into another bin if outdoor temperatures are sufficiently cold to break up hot spots and control infestations.
6. Check the top of binned seeds and remove snow, if present, before a crust of mould develops.
7. If an insect infestation occurs and aeration is not available, seal the bin and fumigate the bulk with phosphine gas.

References

Acasio, U.A. (1997) *Handling and Storage of Soybeans and Soybean Meal*. Technical Bulletin of the American Soybean Association. American Soybean Association, St Louis, MO, USA.

Ahmad, I.S., Reid, J.F. and Paulsen, M.R. (1994) Morphological feature extraction by machine vision of damaged soybean seeds. ASAE Paper No. 94-3014. American Society of Agricultural Engineers, St Joseph, MI, USA.

Ahmad, I.S., Reid, J.F. and Paulsen, M.R. (1999) Color classifier for symptomatic soybean seeds using image processing. *Plant Disease* 83, 320–327.

Alam, A. and Shove, G.C. (1973) Simulation of soybean drying. *Transactions of the ASAE* 16, 134–136.

AOAC (1984) *Official Methods of Analysis*, 14th edn. Association of Official Analytical Chemists, Washington, DC, USA.

ASABE (2008) *ASABE Standard S352.2: Moisture Measurement – Unground Grain and Seeds*. American Society of Agricultural and Biological Engineers, St Joseph, MI, USA.

Bakker-Arkema, F.W., Lerew, L.E., DeBoer, S.F. and Roth, M.G. (1974) Grain dryer simulation. Research Report 224. The MSU Agricultural Experiment Station, East Lansing, MI, USA.

Barrozo, M.A.S., Murata, V.V. and Costa, S.M. (1998) The drying of soybean seeds in countercurrent and concurrent moving bed dryers. *Drying Technology* 16, 2033–2047.

Bartsch, J.A., Haugh, C.G., Athow, K.L. and Peart, R.M. (1986) Impact damage to soybean seed. *Transactions of the ASAE* 29, 582–586.

Berglund, D.R. and Helms, T.C. (2003) Soybean production. Report No. A-250, NDSU Extension Service, North Dakota State University, Fargo, ND, USA.

Brader, B., Lee, R.C., Plarre, R., Burkholder, W., Kitto, G.B., Kao, C., Polston, L., Dorneau, E., Szabo, I., Mead, B. and Rouse, B. (2002) A comparison of screening methods for insect contamination in wheat. *Journal of Stored Products Research* 38, 75–86.

Casady, W.W., Paulsen, M.R., Reid, J.F. and Sinclair, J.B. (1992) A trainable algorithm for inspection of soybean seed quality. *Transactions of the ASAE* 35, 2027–2033.

CGC (2008) *Official Grain Grading Guide*. Canadian Grain Commission, Winnipeg, MB, Canada.

Cotton, R.T. and Wilbur, D.A. (1982) Insects. In: Christenses, C.M. (ed.) *Storage of Cereal Grains and Their Products*. American Association of Cereal Chemists, Inc., St Paul, MN, USA, pp. 304–308.

Dennis, N.M. and Decker, R.W. (1962) A method and machine for detecting living internal infestation in wheat. *Journal of Economic Entomology* 55, 199–203.

Deshpande, S.D. and Bal, S. (1999) Specific heat of soybeans. *Journal of Food Process Engineering* 22, 469–477.

Deshpande, S.D., Bal, S. and Ojha, T.P. (1996) Bulk thermal conductivity and diffusivity of soybean. *Journal of Food Processing and Preservation* 20, 177–189.

Dowell, F.E., Throne, J.E. and Baker, J.E. (1998) Automated nondestructive detection of internal insect infestation of wheat kernels by near-infrared reflectance spectroscopy. *Journal of Economic Entomology* 91, 899–904.

FAO (2009) *FAOSTAT*. Food and Agriculture Organization of the United Nations, Rome, Italy. Available at: http://faostat.fao.org (accessed 3 December 2009).

Felipe, C.A.S. and Barrozo, M.A.S. (2003) Drying of soybean seeds in a concurrent moving bed: Heat and mass transfer and quality analysis. *Drying Technology* 21, 439–456.

Gely, M.G. and Giner, S.A. (2007) Diffusion coefficient relationships during drying of soya bean cultivars. *Biosystems Engineering* 96, 213–222.

Gunasekaran, S., Cooper, T.M. and Berlage, A.G. (1988) Soybean seed coat and cotyledon crack detection by image processing. *Journal of Agricultural Engineering Research* 41, 139–148.

Hou, H.J. and Chang, K.C. (2004) Storage conditions affect soybean color, chemical composition and tofu qualities. *Journal of Food Processing and Preservation* 28, 473–488.

Howe, R.W. and Oxley, T.A. (1952) Detection of insects by their carbon dioxide production. Pest Infestation Research Report of the Great Britain Department of Scientific and Industrial Research. HMSO, London, UK.

Hurburgh, C.R. (2008) Soybean drying and storage. Report No. Pm-1636. Cooperative Extension Service, Iowa State University of Science and Technology, Ames, IA, USA.
Hutchinson, D. and Otten, L. (1983) Thin-layer air drying of soybeans and white beans. *Journal of Food Technology* 18, 507–522.
Ignatowicz, S., Wesolowska, B. and Zaedee, I.H. (1994) Detection of irradiated insect pests in stored products: Locomotor activity of irradiated adult beetles. In: Highley, E., Wright, E.J., Banks, H.J. and Champ, B.R. (eds) *Proceedings of the 6th International Working Conference on Stored Product Protection*, Canberra, Australia. CAB International, Wallingford, UK, pp. 1209–1213.
Isik, E. (2007) Some engineering properties of soybean grains. *American Journal of Food Technology* 2, 115–125.
Karunakaran, C., Jayas, D.S. and White, N.D.G. (2003) Soft X-ray inspection of wheat kernels infested by *Sitophilus oryzae*. *Transactions of the ASAE* 46, 739–745.
Kashaninejad, M., Ahmadi, M., Daraei, A. and Chabra, D. (2008) Handling and frictional characteristics of soybean as a function of moisture content and variety. *Powder Technology* 188, 1–8.
Kitic, D. and Viollaz, P.E. (1984) Comparison of drying kinetics of soybeans in thin layer and fluidized beds. *Journal of Food Technology* 19, 399–408.
Lacerda, A.F., Lisboa, M.H., Oliveira, M.S. and Barrozo, M.A.S. (2004) Two-phase modeling and the quality of soybean seeds dried in a counter-current moving bed drier. *Drying Technology* 27, 432–448.
Lessard, F.F. and Andrieu, A.J. (1986) Development of a rapid method to determine insect infestation in grain bins with electro-acoustic devices. In: Donahaye, E. and Navarro, S. (eds) *Proceedings of the 4th International Working Conference on Stored Product Protection*, Tel Aviv, Israel. Maor-Wallach Press, Jerusalem, Israel, pp. 643.
Li, Z., Kobayashi, N., Watanabe, F. and Hasatani, M. (2002) Sorption drying of soybean seeds with silica gel. *Drying Technology* 20, 223–233.
Loschiavo, S.R. (1974) Detection of insects by traps in grain-filled boxcars during transit. In: Brady, E.U., Brower, J.H., Hunter, P.E., Jay, E.G., Lum, P.T.M., Lund, H.O., Mullen, M.A. and Davis, R. (eds) *Proceedings of the 1st International Working Conference on Stored Product Entomology*, Savannah, Georgia, USA. Savannah, Georgia, USA, pp. 639–650.
Loschiavo, S.R. (1985) Post-harvest grain temperature, moisture and insect infestation in steel granaries in Manitoba. *The Canadian Entomologist* 117, 7–14.
Loschiavo, S.R. and Atkinson, J.M. (1967) A trap for detection and recovery of insects in stored grain. *The Canadian Entomologist* 99, 1160–1163.
Loschiavo, S.R. and Atkinson, J.M. (1973) An improved trap to detect beetles (*Coleoptera*) in stored grain. *The Canadian Entomologist* 105, 437–440.
Manuwa, S.I. (2007) Moisture-dependent physical properties of improved soybean var. TGX 1448-2E. *Journal of Food Science and Technology* 44, 371–374.
Mbuvi, S.W., Litchfield, J.B. and Sinclair, J.B. (1989) Physical properties of soybean seeds damaged by fungi and a virus. *Transactions of the ASAE* 32, 2093–2096.
Mensah, J.K., Nelson, G.L., Herum, F.L. and Richard, T.G. (1984) Mechanical properties related to soybean seed coat cracking during drying. *Transactions of the ASAE* 27, 550–555.
Milani, A.P., Bucklin, R.A., Teixeira, A.A. and Kebeli, H.V. (2000) Soybean compressibility and bulk density. *Transactions of the ASAE* 43, 1789–1793.
Misra, R.N. and Young, J.H. (1980) Numerical solution of simulatenous moisture diffusion and shrinkage during soybean drying. *Transactions of the ASAE* 23, 1277–1282.

Misra, R.N., Young, J.H. and Hamann, D.D. (1981) Finite element procedures for estimating shrinkage stresses during soybean drying. *Transactions of the ASAE* 24, 751–755.

Muir, W.E. (1997) Detection and measurement of insect and mite infestations (Chapter 10). In: *Grain Preservation Biosystems – An incomplete class notes*. Department of Biosystems Engineering, University of Manitoba, Winnipeg, MB, Canada, pp. 10-9–10-11.

Muir, W.E. and Sinha, R.N. (1988) Physical properties of cereal and oilseed cultivars grown in western Canada. *Canadian Agricultural Engineering* 30, 51–55.

Neethirajan, S., Karunakaran, C., Jayas, D.S. and White, N.D.G. (2007) Detection techniques for stored-product insects in grain. *Food Control* 18, 157–162.

Nuh, D.N. and Brinkworth, B.J. (1997) A novel thin-layer model for crop drying. *Transactions of the ASAE* 40, 659–669.

Osella, C.A., Gordo, N.A., Gonzalez, R.J., Tosi, E. and Re, E. (1997) Soybean heat-treated using a fluidized bed. *Lebensmittel Wissenschaft und Technologie* 30, 676–680.

Paliwal, J., Jayas, D.S., White, N.D.G. and Muir, W.E. (1999) Effects of pneumatic conveying of grain on mortality of insects. *Applied Engineering in Agriculture* 15, 65–68.

Parde, S.R., Kausal, R.T., Jayas, D.S. and White, N.D.G. (2002) Mechanical damage to soybean seed during processing. *Journal of Stored Products Research* 38, 385–394.

Paulsen, M.R., Nave, W.R. and Gray, L.E. (1981) Soybean seed quality as affected by impact damage. *Transactions of the ASAE* 24, 1577–1582 and 1589.

Pederson, J.R. (1992) Insects: Identification, damage and detection. In: Sauer, D.B. (ed.) *Storage of Cereal Grains and Their Products*. American Association of Cereal Chemists, St Paul, MN, USA, pp. 472–489.

Polat, R., Atay, U. and Saglam, C. (2006) Some physical and aerodynamic properties of soybean. *Journal of Agronomy* 5, 74–78.

Rafiee, S., Keyhani, A. and Mohammadi, A. (2008) Soybean seeds mass transfer simulation during drying using finite element method. *World Applied Sciences Journal* 4, 284–288.

Reed, C.R., Wright, V.F., Mize, T.W., Pederson, J.R. and Brockschmidt-Evans, J. (1991) Pitfall traps and grain samples as indicators of insects in farm-stored wheat. *Journal of Economic Entomology* 84, 1381–1387.

Sabbah, M.A., Meyer, G.E., Keener, H.M. and Roler, W.L. (1979) Simulation studies of reversed-direction air-flow drying method for soybean seed in a fixed bed. *Transactions of the ASAE* 22, 1162–1166.

Saio, K., Nikkuni, I., Ando, Y., Otsuru, M., Terauchi, Y. and Kito, M. (1980) Soybean quality changes during model storage studies. *Cereal Chemistry* 57, 77–82.

Sangkram, U. and Noomhorm, A. (2002) The effect of drying and storage of soybean on the quality of bean, oil, and lecithin production. *Drying Technology* 20, 2041–2054.

Sauer, D.B., Meronuck, R.A. and Christensen, C.M. (1992) Microflora. In: Sauer, D.B. (ed.) *Storage of Cereal Grains and Their Products*. American Association of Cereal Chemists, St Paul, MN, USA, pp. 313–340.

Shatadal, P. and Tan, J. (2003) Identifying damaged soybeans by color image analysis. *Transactions of the ASAE* 19, 65–69.

Shivhare, U., Raghavan, V., Bosisio, R. and Giroux, M. (1993) Microwave drying of soybean at 2.45 GHz. *Journal of Microwave Power and Electromagnetic Energy* 28, 11–17.

Singh, B.P.N. and Kulshrestha, S.P. (1987) Kinetics of water sorption by soybean and pigeonpea grains. *Journal of Food Science* 52, 1538–1544.

Soponronnarit, S., Swasdesevi, T., Wetchacama, S. and Wutiwiwatchai, W. (2001) Fluidized bed drying of soybeans. *Journal of Stored Products Research* 37, 133–151.

Spencer, M.R. (1976) Effect of shipping on quality of seeds, meals, fats, and oils. *Journal of American Oil Chemists' Society* 53, 238–240.

Sreenarayanan, V.V., Visvanathan, R. and Subramaniyan, V. (1988) Physical and thermal properties of soybean. *Journal of Agricultural Engineering* 25, 76–82.

Subrahmanyan, V., Swaminathan, M., Pingale, S.V. and Kadkol, S.B. (1955) Uric acid as an index of insect filth in cereal and milled products. *Bulletin of Central Food Technology Research Institute* 4, 86–87.

Ting, K.C., White, G.M., Ross, I.J. and Loewer, O.J. (1980) Seed coat damage in deep-bed drying of soybeans. *Transactions of the ASAE* 23, 1293–1300.

Tunde-Akintunde, T.Y., Olajide, J.O. and Akintunde, B.O. (2005) Mass volume-area related and mechanical properties of soybean as a function of moisture and variety. *International Journal of Food Properties* 8, 449–456.

Urbanski, G.E., Wei, L.S. and Nelson, A.I. (1980) Effect of freeze damage on soybean quality and storage stability. *Journal of Food Science* 45, 208–212.

USDA (2007) *U.S. Soybean Inspection*. USDA, GIPSA Technical Services Division, Kansas City, MO, USA.

Wang, D., Ram, M.S. and Dowell, F.E. (2002) Classification of damaged soybean seeds using near-infrared spectroscopy. *Transactions of the ASAE* 45, 1943–1948.

Wang, D., Dowell, F.E., Ram, M.S. and Schapaugh, W.T. (2004) Classification of fungal-damaged soybean seeds using near infrared spectroscopy. *International Journal of Food Properties* 7, 75–82.

Weinzierl, R. and Higgins, R. (2008) Insect pest management for stored grain. In: Overmier, M. (ed.) *2008 Illinois Agricultural Pest Management Handbook*. University of Illinois at Urbana-Champaign, IL, USA, pp. 147–156.

White, G.M., Loewer, O.J., Ross, I.J. and Egli, D.B. (1976) Storage characteristics of soybeans dried with heated air. *Transactions of the ASAE* 19, 306–310.

White, G.M., Bridges, T.C., Loewer, D.J. and Ross, I.J. (1980) Seed coat damage in thin-layer drying of soybeans. *Transactions of the ASAE* 23, 224–227.

White, N.D.G. (2001) Protection of farm-stored grains, oilseeds and pulses from insects, mites, and moulds. In: White, N.D.G. (ed.) *Protection of Farm-Stored Grains, Oilseeds and Pulses from Insects, Mites, and Moulds*. Agriculture and Agri-Food Canada Publication 1851(E) Revised, Cereal Research Centre, Winnipeg, MB, Canada, pp. 5–53.

White, N.D.G., Jayas, D.S. and Demianyk, C.J. (1997) Movement of stored grain to control stored-product insects and mites. *Phytoprotection* 78, 76–84.

Wigger, W.D., Paulsen, M.R., Litchfield, J.B. and Sinclair, J.B. (1988) Classification of fungi-damaged soybeans using color-image processing. ASAE Paper No. 883053. American Society of Agricultural Engineers, St Joseph, MI, USA.

Wiriyampaiwong, S., Soponronnarit, S. and Prachayawarakorn, S. (2003) Soybean drying by two-dimensional spouted bed. *Drying Technology* 21, 1735–1757.

Wongo, L.E. (1990) Factors of resistance in sorghum against *Sitotroga cerealella* (Oliv.) and *Sitophilus oryzae* (L.). *Insect Science and its Applications*, 11, 179–188.

Zisman, U. and Calderon, M. (1990) Early detection of insect infestation in grain samples by measuring the carbon dioxide content. In: Lessard, F.F. and Ducom, P. (eds) *Proceedings of the 5th International Working Conference on Stored Product Protection*, Bordeaux, France. Imprimerie du Médoc, Bordeaux, France, pp. 1465–1466.

13 Diseases of Soybean and Their Management

Glen L. Hartman[1,2] and Curtis B. Hill[1]
[1]*Department of Crop Sciences, National Soybean Research Center, University of Illinois, USA;* [2]*USDA Agricultural Research Service, Urbana, Illinois, USA*

13.1 Introduction

Soybean, the sole domesticated member of 25 known *Glycine* species (Hymowitz, 2008), is the most important oilseed crop worldwide. In 2007, an estimated 220 million t of soybean were produced on 95 million ha worldwide, with the USA, Brazil and Argentina leading in production at 71, 58 and 45 million t, respectively (FAO, 2008). Most of the diseases covered in this chapter are common to these three countries and to most other soybean-producing countries in the world. The origin of soybean domestication is China (Hymowitz, 2008) and most of the pathogens of soybean, with a few exceptions, developed their relationship with soybean in its Asian centre of origin.

More than 300 species of pathogens attack soybean worldwide, although relatively few cause significant economic damage (Hartman *et al.*, 1999). Parasitic microorganisms such as bacteria, fungi, nematodes, Stramenopiles and viruses are responsible for the most economically important soybean diseases. Many abiotic soybean disorders, caused by unfavourable environmental or nutritional conditions, are also important, but are not the focus of this chapter.

In 1994, the estimated worldwide loss due to soybean diseases was 11% (Hartman *et al.*, 1999). From 2001 to 2003, this estimate jumped to 23%: 11% due to plant parasitic bacteria and fungi, 1% to viruses and 11% to animal pests including plant parasitic nematodes (Oerke, 2006). Losses due to diseases could be much higher, but successful management practices, including cultural and seed sanitation techniques, chemical applications and the deployment of disease resistance genes, have played a role in reducing the impact of soybean pathogens. The extent of economic plant damage depends upon the type of pathogen, the plant tissue being attacked, the number of plants affected, disease severity, environmental

Table 13.1. Soybean diseases organized by the most likely time during the season to observe symptoms. The causal organism and general disease management practices are also shown.

Season	Common name	Scientific name	Management
Early	Phomopsis seed decay	*Phomopsis longicolla*	Sanitation; fungicides; resistance
	Rhizoctonia damping-off and root rot	*Rhizoctonia solani*	Seed treatments; partial resistance
	Pythium damping-off and root rot	*Pythium* spp.	Seed treatments; partial resistance
	Phytophthora root and stem rot	*Phytophthora sojae*	Seed treatments; complete resistance; partial resistance
Mid	Bacterial pustule	*Xanthomonas axonopodis* pv. *glycines*	Seed sanitation; resistance
	Bean pod mottle	*Bean pod mottle virus*	Vector control
	Soybean mosaic	*Soybean mosaic virus*	Resistance
	Sclerotinia stem rot	*Sclerotinia sclerotiorum*	Cultural practices; fungicides; partial resistance
	Frogeye leaf spot	*Cercospora sojina*	Fungicides; resistance
	Soybean rust	*Phakopsora pachyrhizi*	Fungicides; resistance
Late	Soybean cyst nematode	*Heterodera glycines*	Crop rotation; resistance
	Stem canker	*Diaporthe* spp., *Phomopsis* spp.	Crop rotation; resistance
	Sudden death syndrome	*Fusarium virguliforme*	Partial resistance
	Cercospora leaf blight	*Cercospora kikuchii*	Fungicides; resistance
	Anthracnose	*Colletotrichum* spp.	Sanitation
	Charcoal rot	*Macrophomina phaseolina*	Partial resistance

conditions, host plant susceptibility, plant stress levels and the stage of plant development.

This chapter takes the reader chronologically through three periods of the soybean growing season – the early, mid and late season – highlighting the major diseases that attack during these periods (Table 13.1). While many soybean pathogens may attack plants multiple times over the season, and some attack multiple plant parts, this chapter emphasizes the time during the growing season that the disease has the most significant economic and visual impact (Fig. 13.1). Each disease includes a description of the causal organism, symptoms, epidemiology and disease management practices.

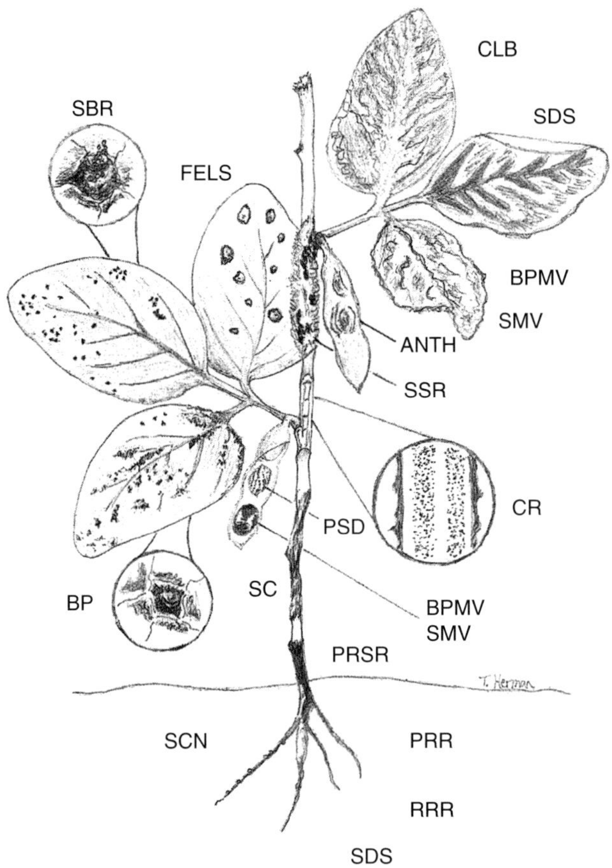

Fig. 13.1. Diagrammatic representation of a diseased soybean plant.
ANTH, anthracnose on pods; BP, bacterial pustule; BPMV, *Bean pod mottle virus*; CLB, Cercospora leaf blight; CR, charcoal rot; FELS, frogeye leaf spot; PRR, Pythium root rot; PRSR, Phytophthora root and stem rot; PSD, Phomopsis seed decay; RRR, Rhizoctonia root rot; SBR, soybean rust; SC, stem canker; SCN, soybean cyst nematode; SDS, sudden death syndrome; SMV, *Soybean mosaic virus*; SSR, Sclerotinia stem rot.

13.2 Management of Early-season Diseases

Seed decay

A soybean seed begins to germinate once it absorbs some amount of water. Any pathogens existing on the surface of the seed coat or that have entered into internal seed tissues, including the embryo, also begin to grow. Some of these pathogens may not have much impact on the plant until later growth stages, but a few can rot the seed during this early growth stage of moisture imbibition and germination.

Seed decay is primarily caused by *Phomopsis longicolla* (Hartman *et al.*, 1999), although soft-rotting bacteria, including *Bacillus subtilis*, and other fungi in the *Diaporthe–Phomopsis* complex of species may also be involved to a lesser extent (Hartman *et al.*, 1999). Symptoms of decay caused by *P. longicolla* are shrivelled, cracked, elongated seeds with a white, chalky appearance; however, sometimes no visible seed symptoms are observed. Infection begins as the seed matures in the seed crop during the previous season. Infected seeds are slow to germinate or may not germinate at all. This can significantly reduce plant stands and may result in a reduced crop yield.

P. longicolla is a pycnidial fungus with no known teleomorph. The fungus forms black pycnidia that produce two kinds of hyaline conidia: ellipsoidal to fusiform α-conidia and filiform β-conidia (Hartman *et al.*, 1999). Conidia produced on infected crop debris are the primary sources of inoculum. Warm, humid air promotes sporulation, while wind and rain spread the conidia over short distances. Infected seed provides long-range dissemination of the pathogen. Stressed plants, for example those already infected with a virus, are more vulnerable than healthy plants to seed infection by *P. longicolla* (Koning *et al.*, 2003).

Seed decay is primarily controlled by seed sanitation. Seed production fields are rotated with a non-host crop to eliminate *P. longicolla*-infested soybean crop debris (Garzonio and McGee, 1983) and are thoroughly inspected for *P. longicolla* infection before seed is harvested (McGee, 1986). Seed lots are tested for infection by enzyme-linked immunosorbent assay (Hartman *et al.*, 1999), polymerase chain reaction (PCR) (Zhang *et al.*, 1999), near infrared radiation (Wang *et al.*, 2004) or visual inspection of plated seed samples (Jackson *et al.*, 2005). Suitable fungicide seed treatments may be applied to eliminate seed infection and maximize seed viability (Munshi *et al.*, 2004); however, the additional input cost and the effectiveness of such treatments depends on whether environmental conditions conducive to seed infection occur. Although not widely advertised or used in commercial soybean varieties, resistance to *P. longicolla* infection is available to limit seed infection (Jackson *et al.*, 2005; Smith *et al.*, 2008b). Dominant and complementary genes control resistance, and seed infection is significantly reduced in plants possessing a resistance gene.

Seedling diseases

The emergence of soybean plants from seeds that have survived seed-decay pathogens normally takes about 5–10 days depending on the temperature, moisture, planting depth and cultivar genetics. At this time, seedling radicles, or primary roots, may be attacked by fungal pathogens soon after seed germination. Radical infections may spread up through the hypocotyl and attack the cotyledons. As emergence continues, lateral roots begin to grow from the radicles. Root hairs appear and provide key nutrient- and water-absorbing functions and the taproot continues to grow and branch. Root hairs are particularly vulnerable to attack from soil-borne pathogens. Loss of both cotyledons at or soon after emergence of soybean seedlings will reduce yields.

Infection that occurs at this stage can result in pre- and post-emergence death of the seedlings or 'damping-off', or the pathogen remains latent with symptoms developing later in the growing season. *Rhizoctonia solani* and *Pythium* species are soil-borne fungi and the most common pathogens causing damping-off of soybean seedlings. *R. solani* is one of the most common soil-borne plant pathogens worldwide and is also a major component of soybean root-rot complex (Hartman *et al.*, 1999). *R. solani* pathogenic on soybean is a multinucleate basidiomycete with the teleomorph *Thanatephorus cucumeris*. Many species of *Pythium* cause damping-off of soybean seedlings and may be responsible for 30% of stand reduction in soybean fields in temperate regions (Hartman *et al.*, 1999; Kirkpatrick *et al.*, 2006). In a recent study in Iowa in the USA, *Pythium* species were more prevalent than *R. solani* and *Phytophthora sojae* in soybean seedlings, and had a greater impact on soybean yields than the other two pathogens (Murillo-Williams and Pedersen, 2008). Cool, moist conditions, combined with minimum tillage and early planting, favour the development of damping-off, resulting in thinner soybean seedling stands (Broders *et al.*, 2007). Because the soybean seedling must often push through crusted soil, deeper planting can decrease survival of seed and final stand number by providing a longer opportunity for pathogens to attack at this stage.

Cultural practices and seed treatments are primarily used to control *Pythium* species and *R. solani* (Hartman *et al.*, 1999). Low-cost seed treatments that control damping-off pathogens can significantly enhance profitability, especially when high-quality treated seed is sown (Poag *et al.*, 2005) and in cool, moist soils (Bradley, 2008). Partial resistance against *R. solani* (Bradley *et al.*, 2005; Zhao *et al.*, 2005) and *Pythium* species (Bates *et al.*, 2008) has also been identified, but is not widely deployed in soybean cultivars.

Phytophthora root and stem rot

P. sojae attacks soybean at any growth stage, but it is most damaging early in the season when it attacks emerging seedlings and rots the roots of young soybean plants (Hartman *et al.*, 1999). Soybean plants surviving damping-off may succumb to *P. sojae* root infection, which can cause wilting, stunting and death of infected plants.

Phytophthora is taxonomically related to *Pythium*. Although both genera resemble fungi, they actually belong to the Stramenopiles kingdom, which also includes algae such as kelp and diatoms (Tyler *et al.*, 2008). They belong to a sub-group of the Stramenopiles called Oomycetes, which includes many other destructive plant pathogens such as the downy mildews (Keeling *et al.*, 2005). Conventional fungal control measures are not effective against these pathogens. *P. sojae* has a gene-for-gene interaction with soybean. There are 12 known avirulence genes in *P. sojae* (Shan *et al.*, 2004), which interact with 14 resistance genes at eight soybean loci (Burnham *et al.*, 2003). This interaction has produced tremendous pathogenic diversity in *P. sojae* worldwide (Dorrance *et al.*, 2003; Sugimoto *et al.*, 2007; Nelson *et al.*, 2008). *Lupinus* species are also known as hosts of *P. sojae* (Hartman *et al.*, 1999).

Pathotype-specific resistance against *P. sojae* has been effective in most soybean-growing regions (Dorrance *et al.*, 2003). All of the pathotype-specific resistance genes are dominant over susceptibility and most provide complete resistance that protects the plants throughout their lifespan against incompatible *P. sojae* pathotypes; they can also encourage the selection and development of compatible pathotypes able to overcome the resistance (Dorrance *et al.*, 2003). Overall, pathotype-resistance genes have remained effective in most soybean regions, possibly because commercial soybean breeders have also increased partial resistance to *P. sojae* (Ferro *et al.*, 2006).

Partial resistance is available for use where pathotype-specific resistance is ineffective and may help to increase the durability of race-specific resistance genes. This type of resistance is quantitative in expression and controlled by multiple genes with small effects, but is highly heritable in soybean (Tyler *et al.*, 2008). The principal mechanism of partial resistance is the ability to reduce the rate of lesion expansion following infection (Mideros *et al.*, 2007), possibly due to higher levels of suberin in tissues of partial resistance genotypes (Thomas *et al.*, 2007). This form of resistance may be more durable against changes in pathogen populations than resistance controlled by pathotype-specific, complete resistance genes. It is non-specific towards *P. sojae* pathotypes and relies on small contributions from multiple genes, making it more difficult for the pathogen to overcome the resistance. Soybean genotypes with either a complete resistance gene or good partial resistance to *P. sojae* have also been found to be tolerant to water-saturated soil conditions (Helms *et al.*, 2007).

New molecular genetic technology used to identify DNA markers linked to resistance genes (Sandhu *et al.*, 2005; Gordon *et al.*, 2007; Weng *et al.*, 2007; Sugimoto *et al.*, 2008) has greatly facilitated the selection of resistant and partially resistant plants in soybean breeding programmes. Overall selection efficiency has increased and the technology has enabled selection of multiple resistance genes controlling resistance to *P. sojae* and other soybean pathogens.

13.3 Management of Mid-season Diseases

During the middle part of the growing season, new vegetative nodes on soybean plants develop approximately every 3–5 days until the fifth vegetative node, and then every 2–3 days until the end of flowering. Flowering racemes develop and progress up and down the plant. Root growth continues until flowering ceases. Later in this part of the growing season, as pods develop, stresses that reduce pod number, number of seeds per pod or seed size greatly impact yield.

Bacterial pustule

Several bacterial diseases occur on soybean. The most common and potentially devastating is bacterial pustule. It is distributed worldwide and is most important in tropical and subtropical regions. The cause of bacterial pustule is *Xanthomonas axonopodis* pv. *glycines*, a Gram-negative, rod-shaped bacterium with a single polar flagellum (Hartman *et al.*, 1999). Wide variability in pathogenicity exists among isolates. The pathogen can also be seed-borne and attacks several legume species. The diagnostic symptom of bacterial pustule is the presence of pustules, which are minute, pale-green spots with elevated centres, usually on the abaxial surface of leaves, formed through hypertrophy and hyperplasia. When foliage is covered with pustules, premature defoliation can occur, reducing yields by reducing seed size and quantity. Frequent hard-driving rains and winds promote pathogen spread. Deployment of genetic resistance is the best method of control. The recessive *rxp* gene provides strong resistance, possibly due to higher peroxidase activity that limits bacterial survival, and requires high inoculum density to overcome. Although the resistance gene has been available for several years and linked DNA markers have been identified (Narvel *et al.*, 2001) to facilitate selection of resistant plants, the gene has apparently not been widely deployed in the central US soybean-growing region (Goradia *et al.*, 2009).

Viral diseases

As temperatures increase during spring and early summer in temperate regions, insects become active, including insects that vector important soybean viruses. The bean leaf beetle, *Cerotoma trifurcata*, is the primary vector of *Bean pod mottle virus* (BPMV), while aphids, including the soybean aphid *Aphis glycines*, are vectors of *Soybean mosaic virus* (SMV) (Hartman *et al.*, 2001).

Bean pod mottle

Bean pod mottle, caused by BPMV, does not appear to be widespread outside of the USA. It can interact synergistically with other soybean viruses, such as SMV, to produce very severe symptoms. Symptoms caused by BPMV infection include green to yellow leaf mottling, leaf puckering, mild to severe leaf distortion and, in severe infections, terminal necrosis. Bean pod mottle can cause seed coat mottling (Hobbs *et al.*, 2003) and delay plant maturity, and can be confused with green stem disorder (Hobbs *et al.*, 2006).

BPMV is a bipartite, spherical RNA virus that belongs to the Comoviruses group of plant viruses. There are two distinct subgroups of BPMV based on nucleic acid hybridization analysis (Gu *et al.*, 2002). Natural reassortant BPMV strains with RNA from both subgroups (Gu *et al.*, 2002) and partial diploid strains (Gu *et al.*, 2007) also occur and can produce severe BPMV symptoms. The virus is seed-borne and its host range is restricted to legumes.

Although bean leaf beetle is the primary vector of BPMV, other beetles, including Japanese beetles (Wickizer and Gergerich, 2007), can transmit it to soybean in a non-persistent manner without a latent period. It is also sap and seed transmissible. The virus overwinters in beetle vectors and perennial weeds (Krell *et al.*, 2003).

Controlling the bean leaf beetle vector is the primary strategy to control BPMV (Krell *et al.*, 2004; Bradshaw *et al.*, 2008). Immunity to BPMV does not seem to exist in soybean (Hartman *et al.*, 1999); however, partial resistance and tolerance remain a possibility and further research to find sources of these traits and utilize them for commercial cultivar development is needed.

Soybean mosaic

Caused by SMV, soybean mosaic occurs in all soybean production areas of the world and is one the most common viral pathogens of soybean (Hartman *et al.*, 1999). Symptoms vary with host genotype, virus strain, plant age at infection and environment. Typical symptoms include a mosaic of light and dark green areas on leaves, plant stunting and seed coat mottling.

SMV is a flexuous RNA rod and a member of the *Potyviridae*. The virus is both aphid- and seed-transmitted (Hartman *et al.*, 1999). Numerous strains of the virus exist based on reactions on differential soybean genotypes. Strain specificity of transmission through seed and induced seed coat mottling has been found (Domier *et al.*, 2007). There are two geographically distinct groups in North America and Asia, based on sequences of part of the coat protein coding regions (Domier *et al.*, 2003). The virus infects many hosts, including many genera in six plant families (Hartman *et al.*, 1999).

Early plant infection reduces pod set, increases seed coat mottling and reduces seed size and weight more so than late-season infection (Hartman *et al.*, 1999). At least 32 aphid species, belonging to 15 genera, transmit SMV in a non-persistent manner. Plants that become infected by seed transmission serve as primary inoculum sources for SMV. SMV incidence in the USA was forecast to increase after the introduction of the soybean aphid in 2000 (Hartman *et al.*, 2001). However, probably because of successful seed sanitation practised by seed producers that minimized sources of primary inoculum, an epidemic of SMV did not develop, despite the expanded distribution of the soybean aphid.

Resistance to SMV is available, but is not widely deployed in soybean production. Three known resistant loci (*Rsv1*, *Rsv3* and *Rsv4*) in soybean interact with different SMV strains (Palmer *et al.*, 2004). *Rsv1* is multi-allelic. *Rsv4* controls resistance to all known SMV strains. In a test of commercial and pre-commercial soybean cultivars in the USA, <10% of the 850 cultivars tested were resistant to two common SMV strains (Wang *et al.*, 2006).

Sclerotinia stem rot

Although restricted on soybean to geographic areas with cooler growing conditions, *Sclerotinia* stem rot is distributed worldwide and has a broad host range, including numerous dicotyledons (Hartman *et al.*, 1999). Symptoms include wilting and death of the upper leaves. Lesions can girdle the stem and block vascular flow, limiting pod and seed development. Any part of the plant that comes in contact with infected tissue can also become infected. The diagnostic feature of *Sclerotinia* stem rot is the white, cottony mycelia present on infected plant parts. Infected stems become shredded and bleached in appearance. Large, black, irregular-shaped sclerotia are usually present on infected stems.

The disease is caused by the fungus *Sclerotinia sclerotiorum*, an apothecial ascomycete. Ascospores are forcibly ejected from cup-shaped apothecia that arise directly from sclerotia. The pathogen can infect seed and infest seed lots.

Beginning with the colonization of spent flower petals, the necrotrophic pathogen launches an attack from infected petals into the node and the stem. With a battery of phytotoxins, including oxalic acid, that augment the activity of endopolygalacturonase (Favaron *et al.*, 2004), the pathogen destroys soybean tissues ahead of the invading hyphae. The attack usually involves only the upper half of the plant. Sclerotia of *S. sclerotiorum* that drop to the soil from infected soybean stems can survive for several years. Apothecial production from surviving sclerotia is promoted by the moist conditions found under the thick, closed canopy, typically found in a healthy soybean field. Ascospores released from the apothecia are the primary source of inoculum. The optimum environmental conditions for ascospore germination, blossom colonization and infection of soybean are not clear; however, the disease is more common in relatively cool, moist northern temperate climates than in warmer and drier climates.

The proper timing of fungicide application before peak vulnerability of blossoms has provided protection against *S. sclerotiorum* in potato (Johnson and Atallah, 2006) and peanut (Smith *et al.*, 2008a). Fungicide application when inoculum levels were low was effective at controlling *Sclerotinia* stem rot in soybean (Mueller *et al.*, 2002a). Partial resistance to *Sclerotinia* stem rot has been found (Diers *et al.*, 2006), but has not been widely deployed in soybean. Several quantitative trait loci (QTLs) that control partial resistance to the disease have been mapped in the soybean genome (Zhao *et al.*, 2006; Guo *et al.*, 2008; Vuong *et al.*, 2008). With the sequencing of the *S. sclerotiorum* genome now complete, new approaches to control *Sclerotinia* stem rot, including genetic engineering, have the potential to provide strong resistance to the disease in the future (Lu, 2003; Dickman, 2007). Detoxification of oxalic acid through genetic engineering has been successful in limiting disease development in canola (Dong *et al.*, 2008).

Frogeye leaf spot

Frogeye leaf spot is more important in warm, humid soybean-production regions, such as tropical areas and the southern USA (Hartman *et al.*, 1999). Circular to angular lesions with dark centres appearing on the adaxial surface of leaves give the appearance of 'frog eyes' and serve as the key diagnostic symptom of the appropriately named disease. The disease can attack other plant parts, including stems and pods, but foliar infection has the largest effect on yields.

The causal fungus, *Cercospora sojina*, is an imperfect fungus. It produces conidia on conidiophores borne in fascicles arising from a thin stroma, which are produced on leaf and stem residues or infested seeds. The conidia are much shorter than those produced by *C. kikuchii*, the cause of soybean purple stain disease. Wide variability in virulence exists among *C. sojina* isolates (Mian *et al.*, 2008). The pathogen is seed-borne and its host range is restricted to soybean. *C. sojina* survives as mycelium in infected seeds of infested soybean residue. Warm, humid weather promotes sporulation and splashing rain helps disseminate conidial inoculum. The disease can rapidly spread in prolonged warm and moist conditions.

Fungicide application and plant resistance have been effective in controlling the disease. Flusilazole, benomyl and the mixture of flusilazole and carbendazim have reduced yield losses caused by frogeye leaf spot, and additionally soybean rust, in Africa. In the southern USA, frogeye symptoms have been significantly reduced with higher rates and multiple applications of the strobilurin fungicides. Several resistance genes have been identified in soybean (Mian *et al.*, 2008). The *Rcs3* gene remains resistant to all known *C. sojina* pathotypes (Mian *et al.*, 2008) and has been mapped in the soybean genome (Mian *et al.*, 1999; Missaoui *et al.*, 2007). A novel toxic soybean protein (SBTX) has been found to have an inhibitory effect on *C. sojina* (Vasconcelos *et al.*, 2008) that may be exploited in soybean.

Soybean rust

Soybean rust is found extensively in most areas of the world where soybean is grown. Its origin is in Asia, but the pathogen has recently spread into the Americas including Brazil, Canada and the USA (Hartman *et al.*, 1999; Isard *et al.*, 2005). Soybean rust is tropical and does not overwinter in cold temperate climates.

The most common symptoms are tan to dark-brown or reddish-brown lesions with one to many erumpent, globose uredinia, particularly on the underside of leaflets (Hartman *et al.*, 1999). Lesions tend to be angular and restricted by leaf veins, and are frequently associated with leaf chlorosis. Under heavy infection, this may result in premature defoliation and early maturity. Soybean is susceptible at any stage, but symptoms usually appear during the mid to late season because of the requirement for a prolonged wet, cool period for infection and sporulation.

Soybean rust is caused by two species of fungi: *Phakopsora meibomiae* (anamorph *Malupa vignae*) and *P. pachyrhizi* (anamorph *Malupa sojae*), which are differentiated by morphological and molecular techniques (Ono *et al.*, 1992; Frederick *et al.*, 2002). Most research and published data pertains to *P. pachyrhizi*, the more aggressive of the two species and with the widest geographic distribution (Hartman *et al.*, 1992a). Soybean rust is a microcyclic fungus with telia (Harmon *et al.*, 2006) that have no known role in the life cycle of the fungus. Virulence among rust isolates is highly variable (Twizeyimana *et al.*, 2009). The pathogen is not seed-borne in soybean, but it has a broad host range that includes 150 species in 53 genera of the legume family *Fabaceae* (Slaminko *et al.*, 2008a,b). The fungus is an obligate pathogen and can only grow on a living plant.

Rust epidemics are most severe during long periods of leaf wetness when the mean daily temperature is <28°C. Urediniospores, the primary means of disease spread, require free water for germination and penetration and can cycle from initial infection to urediniospore production in as little as 9 days. Hurricane Ivan in 2004 was reported as the probable event that transported *P. pachyrhizi* spores from South America to North America (Isard *et al.*, 2005). The quantitative relationship between rust severity and yield showed that this disease has the potential to cause up to 80% yield loss (Hartman *et al.*, 1991).

Primary control of rust is with the application of fungicides. Before 2003, only a few fungicides were registered for controlling soybean rust (Miles *et al.*, 2003), but due to its rapid spread and potential impact, fungicide registration has been accelerated (Miles *et al.*, 2007). The proper timing of fungicide sprays is critical to effective control (Mueller *et al.*, 2009). Since the 2005 growing season, the United States Department of Agriculture (USDA) has provided a tracking system to monitor disease movement using a web site that maps the distribution of soybean rust in North America (USDA, 2008). Major emphasis has been placed on modelling epidemics to produce forecasts to alert soybean producers when to apply fungicides for rust prevention.

Resistance to soybean rust is pathotype-specific. Five dominant resistance genes are known to date (Hartman *et al.*, 2005; Calvo *et al.*, 2008). *Rpp1* has recently been found to be multiallelic (Chakraborty *et al.*, 2009) and has been mapped to soybean LG G (Hyten *et al.*, 2007). The durability of some of rust resistance genes has been short-lived. Additional resistance genes found in perennial wild *Glycine* species from Australia (Burdon, 1988; Hartman *et al.*, 1992b; Schoen *et al.*, 1992) may eventually be introgressed into soybean to improve overall rust resistance.

There are few cultural controls for soybean rust. In Brazil, however, the government has prohibited the planting of soybean during the off season in an attempt to interrupt and then reduce the amount of inoculum available for the peak season. This policy has only been in place for a few years, but by eliminating soybean during the off season, there appears to have been some reduction in rust incidence (Farias Neto, 2007).

13.4 Management of Late-season Diseases

Rapid pod growth and the beginning of seed development typify this part of the soybean growing season. It is the most crucial period for seed yield. Stress during this stage causes a greater yield reduction than at any other time, mainly the result of fewer pods. Seed filling requires plentiful supply of water and nutrients that if in short supply can reduce seed size. Leaf loss of 100% at this stage may reduce yields by 80% by reducing pod numbers and the number of seeds per pod, and to a lesser extent by reducing seed size. In the middle of this stage of the growing season, the 'green bean' stage or beginning full-seed stage occurs, when the total pod weight peaks and leaf yellowing and leaf drop rapidly progress. Later, as dry matter begins to peak in individual seeds and seeds contain about 60% moisture, stress may have little effect on yield except in the case of pod abortion or seed shattering. At this point, the crop is safe from a killing frost while soybeans rapidly lose moisture, especially with warm and dry weather, until they are harvested at the end of the season.

Soybean cyst

Soybean cyst nematode (SCN), caused by the plant parasitic nematode *Heterodera glycines*, occurs in most soybean-growing regions (Hartman *et al.*, 1999). Symptoms on the root system range from a slight discolouration to severe necrosis. Diagnosis of SCN is made when white or yellow females are observed attached to the roots. Above-ground symptoms are rarely apparent, but include slight to severe plant stunting and leaf chlorosis.

H. glycines is a sedentary root endoparasite that invades the root and partially redirects root cell functions to satisfy its nutritional demands for development and reproduction (Riggs, 2004). Each cyst contains 50–200 eggs that hatch and second-stage juveniles (J2s) develop that invade soybean roots. The J2s are equipped with a robust stylet that cuts slits in cell walls to facilitate intracellular migration toward the vascular cylinder, where they induce their permanent feeding structure, the syncytium. Cysts are durable and are disseminated by water, wind, soil peds mixed in seed and machinery, surviving for long periods in the soil. The pathogenicity of the nematode is highly variable and a number of 'HG-type' pathotypes have been identified that differ in their ability to reproduce on a set of soybean differentials (Niblack and Riggs, 2004; Niblack *et al.*, 2008). SCN produces up to four generations in warmer climates and as few as two in cooler climates. Symptoms may be enhanced or repressed in association with other pathogens (Gao *et al.*, 2006).

The most effective and common means of management of SCN include host resistance and crop rotation (Niblack and Chen, 2004; Schmitt *et al.*, 2004). Complete resistance to SCN has not been found and currently deployed resistance genes have not proven to be durable over time due to adaptation of *H. glycines* populations. 'Gene' rotation, along with crop rotation, may improve the durability of SCN resistance.

Stem canker

Northern and southern stem canker are two similar diseases of soybean separated by their geographic occurrence in the USA (Hartman *et al.*, 1999).

Both diseases are capable of killing plants from mid season to maturity. Cankers produced by northern stem canker become dark brown with age and usually girdle the stem causing wilt and plant death; southern stem canker stem lesions become long, but rarely girdle the stem.

The two fungi that cause stem canker are *Diaporthe phaseolorum* var. *caulivora* (northern stem canker) and *D. phaseolorum* var. *meridionalis* (southern stem canker) (Hartman *et al.*, 1999). These are perithecial ascomycetes that have a pycnidial asexual stage. The host range of the two fungi is broad and includes non-legume hosts.

Stem canker fungi over-season in debris and splash-driven rain disseminates conidial inoculum to soybean plants. Seed infection can occur in up to 20% of seeds, but the role of seed infection in stem canker epidemics is unclear. Infections leading to cankers occur early in the growing season, although symptoms do not appear until much later during the reproductive growth stages. Inter-plant spread of the disease is localized to a few plants from the inoculum source.

Stem canker management options include practices such as deep tillage that degrade infected soybean debris and reduce inoculum sources and resistance. Four independent genes have been identified that give resistance to southern stem canker. Their effectiveness against northern stem canker is unknown.

Sudden death syndrome

Sudden death syndrome (SDS) is most visible when it turns foliage yellow late in the soybean-growing season. Prior to that, the root mass of SDS-affected soybean plants has already been reduced, discoloured and rotted (Hartman *et al.*, 1999). Longitudinal sections of taproots and lower stems of diseased plants have discoloured grey to brown xylem vessels that extend up the stem, while the pith remains white. Symptoms first appear on leaves as scattered, interveinal, chlorotic spots, which may become necrotic or enlarge and coalesce into chlorotic streaks that become necrotic, leaving only the mid-vein and major lateral veins green. Severely infected plants prematurely defoliate and have increased pod abortion and reduced seed size.

SDS is caused by the soil-borne fungus *Fusarium virguliforme*, formerly known as *F. solani* f. sp. *glycines* (teleomorph unknown). Another species found in South America, *F. tucumaniae* has also been shown to cause SDS (Aoki *et al.*, 2003; Arruda *et al.*, 2005). *F. virguliforme* does not occur in South America and *F. tucumaniae* does not occur in the USA. To date, most research has been on *F. virguliforme*. *F. virguliforme* produces blue to blue-green macroconidia (Hartman *et al.*, 1999) and chlamydospores (Li *et al.*, 1998). The results of a molecular genetic study indicate that *F. virguliforme* is genetically homogeneous and distinct from other *Fusarium* species (Li *et al.*, 2000). *F. virguliforme* isolates do not produce perithecia, whereas *F. tucumaniae* does (Covert *et al.*, 2007). The fungus is known to produce phytotoxins in culture, and culture filtrates can produce SDS symptoms on soybean stem cuttings (Hartman *et al.*, 2004). Host specialization is not known in *F. virguliforme*, but differences among isolates in their aggressiveness in producing disease symptoms have been observed (Li *et al.*, 2009). The fungus also causes root rot on other legume crops (Hartman *et al.*, 1999).

The pathogen infects soybean roots, but not other plant parts (Hartman *et al.*, 1999). Foliar symptoms may be caused by phytotoxins that are translocated to leaves (Jin *et al.*, 1996). SDS is more prevalent in wet growing seasons than hot, dry seasons and with high soil fertility and soil compaction. PCR-based detection methods developed for the specific detection of *F. virguliforme* DNA from soybean roots and field soil are useful in determining the distribution and levels of inoculum associated with SDS in the field (Li and Hartman, 2003; Gao *et al.*, 2004; Li *et al.*, 2008).

Partial resistance may provide the only effective control of SDS (Hartman *et al.*, 1999) and several sources of partial resistance or reduced susceptibility have been identified (Hartman *et al.*, 1997, 2000; Mueller *et al.*, 2002b, 2003). Partial resistance to SDS may primarily limit the effects of SDS phytotoxins. A number of resistance QTLs have been mapped on several soybean linkage groups (Farias Neto *et al.*, 2007).

Cercospora leaf blight

Cercospora leaf blight is distributed worldwide (Hartman *et al.*, 1999) and can attack leaves, stems, petioles, pods and seeds. Upper leaves exposed to the sun have a light purple appearance, highlighted with bronzing. Reddish purple, angular to irregular lesions appear later on both the upper and lower surfaces of leaves. The lesions vary from pinpoint spots to irregular patches. Numerous infections cause rapid chlorosis and necrosis of leaf tissues, resulting in defoliation starting with the upper leaves. Lesions on petioles and stems are slightly sunken and red to purple. Infection of petioles increases defoliation, with the petioles remaining attached to plants. On seeds, discolouration varies from light red to dark purple, and the discoloured areas range from specks to large, irregular blotches that may cover the entire surface of the seed coat.

The causal fungus, *C. kikuchii*, produces hyaline conidia from conidiophores arising from a stoma in stalks of ≥20. The fungus has no known teleomorph stage. There is high genetic variability among isolates of *C. kikuchii* (Cai and Schneider, 2005; Almeida *et al.*, 2006; Imazaki *et al.*, 2006). High variability in random amplification of polymorphic DNA and microsatellite-primed PCR markers among isolates collected in Louisiana suggests the likely existence of a sexual stage that has yet to be found (Cai and Schneider, 2008). Conidia from infected seeds or infected surface debris are the primary sources of inoculum (Schuh, 2003).

Late-season fungicide applications targeted for frogeye leaf spot or soybean rust additionally help control *Cercospora* leaf blight and seed diseases including purple seed stain. Cultivars with different levels of susceptibility have been reported (Orth and Schuh, 1994).

Anthracnose

Anthracnose occurs wherever soybean is grown, but is most destructive in warm, humid conditions under prolonged rainfall. The disease reduces stands, seed quality and yields. Yield losses are proportional to levels of pod infection (Hartman *et al.*, 1999). Soybean is susceptible at all stages, including early-season pre- and post-emergence damping-off. The most noticeable symptoms occur on older plants during seed maturation, and signs of the pathogen become more evident as the plant senesces and/or individual plant parts such as petioles die off. Foliar symptoms include laminar vein necrosis, leaf rolling, petiole cankering and premature defoliation. The presence of acervuli and/or setae on soybean tissues is diagnostic of anthracnose infection.

The most common pathogen associated with anthracnose is *Colletotrichum truncatum* (teleomorph unknown) (Hartman *et al.*, 1999). Other *Colletotrichum* species can also cause anthracnose (Hartman *et al.*, 1999) with regional or local importance (Chen *et al.*, 2006). Most of the *Colletotrichum* species reported on soybean have wide host ranges that include common weeds in soybean fields.

Colletotrichum mycelia can over-season in infected crop residue or seed, or as sclerotia in soil. Inoculum from infected seeds and residue may cause pre- and post-emergence damping-off of seedlings. Conidia are the primary inocula, disseminated by splashing rain predominately during the reproductive stages in warm, moist weather. Once plant infection is established, the fungus may remain quiescent until the soybean plants begin to mature.

Anthracnose management can be indirectly accomplished with cultural practices that promote the breakdown of soybean debris and reduce inoculum sources. Beneficial organisms also reduce the occurrence of anthracnose in soybean (Tripathi *et al.*, 2006). Cultivars vary in susceptibility and soybean resistance has been reported (Hartman *et al.*, 1999).

Charcoal rot

Charcoal rot manifests itself late in the season (Hartman *et al.*, 1999). It is widely distributed worldwide and affects many crop plants. Soybean yield losses of >70% have been reported. Soybean plants infected with charcoal rot exhibit smaller than normal leaflets, loss of plant vigour, leaf yellowing and leaf flagging. The disease can also alter seed composition (Bellaloui *et al.*, 2008). The diagnostic symptom is light-grey or silvery discolouration of epidermal and sub-epidermal tissues in the taproot and a lower stem speckled with small, dark, black microsclerotia, giving a 'charcoal' appearance.

Charcoal rot is caused by the fungal pathogen *Macrophomina phaseolina* (Hartman *et al.*, 1999). The jet-black microsclerotia are the key taxonomic feature of the fungus. *M. phaseolina* is a pycnidial fungus; however, pycnidia are rarely seen on soybean. Considerable genetic variation exists among charcoal rot isolates (Jana *et al.*, 2005; Purkayastha *et al.*, 2006).

The primary source of charcoal rot inoculum is microsclerotia surviving in soil or host residue (Hartman *et al.*, 1999). Hyphae from germinating microsclerotia attack soybean roots through natural openings and the disease progresses into the internal stem tissues. Damage to soybean plants is primarily caused by the plugging of vascular elements with microsclerotia and mycelia and phytotoxins (Ranezabu *et al.*, 2007). *M. phaseolina* can attack soybean plants at any stage; however, infections beginning during the early season remain latent until environmental stresses, such as drought and high temperatures, allow symptom development.

Effective charcoal rot management strategies have not been found. However, partial resistance has been identified in soybean (Paris *et al.*, 2006; Mengistu *et al.*, 2007). The basis of partial resistance to charcoal rot is not associated with slow wilting or drought tolerance (Wrather *et al.*, 2008). Methods to screen for partial resistance in the greenhouse and to identify *M. phaseolina* using PCR technology (Babu *et al.*, 2007) have been developed and may lead to new, efficient methods of screening for resistance.

13.5 Summary of Management Practices

Overall, the disease management practised in soybean has been highly successful in limiting production losses worldwide. Soybean seed producers employing sound seed sanitation through seed certification and deploying resistance genes in their seed products have had the greatest impact on reducing economic losses. Seed sanitation has helped negate the incidence of seed decay along with viral and bacterial infections. Fungicide application has been effective and is likely to expand where diseases that cannot be adequately controlled with resistance become more important, such as soybean rust in the Americas. New fungicide chemistry has improved the spectrum of effectiveness; however, plant resistance is much more economical and environmentally safe for soybean producers and people living in soybean-production areas. *Phytophthora* root rot, SCN and frogeye leaf spot are diseases that have been effectively checked through resistance in most soybean-growing regions. Resistance to most of the major pathogens has increased in soybean genotypes through the combined efforts of pathologists and breeders. Advanced genetic manipulation shows yet further promise of improving resistance to diseases in soybean and keeping pace with nature's ability to overcome it. With these new techniques and practices, traditional cultural control practices such as crop rotation have become less important in most soybean-growing regions.

Acknowledgements

With thanks to Theresa Herman for providing the illustration and editorial assistance.

References

Almeida, A.M.R., Piuga, F.F., Marin, S.R.R., Binneck, E., Sartori, F., Costamilan, L.M., Teixeira, M.R.O. and Lopes, M. (2006) Pathogenicity, molecular characterization, and cercosporin content of Brazilian isolates of *Cercospora kikuchii*. *Fitopatologia Brasil* 30, 594–602.

Aoki, T., O'Donnell, K., Homma, Y. and Lattanzi, A.R. (2003) Sudden death syndrome of soybean is caused by two morphologically and phylogenetically distinct species within the *Fusarium solani* species complex – *F. virguliforme* in North America and *F. tucumaniae* in South America. *Mycologia* 95, 660–684.

Arruda, G.M.T., Miller, R.N.G., Ferreira, M.A.S.V. and Café-Filho, A.C. (2005) Morphological and molecular characterization of the sudden death syndrome pathogen of soybean in Brazil. *Plant Pathology* 54, 53–65.

Babu, B.K., Saxena, A.K., Srivastava, A.K. and Arora, D.K. (2007) Identification and detection of *Macrophomina phaseolina* by using species-specific oligonucleotide primers and probe. *Mycologia* 99, 797–803.

Bates, G.D., Rothrock, C.S. and Rupe, J.C. (2008) Resistance of the soybean cultivar Archer to *Pythium* damping-off and root rot caused by several *Pythium* spp. *Plant Disease* 92, 763–766.

Bellaloui, N., Mengistu, A. and Paris, R. (2008) Soybean seed composition in cultivars differing in resistance to charcoal rot (*Macrophomina phaseolina*). *Journal of Agricultural Science* 146, 667–675.

Bradley, C.A. (2008) Effect of fungicide seed treatments on stand establishment, seedling disease, and yield of soybean in North Dakota. *Plant Disease* 92, 120–125.

Bradley, C.A., Hartman, G.L., Mueller, D.S., Hoffman, D.D., Nickell, C.D. and Pedersen, W.L. (2005) Genetic analysis of partial resistance to *Rhizoctonia solani* in the soybean cultivar 'Savoy'. *Canadian Journal of Plant Pathology* 27, 137–142.

Bradshaw, J.D., Rice, M.E. and Hill, J.H. (2008) Evaluation of management strategies for bean leaf beetles (Coleoptera: Chrysomelidae) and *Bean pod mottle virus* (Comoviridae) in soybean. *Journal of Economic Entomology* 101, 1211–1227.

Broders, K.D., Lipps, P.E., Paul, P.A. and Dorrance, A.E. (2007) Characterization of *Pythium* spp. associated with corn and soybean seed and seedling disease in Ohio. *Plant Disease* 91, 727–735.

Burdon, J.J. (1988) Major gene resistance to *Phakopsora pachyrhizi* in *Glycine canescens*, a wild relative of soybean. *Theoretical Applied Genetics* 75, 923–928.

Burnham, K.D., Dorrance, A.E., Francis, D.M., Fioritto, R.J. and St. Martin, S.K. (2003) Rps8, a new locus in soybean for resistance to *Phytophthora sojae*. *Crop Science* 43, 101–105.

Cai, G. and Schneider, R.W. (2005) Vegetative compatibility groups in *Cercospora kikuchii*, the causal agent of *Cercospora* leaf blight and purple seed stain in soybean. *Phytopathology* 95, 257–261.

Cai, G. and Schneider, R.W. (2008) Population structure of *Cercospora kikuchii*, the causal agent of *Cercospora* leaf blight and purple seed stain in soybean. *Phytopathology* 98, 823–829.

Calvo, E.S., Kiihl, R.A.S., Garcia, A., Harada, A. and Hiromoto, D.M. (2008) Two major recessive soybean genes conferring soybean rust resistance. *Crop Science* 48, 1350–1354.

Chakraborty, N., Curley, J., Frederick, R.D., Hyten, D.L., Nelson, R.L., Hartman, G.L. and Diers, B.W. (2009) Mapping and confirmation of a new allele at Rpp1 from soybean PI 504538A conferring RB lesion type resistance to soybean rust. *Crop Science* 49, 783–790.

Chen, L.S., Chu, C., Liu, C.D., Chen, R.S. and Tsay, J.G. (2006) PCR-based detection and differentiation of anthracnose pathogens, *Colletotrichum gloeosporioides* and *C. truncatum*, from vegetable soybean in Taiwan. *Journal of Phytopathology* 154, 654–662.

Covert, S.F., Aoki, T., O'Donnell, K., Starkey, D., Holliday, A., Geiser, D.M., Cheung, F., Town, C., Strom, A. and Juba, J. (2007) Sexual reproduction in the soybean sudden death syndrome pathogen *Fusarium tucumaniae*. *Fungal Genetics and Biology* 44, 799–807.

Dickman, M. (2007) Approaches for improving crop resistance to soilborne fungal diseases through biotechnology using *Sclerotinia sclerotiorum* as a case study. *Australasian Plant Pathology* 36, 116–123.

Diers, B., Kopisch-Obuch, F., Hoffman, D., Hartman, G., Pedersen, W., Grau, C. and Wang, D. (2006) Registration of AxN-1-55 soybean germplasm with partial resistance to *Sclerotinia* stem rot. *Crop Science* 46, 1403.

Domier, L.L., Latorre, I.J., Steinlage, T.A., McCoppin, N. and Hartman, G.L. (2003) Variability and transmission of *Aphis glycines* of North American and Asian *Soybean mosaic virus* isolates. *Archives of Virology* 148, 1925–1941.

Domier, L.L., Steinlage, T.A., Hobbs, H.A., Yang, H.A., Herrera-Rodriguez, G., Haudenshield, J.S., McCoppin, N.K. and Hartman, G.L. (2007) Similarities in seed and aphid transmission among *Soybean mosaic virus* isolates. *Plant Disease* 91, 546–550.

Dong, X., Ji, R., Guo, X., Foster, S., Chen, H., Dong, C., Liu, Y., Hu, Q. and Liu, S. (2008) Expressing a gene encoding wheat oxalate oxidase enhances resistance to *Sclerotinia sclerotiorum* in oilseed rape (*Brassica napus*). *Planta* 228, 331–340.

Dorrance, A., McClure, S. and Desilva, A. (2003) Pathogenic diversity of *Phytophthora sojae* in Ohio soybean fields. *Plant Disease* 87, 139–146.

FAO (2008) *FAOSTAT*. Food and Agriculture Organization of the United Nations, Rome, Italy. Available at: http://faostat.fao.org (accessed 15 December 2008).

Farias Neto, A.F. (2007) *Development of disease resistant varieties in Brazil*. Available at: http://www.plantmanagementnetwork.org/infocenter/topic/soybeanrust/2007/program (accessed 15 December 2008).

Farias Neto, A.F., Hashmi, R., Schmidt, M., Carlson, S.R., Hartman, G.L., Li, S., Nelson, R.L. and Diers, B.W. (2007) Mapping and confirmation of a new sudden death syndrome resistance QTL on linkage group D2 from the soybean genotypes PI 567374 and 'Ripley'. *Molecular Plant Breeding* 20, 53–62.

Favaron, F., Sella, L. and D'Ovidio, R. (2004) Relationships among endo-polygalacturonase, oxalate, pH, and plant polygalacturonase-inhibiting protein (PGIP) in the interaction between *Sclerotinia sclerotiorum* and soybean. *Molecular Plant-Microbe Interactions* 17, 1402–1409.

Ferro, C.R., Hill, C.B., Miles, M.R. and Hartman, G.L. (2006) Evaluation of soybean cultivars with *Rps1k* gene for partial resistance or field tolerance to *Phytophthora sojae*. *Crop Science* 46, 2427–2436.

Frederick, R.D., Snyder, C.L., Peterson, G.L. and Bonde, M.R. (2002) Polymerase chain reaction assays for the detection and discrimination of the soybean rust pathogens *Phakopsora pachyrhizi* and *P. meibomiae*. *Phytopathology* 92, 217–227.

Gao, X., Jackson, T.A., Lambert, K.N., Li, S., Hartman, G.L. and Niblack, T.L. (2004) Detection and quantification of *Fusarium solani* f. sp. *glycines* in soybean roots with real-time quantitative polymerase chain reaction. *Plant Disease* 88, 1372–1380.

Gao, X., Jackson, T.A., Hartman, G.L. and Niblack, T.L. (2006) Interactions between the soybean cyst nematode and *Fusarium solani* f. sp. *glycines* based on greenhouse factorial experiments. *Phytopathology* 96, 1409–1415.

Garzonio, D.M. and McGee, D.C. (1983) Comparison of seeds and crop residues as sources of inoculum for pod and stem blight of soybeans. *Plant Disease* 67, 1374–1376.

Goradia, L., Hartman, G.L. and Daniel, S.L. (2009) Evaluation of glyphosate-resistant soybean cultivars for resistance to bacterial pustule. *European Journal of Plant Pathology* 124, 331–335.

Gordon, S., Kowitwanich, K., Pipatpongpinyo, W., St. Martin, S. and Dorrance, A. (2007) Molecular marker analysis of soybean plant introductions with resistance to *Phytophthora sojae*. *Phytopathology* 97, 113–118.

Gu, H., Clark, A.J., De Sa, P.B., Pfeiffer, T.W., Tolin, S. and Ghabrial, S.A. (2002) Diversity among isolates of *Bean pod mottle virus*. *Phytopathology* 92, 446–452.

Gu, H., Zhang, C. and Ghabrial, S.A. (2007) Novel naturally occurring *Bean pod mottle virus* reassortants with mixed heterologous RNA1 genomes. *Phytopathology* 97, 79–86.

Guo, X., Wang, D., Gordon, S., Helliwell, E., Smith, T., Berry, S., St. Martin, S. and Dorrance, A. (2008) Genetic mapping of QTLs underlying partial resistance to *Sclerotinia sclerotiorum* in soybean PI 391589A and PI 391589B. *Crop Science* 48, 1129–1139.

Harmon, C.L., Harmon, P.F., Mueller, T.A., Marois, J.J. and Hartman, G.L. (2006) First report of *Phakopsora pachyrhizi* telia on kudzu in the United States. *Plant Disease* 90, 380.

Hartman, G.L., Wang, T.C. and Tschanz, A.T. (1991) Soybean rust development and the quantitative relationship between rust severity and soybean yield. *Plant Disease* 75, 596–600.

Hartman, G.L., Saddoui, E.M., Tschanz, A.T., Macintyre, R. and Lopez, K. (1992a) *Annotated Bibliography of Soybean Rust* (*Phakopsora pachyrhizi* Syd.). Asian Vegetable Research and Development Center, Taipei, Taiwan.

Hartman, G.L., Wang, T.C. and Hymowitz, T. (1992b) Sources of resistance to soybean rust in perennial *Glycine* species. *Plant Disease* 76, 396–399.

Hartman, G.L., Huang, Y.H., Nelson, R.L. and Noel, G.R. (1997) Germplasm evaluation of *Glycine max* for resistance to *Fusarium solani*, the causal organism of sudden death syndrome. *Plant Disease* 81, 515–518.

Hartman, G.L., Sinclair, J.B. and Rupe, J.C. (1999) *Compendium of Soybean Diseases*, 4th edn. American Phytopathological Society Press, St Paul, MN, USA.

Hartman, G.L., Gardner, M.E., Hymowitz, T. and Naidoo, G.C. (2000) Evaluation of perennial *Glycine* species for resistance to soybean fungal pathogens that cause *Sclerotinia* stem rot and sudden death syndrome. *Crop Science* 40, 545–549.

Hartman, G.L., Domier, L.L., Wax, L.M., Helm, C.G., Onstad, D.W., Shaw, J.T., Solter, L.F., Voegtlin, D.J., d'Arcy, C.J., Gray, M.E., Steffey, K.L., Isard, S.A. and Orwick, P.L. (2001) Occurrence and distribution of *Aphis glycines* on soybeans in Illinois in 2000 and its potential control. *Plant Health Progess*. Available at: http://www.plantmanagementnetwork.org/sub/php/brief/aphisglycines (accessed 15 December 2008).

Hartman, G.L., Huang, Y.H. and Li, S. (2004) Phytotoxicity of *Fusarium solani* culture filtrates from soybean and other hosts assayed by stem cuttings. *Australasian Plant Pathology* 33, 9–15.

Hartman, G.L., Miles, M.R. and Frederick, R.D. (2005) Breeding for resistance to soybean rust. *Plant Disease* 89, 664–666.

Helms, T.C., Werk, B.J., Nelson, B.D. and Deckard, E. (2007) Soybean tolerance to water-saturated soil and role of resistance to *Phytophthora sojae*. *Crop Science* 47, 2295–2302.

Hobbs, H.A., Hartman, G.L., Wang, Y., Hill, C.B., Bernard, R.L., Pedersen, W.L. and Domier, L.L. (2003) Occurrence of seed coat mottling in soybean plants inoculated with *Bean pod mottle virus* and *Soybean mosaic virus*. *Plant Disease* 87, 1333–1336.

Hobbs, H., Hill, C., Grau, C., Koval, N., Wang, Y., Pedersen, W., Domier, L. and Hartman, G. (2006) Green stem disorder of soybean. *Plant Disease* 90, 513–518.

Hymowitz, T. (2008) The history of the soybean. In: Johnson, L.A., White, P.J. and Galloway, R. (eds) *Soybeans: Chemistry, Production, Processing, and Utilization*. AOCS Press, Urbana, IL, USA, pp. 1–31.

Hyten, D.L., Hartman, G.L., Nelson, R.L., Frederick, R.D., Concibido, V.C. and Cregan, P.B. (2007) Map location of the *Rpp1* locus that confers resistance to *Phakopsora pachyrhizi* (soybean rust) in soybean. *Crop Science* 47, 837–838.

Imazaki, I., Homma, Y., Kato, M., Vallone, S., Yorinori, J.T., Henning, A.A., Iizumi, H. and Koizumi, S. (2006) Genetic relationships between *Cercospora kikuchii* populations from South America and Japan. *Phytopathology* 96, 1000–1008.

Isard, S.A., Gage, S.H., Comtois, P. and Russo, J.M. (2005) Principles of the atmospheric pathway for invasive species applied to soybean rust. *BioScience* 55, 851–861.

Jackson, E.W., Fenn, P. and Chen, P. (2005) Inheritance of resistance to *Phomopsis* seed decay in soybean PI 80837 and MO/PSD-0259 (PI 562694). *Crop Science* 45, 2400–2404.

Jana, T.K., Singh, N.K., Koundal, K.R. and Sharma, T.R. (2005) Genetic differentiation of charcoal rot pathogen, *Macrophomina phaseolina*, into specific groups using URP-PCR. *Canadian Journal of Microbiology* 51, 159–164.

Jin, H., Hartman, G.L., Nickell, C.D. and Widholm, J.M. (1996) Characterization and purification of a phytotoxin produced by *Fusarium solani*, the causal agent of soybean sudden death syndrome. *Phytopathology* 86, 277–282.

Johnson, D. and Atallah, Z. (2006) Timing fungicide applications for managing *Sclerotinia* stem rot of potato. *Plant Disease* 90, 755–758.

Keeling, P.J., Burger, G., Durnford, D.G., Lang, B.F., Lee, R.W., Pearlman, R.E., Roger, A.J. and Gray, M.W. (2005) The tree of eukaryotes. *Trends in Ecology & Evolution* 20, 670–676.

Kirkpatrick, M.T., Rupe, J.C. and Rothrock, C.S. (2006) Soybean response to flooded soil conditions and the association with soilborne plant pathogenic genera. *Plant Disease* 90, 592–596.

Koning, G., Tekrony, D.M. and Ghabrial, S.A. (2003) Soybean seedcoat mottling: Association with *Soybean mosaic virus* and *Phomopsis* spp. seed infection. *Plant Disease* 87, 413–417.

Krell, R.K., Pedigo, L.P., Hill, J.H. and Rice, M.E. (2003) Potential primary inoculum sources of *Bean pod mottle virus* in Iowa. *Plant Disease* 87, 1416–1422.

Krell, R.K., Pedigo, L.P., Hill, J.H. and Rice, M.E. (2004) Bean leaf beetle (Coleoptera: Chrysomelidae) management for reduction of bean pod mottle virus. *Journal of Economic Entomology* 97, 192–202.

Li, S. and Hartman, G.L. (2003) Molecular detection of *Fusarium solani* f. sp. *glycines* in soybean roots and soil. *Plant Pathology* 52, 74–83.

Li, S., Hartman, G.L. and Gray, L.E. (1998) Chlamydospore formation, production, and nuclear status in *Fusarium solani* f. sp. *glycines* soybean sudden death syndrome-causing isolates. *Mycologia* 90, 414–421.

Li, S., Tam, Y.K. and Hartman, G.L. (2000) Molecular differentiation of *Fusarium solani* f. sp *glycines* from other *F. solani* based on mitochondrial small subunit rDNA sequences. *Phytopathology* 90, 491–497.

Li, S., Hartman, G.L., Domier, L.L. and Boykin, D. (2008) Quantification of *Fusarium solani* f. sp. *glycines* isolates in soybean roots by colony-forming unit assays and real-time quantitative PCR. *Theoretical and Applied Genetics* 117, 343–352.

Li, S., Hartman, G.L. and Chen, Y. (2009) Evaluation of aggressiveness of *Fusarium virguliforme* isolates that cause sudden death syndrome. *Journal of Plant Pathology,* 91, 77–86.

Lu, G. (2003) Engineering *Sclerotinia sclerotiorum* resistance in oilseed crops. *African Journal of Biotechnology* 2, 509–516.

McGee, D.C. (1986) Prediction of *Phomopsis* seed decay by measuring soybean pod infection. *Plant Disease* 70, 329–333.

Mengistu, A., Ray, J.D., Smith, J.R. and Paris, R.L. (2007) Charcoal rot disease assessment of soybean genotypes using a colony-forming unit index. *Crop Science* 47, 2453–2461.

Mian, M.A.R., Wang, T., Phillips, D.V., Alvernaz, J. and Boerma, H.R. (1999) Molecular mapping of the *Rcs3* gene for resistance to frogeye leaf spot in soybean. *Crop Science* 39, 1687–1691.

Mian, M.A.R., Missaoui, A.M., Walker, D.R., Phillips, D.V. and Boerma, H.R. (2008) Frogeye leaf spot of soybean: A review and proposed race designations for isolates of *Cercospora sojina* Hara. *Crop Science* 48, 14–24.

Mideros, S., Nita, M. and Dorrance, A.E. (2007) Characterization of components of partial resistance, *Rps2,* and root resistance to *Phytophthora sojae* in soybean. *Phytopathology* 97, 655–662.

Miles, M.R., Hartman, G.L., Levy, C. and Morel, W. (2003) Current status of soybean rust control by fungicides. *Pesticide Outlook* 14, 197–200.

Miles, M.R., Levy, C., Morel, W., Mueller, T., Steinlage, T., Van Rij, N., Frederick, R.D. and Hartman, G.L. (2007) International fungicide efficacy trials for the management of soybean rust. *Plant Disease* 91, 1450–1458.

Missaoui, A., Ha, B., Phillips, D. and Boerma, H. (2007) Single nucleotide polymorphism detection of the *Rcs3* gene for resistance to frogeye leaf spot in soybean. *Crop Science* 47, 1681–1690.

Mueller, D.S., Dorrance, A.E., Derksen, R.C., Ozkan, E., Kurle, J.E., Grau, C.R., Gaska, J.M., Hartman, G.L., Bradley, C.A. and Pedersen, W.L. (2002a) Efficacy of fungicides on *Sclerotinia sclerotiorum* and their potential for control of *Sclerotinia* stem rot on soybean. *Plant Disease* 86, 26–31.

Mueller, D.S., Hartman, G.L., Nelson, R.L. and Pedersen, W.L. (2002b) Evaluation of *Glycine max* germ plasm for resistance to *Fusarium solani* f. sp. *glycines*. *Plant Disease* 86, 741–746.

Mueller, D.S., Nelson, R.L., Hartman, G.L. and Pedersen, W.L. (2003) Response of commercially developed cultivars and ancestral soybean lines to *Fusarium solani* f. sp. *glycines*. *Plant Disease* 87, 827–831.

Mueller, T.A., Miles, M.R., Morel, W., Marios, J.J., Wright, D.L., Kemerait, R.C., Levy, C. and Hartman, G.L. (2009) Effect of fungicide and timing of application on soybean rust severity and yield. *Plant Disease* 93, 243–248.

Munshi, G.D., Randhawa, M. and Sharma, R.C. (2004) Effect of fungicidal seed treatments on *Phomopsis* and other seed mycoflora of soybean. *Journal of Research, Punjab Agricultural University* 41, 352–355.

Murillo-Williams, A. and Pedersen, P. (2008) Early incidence of soybean seedling pathogens in Iowa. *Agronomy Journal* 100, 1481–1487.

Narvel, J.M., Jakkula, L.R., Phillips, D.V., Wang, T., Lee, S.H. and Boerma, H.R. (2001) Molecular mapping of *Rxp* conditioning reaction to bacterial pustule in soybean. *Journal of Heredity* 92, 267–270.

Nelson, B.D., Mallik, I., McEwen, D. and Christianson, T. (2008) Pathotypes, distribution, and metalaxyl sensitivity of *Phytophthora sojae* from North Dakota. *Plant Disease* 92, 1062–1066.

Niblack, T.L. and Chen, S. (2004) Cropping systems. In: Schmitt, D.P., Wrather, J.A. and Riggs, R.D. (eds) *Biology and Management of the Soybean Cyst Nematode*, 2nd edn. Schmitt & Associates, Marceline, Missouri, USA, pp. 181–206.

Niblack, T.L. and Riggs, R.D. (2004) Variation in virulence phenotypes. In: Schmitt, D.P., Wrather, J.A. and Riggs, R.D. (eds) *Biology and Management of Soybean Cyst Nematode*, 2nd edn. Schmitt & Associates, Marceline, Missouri, USA, pp. 57–72.

Niblack, T.L., Colgrove, A.L., Colgrove, K. and Bond, J.P. (2008) Shift in virulence of soybean cyst nematode is associated with use of resistance from PI 88788. *Plant Health Progress*. Available at: http://www.plantmanagementnetwork.org/php/elements/sum2.aspx?id=6798 (accessed 15 December 2008).

Oerke, E.C. (2006) Crop losses to pests. *Journal of Agricultural Science* 144, 31–43.

Ono, Y., Buritica, P. and Hennen, J.F. (1992) Delimitation of *Phakopsora*, *Physopella* and *Cerotelium* and their species on Leguminosae. *Mycological Research* 96, 825–850.

Orth, C.E. and Schuh, W. (1994) Resistance of 17 soybean cultivars to foliar, latent, and seed infection by *Cercospora kikuchii*. *Plant Disease* 78, 661–664.

Palmer, R.G., Pfeiffer, T.W., Buss, G.R. and Kilen, T.C. (2004) Qualitative genetics. In: Boerma, H.R. and Specht, J.E. (eds) *Soybeans: Improvement, Production, and Uses*, 3rd edn. Agronomy Monograph 16. American Society of Agronomy, Crop Science Society of America, and Soil Science Society of America, Madison, WI, USA, pp. 137–233.

Paris, R.L., Mengistu, A., Tyler, J.M. and Smith, J.R. (2006) Registration of soybean germplasm line DT97-4290 with moderate resistance to charcoal rot. *Crop Science* 46, 2324.

Poag, P.S., Popp, M., Rupe, J., Dixon, B., Rothrock, C. and Boger, C. (2005) Economic evaluation of soybean fungicide seed treatments. *Agronomy Journal* 97, 1647–1657.

Purkayastha, S., Kaur, B., Dilbaghi, N. and Chaudhury, A. (2006) Characterization of *Macrophomina phaseolina*, the charcoal rot pathogen of cluster bean, using conventional techniques and PCR-based molecular markers. *Plant Pathology* 55, 106–116.

Ranezabu, M., Shier, W.T., Abbas, H.K., Tonos, J.L., Baird, R.E. and Sciumbato, G. (2007) Soybean charcoal rot disease fungus *Macrophomina phaseolina* in Mississippi produces the phytotoxin (-)-botryodiplodin, but no detectable phaseolinone. *Journal of Natural Products* 70, 128–129.

Riggs, R.D. (2004) History and distribution. In: Schmitt, D.P., Wrather, J.A. and Riggs, R.D. (eds) *Biology and Management of Soybean Cyst Nematode*, 2nd edn. Schmitt & Associates, Marceline, MO, USA, pp. 41–56.

Sandhu, D., Schallock, K., Rivera-Velez, N., Lundeen, P., Cianzio, S. and Bhattacharyya, M. (2005) Soybean *Phytophthora* resistance gene *Rps8* maps closely to the *Rps3* region. *Journal of Heredity* 96, 536–541.

Schmitt, D.P., Barker, K.R. and Riggs, R.D. (2004) Potential means of management. In: Schmitt, D.P., Wrather, J.A. and Riggs, R.D. (eds) *Biology and Management of Soybean Cyst Nematode*, 2nd edn. Schmitt & Associates, Marceline, MO, USA, pp. 57–72.

Schoen, D.J., Burdon, J.J. and Brown, A.H.D. (1992) Resistance of *Glycine tomentella* to soybean leaf rust *Phakopsora pachyrhizi* in relation to ploidy level and geographic distribution. *Theoretical and Applied Genetics* 83, 827–832.

Schuh, W. (2003) Influence of interrupted dew periods, relative humidity, and light on disease severity and latent infections caused by *Cercospora kikuchii* on soybean. *Phytopathology* 83, 109–113.

Shan, W., Cao, M., Leung, D. and Tyler, B.M. (2004) The Avr1b locus of *Phytophthora sojae* encodes an elicitor and a regulator required for avirulence on soybean plants carrying resistance gene *Rps1b*. *Molecular Plant-Microbe Interactions* 17, 394–403.

Slaminko, T.L., Miles, M.R., Frederick, R.D., Bonde, M.R. and Hartman, G.L. (2008a) New legume hosts of *Phakopsora pachyrhizi* based on greenhouse evaluations. *Plant Disease* 92, 767–771.

Slaminko, T.L., Miles, M.R., Marios, J.J., Wright, D.L. and Hartman, G.L. (2008b) Hosts of *Phakopsora pachyrhizi* identified in field evaluations in Florida. *Plant Health Progress*. Available at: http://www.plantmanagementnetwork.org/php/elements/sum2.aspx?id=7078 (accessed 15 December 2008).

Smith, D., Garrison, M., Hollowell, J., Isleib, T. and Shew, B. (2008a) Evaluation of application timing and efficacy of the fungicides fluazinam and boscalid for control of *Sclerotinia* blight of peanut. *Crop Protection* 27, 823–833.

Smith, S., Fenn, P., Chen, P. and Jackson, E. (2008b) Inheritance of resistance to *Phomopsis* seed decay in PI 360841 soybean. *Journal of Heredity* 9, 588–592.

Sugimoto, T., Watanabe, K., Yoshida, S., Aino, M., Matsuyama, M., Maekawa, K. and Irie, K. (2007) The effects of inorganic elements on the reduction of *Phytophthora* stem rot disease of soybean, the growth rate and zoospore release of *Phytophthora sojae*. *Journal of Phytopathology* 155, 97–107.

Sugimoto, T., Yoshida, S., Watanabe, K., Aino, M., Kanto, T., Maekawa, K. and Irie, K. (2008) Identification of SSR markers linked to the *Phytophthora* resistance gene Rps1-d in soybean. *Plant Breeding* 127, 154–159.

Thomas, R., Fang, X., Ranathunge, K., Anderson, T.R., Peterson, C.A. and Bernards, M.A. (2007) Soybean root suberin: Anatomical distribution, chemical composition, and relationship to partial resistance to *Phytophthora sojae*. *Plant Physiology* 144, 299–311.

Tripathi, M., Johri, B.N. and Sharma, A. (2006) Plant growth-promoting *Pseudomonas* sp. strains reduce natural occurrence of anthracnose in soybean (*Glycine max* L.) in central Himalayan region. *Current Microbiology* 52, 390–394.

Twizeyimana, M., Ojiambo, P.S., Sonder, K., Ikotun, T., Hartman, G.L. and Bandyopadhyay, R. (2009) Pathogenic variation of *Phakopsora pachyrhizi* infecting soybean in Nigeria. *Phytopathology* 99, 353– 361.

Tyler, B.M., Jiang, R.H.Y., Zhou, L., Tripathy, S., Dou, D., Torto-Alalibo, T., Li, H., Mao, Y., Liu, B. and Vega-Sanchez, M. (2008) Functional genomics and bioinformatics of the *Phytophthora sojae* soybean interaction. In: Gustafson, J.P., Tayler, J. and Stacey, G. (eds) *Genomics of Disease*. Springer, New York, USA, pp. 67–78.

USDA (2008) Pest information platform for extension and education. Online publication. Available at: www.sbr.usa.net (accessed 15 December 2008).

Vasconcelos, I., Morais, J., Siebra, E., Carlini, C., Sousa, D., Beltramini, L., Melo, V. and Oliveira, J. (2008) SBTX, a new toxic protein distinct from soyatoxin and other toxic soybean [*Glycine max*] proteins, and its inhibitory effect on *Cercospora sojina* growth. *Toxicon* 51, 952–963.

Vuong, T., Diers, B.W. and Hartman, G.L. (2008) Identification of QTL for resistance to *Sclerotinia* stem rot (*Sclerotinia sclerotiorum*) in plant introduction 194639. *Crop Science* 48, 2209–2214.

Wang, D., Dowell, F.E., Ram, M.S. and Schapaugh, W.T. (2004) Classification of fungal-damaged soybean seeds using near-infrared spectroscopy. *International Journal of Food Properties* 7, 75–82.

Wang, Y., Hobbs, H.A., Bowen, C.R., Bernard, R.L., Hill, C.B., Haudenshield, J.S., Domier, L.L. and Hartman, G.L. (2006) Evaluation of soybean cultivars, 'Williams' isogenic lines, and other selected soybean lines for resistance to two *Soybean mosaic virus* strains. *Crop Science* 46, 2649–2653.

Weng, C., Yu, K., Anderson, T.R. and Poysa, V. (2007) A quantitative trait locus influencing tolerance to *Phytophthora* root rot in the soybean cultivar 'Conrad'. *Euphytica* 158, 81–86.

Wickizer, S.L. and Gergerich, R.C. (2007) First report of Japanese beetle (*Popillia japonica*) as a vector of *Southern bean mosaic virus* and *Bean pod mottle virus*. *Plant Disease* 91, 637–637.

Wrather, A., Shannon, G., Carter, T.E., Bond, J., Rupe, J. and Almeida, A. (2008) Reaction of drought tolerant soybean genotypes to *Macrophomina phaseolina*. *Plant Health Progress*. Available at: http://www.plantmanagementnetwork.org/php/elements/sum.aspx?id=6948&photo=4148 (accessed 15 December 2008).

Zhang, A.W., Hartman, G.L., Curio-Penny, B., Pedersen, W.L. and Becker, K.B. (1999) Molecular detection of *Diaporthe phaseolorum* and *Phomopsis longicolla* from soybean seeds. *Phytopathology* 89, 796–804.

Zhao, G., Ablett, G.R., Anderson, T.R., Rajcan, I. and Schaafsma, A.W. (2005) Inheritance and genetic mapping of resistance to *Rhizoctonia* root and hypocotyl rot in soybean. *Crop Science* 45, 1441–1447.

Zhao, J., Udall, J., Quijada, P., Grau, C., Meng, J. and Osborn, T. (2006) Quantitative trait loci for resistance to *Sclerotinia sclerotiorum* and its association with a homeologous non-reciprocal transposition in *Brassica napus* L. *Theoretical and Applied Genetics* 112, 509–516.

14 Insect Pests of Soybean and Their Management

Matthew E. O'Neal and Kevin D. Johnson
Department of Entomology, Iowa State University, Ames, Iowa, USA

14.1 Introduction

The ecology and management of arthropods in soybean (*Glycine max* (L.) Merrill) has been thoroughly reviewed three times in the last 32 years (Turnipseed and Kogan, 1976; Kogan and Turnipseed, 1987; Sinclair *et al.*, 1997). The majority of the first review (Turnipseed and Kogan, 1976) described the community of insects associated with soybean. In that initial review the three main insect pest management tools (insecticides, host plant resistance [HPR] and biological control) comprised a little over three pages of a 33-page review. This is noted not as a criticism, but as a reflection of the state of pest management at that time. The next review focused more on the practice of pest management in soybean (Kogan and Turnipseed, 1987); nine out of 29 pages reviewed not only improvements in pest management tools, but also detailed case studies that incorporated multiple tools for the effective management of defoliators, pod-feeders and stem-borers identified as significant pests. The main tools were insecticides recommended at times when their application can limit yield loss (i.e. at an economic threshold), HPR and biological control agents. Additional reviews and compilations of insect soybean pests have been written since. Almost 11 years after the third review (Sinclair *et al.*, 1997), significant advancements have been made to the most common tools that can be used against insect pests of soybean. For North America, Higley and Boethel (1994) compiled an extensive description of 39 different insect pests of soybean. Within a worldwide perspective, Sinclair *et al.* (1997) provided a detailed explanation of how multiple pest types (weeds, pathogens and arthropods) can be managed.

The current chapter reviews the most significant insect pests of soybean and gives an overview of integrated pest management (IPM) theory and the consequences that can occur with greater frequency when IPM is not practised. The main tactics used in IPM will be briefly reviewed and their potential and limitations for soybean production will be discussed. This chapter also includes an update of the recent invasion of the soybean aphid (*Aphis glycines*) to North America and Australia from its native range in Asia and discusses how multiple tactics are employed to limit this economically important pest of soybean.

14.2 Insect Pests by Feeding Guild

Insects are the most diverse class of organisms, with over a million species identified and an estimated 10 million more yet to be discovered. Fortunately only a small portion (<1%) are considered pests and many are beneficial for crop production. The type of feeding varies by the type of insect, based on its ecology, morphology and phenology. These factors help to determine the feeding guild in which a herbivore resides. Guilds are comprised of organisms that share similar feeding strategies and life history traits. For soybean, these feeding guilds include defoliators (Lepidoptera, Coleoptera, Orthoptera), phloem feeders (Hemiptera) and seed feeders (Coleoptera, Diptera). The first step in preventing yield loss due to these forms of herbivory is the identification of the pest. From there, the appropriate management tactic can be selected. Several insect pests from different orders feed on soybean throughout the season. The most significant pests, as noted by Kogan and Turnipseed (1987) and Sinclair *et al.* (1997), have been highlighted in Table 14.1. In the following sections the type of herbivores that feed on soybean, based on their feeding guild, are briefly discussed.

Defoliators

Several insect orders – Orthoptera (grasshoppers), Coleoptera (beetles) and Lepidoptera (moths and butterflies) – are capable of feeding on soybean leaves. These insects typically have chewing mouthparts that either remove leaf area or destroy the leaf surface. Members of the last two orders have been identified as some of the more significant global soybean pests. The immature (larval, caterpillar) stages of Lepidoptera participate in leaf feeding. Adult Lepidoptera do not have chewing mouthparts. Both the adult and immature stages of Coleopteran soybean pests possess chewing mouthparts and can damage leaves. Kogan and Turnipseed (1987) identified the following as major or frequent defoliators of soybean: *Helicoverpa zea* (corn earworm), *Heliothis armigera* (America bollworm), *Heliothis punctigera* (Australian bollworm), *Pseudoplusia includens* (soybean looper), *Anticarsia gemmatalis* (velvetbean caterpillar) and *Spilosoma obliqua* (Bihar hairy caterpillar). Both adult and immature stages of Orthoptera are capable of feeding on soybean leaves. Although grasshoppers often feed on soybean, they rarely reach pest status (DeGooyer and Browde, 1994).

Although some defoliators leave very distinct patterns of feeding (e.g. *S. obliqua*), it is often difficult to determine which species is responsible for

Table 14.1. Arthropod pests of soybeans based on plant growth stage (adapted and updated from Sinclair *et al.*, 1997).

	Early season (V2 to V3–V5)	Mid-season (V8–R2 to V1–R5)	Late-season (R7–R8)
North America	Seedlings *Delia platura*^{D}	Defoliators *Helicoverpa zea*^{L}, *Heliothis virescens*^{L}, *Pseudoplusia includens*^{L}, *Anticarsia gemmatalis*^{L}, *Spodoptera* spp.^{L}, *C. trifurcata*^{C}	Seed feeders *H. zea*^{L}, *Nezara viridula*^{H}, *C. trifurcata*^{C}, *Acrosternum hilare*^{H}
	Leaf feeders *Cerotoma trifurcata*^{D}, *Agrotis* spp.^{D}	Leaf feeders *Tetranychus urticae*^{A}	
		Phloem feeders *A. glycines*^{H}	
South America	Leaf feeders *Agrotis* spp.^{D}	Defoliators *P. includens*^{L}, *Spodoptera* spp.^{L}, *Anticarsia aporema*^{L}, *Cerotoma* spp.^{C}	Seed feeders *Etiella zinckenella*^{L}, *Maruca testulalis*^{L}, *Laspeyresia fabivora*^{D}, *N. viridula*^{H}, *Piezodorus guildinii*^{H}, *Euschistus* spp.^{H}
	Stem and stalk borers *Elasmopalpus lignosellus*^{D}	Phloem feeders *Bemisia tabaci*^{H}	
Asia	Seedlings *D. platura*^{D}	Defoliators *Heliothis viriplaca*^{L}, *Matsumuraeses phasoli*^{L}, *Agrotis ipsilon*^{L}, *Melanagromyza* spp.^{D}, *Anomala rufocuprea*^{C}	Seed feeders *H. viriplaca*^{L}, *Etiella zinckenella*^{L}, *Hedylepta indicata*^{L}, *Piezodorus hybneri*^{H}, *Dolycoris baccarum*^{H}, *Riptortus clavatus*^{H}, *Asphondylia* spp.^{D}
	Phloem feeders *Aphis glycines*^{H}	Phloem feeders *A. glycines*^{H}, *Aulacorthum solani*^{H}	
	Stem and stalk borers *Melanagromyza* spp.^{D}		

A, Arachnid; C, Coleoptera (beetles); D, Diptera (fly); H, Hemiptera (aphids and true bugs); L, Lepidoptera (butterflies and moths).

the loss of leaf tissue. Efforts to quantify the effect of leaf tissue removal on yield have typically relied on manual leaf removal (Stone and Pedigo, 1972). This has aided the development of a model for predicting crop yields for soybean (CROPGRO, SOY-GRO). Timsina *et al.* (2007) observed that the model's predictions follow the empirical data that outlines a general pattern of soybean sensitivity to defoliation: low during vegetative growth, increasing until the beginning of seed growth and decreasing thereafter. Several factors can influence the impact of defoliation on soybean yield, including water stress (Li *et al.*, 2005), salinity stress (Li *et al.*, 2006) and competition with weeds (Grymes *et al.*, 1999), although the last interaction only contributes in an additive fashion to yield reduction. Efforts to estimate the impact of insect defoliation are often confounded by environmental interactions. Notably, Haile *et al.* (1998) observed no yield reductions with defoliation as high as 60% when ample soil moisture was available. Additionally, the sensitivity of soybean to defoliation may vary based on the phenology of the crop; that is, when the sensitive periods of soybean growth occur in relationship to the occurrence of defoliating insects. By altering the planting date or selecting earlier-maturing varieties, growers can alter the timing of key growth stages in the life of the plant, depending on when the impact of defoliation could be greatest (McPherson *et al.*, 1996; Malone *et al.*, 2002) and to what community of herbivores the plants are exposed (Lourencao *et al.*, 2000; McPherson *et al.*, 2001).

The impact of defoliation is not limited to loss of leaf area. In North America, several species of Coleoptera are vectors of *Bean pod mottle virus* (Mabry *et al.*, 2003; Werner *et al.*, 2003). The most efficient transmission of *Bean pod mottle virus* is by the bean leaf beetle (*Cerotoma trifurcata*) (Giesler *et al.*, 2002). The larval stages feed on the roots of soybean and the adults feed on leaves and pods, depending upon availability (Pedigo, 1994). Reductions in yield and seed quality are highest when transmission occurs in the early vegetative stages.

Phloem feeders

Insects that feed on phloem are generally found in the order Hemiptera and possess piercing–sucking mouthparts that are capable of entering vascular tissue. Both immature (nymph) and adult stages are capable of this feeding. Damage is often not visibly apparent to the same degree as defoliation. Some phloem feeders, such as *Spissistilus festinus* (three-cornered alfalfa hopper), can girdle the stems of plants, resulting in lodging. This can reduce yields, especially if it occurs late in the growing season. Kogan and Turnipseed (1987) identified *Bemisia tabaci* (sweet-potato whitefly) as a major or frequent phloem feeder of soybean. Within the native range of soybean, three aphid species – *Aulacorthum solani*, *Aphis craccivora*, and *Aphis glycines* – have been identified as soybean pests. As demonstrated by *A. glycines* outbreaks in North America, aphids can have a significant impact on soybean yield. One concern with phloem feeders such as *B. tabaci* and *A. glycines* is that the impact to the plant may go undetected without a close visual inspection. The visual impact to the plant from their feeding may only become apparent after large, yield-reducing populations have developed. Both *B. tabaci* and *A. glycines* are capable of exponential population growth during a growing season. Like most other phloem-feeding insects, aphids and whiteflies excrete honeydew (incompletely digested phloem), which contains a high concentration of sugars. When populations of these insects are large, so much honeydew is excreted that the leaf surface is darkened from the growth of mould growing on this sugary substance. Therefore, in addition to reducing yield by direct feeding on the plant, the build up of this sooty mould can further reduce photosynthesis (Macedo *et al.*, 2003).

Many phloem-feeding insects are efficient vectors for transmitting plant viruses. *B. tabaci* is a vector of several plant viruses (Jones, 2003; Varma and Malathi, 2003), as is *A. glycines* (Clark and Perry, 2002). Of particular concern is the potential for a combination of viruses from different vectors to act synergistically (Anjos *et al.*, 1992). As an additional complication, the transmission of soybean plant viruses does not require the presence of insects that colonize and feed for extended periods. The transmission of plant viruses can occur via insects that are transients within a soybean field (Perring *et al.*, 1999; Pedersen *et al.*, 2007), thus complicating the management of insect-vectored diseases. Because of the difficulty of keeping plants free of insects, the management of insect-vectored diseases is limited to HPR targeting the disease.

Seed and pod feeders

Another feeding guild that routinely causes yield loss is the pod and seed feeders. Many of the same orders of insects (Coleoptera, Lepidoptera) and species (*H. armigera, H. zea, C. trifurcata*) that feed on foliage also feed on pods and seeds. Some species feed almost entirely within pods during the immature stages (*Etiella zinckenella*), causing moderate yield loss (Van den Berg *et al.*, 1998a). The most widespread and significant pests of soybean during the later growth stages are hemipterans, specifically the pentatomid stinkbugs (Kogan and Turnipseed, 1987) such as *Nezara viridula* and *Piezodorus guildinii*. The most ubiquitous and possibly damaging is *N. viridula*, which is found in all soybean-producing areas of the world. Knight and Gurr (2007) reviewed the many management strategies targeting *N. viridula*. Like all hemipterans, *N. viridula* has a piercing–sucking mouthpart that can result in flower bud abscission, seed deformation and, if sufficiently extensive, abortion of pods. The relationship between yield reduction and stinkbug pod feeding is related to the growth stage of the plant, with later stages suffering greater yield reductions (Sinclair *et al.*, 1997). In addition to directly reducing yield, stinkbugs transmit yeast spot disease (Kimura *et al.*, 2008a,b).

Stem borers and root feeders

As with the seed- and pod-feeding guild, many of the same orders of insects (Coleoptera, Lepidoptera, Diptera) that feed on foliage also feed on stems. Within Asia, members of the leaf-miner fly family (Agromyzids, specifically the genus *Melanagromyza*) are thought to have a co-evolved relationship with soybean (Chiang and Norris, 1983). These stem-boring flies can interfere with growth (Talekar, 1989), but this damage does not often result in yield loss (Van den Berg *et al.*, 1998b). In general, the damage observed by growers to stem-feeding maybe at harvest when the weakened stems are prone to falling, often referred to as lodging, making harvesting difficult. Outside the native range of soybean, novel associations have formed between stem-boring insects and soybean. Such insects include *Elasmopalpus lignosellus* (the lesser cornstalk borer) and *Dectes texanus*. The latter has expanded its host range from sunflower (*Helianthus annuus*) to include soybean (Michaud and Grant, 2005). There are few known soybean obligate root feeders. Several generalist root herbivores, including various species of grubs (*Scarabaeidae* species), wireworms (*Elateridae* species) and the seedcorn maggot (*Delia platura*), have been reported to cause stand loss (Turnipseed and Kogan, 1976; Hammond, 1995), especially when soybean is planted in high-residue situations (mature hay stands, some no-till fields and following cover crops). The larval stages of *C. trifurcata* (bean leaf beetle) have been recorded feeding on the roots of soybean (Pedigo, 1994), although the impact of this feeding is not considered to be great.

14.3 Pest Management Principles

IPM programmes provide efficient and economical pest management (Kogan, 1998). A cost–benefit analysis (Poston *et al.*, 1983) is the essential foundation of any IPM programme (Stone and Pedigo, 1972). In order for the cost–benefit analysis to be effective it should include not only the control costs, costs associated with implementation and crop value, but also crop response to pest injury and the proportionate injury per individual pest (Poston *et al.*, 1983). Pedigo *et al.* (1986) defined injury as the physiological response of a plant to a pest and damage as the measurable injury caused by a pest. This relationship between injury and damage (yield) is often referred to as the damage-response curve (Fig. 14.1). Not all components of the damage curve are observable, but all are possible plant responses to injury. A key component in this relationship is the damage boundary. The damage boundary occurs where yield loss is first detectable. Determining the pest density where the damage boundary occurs is critical to determining if the injury caused by an arthropod is producing meaningful damage.

Once the damage-response curve has been characterized and effective sampling protocols for the pest in question have been developed, the economic injury level (EIL) and an economic threshold (ET) can be calculated. The EIL is defined as the lowest pest density that will cause economic damage (Stern *et al.*, 1959; Poston *et al.*, 1983; Pedigo *et al.*, 1986). The ET is defined as the lowest pest density that should trigger a management action to prevent the EIL being reached (Pedigo *et al.*, 1986; Pedigo and Rice, 2006).

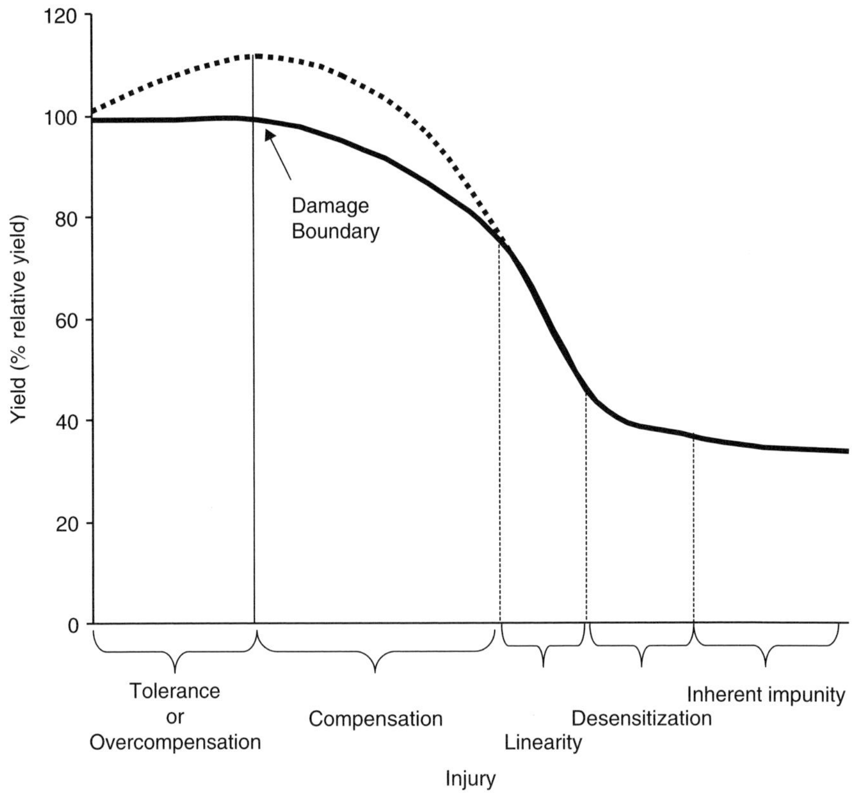

Fig. 14.1. The general relationship between increasing insect-derived injury and the relative yield, calculated as a percentage of yield in the presence of the insect divided by yield in its absence (i.e. the 'damage curve'). Major regions of the damage curve include the damage boundary (DB, the injury level at which yield loss is first detectable), tolerance (no damage per unit injury), overcompensation (negative damage per unit injury, noted with the dotted line), compensation (increasing damage per unit injury, this is where the DB is first crossed), linearity (constant damage per unit injury), desensitization (decreasing damage per unit injury) and inherent impunity (no additional damage per unit injury) (modified from Pedigo *et al.*, 1986; Peterson and Higley, 2001).

Pest management practices within an organic production context

Organic production, in general and of soybeans specifically, is a small but growing portion of agriculture. Many countries and trade organizations require producers to grow crops under defined conditions to label their crops as organic. Although the requirements for certification may vary by agency, a common constraint is the limitation on synthetic insecticides or genetically modified (GM) crops. Note that the use of non-transgenic HPR can be confounded by soybean varieties that contain a GM-based herbicide resistance. Therefore, even though the resistance to an insect pest may have been derived from within the soybean germplasm through conventional breeding technology, it would not be available to an organic grower. Beyond these two tools, all other pest management tactics are applicable for organic production. However, these limitations are not trivial and, as a result, organic growers must take a more holistic systems approach to managing insect pests.

In a review of arthropods pest management for organic crop production, Zehnder *et al.* (2007) outlined four phases of strategies (Table 14.2). These methods begin with cultural practices that will limit pest outbreaks. The second and third phases are biological control tactics (conservation, inundative and inoculative release of biocontrol agents). The fourth phase is the use of organic-approved behavioural modifiers and insecticides. The former includes repellents and attractants that are often based on insect-derived semiochemicals, such as sex pheromones. The latter cannot be of a synthetic origin. Despite this limitation a wide variety of insecticides is available, including those derived from plants (neem extract, pyrethrum), microbes (*Bacillus thuringiensis*, granulosis viruses) and fungal pathogens. As noted by Zehnder *et al.* (2007), these products typically cost more and are not as efficacious as synthetic, broad-spectrum insecticides. Part of this lack of efficacy is due to reduced residual activity when applied; many products quickly break down in sunlight. Growers who produce soybeans organically can still use the tactics and concepts that have been outlined in this chapter; however, they should be aware of the limitations based on the requirements of their certifying agency. It is interesting to note that a complaint of IPM is that it often inverts the four phases outlined by Zehnder *et al.* (2007), relying first

Table 14.2. Organization of arthropod pest management strategies from preventative to more direct tactics (adapted from Zehnder *et al.*, 2007).

Phase	Pest management tactic	Considerations for the organic grower
1	Cultural practices compatible with natural processes and ecosystem services, such as crop rotation, soil management and host plant resistance	Transgenic-based host plant resistance is prohibited (only non-GMO allowed)
2	Land and vegetation management to enhance natural enemy impact (i.e. conservation biological control)	This phase may be facilitated by altering buffers that separate organic from conventional farms with habitats that conserve the natural enemies of insect pests
3	Inundative and inoculative releases of biological control agents, use of mating disruption	–
4	Use of insecticides as pest densities reach threshold levels	Synthetic insecticides are prohibited, but insecticides of a biological or mineral origin are permitted

GMO, genetically modified organism.

on insecticides and then developing more ecologically based approaches to mitigate the subsequent ecological backlash that results from their overuse (Lewis *et al.*, 1997). Therefore, it is suggested that the approach outlined in Table 14.2 is not limited to organic growers, but referred to by all soybean growers. Note that many aspects of this approach are explored in the case study in the final section of this chapter.

Ecological backlash

Pest management practices can result in three forms of ecological backlash (Pedigo and Rice, 2006): resistance, resurgence and replacement. Resistance is defined as a heritable change in pest sensitivity to a control technique (Pedigo and Rice, 2006). Resistance to insecticides is likely the most significant of these due to both its frequency and implications for future pest management (Nauen *et al.*, 2001; Nauen and Denholm, 2005; Pedigo and Rice, 2006; Gorman *et al.*, 2008; Karunker *et al.*, 2008; Nauen *et al.*, 2008). The most common form of insecticide resistance is physiological (Nauen and Elbert, 2003; Pedigo and Rice, 2006; Karunker *et al.*, 2008; Nauen *et al.*, 2008), including detoxification through enzymatic actions and modifications that prevent insecticides from binding to the target site. Since the reproductive potential of herbivores is often greater than that of their natural enemies (predators and parasitoids) and the insecticides commonly used are toxic to both, the use of insecticides can often exacerbate a pest problem. The loss of biological control can lead to either resurgence of a target pest or replacement by a herbivore that was not considered a pest before the pest management programme was begun. Resurgence occurs following the implementation of a tactic when a pest population rebounds to a population higher than that before the tactic was used (Pedigo *et al.*, 1986; Bozsik, 2006). Replacement is when a primary pest's population is controlled, allowing a secondary pest that is not controlled to fill a now-vacant ecological niche that was once filled by the target pest or to escape mortality from natural enemies. In the future, soybean IPM programmes will include components to prevent ecological backlash before new tools, such as GM soybeans resistant to insects, are allowed to be marketed.

GM soybeans are commercially available (i.e. herbicide-resistant soybeans) and have been widely adopted by growers, with nearly 70% of worldwide hectares planted to herbicide-resistant soybeans (James, 2008). GM soybeans have been developed that are resistant to herbivores (Stewart *et al.*, 1996). However, GM soybeans that are resistant to insects are not yet commercially available. If these products are released then growers may need to adopt resistance management plans from other GM crops (e.g. corn and cotton; Tabashnik, 2008). To date, the strategy for resistance management of GM crops includes a high dose of the toxin coupled with a refuge (Gould, 1998). The combination of these tactics is thought to limit the occurrence of resistant alleles in a homozygous form and thus prevent the occurrence of a population of insects resistant to the GM-based trait.

Managing resurgence and replacement requires a systems approach, with an understanding of how the ecology of the soybean ecosystem is altered due to a given pest management tactic. In North America, the two-spotted spider mite (*Tetranychus urticae*) is rarely a pest unless abiotic conditions favour its development (dry conditions). However, the application of pyrethroids can lead to outbreaks even in unfavourable conditions due to the loss of mite predators. In North America (Johnson *et al.*, 2008) and Indonesia (Van den Berg *et al.*, 1997), the biological control of *A. glycines* can also be disrupted by broad-spectrum insecticides that remove predators. With the spread of Asian soybean rust (*Phakopsora pachyrhizi*) and the accompanying increase in insecticide and fungicide use, pest managers may exacerbate the pest status of pre-existing and invasive pests. Endemic fungal pathogens contribute to the limited pest status of some soybean arthropods in North America (such as *T. urticae*, see Klubertanz *et al.*, 1991); however, these pathogens may be negatively affected by the active ingredients in many commercially available fungicides (Latteur and Jansen, 2002).

The use of insecticides within the context of an IPM programme is thought to reduce the potential of an ecological backlash from occurring (Stern *et al.*, 1959). However, a robust IPM programme should incorporate multiple control tactics and vigilance in order to prevent target pests from experiencing strong selection pressures (Pedigo *et al.*, 1986; Kogan, 1998). The three main tactics for insect pest management that can be incorporated into an IPM scheme are reviewed below.

Insecticides

Insecticidal management of insect pests has and continues to be one of the most effective means of quickly reducing pest populations in soybean. Previous reviews of insect pest management have discussed the pyrethroid and organophosphate classes of chemistry (Turnipseed and Kogan, 1976; Kogan and Turnipseed, 1987). In recent years, several insecticides with new modes of action have come to the market with multiple benefits, including reduced human toxicity, increased efficacy, plant mobility and insect selectivity (Harrewijn and Kayser, 1997; Kunkel *et al.*, 1999; Elbert and Nauen, 2000; Mukherjee and Gopal, 2000; Bostanian *et al.*, 2001; Elzen, 2001; Sechser *et al.*, 2002; Koppenhofer *et al.*, 2003; Ako *et al.*, 2004; Ohnesorg *et al.*, 2009).

Insecticides capable of plant systemic translocation allow growers to manage insect pests such as *Dectes texanus* (soybean stem borer) and *Odontota horni* (soybean leafminer), in which life history traits minimize exposure to contact insecticides. Plant systemic insecticides may move through either or possibly both of the xylem (apoplastic movement) and phloem (symplastic movement). The available plant systemic insecticides registered for use on soybean primarily consist of two modes of action: nicotinic acetylcholine receptor agonists (neonicotinoids) and lipid synthesis inhibitors. However, this list is expected to grow quickly in the coming years. Neonicotinoids were first commercialized in the 1990s and were one of the first insecticides with a plant systemic mode of action, exhibiting apoplastic plant mobility. They may be applied either to the soybean seed at planting or as a foliar formulation (Mukherjee and Gopal, 2000; Elzen, 2001; Buchholz and Nauen, 2002; Weichel and Nauen, 2003). Common neonicotinoid insecticides include thiamethoxam, imidacloprid and clothianidin. However, the impact of a seed-applied insecticide is limited by how active it remains throughout the season. Neonicotinoid seed treatments are toxic against both leaf-feeding (bean leaf beetle) and phloem feeding insects (*A. glycines*, *B. tabaci*), and their impact on insect pests that colonize soybean later in the season is limited. For example, bean leaf beetles colonize soybean fields in North America as the plants emerge, and seed treatments have been very effective in reducing defoliation in the plant's early vegetative stages (Bradshaw *et al.*, 2008). As the plant matures and gains vegetative mass, however, the amount of toxin declines and pest management effectiveness decreases. In much of North America, soybean fields are not colonized by *A. glycines* until nearly 2 months after plants emerge, and the utility of a seed treatment is limited for this pest (McCornack and Ragsdale, 2006; Johnson *et al.*, 2008).

An ecological backlash in the form of resurgence and replacement is a major concern of any pest management programme (Stern *et al.*, 1959). Ecological backlash can be caused by insecticides that remove beneficial insects from the ecosystem (Stern *et al.*, 1959; Bozsik, 2006). Insecticides that have a limited impact on beneficial insecticides such as natural enemies and pollinators have long been sought. Systemic insecticides may limit resurgence and replacement, as the systemic activity reduces exposure to non-target organisms. *In vitro* assays have shown that neonicotinoids have a low degree of selectivity; however, when neonicotinoid insecticides are applied as seed treatments, non-target impacts are limited to insects that feed on treated plants (Maienfisch *et al.*, 2001; Nauen *et al.*, 2002, 2003; Bozsik, 2006). Other studies have demonstrated the efficacy of pest-specific modes of action that have a reduced effect on beneficial insect communities (Kraiss and Cullen, 2008; Ohnesorg *et al.*, 2009). Such insecticides, sometimes referred to as biopesticides or reduced-risk insecticides, that are effective against pests but have limited impact on natural enemies, may expand the durability of IPM programmes in soybean (for a more detailed example, see the Knight and Gurr 2007 review of *N. viridula* management in soybean).

Host plant resistance

Since 1978 there has been increasing interest in HPR for the management of insect pests. This includes both conventional and GM-derived resistance. Although one of the most successful breeding programmes for insect resistance in soybean targets a phloem feeder (potato leafhopper, *Empoasca fabae*), the majority have focused on leaf-feeding insects (specifically Lepidoptera and Coleoptera herbivores). Since the invasion of the soybean aphid in the USA, efforts to locate aphid resistance in soybean germplasm have resulted in several successes. As mentioned earlier, GM soybeans engineered for resistance against insect pests have been developed (Walker *et al.*, 2000). However, these are not yet commercially available. Therefore, only forms of resistance developed through conventional (non-GM) methods are reviewed in this chapter.

From the perspective of an entomologist, HPR against insects comes in three different forms: antixenosis, antibiosis and tolerance. The first, antixenosis, is the inability of an insect pest to find and feed on the plant. This can involve breeding for greater pubescence on leaves and stems, which 40 years ago resulted in a successful reduction in leafhopper injury to soybeans. *Empoasca fabae* (potato leafhopper) can greatly reduce the ability of soybean to grow in the USA. By increasing soybean pubescence, this once formidable pest was reduced to a non-issue (Metcalf and Luckmann, 1994).

Antixenosis has been observed in some forms of resistance that are currently being investigated by US soybean breeders for *A. glycines* (Diaz-Montano *et al.*, 2006; Hesler *et al.*, 2007). Complicating this is the presence of the second form of resistance, antibiosis, which is the inability for the pest to grow and reproduce while feeding on a resistant plant. Evidence for antibiosis has been reported in soybean germplasm by several groups of soybean breeders (Mensah *et al.*, 2005; Hill *et al.*, 2006a; Hesler *et al.*, 2007). When aphids are placed on these plants they produce fewer offspring; it is not clear if resistant plants produce a toxin or are just less nutritious for aphids than susceptible soybean. A source of antibiosis in soybean is attributed to a single dominant gene (Hill *et al.*, 2006a,b). Beginning in 2009, this gene (*Rag1*) has been made commercially available on a limited basis in North America. Interestingly, a biotype of soybean aphid that is capable of surviving on *Rag1*-containing soybean was discovered in North America before the commercial release of this resistance in North America (Kim *et al.*, 2008).

The last form of resistance is tolerance – the ability of a plant to produce high yields despite the feeding of an insect pest. Tolerance is difficult to test in the laboratory because tolerant plants will continue to support large insect populations. Screening for this form of resistance is, therefore, not possible within greenhouses where plants are not taken to yield for verification. Some researchers have found pseudo-tolerance to defoliation, as soybean planted in narrow rows was able to tolerate greater leaf area removal than plants in wide rows (Hammond *et al.*, 2000).

Biological control

Biological control is the intentional manipulation of predators, parasitoids and pathogens to suppress the population of a specific pest, resulting in reduced abundance or damage. The trio of predators, parasitoids and pathogens is commonly referred to as 'natural enemies'. There are three approaches to employing natural enemies in a biological control programme: importation, augmentation and conservation. As an annual crop, a soybean field is an ephemeral habitat that is often difficult for natural enemies to colonize and establish with enough time to limit herbivore populations from building to economically damaging levels. Therefore, it is not appropriate to expect a similar level of pest control from biological control programmes as compared to insecticide-based programmes. Rather, biological control programmes can delay pest outbreaks or shift outbreaks to periods when crop susceptibility is less (Wiedemann and Smith, 1997). Despite this inherent limitation, biological control programmes have been successful. A very brief review of the three methods and their relevance for soybean pest management is given below.

Possibly the most-studied approach is importation biological control, in which an exotic (i.e. non-native) natural enemy is released against an invasive pest. This is sometimes called classical biological control because one of the first documented successes using this approach occurred in 1888 with the release of the vedalia beetle (*Rodolia cardinalis*) in the US state of California for the management of cottony cushion scale (*Icerya purchasi*) on citrus. Typically, classical biological control programmes require an understanding of the origin of the insect pest, identifying natural enemies that contribute to its mortality and releasing them in exotic areas where it is a pest (Hajek, 2004). The natural-enemy-release hypothesis is often invoked to justify this approach. This hypothesis proposes that it is the absence of natural enemies from the native range of the exotic herbivore that allow it to exceed the population levels (carrying capacity) typically observed in the native range. The goal of a classical biological control programme is the prolonged suppression of the exotic herbivore, such that the need for future management practices (i.e. chemical inputs) is reduced because the carrying capacity of the pest is reduced. Such programmes are not without a certain degree of risk, and the released natural enemy may fail to establish and impact the target pest. An additional risk associated with this approach is the potential attack of native insects by the exotic natural enemy (Louda *et al.*, 2003).

Several classical biological control programmes have been conducted in soybeans (Kogan and Turnipseed, 1987). Recently, a classical biological control programme targeting the soybean aphid in North America has resulted in the release of a parasitoid wasp (*Binodoxys communis*).

A more interactive approach to biological control is the release of commercially raised insect predators, often called augmentation or inoculation biological control. Such programmes involve the release of native or endemic natural enemies, and require repeated releases to ensure the natural enemy is sufficiently present to have an impact on pests. The use of baculoviruses for the control of the lepidopteran leaf feeder (*Anticarsia gemmatalis*) in Brazil (Moscardi, 1999) is an example where repeated inoculations of a natural enemy have provided significant control, resulting in a reduction in insecticide use. Although there are several commercial sources of predators (such as ladybirds/ladybeetles) that feed on soybean insect pests such as aphids, there is no evidence that purchasing these predators for release in soybean fields has any impact on soybean aphid populations. This approach is not currently recommended. As demonstrated in cotton, it is very difficult to release enough predators to increase the background population of predators. Furthermore, annual cropping systems such as soybean may be a sink for natural enemies that themselves become victims to predation through intra-guild predation (Rosenheim *et al.*, 1993).

Of the three approaches to biological control, Conservation may be the least studied (Hajek, 2004). However, there is a growing literature on how the use of conservation and habitat management techniques can enhance the impact of currently occurring natural enemies (Landis *et al.*, 2000). As many have noted, land-use practices within and around agricultural fields affect the ability of natural enemies to colonize and maintain a sufficient carrying capacity to reduce herbivore colonization and population growth (Thies *et al.*, 2005; Brewer *et al.*, 2008). The spectrum of conservation practices that promote greater biological control of existing natural enemies include using ET and selective insecticides that have a reduced impact on natural enemies (Stern *et al.*, 1959; Ohnesorg *et al.*, 2009), incorporating within-field refuges (cover crops and living mulches) from agricultural disturbances (e.g. Schmidt *et al.*, 2007) and developing extra-field habitats that can improve the biodiversity and abundance of natural enemies (Hickman and Wratten, 1996). Although soybean growers may be unable to apply such a practice within their farms, the impact of land use surrounding a soybean field can help to inform farmers about the risk of insect pest outbreaks to their fields (Gardiner *et al.* 2009a).

14.4 Combining Multiple Management Tools for Effective Pest Management: a Case Study of *Aphis glycines* in North America

Soybean was first domesticated in China and many of the insects that co-evolved with the plant did not travel to the regions that currently produce the majority of soybeans (e.g. USA, Brazil and Argentina). Therefore, the majority of the insects that colonize soybean have done so through novel associations.

Kogan and Turnipseed (1987) noted: 'Entire guilds are missing from the Western Hemisphere: foremost among them are such soybean-colonizing aphids as the soybean specialist *A. glycines*, a common pest in East Asia.' Recently, that niche has been filled as populations of *A. glycines* have established outside of China, in Indonesia, Australia and North America (Van den Berg *et al.*, 1997; Venette and Ragsdale, 2004; Brier *et al.*, 2008). In North America, the invasion of *A. glycines* has altered the pest management practices of soybean growers, especially in the leading soybean-producing states (Iowa, Minnesota and Illinois). Soybean had received very few, if any, insecticide applications (Fernandez-Cornejo and Jans, 1999).With the arrival of *A. glycines* and its rapid spread across the north-central region of the USA, however, growers have dramatically increased the use of insecticides over an area ranging from 4 to 14 million ha year^{-1} (M.E. O'Neal, unpublished data). Although *A. glycines* was noted as an important pest of soybean in its native range by Kogan and Turnipseed (1987), its impact in the USA has been markedly greater than that within its native range (Liu *et al.*, 2004; Ragsdale *et al.*, 2004; Wu *et al.*, 2004). It is not surprising that invasive species, such as *A. glycines*, that escape mortality factors from their native range have remarkably different population levels within exotic ranges (Elton, 2000). *A. glycines* is not the only invasive species to attack soybean. For example, *B. tabaci* is established in many soybean-growing regions and

IPM for this well-studied pest (Byrne and Bellows, 1991) has been recently reviewed (Brier *et al.*, 2008; Takahshi *et al.*, 2008) for soybean. *B. tabaci* is particularly difficult to manage as it has quickly developed resistance to several insecticide classes, making true IPM (i.e. the incorporation of multiple tactics) necessary for its successful management (Naranjo, 2001). The challenge for developing management plans for *A. glycines*, then, is to produce recommendations that do not result in insecticide resistance.

The arrival and spread of *A. glycines* across the north-central region of the USA was rapid, with the first observations made during the summer of 2000. Since very few, if any, native aphids feed on soybeans, the spread of this invasive soybean specialist was easily documented. Within 4 years of its discovery in the USA, *A. glycines* had spread across 22 US states and three Canadian provinces (Venette and Ragsdale, 2004). Its establishment across a large area of the US soybean-growing region is facilitated by the previous establishment and spread of its overwintering host *Rhamnus cathartica* (Ragsdale *et al.*, 2004). Field studies of soybean have revealed a diverse community of natural enemies (Rutledge *et al.*, 2004; Nielsen and Hajek, 2005; Mignault *et al.*, 2006). The community of insect predators within soybean has apparently been altered by the arrival of the aphid to include more species that are noted aphid-feeding specialists (Schmidt *et al.*, 2008). Included in this community is *Harmonia axyridis*, an Asian coccinellid that has been observed to contribute to the biological control of *A. glycines* in Indonesia (Van den Berg *et al.*, 1997).

With the invasion of *A. glycines* into North America, it has been noted that this aphid is not considered a pest in its native range (Wu *et al.*, 2004). Liu *et al.* (2004) revealed that a community of natural enemies that include both predators and parasitoids suppresses *A. glycines* populations in China. Surveys of the natural enemy community within North American soybean reveal the presence of some members of the predatory community (*Harmonia axyridis*, Orius, syrphids), but an absence of these parasitoids (Rutledge *et al.*, 2004; Schmidt *et al.*, 2008). After several years of host range studies that indicated a high degree of specificity for *A. glycines* (see Wyckhuys and Heimpel, 2007, but also Wyckhuys *et al.*, 2007 for a description of how ecological factors may further limit the attack of an exotic parasitoid on native aphids), federal approval was received for the release of a parasitoid wasp in the US in 2007. Releases of this wasp are ongoing; the impact on *A. glycines* has not been measured and is unlikely until populations of the parasitoid have been established within the new range. Additional candidate wasps are in quarantine and maybe released in the near future (Heimpel *et al.*, 2004).

Despite the absence of parasitoid wasps in North America (Kaiser *et al.*, 2007; Noma and Brewer, 2008; Schmidt *et al.*, 2008), several field studies have revealed that the existing community of natural enemies contributes to the suppression of soybean aphid populations (Fox *et al.*, 2004; Rutledge and O'Neil, 2005; Costamagna and Landis, 2006, 2007) in soybean. In the absence of predation, soybean aphid population growth is significantly faster (two to seven times). Despite the community of natural enemies currently found in the soybean fields of North America, economic outbreaks of

A. glycines continue to occur. Gardiner *et al.* (2009a) observed that a contributing factor to the occurrence of these outbreaks is the land-use patterns around soybean fields. As the amount of non-crop, perennial habitat increases within a 2.5-km region around a soybean field, the risk of aphid outbreaks increases. This risk is mitigated by natural enemies, predominately the coccinellid community and specifically *H. axyridis* (Gardiner *et al.*, 2009b), which may require an overwintering habitat and alternative prey provided by arboreal habitats (Koch and Galvan, 2008). Although temperature-based models for predicting *A. glycines* population growth exist (McCornack *et al.*, 2004, 2005), they currently do not incorporate mortality due to predation. Improvements in predicting temperature-based models of *A. glycines* population growth, which currently overestimate population growth (M.E. O'Neal, 2008, personal observation), require a component that incorporates natural-enemy-derived mortality. To fully explain the potential of the natural enemy community to suppress *A. glycines*, such a model may need to be spatially explicit, such that it factors in the negative impact that a landscape dominated by annual crop production may have on this ecosystem service (Landis *et al.*, 2008).

Given the impact of predation on *A. glycines*, the development of EIL and ET have been calculated with naturally occurring *A. glycines* populations (Ragsdale *et al.*, 2007). In this way, the subsequent recommendations for insecticide use incorporated the impact of existing natural enemies. Growers are currently recommended to scout fields during the summer months, during which aphids colonize soybean fields, applying a foliar insecticide when populations exceed 250 aphids per plant during the early to mid reproductive stages of soybean. Sampling protocols that aid in the rapid assessment of *A. glycines* populations have been developed based on this ET (Hodgson *et al.*, 2007). This recommendation (scout and apply insecticide only when populations reach the ET) has been shown to be more cost-effective than a preventative approach that applies insecticide based on the growth stage of the plant, regardless of the population density of the aphid (O'Neal, 2008, unpublished data). Due in part to efforts by university scientists and soybean-grower associations, most growers (70%) in the north-central region of the USA are aware of and use these recommendations (Olson *et al.*, 2007).

14.5 Summary: Opportunities for Sustainable Production

In summary, both emerging (i.e. *A. glycines*) and established (i.e. *N. viridula*, *Anticarsia gemmatalis*) pests can potentially be managed with IPM that incorporates multiple tools. Although the potential utility of these tools has not been fully realized, IPM for soybean production can draw from multiple tools to develop a robust, sustainable programme. The challenge for any given pest management programme will be the incorporation of novel approaches into pre-existing practices. Using the example of *A. glycines*, one can see how advances in HPR and importation biological control will have to address how the current use of broad-spectrum insecticides will limit

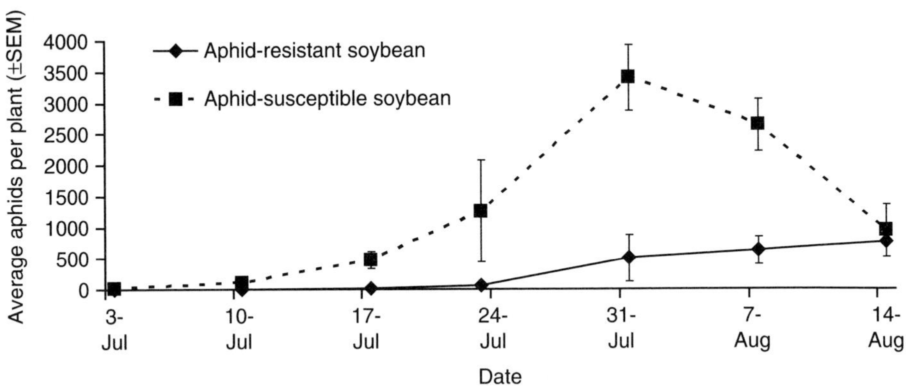

Fig. 14.2. Comparison of a soybean variety that contains the *Rag1* gene, which confers antibiosis against *A. glycines* (aphid-resistant), and a parental line that does not have this gene (aphid-susceptible). Both varieties were grown in replicated plots in which half of the plot was kept free from aphids using a foliage-applied insecticide (insecticide) and the other half was left untreated (no insecticide). The data presented here are from the untreated section. Twenty plants per plot were sampled and the number of *A. glycines* was counted on each. This experiment was conducted in Ames, Iowa, during 2007. Different letters indicate statistically significant differences in yield ($P = 0.05$) (O'Neal, 2008, unpublished data).
SEM, standard error of the mean.

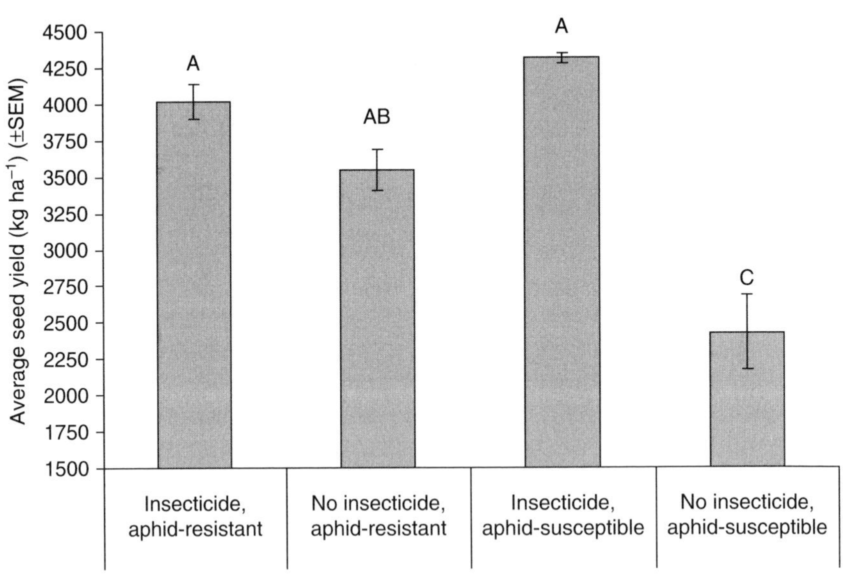

Fig. 14.3. Comparison of a soybean variety that contains the *Rag1* gene, which confers antibiosis against *A. glycines* (aphid-resistant), and parental line that does not have this gene (aphid-susceptible). Both varieties were grown in replicated plots in which half of the plot was kept free from aphids using a foliage-applied insecticide (insecticide) and the other half was left untreated (no insecticide). This experiment was conducted in Ames, Iowa, during 2007, when *A. glycines* populations exceeded the ET. Different letters indicate statistically significant differences in yield ($P = 0.05$) (O'Neal, 2008, unpublished data).
SEM, standard error of the mean.

their effectiveness. As the *Rag1* gene becomes commercially available, it will not result in plants that are aphid-free (Fig. 14.2); *Rag1*-containing plants are likely to require additional pest management tools for the optimal yield to be produced (Fig. 14.3). ETs and EILs will have to be recalculated based on the reduced capacity of *A. glycines* to grow on these plants. As additional pest management tools become available to soybean growers, there will be a continued need for researchers to determine the best approach for integration of and subsequent communication of these tools to growers.

Clearly there is room for improvement on the current status of the pest management tools outlined in this brief review. As noted above, conventionally developed HPR for soybean aphids has yet to produce a variety that is free from aphids to the same degree as GM crops (e.g. Bt-corn). To date, the use of GM soybeans for insect pest management has been limited. However, the overuse of any pest management tool will result in ecological backlash (resistance, resurgence or replacement) that can limit the usefulness of that tool. The integrated use of the main tools of IPM has the capacity to prevent soybean yield loss. However, the challenge will be to develop recommendations that employ these tools sustainably.

References

Ako, M., Borgemeister, C., Poehling, H.M., Elbert, A. and Nauen, R. (2004) Effects of neonicotinoid insecticides on the bionomics of twospotted spider mite (Acari: Tetranychidae). *Journal of Economic Entomology* 97, 1587–1594.

Anjos, R.J., Jarlfors, U. and Ghabrial, S.A. (1992) Soybean mosaic potyvirus enhances the titer of two comoviruses in dually infected soybean plants. *Phytopathology* 82, 1022–1027.

Bostanian, N.J., Larocque, N., Chouinard, G. and Coderre, D. (2001) Baseline toxicity of several pesticides to *Hyaliodes vitripennis* (Say) (Hemiptera: Miridae). *Pest Management Science* 57, 1007–1010.

Bozsik, A. (2006) Susceptibility of adult *Coccinelia septempunctata* (Coleoptera: Coccinellidae) to insecticides with different modes of action. *Pest Management Science* 62, 651–654.

Bradshaw, J.D., Rice, M.E. and Hill, J.H. (2008) Evaluation of management strategies for bean leaf beetles (Coleoptera: Chrysomelidae) and *Bean pod mottle virus* (Comoviridae) in soybean. *Journal of Economic Entomology* 101, 1211–1227.

Brewer, M.J., Noma, T., Elliott, N.C., Kravchenko, A.N. and Hild, A.L. (2008) A landscape view of cereal aphid parasitoid dynamics reveals sensitivity to farm- and region-scale vegetation sructure. *European Journal of Entomology* 105, 503–511.

Brier, H.B., Murray, D.A.H., Wilson, L.J., Nicholas, A.H., Miles, M.M., Grundy, P.R. and McLennan, A.J. (2008) An overview of integrated pest management (IPM) in north-eastern Australian grain farming systems: Past, present and future prospects. *Australian Journal of Experimental Agriculture* 48, 1574–1593.

Buchholz, A. and Nauen, R. (2002) Translocation and translaminar bioavailability of two neonicotinoid insecticides after foliar application to cabbage and cotton. *Pest Management Science* 58, 10–16.

Byrne, D.N. and Bellows, T.S. (1991) Whitefly biology. *Annual Review of Entomology* 36, 431–457.

Chiang, H.S. and Norris, D.M. (1983) Physiological and anatomical stem parameters of soybean resistance to agromyzid beanflies. *Entomologia Experimentalis et Applicata* 33, 203–212.
Clark, A.J. and Perry, K.L. (2002) Transmissibility of field isolates of soybean viruses by *Aphis glycines*. *Plant Disease* 86, 1219–1222.
Costamagna, A.C. and Landis, D.A. (2006) Predators exert top-down control of soybean aphid across a gradient of agricultural management systems. *Ecological Applications* 16, 1619–1628.
Costamagna, A.C. and Landis, D.A. (2007) Quantifying predation on soybean aphid through direct field observation. *Biological Control* 42, 16–24.
DeGooyer, T.A. and Browde, J.A. (1994) Grasshoppers. In: Higley, L.G. and Boethel, D.J. (eds) *Handbook of Soybean Insect Pests*. Entomological Society of America, Lanham, MD, USA, pp. 57–59.
Diaz-Montano, J., Reese, J.C., Schapaugh, W.T. and Campbell, L.R. (2006) Characterization of antibiosis and antixenosis to the soybean aphid (Hemiptera: Aphididae) in several genotypes. *Journal of Economic Entomology* 99, 1884–1889.
Elbert, A. and Nauen, R. (2000) Resistance of *Bemisia tabaci* (Homoptera: Aleyrodidae) to insecticides in southern Spain with special reference to neonicotinoids. *Pest Management Science* 56, 60–64.
Elton, C.S. (2000) *The Ecology of Invasions by Animals and Plants*. University of Chicago Press, Chicago, IL, USA.
Elzen, G.W. (2001) Lethal and sublethal effects of insecticide residues on *Orius insidiosus* (Hemiptera: anthocoridae) and *Geocoris punctipes* (Hemiptera: lygaeidae). *Journal of Economic Entomology* 94, 55–59.
Fernandez-Cornejo, J. and Jans, S. (1999) Pest management in U.S. agriculture. Agriculture Handbook No. 717. Resource Economics Division, U.S. Department of Agriculture Economic Research Service, USA.
Fox, T.B., Landis, D.A., Cardoso, F.F. and DiFonzo, C.D. (2004) Predators suppress *Aphis glycines* matsumura population growth in soybean. *Environmental Entomology* 33, 608–618.
Gardiner, M.M., Landis, D.A., Gratton, C., DiFonzo, C.D., O'Neal, M.E., Chacon, J., Wayo, M.T., Schmidt, N.P., Mueller, E.E. and Heimpel, G.E. (2009a) Landscape diversity impacts biocontrol services in north-central U.S. soybean. *Ecological Applications* 9, 143–154.
Gardiner, M.M., Landis, D.A., Gratton, C., Schmidt, N.P., O'Neal, M.E., Mueller, E., Chacon, J., Heimpel, G.E. and DiFonzo, C.D. (2009b) Landscape composition mediates coccinellid community structure. *Diversity and Distributions* 15, 554–564.
Giesler, L.J., Ghabrial, S.A., Hunt, T.E. and Hill, J.H. (2002) *Bean pod mottle virus*: A threat to US soybean production. *Plant Disease* 86, 1280–1289.
Gorman, K., Liu, Z., Denholm, I., Bruggen, K.U. and Nauen, R. (2008) Neonicotinoid resistance in rice brown planthopper, *Nilaparvata lugens*. *Pest Management Science* 64, 1122–1125.
Gould, F. (1998) Sustainability of transgenic insecticidal cultivars: Integrating pest genetics and ecology. *Annual Review of Entomology* 43, 701–726.
Grymes, C.F., Griffin, J.L., Boethel, D.J., Leonard, B.R., Jordan, D.L. and Russin, J.S. (1999) Soybean response to weed interference and defoliation. *Weed Science* 47, 90–94.
Haile, F.J., Higley, L.G. and Specht, J.E. (1998) Soybean cultivars and insect defoliation: Yield loss and economic injury levels. *Agronomy Journal* 90, 344–352.
Hajek, A. (2004) *Natural Enemies: An Introduction to Biological Control*. Cambridge University Press, Cambridge, UK.

Hammond, R.B. (1995) Timing of plowing and planting: Effects on seedcorn maggot populations in soybean. *Crop Protection* 14, 471–477.

Hammond, R.B., Higley, L.G., Pedigo, L.P., Bledsoe, L., Spomer, S.M. and DeGooyer, T.A. (2000) Simulated insect defoliation on soybean: Influence of row width. *Journal of Economic Entomology* 93, 1429–1436.

Harrewijn, P. and Kayser, H. (1997) Pymetrozine, a fast-acting and selective inhibitor of aphid feeding. *In-situ* studies with electronic monitoring of feeding behaviour. *Pesticide Science* 49, 130–140.

Heimpel, G.E., Ragsdale, D.W., Venette, R., Hopper, K.R., O'Neil, R.J., Rutledge, C.E. and Wu, Z.S. (2004) Prospects for importation biological control of the soybean aphid: Anticipating potential costs and benefits. *Annals of the Entomological Society of America* 97, 249–258.

Hesler, L.S., Dashiell, K.E. and Lundgren, J.G. (2007) Characterization of resistance to *Aphis glycines* in soybean accessions. *Euphytica* 154, 91–99.

Hickman, J.M. and Wratten, S.D. (1996) Use of *Phacelia tanacetifolia* strips to enhance biological control of aphids by hoverfly larvae in cereal fields. *Journal of Economic Entomology* 89, 832–840.

Higley, L.G. and Boethel, D.J. (1994) *Handbook of Soybean Insect Pests*. Entomological Society of America, Lanham, MD, USA.

Hill, C.B., Li, Y. and Hartman, G.L. (2006a) Soybean aphid resistance in soybean Jackson is controlled by a single dominant gene. *Crop Science* 46, 1606–1608.

Hill, C.B., Li, Y. and Hartman, G.L. (2006b) A single dominant gene for resistance to the soybean aphid in the soybean cultivar Dowling. *Crop Science* 46, 1601–1605.

Hodgson, E.W., McCornack, B.P., Koch, K.A., Ragsdale, D.W., Johnson, K.D., O'Neal, M.E., Cullen, E.M., Kraiss, H.J., DiFonzo, C.D. and Behnken, L.M. (2007) Field validation of speed scouting for soybean aphid. *Crop Mangement*. Available at: http://www.plantmanagementnetwork.org/pub/cm/research/2007/speed (accessed 29 June 2009).

James, C. (2008) *Global Status of Commercialized Biotech/GM Crops: 2008*. ISAAA Brief No. 39. ISAAA, Ithaca, NY, USA.

Johnson, K.D., O'Neal, M.E., Bradshaw, J.D. and Rice, M.E. (2008) Is preventative, concurrent-management of the soybean aphid (Hemiptera: Aphididae) and bean leaf beetle (Coleoptera: Chrysomelidae) possible? *Journal of Economic Entomology* 101, 801–809.

Jones, D.R. (2003) Plant viruses transmitted by whiteflies. *European Journal of Plant Pathology* 109, 195–219.

Kaiser, M.E., Noma, T., Brewer, M.J., Pike, K.S., Vockeroth, J.R. and Gaimari, S.D. (2007) Hymenopteran parasitoids and dipteran predators found using soybean aphid after its midwestern United States invasion. *Annals of the Entomological Society of America* 100, 196–205.

Karunker, I., Benting, J., Lueke, B., Ponge, T., Nauen, R., Roditakis, E., Vontas, J., Gorman, K., Denholm, I. and Morin, S. (2008) Over-expression of cytochrome P450 CYP6CM1 is associated with high resistance to imidacloprid in the B and Q biotypes of *Bemisia tabaci* (Hemiptera: Aleyrodidae). *Insect Biochemistry and Molecular Biology* 38, 634–644.

Kim, S.K., Hill, C.B., Hartman, G.L., Rouf-Mian, M.A. and Diers, B.W. (2008) Discovery of soybean aphid biotypes. *Crop Science* 48, 923–928.

Kimura, S., Tokumaru, S. and Kuge, K. (2008a) *Eremothecium ashbyi* causes soybean yeast-spot and is associated with stink bug, *Riptortus clavatus*. *Journal of General Plant Pathology* 74, 275–280.

Kimura, S., Tokumaru, S. and Kikuchi, A. (2008b) Carrying and transmission of *Eromothecium coryli* (Peglion) Kurtzman as a causal pathogen of yeast-spot diesase in soybeans by *Riptortus clavatus* (Thunberg), *Nezara antennata* Scott, *Piezodorus hybneri* (Gmelin) and *Dolycoris baccarum* (Linnaeus). *Japanese Journal of Applied Entomology and Zoology* 52, 13–18.

Klubertanz, T.H., Pedigo, L.P. and Carlson, R.E. (1991) Impact of fungal epizootics on the biology and mangement of the 2-spotted spider-mite (Acari, Tetranychidae) in soybean. *Environmental Entomology* 20, 731–735.

Knight, K.M.M. and Gurr, G.M. (2007) Review of *Nezara viridula* (L.) management strategies and potential for IPM in field crops with emphasis on Australia. *Crop Protection* 26, 1–10.

Koch, R.L. and Galvan, T.L. (2008) Bad side of a good beetle: The North American experience with *Harmonia axyridis*. *Biocontrol* 53, 23–35.

Kogan, M. (1998) Integrated pest management: Historical perspectives and contemporary developments. *Annual Review of Entomology* 43, 243–270.

Kogan, M. and Turnipseed, S.G. (1987) Ecology and management of soybean arthropods. *Annual Review of Entomology* 32, 507–538.

Koppenhofer, A.M., Cowles, R.S., Cowles, E.A., Fuzy, E.M. and Kaya, H.K. (2003) Effect of neonicotinoid synergists on entomopathogenic nematode fitness. *Entomologia Experimentalis et Applicata* 106, 7–18.

Kraiss, H. and Cullen, E.M. (2008) Efficacy and nontarget effects of reduced-risk insecticides on *Aphis glycines* (Hemiptera: Aphididae) and its biological control agent *Harmonia axyridis* (Coleoptera: Coccinellidae). *Journal of Economic Entomology* 101, 391–398.

Kunkel, B.A., Held, D.W. and Potter, D.A. (1999) Impact of halofenozide, imidacloprid, and bendiocarb on beneficial invertebrates and predatory activity in turfgrass. *Journal of Economic Entomology* 92, 922–930.

Landis, D., Wratten, S. and Gurr, G. (2000) Habitat management to conserve natural enemies of arthropod pests in agriculture. *Annual Review of Entomology* 45, 175–201.

Landis, D.A., Gardiner, M.M., van der Werf, W. and Swinton, S.M. (2008) Increasing corn for biofuel production reduces biocontrol services in agricultural landscapes. *Proceedings of the National Academy of Sciences of the United Sates of America* 105, 20552–20557.

Latteur, G. and Jansen, J.P. (2002) Effects of 20 fungicides on the infectivity of conidia of the aphid entomopathgenic fungus *Erynia neoaphidis*. *BioControl* 47, 435–444.

Lewis, W.J., van Lenteren, J.C., Phatak, S.C. and Tumlinson, J.H. (1997) A total system approach to sustainable pest management. In: *Proceedings of the National Academy of Sciences of the United States of America* 94, 12243–12248.

Li, X.J., An, P., Eneji, A.E., Hamamura, K., Lux, A. and Inanaga, S. (2005) Growth and yield responses of two soybean cultivars to defoliation and water stress. *Biologia* 60, 467–472.

Li, X.J., An, P., Inanaga, S., Eneji, A.E. and Tanabe, K. (2006) Salinity and defoliation effects on soybean growth. *Journal of Plant Nutrition* 29, 1499–1508.

Liu, J., Wu, K.M., Hopper, K.R. and Zhao, K.J. (2004) Population dynamics of *Aphis glycines* (Homoptera: Aphididae) and its natural enemies in soybean in northern China. *Annals of the Entomological Society of America* 97, 235–239.

Louda, S.M., Pemberton, R.W., Johnson, M.T. and Follett, P.A. (2003) Nontarget effects – the Achilles' heel of biological control? Retrospective analyses to reduce risk associated with biocontrol introductions. *Annual Review of Entomology* 48, 365–396.

Lourencao, A.L., Pereira, J.C.V.N.A., De Miranda, M.A.C. and Ambrosano, G.M.B. (2000) Evaluation of damage to early maturity soybean cultivars and lines caused by stink bugs and caterpillars. *Pesquisa Agropecuaria Brasileira* 35, 879–886.

Mabry, T.R., Hobbs, H.A., Steinlage, T.A., Johnson, B.B., Pedersen, W.L., Spencer, J.L., Levine, E., Isard, S.A., Domier, L.L. and Hartman, G.L. (2003) Distribution of leaf-feeding beetles and *Bean pod mottle virus* (BPMV) in Illinois and transmission of BPMV in soybean. *Plant Disease* 87, 1221–1225.

Macedo, T.B., Bastos, C.S., Higley, L.G., Ostlie, K.R. and Madhavan, S. (2003) Photosynthetic responses of soybean to soybean aphid (Homoptera: Aphididae) injury. *Journal of Economic Entomology* 96, 188–193.

Maienfisch, P., Angst, M., Brandl, F., Fischer, W., Hofer, D., Kayser, H., Kobel, W. Rindlisbacher, A., Senn, R., Steinemann, A. and Widmer, H. (2001) Chemistry and biology of thiamethoxam: A second generation neonicotinoid. *Pest Management Science* 57, 906–913.

Malone, S., Herbert, D.A. and Holshouser, D.L. (2002) Relationship between leaf area index and yield in double-crop and full-season soybean systems. *Journal of Economic Entomology* 95, 945–951.

McCornack, B.P. and Ragsdale, D.W. (2006) Efficacy of thiamethoxam to suppress soybean aphid populations in Minnesota soybean. *Crop Mangement.* Available at: http://www.plantmanagementnetwork.org/pub/cm/research/2006/aphid (accessed 29 June 2009).

McCornack, B.P., Ragsdale, D.W. and Venette, R.C. (2004) Demography of soybean aphid (Homoptera: Aphididae) at summer temperatures. *Journal of Economic Entomology* 97, 854–861.

McCornack, B.P., Carrillo, M.A., Venette, R.C. and Ragsdale, D.W. (2005) Physiological constraints on the overwintering potential of the soybean aphid (Homoptera: Aphididae). *Environmental Entomology* 34, 235–240.

McPherson, R.M., Ruberson, J.R., Hudson, R.D. and Jones, D.C. (1996) Soybean maturity group and incidence of velvetbean caterpillars (Lepidoptera: Noctuidae) and Mexican bean beetles (Coleoptera: Coccinellidae). *Journal of Economic Entomology* 89, 1601–1607.

McPherson, R.M., Wells, M.L. and Bundy, C.S. (2001) Impact of the early soybean production system on arthropod pest populations in Georgia. *Environmental Entomology* 30, 76–81.

Mensah, C., DiFonzo, C.D., Nelson, R.L. and Wang, D.C. (2005) Resistance to soybean aphid in early maturing soybean germplasm. *Crop Science* 45, 2228–2233.

Metcalf, R.L. and Luckmann, W.H. (1994) *Introduction to Insect Pest Management*, 3rd edn. Wiley & Sons, New York, USA.

Michaud, J.P. and Grant, A.K. (2005) The biology and behavior of the longhorned beetle, *Dectes texanus* on sunflower and soybean. *Journal of Insect Science* 5, 1–15.

Mignault, M.P., Roy, M. and Brodeur, J. (2006) Soybean aphid predators in Quebec and the suitability of *Aphis glycines* as prey for three Coccinellidae. *BioControl* 51, 89–106.

Moscardi, F. (1999) Assessment of the application of baculoviruses for control of lepidoptera. *Annual Review of Entomology* 44, 257–289.

Mukherjee, I. and Gopal, M. (2000) Environmental behaviour and translocation of imidacloprid in eggplant, cabbage and mustard. *Pest Management Science* 56, 932–936.

Naranjo, S.E. (2001) Conservation and evaluation of natural enemies in IPM systems for *Bemisia tabaci*. *Crop Protection* 20, 835–852.

Nauen, R. and Denholm, I. (2005) Resistance of insect pests to neonicotinoid insecticides: Current status and future prospects. *Archvies of Insect Biochemistry and Physiology* 58, 200–215.

Nauen, R. and Elbert, A. (2003) European monitoring, of resistance to insecticides in *Myzus persicae* and *Aphis gossypii* (Hemiptera: Aphididae) with special reference to imidacloprid. *Bulletin of Entomological Research* 93, 47–54.

Nauen, R., Ebbinghaus-Kintscher, U. and Schmuck, R. (2001) Toxicity and nicotinic acetylcholine receptor interaction of imidacloprid and its metabolites in *Apis mellifera* (Hymenoptera: Apidae). *Pest Management Science* 57, 577–586.

Nauen, R., Stumpf, N. and Elbert, A. (2002) Toxicological and mechanistic studies on neonicotinoid cross resistance in Q-type *Bemisia tabaci* (Hemiptera: Aleyrodidae). *Pest Management Science* 58, 868–875.

Nauen, R., Ebbinghaus-Kintscher, U., Salgado, V.L. and Kaussmann, M. (2003) Thiamethoxam is a neonicotinoid precursor converted to clothianidin in insects and plants. *Pesticide Biochemistry and Physiology* 76, 55–69.

Nauen, R., Bielza, P., Denholm, I. and Gorman, K. (2008) Age-specific expression of resistance to a neonicotinoid insecticide in the whitefly *Bemisia tabaci*. *Pest Management Science* 64, 1106–1110.

Nielsen, C. and Hajek, A.E. (2005) Control of invasive soybean aphid, *Aphis glycines* (Hemiptera: Aphididae), populations by existing natural enemies in New York State, with emphasis on entomopathgenic fungi. *Environmental Entomology* 34, 1036–1047.

Noma, T. and Brewer, M.J. (2008) Seasonal abundance of resident parasitoids and predatory flies and corresponding soybean aphid densities, with comments on classical biological control of soybean aphid in the US Midwest. *Journal of Economic Entomology* 101, 278–287.

Ohnesorg, W.J., Johnson, K.D. and O'Neal, M.E. (2009) Impact of reduced-risk insecticides on soybean aphid and associated natural enemies. *Journal of Economic Entomology* 102, 1816–1826.

Olson, K.D., Badibanga, T. and DiFonzo, C. (2007) Farmers awareness and use of IPM for soybean aphid control: Survey results for the 2004, 2005, and 2006 crop years. University of Minnesota, Department of Applied Economics. Staff Paper P07-13. Available at: http://purl.umn.edu/7355 (accessed 29 June 2009).

Pedersen, P., Gran, C., Cullen, E. and Hill, J. (2007) Potential for integrated management of soybean virus disease. *Plant Disease* 91, 1255–1259.

Pedigo, L.P. (1994) Bean leaf beetle. In: Higley, L.G. and Boethel. D.J. (eds) *Handbook of Soybean Insect Pests*. Entomological Society of America, Lanham, MD, USA, pp. 57–59.

Pedigo, L.P. and Rice, M.E. (2006) *Entomology and Pest Management*. Pearson Prentice Hall, Columbus, OH, USA.

Pedigo, L.P., Hutchins, S.H. and Higley, L.G. (1986) Economic injury levels in theory and practice. *Annual Review of Entomology* 31, 341–368.

Perring, T.M., Gruenhagen, N.M. and Farrar, C.A. (1999) Management of plant viral diseases through chemical control insect vectors. *Annual Review of Entomology* 44, 457–481.

Peterson, R.K.D. and Higley, L.G. (2001) *Biotic Stress and Yield Loss*. CRC Press, Baton Rouge, FL, USA.

Poston, F.L., Pedigo, L.P. and Welch, S.M. (1983) Economic injury levels: Reality and practicality. *Bulletin of the Entomological Society of America* 29, 49–53.

Ragsdale, D.W., Voegtlin, D.J. and O'Neil, R.J. (2004) Soybean aphid biology in North America. *Annals of the Entomological Society of America* 97, 204–208.

Ragsdale, D.W., McCornack, B.P., Venette, R.C., Potter, B.D., Macrae, I.V., Hodgson, E.W., O'Neal, M.E., Johnson, K.D., O'Neil, R.J., DiFonzo, C.D., Hunt, T.E., Glogoza, P.A. and Cullen, E.M. (2007) Economic threshold for soybean aphid (Hemiptera: Aphididae). *Journal of Economic Entomology* 100, 1258–1267.

Rosenheim, J.A., Wilhoit, L.R. and Armer, C.A. (1993) Influence of intraguild predation among generalist insect predators on the suppression of an herbivore population. *Oecologia* 96, 439–449.

Rutledge, C.E. and O'Neil, R.J. (2005) *Orius insidiosus* (Say) as a predator of the soybean aphid, *Aphis glycines* Matsumura. *Biological Control* 33, 56–64.

Rutledge, C.E., O'Neil, R.J., Fox, T.B. and Landis, D.A. (2004) Soybean aphid predators and their use in integrated pest management. *Annals of the Entomological Society of America* 97, 240–248.

Schmidt, N.P., O'Neal, M.E. and Singer, J.W. (2007) Alfalfa living mulch advances biological control of soybean aphid. *Journal of Environmental Entomology* 36, 416–424.

Schmidt, N.P., O'Neal, M.E. and Dixon, P.M. (2008) Aphidophagous predators in Iowa soybean: A community comparison across multiple sampling methods. *Annals of the Entomological Society of America* 101, 341–350.

Sechser, B., Reber, B. and Bourgeois, F. (2002) Pymetrozine: Selectivity spectrum to beneficial arthropods and fitness for integrated pest management. *Anzeiger für Schadlingskunde-Journal of Pest Science* 75, 72–77.

Sinclair, J.B., Kogan, M. and McGlamery, M.D. (1997) *Guidelines for the Integrated Management of Soybean Pests*. National Soybean Research Laboratory Publication No. 2, College of Agricultural, Consumer and Environmental Sciences, University of Illinois at Urbana-Champaign, Urbana, IL, USA.

Stern, V.M., Smith, R.F., Bosh, R.V. and Hagen, K.S. (1959) The integrated control concept. *Hilgardia* 29, 81–101.

Stewart, C.N., Adang, M.J., All, J.N., Boerma, H.R., Cardineau, G., Tucker, D. and Parrott, W.A. (1996) Genetic transformation, recovery, and characterization of fertile soybean transgenic for a syntheic *Bacillus thuringiensis* cry1Ac gene. *Plant Physiology* 112, 121–129.

Stone, J.D. and Pedigo, L.P. (1972) Development and economic injury level of green cloverworm (Lepidoptera: Nocutidae) on soybean in Iowa. *Journal of Economic Entomology* 65, 197–201.

Tabashnik, B.E. (2008) Delaying insect resistance to transgenic crops. *Proceedings of the National Academy of Science* 105, 19029–19030.

Takahshi, K.M., Berti, E. and Lourencao, A.L. (2008) Biology of *Bemisia tabaci* (GENN.) B-biotype and parasitism by *Encarsia formosa* (GAHAN) on collard, soybean and tomato plants. *Scientia Agricola* 65, 639–642.

Talekar, N.S. (1989) Characteristics of *Melanagromyza sojae* (Diptera, Agromyzidae) damage in soybean. *Journal of Economic Entomology* 82, 584–588.

Thies, C., Roschewitz, I. and Tscharntke, T. (2005) The landscape context of cereal aphid-parasitoid interactions. *Proceedings of the Royal Society B* 272, 203–210.

Timsina, J., Boote, K.J. and Duffield, S. (2007) Evaluating the CROPGRO soybean model for predicting impacts of insect defoliation and depodding. *Agronomy Journal* 99, 148–157.

Turnipseed, S.G. and Kogan, M. (1976) Soybean entomology. *Annual Review of Entomology* 21, 247–282.

Van den Berg, H., Ankasah, D., Muhammad, A., Rusli, R., Widayanto, H.A., Wirasto, H.B. and Yully, I. (1997) Evaluating the role of predation in population fluctuations of the soybean aphid, *Aphis glycines* in farmers' fields in Indonesia. *Journal of Applied Ecology* 34, 971–984.

Van den Berg, H., Shepard, B.M. and Nasikin (1998a) Damage incidence by *Etiella zinckenella* in soybean in East Java, Indonesia. *International Journal of Pest Management* 44, 153–159.

Van den Berg, H., Shepard, B.M. and Nasikin (1998b) Response of soybean to attack by stemfly *Melanagromyza sojae* in farmers' fields in Indonesia. *Journal of Applied Ecology* 35, 514–522.

Varma, A. and Malathi, V.G. (2003) Emerging geminivirus problems: A serious threat to crop production. *Annals of Applied Biology* 142, 145–164.

Venette, R.C. and Ragsdale, D.W. (2004) Assessing the invasion by soybean aphid (Homoptera: Aphididae): Where will it end? *Annals of the Entomological Society of America* 97, 219–226.

Walker, D.R., All, J.N., McPherson, R.M., Boerma, H.R. and Parrott, W.A. (2000) Field evaluation of soybean engineered with a synthetic cry1Ac transgene for resistance to corn earworm, soybean looper, velvetbean caterpillar (Lepidoptera: Noctuidae) and lesser cornstalk borer (Lepidoptera: Pyralidae). *Journal of Economic Entomology* 93, 613–622.

Weichel, L. and Nauen, R. (2003) Uptake, translocation and bioavailability of imidacloprid in several hop varieties. *Pest Management Science* 60, 440–446.

Werner, B.J., Krell, R.K., Pedigo, L.P. and Hill, J.H. (2003) The soybean leaf miner (Coleoptera: Chrysomelidae) as a vector of Bean pod mottle virus. *Journal of the Kansas Entomological Society* 76, 643–644.

Wiedenmann, R.N. and Smith, J.W. Jr (1997) Attributes of natural enemies in ephemeral crop habitats. *Biological Control* 10, 16–22.

Wu, Z.S., Schenk-Hamlin D., Zhan, W.Y., Ragsdale, D.W. and Heimpel, G.E. (2004) The soybean aphid in China: A historical review. *Annals of the Entomological Society of America* 97, 209–218.

Wyckhuys, K.A.G. and Heimpel, G.E. (2007) Response of the soybean aphid parasitoid *Binodoxys communis* to olfactory cues from target and non-target host-plant complexes. *Entomologia Experimentalis et Applicata* 123, 149–158.

Wyckhuys, K.A.G., Koch, R.L. and Heimpel, G.E. (2007) Physical and ant-mediated refuges from parasitism: Implications for non-target effects in biological control. *Biological Control* 40, 306–313.

Zehnder, G., Gurr, G.G., Kuhne, S., Wade, M.R., Wratten, S.D. and Wyss, E. (2007) Arthropod pest management in organic crops. *Annual Review of Entomology* 52, 57–80.

15 Nematodes of Soybean and Their Management

Edward O. Oyekanmi[1,2,3] and B. Fawole[1]
[1]*Crop Protection and Environmental Biology Department, Faculty of Agriculture and Forestry, University of Ibadan, Ibadan, Nigeria;* [2]*Biological Sciences Department, Wesley University of Science and Technology, Ondo, Nigeria;* [3]*Nematology Unit, International Institute of Tropical Agriculture, Oyo Road, Ibadan, Nigeria*

15.1 Introduction

Soybean (*Glycine max* (L.) Merrill) is a globally important oilseed crop and source of high-quality protein (Sinclair and Backman, 1989). In many parts of the world soybean has in recent years become increasingly popular as a source of oil and protein for human consumption (Zarkadas *et al.*, 1993) and animal fodder (Manyong *et al.*, 1996) and an important component in improved crop rotation systems (Carsky *et al.*, 1997). Soybean is a promising crop that will help agricultural scientists and other stakeholders to achieve the goal of sustainable agriculture and food production for the world's ever-increasing population. Thus, looking into factors that can reduce the agronomic productivity of soybean, such as the menace of plant parasitic nematodes on the crop, is a worthwhile task.

Plant parasitic nematodes are mostly microscopic worms that attack crops. They are known to be a major biotic constraint to crop production on various continents of the world (Lowe, 1992; Jones and Perry, 2004). About 24 genera and many species of plant parasitic nematodes are associated with soybean. It has been estimated that about 11% of world crop production is lost as a result of plant nematode damage, which accounts for one third of all losses attributed to pests (Whitehead, 1998; Agrios, 2005). These losses vary from negligible to crop failure. If nematodes are left unchecked, severe financial losses are incurred by growers (Schmitt, 1985; Whitehead, 1998).

15.2 Plant Parasitic Nematodes Associated with Soybean Production

Among the various constraints to soybean production are plant parasitic nematodes (Sikora and Greco, 1990). The amount of damage caused by plant parasitic nematodes is related to many variables, including the nematode species, the size of the nematode population, the susceptibility of the host plant and various environmental factors such as temperature, duration of the growing season, availability of water and nutrients to the plant and the presence of other organisms contributing to the total damage inflicted upon the crop (Zirakparvar, 1985; Back *et al.*, 2002). The following nematode genera, among others, are associated with soybean in various agroecological zones: *Meloidogyne*, *Heterodera*, *Pratylenchus*, *Rotylenchulus*, *Hoplolaimus*, *Belonolaimus*, *Helicotylenchus*, *Tylenchorhynchus*, *Scutellonema*, *Paratrichodorus*, *Criconema* and *Xiphinema* (Schmitt, 1985; Lowe, 1992; Sikora *et al.*, 2005a). Although a large number of nematodes are associated with soybean, the sedentary nematodes of *Meloidogyne* species (root-knot nematodes) and *Heterodera glycines* (Ichinohe) are particularly serious pests (Schmitt, 1985; Fourie *et al.*, 2001; Sikora *et al.*, 2005a).

When the relationship between nematodes and a single plant is examined, the effect on the latter can range from death in some instances to absence of detectable damage in others (Zirakparvar, 1985). Depending on specific life cycles, plant parasitic nematodes are able to cause a variety of types of wound on host plant roots while entering or feeding. For example, ectoparasitic nematodes such as *Trichodorus* and *Tylenchorhynchus* species feed on roots while entering. Endoparasitic nematodes such as *Meloidogyne* and *Heterodera* species are more disruptive to their host's roots (Von Mende *et al.*, 1998; Back *et al.*, 2002).

As nematodes steal nutrients from the roots, the plants are weakened and do not grow well. Subsequently, plants may be more vulnerable to attack by other stresses such as insects, diseases and drought. Part of the damage done by plant parasitic nematodes to soybean includes the damage to the root system, which causes sloughing off of the root and severe necrosis. Root pruning as well as proliferation of lateral roots may occur, and a reduction of the grain yield is normally seen (Zirakparvar, 1980; Lowe, 1992).

In addition to *Meloidogyne* and *Heterodera* genera, the parasitism of *Pratylenchus* and *Rotylenchulus* is discussed in this chapter. These four genera are important to soybean on a global basis and may cause high economic loss (Schmitt, 1985).

Root-knot nematodes

Root-knot nematodes (*Meloidogyne* species) are very important pests of soybean. Species include *M. arenaria*, *M. hapla*, *M. incognita* and *M. javanica*, which limit soybean grain yield, symbiotic nitrogen fixation (SNF) and growth (Schmitt and Noel, 1984; Sikora *et al.*, 2005a; Oyekanmi *et al.*, 2007; Coyne and Oyekanmi, 2007).

Epidemiology and biology

The epidemiology of plant disease deals with the dynamic process of host–pathogen interactions, which determine the prevalence and severity of disease. Root-knot nematodes are worldwide in distribution and are able to adapt to various environmental conditions and parasitize wide host ranges (Schmitt, 1985; Whitehead, 1998). Although races with differing host preferences have been demonstrated, *M. hapla* is limited to temperate soil and to cool soil in the tropics, *M. arenaria* is found in tropical, subtropical and warm temperate soils and *M. incognita* and *M. javanica* are typically tropical and subtropical (Whitehead, 1998).

Every succulent soybean tissue in the soil, especially the roots, is subject to attack by root-knot nematodes. In locations where soil temperatures at the time of soybean maturity are suitable for egg hatching (>15°C), hatching occurs. Root-knot nematodes survive the winter in soil as infective juveniles. The matured females lay eggs in the roots. A female lays several hundred eggs, which are deposited in a gelatinous matrix through a rupture in the galled surface. These give rise to the first-stage juveniles, which are enclosed within the egg; second-stage juveniles emerge after about 10 days. In *M. incognita*, the second-stage juveniles migrate through the soil to reach young roots, which are usually invaded just behind and along the sides of the roots. They normally settle at a feeding site close to the point of entry and in most cases penetrate the pericycle. Initiation of galling occurs within 2–3 days of penetration to the pericycle. The nematodes then become sedentary and moult three times to become pear-shaped or saccate females (which are the majority) or vermiform males. The males are few, elongate (1.0–1.5 mm and migratory (Whitehead, 1998). Some stages of *M. incognita* are shown in Fig. 15.1.

Depending on the host plant, nematode species, soil temperatures and length of maturity of the genotypes, there may be up to three generations per soybean growing cycle. However, several overlapping generations, each of 3–4 weeks' duration are completed during the season. Reproduction is often through parthenogenesis, although sexual reproduction occurs in

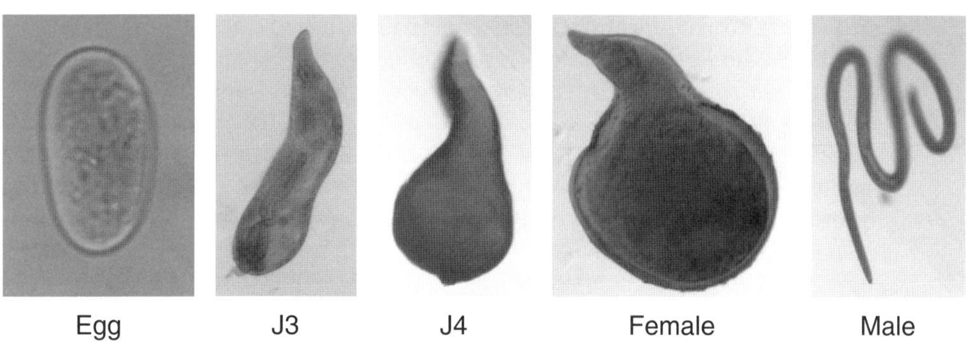

Fig. 15.1. Some stages of root-knot nematode (*Meloidogyne incognita*) (photo by Edward Oyekanmi).

several species, especially when infestations become large and many juveniles crowded together in limited plant space develop as males (Whitehead, 1998). The presence of males in a population, however, is not proof of sexual reproduction. *M. arenaria* reproduces mainly by meiotic parthenogenesis and sexually. *M. incognita* reproduces by parthenogenesis, although males are often common and may aggregate at the posterior end of attractive females. *M. javanica* also reproduces by parthenogenesis (Whitehead, 1998).

The number of nematodes in the soil reaches a maximum at soybean maturity. The population declines slowly through the winter and then precipitously as soil temperatures increase in the spring. The rapid decrease in the number of infective nematodes in the soil can occur several weeks before soybean planting due to factors such as increased activity of the nematodes and depletion of their food reserves, along with an increase in the predatory and parasitic activities of the soil arthropod and microbial communities, which are detrimental to the survival of the nematodes. At planting time, the number of nematodes in the soil is at its lowest, usually <10% of that at soybean maturity (Sinclair and Backman, 1989).

Root-knot nematodes form disease complexes with other pathogens such as bacteria and fungi. The nematodes predispose plants to other infections by puncturing the host plant tissue in the process of feeding. Pathogens enter the plant tissue through such entries and become established. Consequently the plants are weakened, making them prone to other pathogens. In soybean, nodulation by *Bradyrhizobium japonicum* has been suppressed by infestation of the roots with *M. hapla* (Whitehead, 1998). Infestation by *Meloidogyne* species can also lead to reduced nodulation or inhibit nodulation by symbiotic nitrogen-fixing bacteria (Musarrat and Haseeb, 2000).

Damage threshold

Soybean roots infected with root-knot nematodes show swelling or deformation, known as galls. Galls are initiated when the tissue of roots are infected. They are formed when large multinuclear syncytia (giant or transfer cells) form around the anterior ends of the cell cytoplasm. Soybean roots infected with root-knot are deformed and galled. They are unable to effectively absorb nutrients and water uptake is also negatively affected (Sikora *et al.*, 2005a).

The ability of soybean to carry out effective SNF is undoubtedly severely affected as a result of *M. incognita* infection (Coyne and Oyekanmi, 2007). *M. incognita* affects SNF differently, depending on the genotype, while the effect on SNF is also known to depend on specific nematodes species (Carneiro *et al.*, 2002). *Meloidogyne* species cause varying degrees of stunting, sclerosis and in some cases early senescence, depending on the initial population density. Losses can often be related to the intensity of galling, which is also dependent on the initial population density (Sikora *et al.*, 2005a). Losses of 90% due to *M. incognita* have been reported from Florida, USA (Kinloch, 1982). The damage caused to soybean is also influenced by environmental factors, especially moisture, soil fertility status, soil compaction and soil pH. Likewise, the presence of other pathogens such as *Fusarium oxysporum* exacerbates losses due to root-knot infection. Akinsanmi and Adekunle (2003) reported that when *M. incognita* race 2 and *F. oxysporum* were present as a concomitant inoculation to soybean, both the growth and yield of soybean were significantly reduced, compared to when *M. incognita* alone was applied.

Management measures

As root-knot nematode density in the field increases, both prophylactic (preventive) and curative (corrective) management measures are required to avoid yield loss. Several approaches have been initiated with varying degrees of success in the management of root-knot nematodes associated with soybean. These are host plant resistance, cultural practices, biological control and nematicide application.

Screening for root-knot resistance is acceptable to growers and scientists because it is cheap and environment-friendly (Oyekanmi and Coyne, 2009). Host plant resistance utilization is rated the best in soybean nematode pest management (Sikora *et al.*, 2005b). In host plant resistance utilization research conducted on selected soybean cultivars by Oyekanmi and Coyne (2009), the tropical *Glycine max* (L.) Merrill TGx 1485-ID was moderately resistant to *M. incognita* and is promising in the management of root-knot nematode. Fig. 15.2 shows a part of the results obtained from the screening work.

Soybean breeders are breeding for nematode resistance in order to make sustainable agriculture available to farmers. This practice is in agreement with the global call to reduce agrochemical inputs into crop production (Holderness *et al.*, 2000; Khan *et al.*, 2002; Oyekanmi *et al.*, 2008). Resistance to *M. incognita* has received the most attention. In crosses between resistant and susceptible parents, resistance levels approximately equal to those of resistant parents have been recovered in moderate-size populations. This indicates that very few major genes control resistance. The absence of discrete classes suggests that parents also differ in other genes with smaller effects (Hinson, 1985).

Cultural practices such as planting a non-suitable host of root-knot nematodes in a crop rotation cycle may prevent build-up of root-knot nematodes. A number of grasses, cereals and Compositae are not very suitable hosts for *Meloidogyne* species, which can dramatically lessen soil infestations. Murphy *et al.* (1974) found that *M. incognita* decreased below detectable levels in the soil after 4, 5 or more years of Bahia grass (*Paspalum notation*) or Bermuda grass (*Cynodon dactylon*), but other important nematodes (*Paratrichodorus minor*) were increased after either grass.

Among the Compositae, the species of *Crotalaria* (notably *C. spectabilis* and *C. juncea*) and *Tagetes* (particularly *T. erecta*, *T. patula* and *T. minuta*) are resistant or immune to root-knot nematodes. These plants react hypersensitively to invasion by juvenile root-knot nematodes, producing nematoxins (e.g. α-terthienyl) that prevent the nematodes feeding in the root. As a result, the nematodes die and the soil infestation is greatly decreased. In this way, *M. incognita* is controlled by the utilization of a non-suitable host in the soybean rotation plan (Whitehead, 1998).

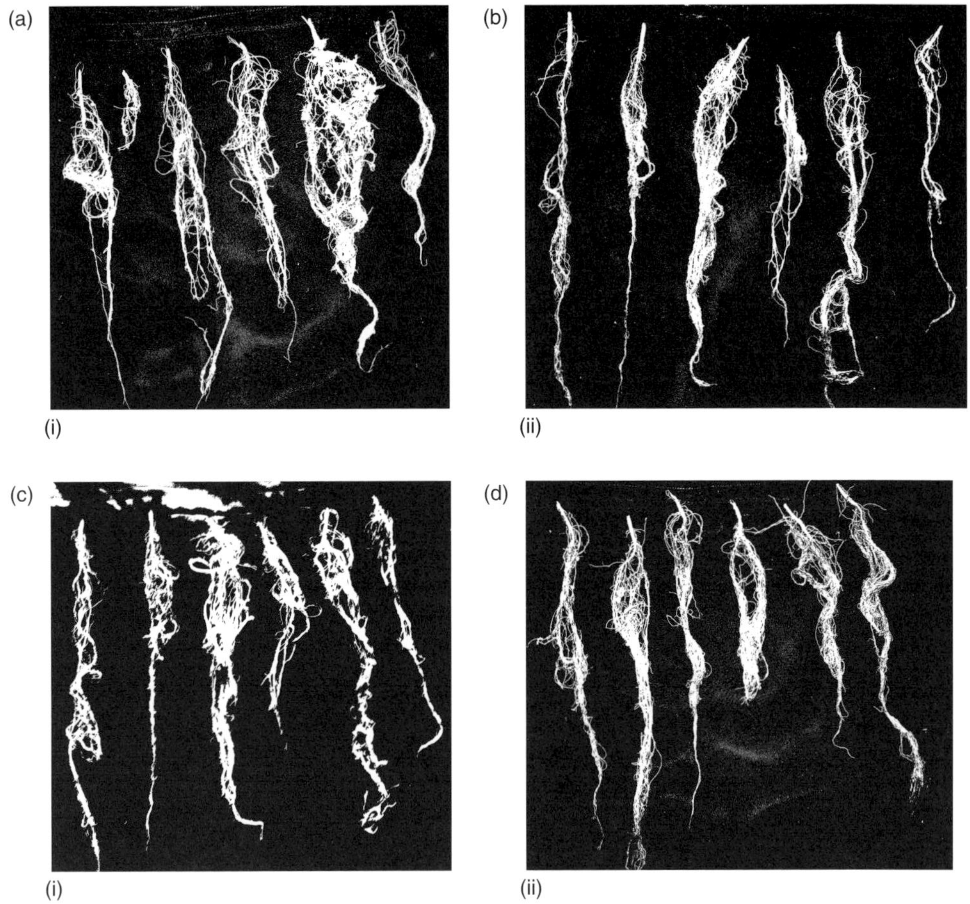

Fig. 15.2. (a) and (b): Roots of six soybean cultivars harvested from soil inoculated with and without *Meloidogyne incognita*, respectively. (c) and (d): Roots of six soybean cultivars harvested from soil inoculated with and without *Pratylenchus* species, respectively. (i) represents a set of roots with nematode infection, while (ii) represents another set of roots as in (i), but without nematode infection (photo by Edward Oyekanmi).

However, the efficacy of the plants may vary with the cultivar and nematode isolate or isolates. A non-host in one region may, therefore, be less effective or ineffective in another region. In addition, nematicides are commonly used in the control of root-knot nematodes, but they are rarely cost-effective (Sinclair and Backman, 1989).

Soybean cyst nematode

One of the major limiting nematode pests to soybean production is the soybean cyst nematode (SCN) (*Heterodera glycines* Ichinohe). It is a dynamic parasite that reproduces by amphimixis. Through its large number of genes for parasitic capability or through genetic recombination, or both, this nematode adapts to a wider host range than most species of cyst nematodes (Riggs, 1985). After a period of consecutive cropping of a farmland with a soybean cultivar that is resistant, the SCN has been able to adapt and eventually parasitize those cultivars.

Epidemiology and biology

The SCN is an important nematode pest in temperate soil, but also occurs in tropical regions where soybean is grown. The nematode causes severe stunting and yellowing of the foliage and in extreme cases plant death (Sikora *et al.*, 2005a). Five races of SCN have been described (Golden *et al.*, 1970; Inagaki, 1979) based on differential host ranges. Other races have also been observed, and the number of races is increasing (Riggs, 1985).

The SCN has an egg stage, four juvenile stages and an adult stage (male or female). First-stage juveniles develop within the egg and moult once to become second-stage juveniles, which emerge from the egg (in the gelatinous matrix or in the cyst). Factors such as temperature and moisture affect hatching, but considerable spontaneous hatching occurs when eggs are not in diapause (Sinclair and Backman, 1989). Second-stage juveniles are approximately 450 nm long. About half of the tail of the second-stage juvenile is hyaline. Second-stage juvenile penetrate roots approximately ≥1 cm behind the root tip, migrate to the vascular tissue and place their tip region adjacent to the stele (endodermis). Enlarged, multinucleate (syncytial) cells form around the head of these sedentary juveniles in response to nematode stylet probing and feeding (Whitehead, 1998). The nematodes begin to enlarge and become sedentary when feeding begins. Three more moults occur, resulting in third- and fourth-stage juveniles and adults (Sinclair and Backman, 1989).

Adult females are white when young and turn yellow with age. Upon death the body wall hardens and becomes a dark brown cyst. The cyst wall has a pattern of irregular, short zigzag lines. The female produces a gelatinous matrix at the vulva cone, which usually contains some eggs. The female body is also filled with eggs. *H. glycine* cysts are lemon-shaped with dimensions of 560–850 × 350–590 nm. Brown bullae (internal knobs in the anal area) are present.

Males are vermiform and 1.0–1.5 mm long. The male SCN matures faster than the female; the males are required for reproduction. Development occurs at temperatures of 18–32°C, although 24–28°C is optimal. Development does not occur at temperatures >33°C and is very slow at <16°C. Eggs within the cyst may survive for ≥11 years. Their survival may be due in part to protection by the cyst wall and in part to genetically controlled diapauses. The SCN possesses a high degree of genetic variability that is expressed phenotypically.

H. glycines is reported to have a diapauses stage that may reduce spontaneous emergence at a given time of the year. The SCN is also susceptible to desiccation. The percentage survival of eggs and juveniles decreases with increasing temperature from northern to southern soybean-growing regions of the USA (Noel, 1985). This reduction is thought to be due to the influence of temperature on nematode activity and increased biological control through soil pathogens and parasites (Sikora *et al.*, 2005a). The SCN completes six to seven generations per season in temperate growing areas, with the greatest increase in density occurring in the first generation (Lawn and Noel, 1986).

Damage threshold

Foliar symptoms on seedlings vary from slight stunting to severe chlorosis and death. The mature soybean plant may be stunted, chlorotic or both when infected with SCN; the symptoms manifested are not entirely diagnostic because they are similar to those caused by nitrogen and potassium deficiencies. The root system has symptoms ranging from slight discolouration to severe necrosis. Some populations of the nematode, especially race 1, also affect nitrogen fixation (Sinclair and Backman, 1989). Nodulation may be slightly to completely inhibited and the nitrogen fixation efficiency of the remaining nodules may be reduced. Due to the misleading symptoms, diagnosis of the disease must be based on signs, namely the white to yellow females and eruption from the roots (Sinclair and Backman, 1989).

In soybean, yield losses depend on the numbers of hatchable nematodes in the soil at planting, the cultivar grown, soil moisture stress, the virulence of nematode population and the presence of interacting organisms (e.g. *Verticillium* species) (Whitehead, 1998). Yield losses can range from 10% to 80% depending on rainfall, soil fertility, the presence of other diseases and nematode density (Jacobsen *et al.*, 1983).

According to Noel (1984) in a study on silt loam soils with 2% organic matter, economic losses were incurred when densities were ≥699 eggs and juveniles, or 12 cysts with viable eggs in 250 cm^3 of soil.

Management measures

Measures of control of SCN include cultural methods, such as the use of crop rotation and trap crops, and a chemical control approach such as the application of nematicides. The use of resistant soybean varieties is also important in the management of SCN.

Crop rotation is effective in controlling the SCN because few crops are susceptible to it. Growing a non-host for 2 years is generally adequate to allow a susceptible soybean cultivar to be grown. However, additional benefits may be achieved by growing a non-host for an additional year if the population density of the nematode is extremely high. Occasionally, a resistant soybean cultivar may be used in place of a non-host.

Soybean varieties with resistance to the SCN are available. In most cases field populations appear to be mixtures of nematode genera, which affects nematode management strategies. Selection force imposed by a resistant cultivar may result in changes in the gene frequency of the nematode population. Continuous or frequent use of resistant cultivars results in race shift and 'resistance-breaking' types increase (Sinclair and Backman, 1989).

The use of nematicides provides some control of this nematode; however, the approach may not be economically viable. For effective long-term control of the SCN, integration of control practices is the best option. This approach includes the use of resistant cultivars, crop rotation and crop management. However, the particular integration of control tactics is dependent on the crop, available cultivars and the ability to manage soil water (Sinclair and Backman, 1989).

Lesion nematodes

Some species of lesion nematodes such as *Pratylenchus agilus, P. alleni, P. brachyurus, P. coffeae, P. crenatus, P. hexincisus, P. neglectus, P. penetrans, P. sefaensis, P. scribneri, P. vulnus* and *P. zeae* attack soybean. Lesion nematodes are found in soils throughout the world and attack a wide range of crops and weeds.

Epidemiology and biology

Lesion nematodes are endoparasites that attack the root cortex. Their feeding generally causes dark lesions and an overall browning of the roots, and may decrease root growth by 25% (Sinclair and Backman, 1989). The epidermis and cortex of several infected roots may slough away from the stele. The degree of damage is generally influenced by the nematode population at the time of planting and by temperature. If water and nutrients become limiting, plants become yellowish and stunted, and yields may be adversely affected when nematodes feed on plants.

A female and male of *Pratylenchus* species are shown in Fig. 15.3. The adults range from 500–800 µm in length. Species such as *P. brachyurus* and *P. zeae* are found primarily in warm regions. Others such as *P. hexincisus* and *P. penetrans* are commonly found in heavy-textured soils and are abundant in temperate regions (Sinclair and Backman, 1989).

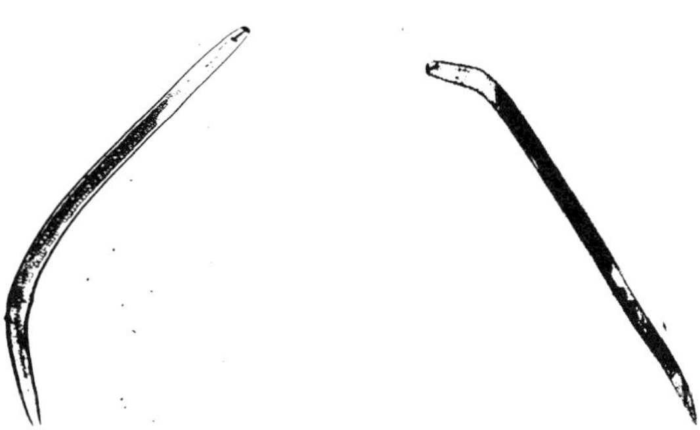

Female Male

Fig. 15.3. A female and male *Pratylenchus* species (photo by Edward Oyekanmi).

Damage threshold

Pratylenchus species infect soybean in most growing areas. Lesion nematodes cause stunting, leaf yellowing and yield losses, depending on soil densities at planting. Yield losses are linearly related to *P. brachyurus* densities in a sandy-clay loam soil (Schmitt and Baker, 1981).

The lesion nematode is normally a parasite of the root cortex, but it may also harm phloem and sometimes xylem. Reproduction is sexual. Although it is not possible to give precise threshold populations above which losses occur, five to ten nematodes $100\,g^{-1}$ of soil may be considered damaging (Whitehead, 1998). For most *Pratylenchus* species infections, infested roots show typical lesions that may eventually girdle the root. These root lesion nematodes are obligate root endoparasites and are injurious to many crops. They feed primarily on the root cortex, through which they migrate, creating cavities and channels in which their eggs are deposited singly or in small groups. Before penetrating the roots, *Pratylenchus* species sometimes browse on the root surfaces and root hairs. Small lesions, at first yellow and then brown to black, develop where the nematodes enter and feed in the roots. These often enlarge and may eventually girdle the roots. Cells on which the nematodes feed and cavities and channels through which they pass become necrotic and secondary invading bacteria and fungi may rot the root. The loss of functional feeder and sometimes main roots results in leaf chlorosis, twig die-back and yield loss. *Pratylenchus* species are known to increase damage caused by root-rotting fungi, which may further reduce yield (Sikora *et al.*, 2005a).

Management measures

The management of *Pratylenchus* species is affected by wide host ranges and the presence of multiple species in a field. However, the use of resistant varieties in the management of lesion nematodes is an effective means of management. Most of these cultivars are also resistant to other nematodes, such as reniform and root-knot nematodes.

Cultural methods such as crop rotation do not adequately control *Pratylenchus* species, although some reduction in soil infestation may be achieved by growing a poor host (Whitehead, 1998). Nematicides may be used as a control measure for *Pratylenchus* species. If this is being considered, an economic analysis should first be conducted to ascertain profitability (Zirakparvar, 1985).

Reniform nematodes

Rotylenchulus reniformis (Linford & Olivera), the reniform nematode, is widespread in the tropics and subtropics, where it attacks and multiplies on a wide range of cultivated plants. Infection of soybean roots by the reniform nematode was reported in 1956 on the Gold Coast (now Ghana), West Africa, and in 1967 in South Carolina, USA (Sinclair and Backman, 1989).

Epidemiology and biology

R. reniformis is now distributed in the south-eastern and Gulf Coast areas of the USA and in most tropical regions of the world. It is a semi-endoparasite that partially embeds itself in the root. Because it is small and soil particles adhere to the portion of the body outside the root, the nematode is easily overlooked. The eggs are elongate (72–100 × 33–44 μm) and hatch within 24 h, yielding a sex ratio of about 1:1. Larvae are 330–445 μm long and have a well-developed stylet of 14–18 μm long. They moult three times in the soil with little or no root feeding and become adults (Sinclair and Backman, 1989).

The mature females of *R. reniformis* are sedentary, semi-endoparasites of roots. They feed on cortical parenchyma, the pericycle or even the phloem, depending on the host species. The male has a poorly developed stylet and median oesophageal bulb and does not feed. Sexual reproduction is the norm, but parthogenesis has also been reported (Sinclair and Backman, 1989). The kidney-shaped females produce a gelatinous matrix (the egg sac) that covers the female body and in which about 50–70 eggs (depending on the host plant) are laid. Soil adhering to this matrix often can hamper detection of the female on the root surface (Whitehead, 1998; Sikora *et al.*, 2005a). All juveniles are passed into the soil in as little as 10 days, with the complete life cycle taking up to about 30 days, depending on the host and soil conditions. *R. reniformis* survives in the soil in the juvenile and adult male stages. Immature females penetrate the root and establish in the endodermis.

Damage threshold

The reniform nematode can cause stunting and chlorosis on soybean. *R. reniformis* has been found to be associated with soybean damage in tropical and subtropical countries (Schmitt and Noel, 1984). Significant yield losses may result from such attacks, especially when other root pathogens are also involved.

Management measures

The use of resistant varieties (Birchfield *et al.*, 1971; Lim and Castillo, 1979) and a sound rotation plan with non-host crops for ≥2 years are good management options. Soybean cultivars that are resistant to *H. glycines* can also be resistant to *R. reniformis* (Sikora *et al.*, 2005a).

15.3 Recent Advances in Soybean Nematode Management

Stringent exclusion and quarantine strategy

Phytosanitary regulations guide the transfer of plant materials. These regulations are becoming more stringent, now that there is worldwide awareness concerning the transfer of pests from one particular endemic location to a relatively safe location (Biosecurity, 2003). Plant parasitic nematodes are commonly known as the farmer's hidden enemy. They are some of the pests looked for in phytosanitary investigations. Grains contaminated with seed-borne nematodes can be a threat to high soybean-producing countries, if such infected materials are conveyed across borders. In addition, plant materials contaminated with *Heterodera* species or root-knot nematodes can be devastating to soybean-producing areas that do not already have such nematode problems. Thus, a phytosanitary certificate showing a clean bill of no nematode infection is required when plant materials are imported or exported. Likewise, care should be taken to confine (quarantine) infected or suspected plant materials to avoid the spread of such pests. According to Sikora *et al.* (2005b), exclusion is the most effective and economical means of preventing the introduction of important pests. In addition, the prevention of local spread within a country or region has been effective in the past and should be strengthened in the future. An example of the transfer of an economically important nematode is that of SCN *H. glycine* to Brazil. Nematologists and other plant protection experts should be involved in phytosanitary and quarantine services at both national and international levels.

Nematode-suppressive soil

Suppressive soils prevent nematodes from establishing and causing disease, and they diminish disease severity after initial nematode damage in continuous culturing of a host. Identification of soils that inhibit a tiny soybean-destroying nematode is an important tool in reducing yield losses (Westphal, 2005). Using plants bred to resist pests is not the complete answer, so it is important to find suppressive mechanisms. This method of nematode management is much more desirable than using chemicals in order to limit damage to the environment.

Currently, efforts are being intensified to extend this research to finding ways to increase nematode suppression in soil. This is important because nematode populations constantly change, enabling them to overcome certain types of resistance, including that of plants bred to be resistant to the organisms. A range of non-specific and specific soil treatments, followed by infestation with a target nematode, have been employed to identify nematode-suppressive soils. Biocidal treatments, soil transfer tests and baiting approaches, together with observations of the plant parasitic nematodes in the root zone of susceptible host plants, have improved the understanding of nematode-suppressive soils (Westphal, 2005). Efforts should be further intensified to locate more suppressive soils, to make this tool more widely available and to investigate further the mechanisms that create its effectiveness against the cyst nematode.

Breeding for nematode resistance and crop performance

There has been tremendous research in the area of breeding for resistance in grain legumes, and it is worth making use of research results in this area of science. Prospects for major contributions to protecting world food and grain production through crop resistance and tolerance to nematodes are fundamental. Agricultural scientists consistently identify plant resistance as the highest research priority for nematode pest management. The advantages of breeding crop plants that are resistant to injurious parasitic nematodes, and growing them on infested land, are many and varied (Ferris *et al.*, 1992). Host resistance is a viable alternative because it is cheap and poses no technical difficulties to the farmer (Trudgill, 1991).

Modern biotechnology is based on scientific advances that make it possible to isolate and clone specific pieces of DNA-containing genes, and to sequence the nucleotides in a DNA molecule (the genetic code) so that the precise location and structure of genes can be studied at the molecular level (Goodman and Kiser, 1985). Numerous disease-resistance (*R*) gene homologues have been identified in legume species by sequence identity of conserved motifs with known *R* genes (Kanazin *et al.*, 1996; Yu *et al.*, 1996; Zhu *et al.*, 2002). The challenge is to identify the specific genes conferring resistance to a particular pathogen. One example of this is the *Rpg1-b* gene from soybean, which confers resistance to *Pseudomonas syringae* pv. *glycinea* (causing bacterial blight) carrying the *avrB* gene in a classic gene-for-gene specific manner (Ashfield *et al.*, 2004). Molecular breeding techniques may be extended to nematodes of international interest such as *Meloidogyne* and *Heterodera*.

Remote-sensing utilization and host plant resistance

Geospatial facilities linked with pathological information systems could be exploited in soybean pest management (Nutter *et al.*, 2002; Schmitz *et al.*, 2004). Tremendous progress has been made in the use of remote sensing, using infrared and digital thermography, to detect areas in a field that are infested with nematodes (Sikora *et al.*, 2005b). This approach can give exact location where the plant resistance approach can be utilized, so that susceptible and non-host crops can be established in the same plot. This methodology has been used for cyst nematode management.

Biofertilizer and biopesticide utilization

The ability of legumes to form symbiotic mutualistic relationships with certain bacteria in the *Rhizobiales* (collectively called rhizobia) and to harness the ability of the bacteria to fix atmospheric nitrogen into ammonia has a tremendous impact on natural and agricultural ecosystems (Coyne and Oyekanmi, 2007). The interaction enables legumes to produce protein-rich seeds and foliage that are critical to many human and animal diets. Past research has illuminated many of the facets of plant–bacterium recognition, nodule formation, nitrogen fixation and ammonia assimilation. Legumes are a rich source of flavonoids, notably isoflavones and isoflavanones, which are

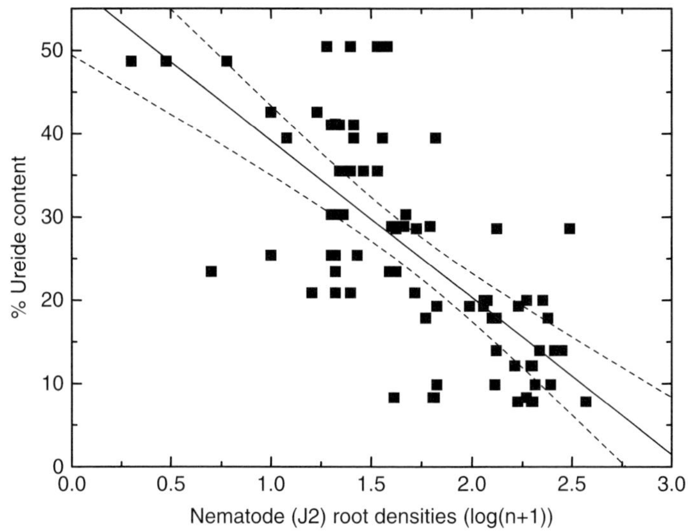

Fig. 15.4. Ureide content (symbiotic nitrogen fixation) of shoots regressed against second-stage juvenile density in two cultivars of soybean, TGx 1485-1D and TGx 1448-E (Coyne and Oyekanmi, 2007; reprinted with permission).

not found in *Arabidopsis* (*Arabidopsis thaliana*). Legume nodules are also rich sources of cysteine cluster proteins, some of which have been shown to have antimicrobial activity and may play a role in protecting nodules from pathogens (Yu *et al.*, 1996; Zhu *et al.*, 2002). SNF and atmospherically fixed nitrogen are important alternative sources of usable nitrogen for legume (and other) crops for improved yields, especially on degraded soils and under low external input cropping systems (Vincent, 1982; FAO, 1983; Buttery *et al.*, 1992). SNF is dependent on numerous factors, including host crop, genotype and microsymbiont, but may be limited by pedoclimatic factors and pests, especially those that reduce effective nodulation such as root-knot nematodes (Sasser and Carter, 1985; Brockwell *et al.*, 1991; Graham, 1992).

According to Coyne and Oyekanmi (2007) *Glomus mosseae*, *Bradyrhizobium japonicum* and *Trichoderma pseudokoningii* act as biocontrol agents against *M. incognita*. The capacity of *M. incognita* to reduce the SNF potential of soybean is highly related to soybean's ability to fix nitrogen when infected (Fig. 15.4). However, it is noteworthy that the effect on SNF is not necessarily related to the extent of visual galling damage observed. Therefore, galling damage may not be the most useful tool as an estimate or indicator of crop damage, except to provide evidence of root-knot nematode infection and likely loss of production.

The arbuscular mycorrhizal fungus microbial agents *G. mosseae*, *B. japonicum* and *T. pseudokoningii* have been found to act on soybean nematodes as biopesticides (Oyekanmi *et al.*, 2008). The treatments (biocontrol agents in various factorial combinations and nematicides) reduced *M. incognita* density (Fig. 15.5). This supports the idea that bio-enhancement agents,

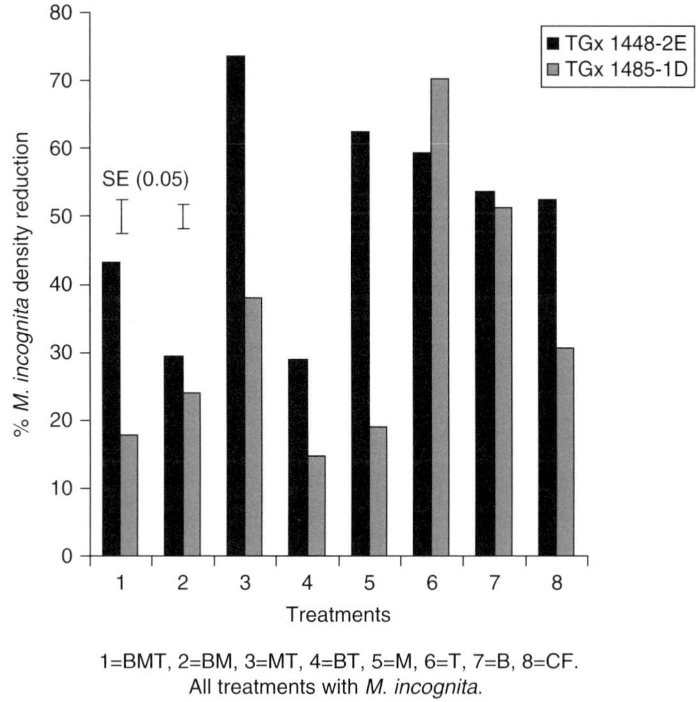

Fig. 15.5. Percentage reduction of *Meloidogyne incognita* density in soybean root harvested at 8 weeks after planting following various treatments in *M. incognita*-infested soil (Oyekanmi *et al.*, 2008; reprinted with permission). B, *Bradyrhizobium japonicum*; CF, carbofuran; M, *Glomus mosseae*; SE, standard error; T, *Trichoderma pseudokoningii*.

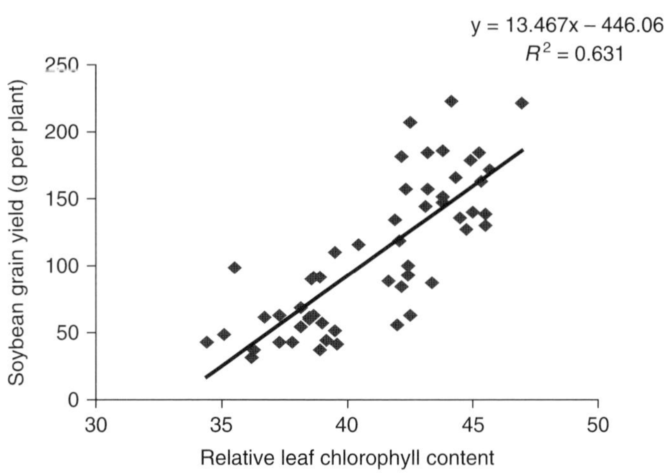

Fig. 15.6. Soybean grain yield regressed with leaf chlorophyll content under root-knot nematode infection (reprinted with permission from Oyekanmi *et al.*, 2007).

which are again environmentally friendly, may be very promising in the production of soybean in soil infested with soybean nematodes.

In a similar study by Oyekanmi *et al.* (2007), the biocontrol agents mentioned above responded by improving the yield of soybean in the presence of *M. incognita* through enhancing the chlorophyll content of the leaves (Fig. 15.6). However, this was also dependent on the host plant's response to *M. incognita*; in this work, the microbial agents responded as a biofertilizer. With the global call to reduce agrochemical inputs in agricultural production, the application of such microbial agents is an option worth exploring. In addition, health hazards loom with over-dependence on agrochemicals, not to mention risks to the environment. Environmentally sound control strategies will receive legislative, consumer and grower acceptance, providing the cost implications of such biological tools are economically viable.

15.4 Conclusions and Future Prospects

With our knowledge of selected soybean nematodes and the economic losses they cause, as discussed in this chapter, fellow nematologists and other plant protection experts must be armed with basic, applied and advanced knowledge of soybean nematode management. We are expected to rise to the challenges posed by these insidious pests.

The knowledge shared, if well utilized, will make the soybean economy and soybean production more robust, enabling farmers to meet the ever-increasing demands of both national and international markets. The ideas shared will enable all stakeholders in soybean production and the value chain to comply with the global clarion call to reduce agrochemical inputs into agricultural production and utilize eco-friendly strategies to overcome the challenges posed by pests. In this regard, the recent advances in soybean nematode management – such as stringent exclusion and quarantine strategies, nematode-suppressive soil, breeding for nematode resistance and crop performance, remote-sensing utilization and host plant resistance, biofertilizers and biopesticides – could form a viable roadmap for soybean nematode management. Networking among different specialists in specialized areas of nematology, such as nematode taxonomy and nematode management, is essential to enhance the quality of research data. The holistic, interdisciplinary approach to crop science research should be further strengthened so that experts in plant breeding, agronomy, nematology and other related disciplines form formidable research teams. Such teams may deliver the expected results in science, technology and agriculture.

In conclusion, further research for developments in soybean nematode management and information-sharing should be encouraged and supported. This effort, if consolidated and well funded, will give birth to promising eco-friendly strategies in sustainable agriculture.

References

Agrios, G.N. (2005) *Plant Pathology*, 5th edn. Academic Press, San Diego, CA, USA.

Akinsanmi, O.A. and Adekunle, O.K. (2003) Effect of *Fusarium oxysporium* f. sp. *glycines* and *Sclerotium rolfsii* on the pathogenicity of *Meloidogyne incognita* race 2 to soybean. *Plant and Soil* 253, 429–435.

Ashfield, T., Ong, L.E., Nobuta, K., Schneider, C.M. and Innes, R.W. (2004) Convergent evolution of disease resistance gene specificity in two flowering plant families. *Plant Cell* 16, 309–318.

Back, M.A., Haydock, P.P.J. and Jenkinson, P. (2002) Disease complexes involving plant parasitic nematodes and soilborne pathogens. *Plant Pathology* 51, 683–697.

Biosecurity (2003) Biosecurity New Zealand importation countries phytosanitary requirements. Available at: http://www.biosecurity.govt.nz (accessed 22 May 2009).

Birchfield, W., Williams, C., Hartwig, E.E. and Brister, L.R. (1971) Reniform nematode resistance in soybeans. *Plant Disease Reporter* 55, 1043–1045.

Brockwell, J., Pilka, A. and Holliday, R.A. (1991) Soil pH is a major determinant of the numbers of naturally occurring *Rhizobium meliloti* in non-cultivated soils in New South Wales. *Australian Journal of Experimental Agriculture* 31, 211–219.

Buttery, B.R., Park, S.J. and Hume, D.J. (1992) Potential for increasing nitrogen fixation in grain legumes. *Canadian Journal of Plant Science* 72, 323–349.

Carneiro, R.G., Shimizu, M.M., Goncalves, K.V., Mazzafera, P. and Ferraz, L.C. (2002) Transport of ureides and amino acids in nodulated soybean infected by *Meloidogyne incognita* and *M. javanica*. *Nematologica Mediterranea* 30, 17–21.

Carsky, R.J., Abiadoo, R., Dashiell, K.E. and Sanginga, N. (1997) Effect of soybean on subsequent maize grain yield in Guinea savannah of West Africa. *African Crop Science Journal* 5, 31–39.

Coyne, D.L. and Oyekanmi, E.O. (2007) Symbiotic nitrogen fixation of two soybean genotypes as affected by root-knot nematode and microsymbionts. *Journal of Biological Sciences* 7, 1221–1226.

FAO (1983) *Technical Handbook on Symbiotic Nitrogen Fixation*. Food and Agriculture Organization of the United Nations, Rome, Italy.

Ferris, H., Castro, C.E., Caswell, E.P., Jaffee, B.A., Roberts, P.A., Westerdahl, B.B. and Williamson, V.M. (1992) Biological approaches to the management of plant-parasitic nematodes. In: Madden, J.P. (ed.) *Beyond Pesticides: Biological Approaches to Pest Management in California*. University of California Press, CA, USA, pp. 68–101.

Fourie, H., McDonald, A.H. and Loots, G.C. (2001) Plant parasitic nematodes in field crops in South Africa. 6. Soybean. *Nematology* 3, 447–454.

Golden, A.M., Epps, J.M., Riggs, R.D., Duclos, L.A., Fox, J.A. and Bernard, R.L. (1970) Terminology and identity of infraspecific forms of the soybean-cyst nematode (*Heterodera glycines*). *Plant Disease Reporter* 54, 544–546.

Goodman, R.M. and Kiser, J. (1985) Biotechnology and its impact on future developments in soybean production and use. In: Shibles, R. (ed.) *Proceedings* of *World Soybean Research Conference III*. Westview Press, Boulder, CO, USA and London, UK, pp. 261–272.

Graham, P.H. (1992) Stress tolerance in *Rhizobium* and *Bradyrhizobium* and nodulation under adverse soil conditions. *Canadian Journal of Microbiology* 38, 475–484.

Hinson, K. (1985) Breeding for resistance to root-knot nematodes. In: Shibles, R. (ed.) *Proceedings* of *World Soybean Research Conference III*. Westview Press, Boulder, CO, USA and London, UK, pp. 387–393.

Holderness, M., Sharrock, S. Frison, E. and M. Kairo (2000) Organic banana 2000. Towards an organic banana initiative in the Caribbean. Report of the International workshop on the production and marketing of organic bananas by smallholders farmers. International Network for the Improvement of Banana and Plantain, Montpellier, France.

Inagaki, H. (1979) Race status of five Japanese populations of *Heterodera glycines*. *Japanese Journal of Nematology* 9, 1–4.

Jacobsen, B.J., Edwards, D.F., Noel, G.R. and Shurtleff, M.C. (1983) The soybean cyst nematode problem. Report on Plant Disease, University of Illinois Cooperative Extension Service, No. 501, IL, USA.

Jones, T.J. and Perry, N.R. (2004) Plant parasitic nematodes, small animals, big impact. *Biologist* 51, 3.

Kanazin, V., Marek, L. and Shoemaker, R. (1996) Resistance gene analogs are conserved and clustered in soybean. *Proceedings of National Academy of Science, USA* 93, 11746–11750.

Khan, M.R., Kounsar, K. and Hamid, A. (2002) Effect of certain rhizobacteria and antagonistic fungi on root-nodulation and root-knot nematode disease of green gram. *Nematology Mediterranean* 30, 85–89.

Kinloch, R.A. (1982) The relationship between soil populations of *Meloidogyne incognita* and yield reduction of soybean in the Coastal Plains. *Journal of Nematology* 14, 162–167.

Lawn, D.A. and Noel, G.R. (1986) Field interrelationships among *Heterodera glycines*, *Pratylenchus scribneri* and three other nematode species associated with soybean. *Journal of Nematology* 18, 98–105.

Lim, B.K. and Castillo, M.B. (1979) Screening soybeans for resistance to reniform nematode disease in the Philippines. *Journal of Nematology* 11, 275–282.

Lowe, J. (ed.) (1992) *Nematological Research at IITA, 1969–1988: A Summary of Investigations Conducted by F.E. Cavenees. Plant Health Research Monograph 2.* International Institute of Tropical Agriculture, Ibadan, Nigeria.

Manyong, V.M., Dashiell, K.E., Oyewole, B. and Blahut G. (1996) Spread of new soybean varieties in a traditional soybean growing area of Nigeria. A seminar paper in second symposium of the African Association of Farming Systems Research, Extension and Training (AAFSRET), 20–23 August 1996, Ouagadougou, Burkina Faso, pp. 151–162.

Murphy, W.S., Brodie, B.B. and Good, J.M. (1974) Population dynamics of plant nematodes in cultivated soil: Effects of combinations of cropping systems and nematicides. *Journal of Nematology* 6, 103–107.

Musarrat, J. and Haseeb, A. (2000) Agrichemicals as antagonist of lectin-mediated *Rhizobium*-legume symbiosis: Paradigms and prospects. *Current Science* 78, 793–797.

Noel, G.R. (1984) Relating numbers of soybean cyst nematode to crop damage. In: *Proceedings of the 5th Cyst Nematode Workshop*, Raleigh, NC, USA, pp. 17–19.

Noel, G.R. (1985) The soybean cyst nematode. In: Lamberti, F. and Taylor, C.E. (eds) *Cyst Nematodes*. NATO ASI Series, Series A: Life Sciences, Volume 121, Plenum Press, New York, USA, pp. 257–268.

Nutter, F.W. Jr, Tylka, G.L., Guan, J., Morreira, A.J.D., Marett, C.C., Rosburg, T.R., Basart, J.P. and Chong, C.S. (2002) Use of remote sensing to detect soybean cyst nematode-induced plant stress. *Journal of Nematology* 34, 222–231.

Oyekanmi, E.O. and Coyne, D.L. (2009) Screening of six soybean cultivars for *M. incognita* and *Pratylenchus* sp. resistance. Preliminary report, Research for Development, Nematology Unit, IITA-Ibadan, Nigeria.

Oyekanmi, E.O., Coyne, D.L., Fagade, O.E. and Osonubi, O. (2007) Improving root-knot nematode management of two soybean genotypes through the application of *Bradyrhizobium japonicum, Trichoderma pseudokoningii* and *Glomus mosseae* in full factorial combinations. *Crop Protection* 26, 1006–1012.

Oyekanmi, E.O., Coyne, D.L. and Fawole, B. (2008) Utilization of the potentials of selected microorganisms as biocontrol and biofertilizer for enhanced crop improvement. *Journal of Biological Sciences* 8, 746–752.

Riggs, R.D. (1985) Strategies for race stabilization in soybean cyst nematode. In: Shibles, R. (ed.) *Proceedings* of *World Soybean Research Conference III*. Westview Press, Boulder, CO, USA and London, UK, pp. 528–531.

Sasser, J.N. and Carter, C.C. (1985) *An Advanced Treatise on Meloidogyne*, Volumes 1 and 2. North Carolina University, Raleigh, NC, USA.

Schmitt, D.P. (1985) Plant parasitic nematodes associated with soybeans. In: Shibles, R. (ed.) *Proceedings* of *World Soybean Research Conference III*. Westview Press, Boulder, CO, USA and London, UK, pp. 541–546.

Schmitt, D.P. and Baker, K.R. (1981) Damage and reproduction potentials of *Pratylenchus brachyurus* and *P. penetrans* on soybean. *Journal of Nematology* 13, 327–332.

Schmitt, D.P. and Noel, G.R. (1984) Nematode parasites of soybean. In: Nickle, W.R. (ed.) *Plant and Insect Nematodes*. Marcel Dekker, New York, USA, pp. 13–59.

Schmitz, A., Kiewnick, S., Schlang, J. and Sikora, R.A. (2004) Use of high resolution digital thermography to detect *Heterodera schachtii* infestation in sugar beets. *Communications in Agriculture and Applied Biological Sciences* 69, 359–363.

Sikora, R.A. and Greco, N. (1990) Nematode parasites of food legumes In: Luc, M., Sikora, R.A. and Bridge, J. (eds) *Plant Parasitic Nematodes in Subtropical and Tropical Agriculture*. CAB International, Wallingford, UK, pp. 181–235.

Sikora, R.A., Greco, N. and Silva, J.F.V. (2005a) Nematode parasites of food legumes. In: Luc, M., Sikora, R.A. and Bridge, J. (eds) *Plant Parasitic Nematodes in Subtropical and Tropical Agriculture*, 2nd edn. CAB International, Wallingford, UK, pp. 259–318.

Sikora, R.A., Bridge, J. and Starr, J.L. (2005b) Management practices: An overview of integrated nematode management technologies. In: Luc, M., Sikora, R.A. and Bridge, J. (eds) *Plant Parasitic Nematodes in Subtropical and Tropical Agriculture*, 2nd edn. CAB International, Wallingford, UK, pp. 793–825.

Sinclair, J.B. and Backman, P.A. (eds) (1989) *Compendium of Soybean Diseases*. American Phytopathology Society, St Paul, MN, USA.

Trudgill, D.L. (1991) Resistance to and tolerance of plant parasitic nematodes in plants. *Annual Review of Phytopathology* 29, 167–192.

Vincent, J.M. (1982) *Nitrogen Fixation in Legumes*. Academic Press, Sydney, Australia.

Von Mende, N., Gravanto Nobre, M.J. and Perry, R.N. (1998) Host finding, invasion and feeding. In: Sharma, S.B. (ed.) *The Cyst Nematodes*. Chapman and Hall, London, UK, pp. 215–238.

Westphal, A. (2005) Detection and description of soils with specific nematode suppressiveness. *Journal of Nematology* 37, 121–132.

Whitehead, A.G. (1998) *Plant Nematode Control*. CAB International, Wallingford, UK.

Yu, Y., Buss, G. and Maroof, M. (1996) Isolation of a superfamily of candidate disease-resistance genes in soybean based on a conserved nucleotide-binding site. *Proceedings of National Academy of Science, USA* 93, 11751–11756.

Zarkadas, C.G., Yu, Z., Voldeng, H.D. and Minero-Amador, A. (1993) Assessment of the protein quality of a new high-protein soybean cultivar by amino acid analysis. *Journal of Agriculture, Food and Chemistry* 41, 616–623.

Zhu, H., Cannon, S.B., Young, N.D. and Cook, D.R. (2002) Phylogeny and genomic organization of the TIR and non-TIR NBS-LRR resistance gene family in *Medicago truncatula*. *Molecular Plant-Microbe Interactions* 15, 529–539.

Zirakparvar, M.E. (1980) Host range of *Pratylenchus hexincisus* and its pathogenicity on corn, soybean and tomato. *Phytopathology* 70, 549–573.

Zirakparvar, M.E. (1985) Chemical control of nematodes attacking soybeans. In: Shibles, R. (ed.) *Proceedings* of *World Soybean Research Conference III*. Westview Press, Boulder, CO, USA and London, UK, pp. 523–527.

16 Soybean Processing and Utilization

Nawab Ali
Indian Council of Agricultural Research, KAB-II, Pusa, New Delhi, India

16.1 Introduction

Soybean (*Glycine max* (L.) Merrill) is an environment-friendly grain legume. Globally, it is a major source of protein, oil and health-promoting phytochemicals for human nutrition and livestock feed. Soybean cultivation also improves soil health because of its ability to fix atmospheric nitrogen and its deep root system. Soybean has now become an important world commodity because of its wide range of geographical adaptation, unique chemical composition, good nutritional value, functional health benefits and industrial applications. Soybean is, therefore, a human-, animal- and soil-friendly crop and its production and utilization should be encouraged and promoted for the betterment of our planet and its inhabitants. There is a great potential for the production and utilization of soybean and its derivatives for food, feed, pharmaceutical and industrial applications throughout the world.

Soybean originated in Manchuria, China, and from there it spread to Korea and Japan. Later, because of economic, nutritional and ecological benefits, soybean cultivation spread to many countries of the world, especially in Asia, Central America, South America, Europe and Africa. The present world production of soybean is about 223 million t, accounting for nearly 57% of the total global oilseed production. It provides approximately 60% of vegetable protein and 30% of vegetable oil in the world.

Soybean is an excellent source of nutrition and health-promoting phytochemicals. People also know soybean as a 'miracle bean', 'wonder bean' and 'golden bean' and say that it is a golden gift of nature to humanity for health and happiness (Holt, 1998; Ali, 2000, 2008; SOPA, 2002). Looking at the wide versatility of soybean uses and its production potential, it may not be out of place to call it the golden grain of the globe.

The production and utilization of soybean is growing. Global soybean production in the first decade of the 20th century was about 22 million t;

this has now, in the first decade of the 21st century, increased more than ten times to about 223 million t. Soybean has been and continues to be a major source of well-being for people in different regions of the world. Soybean use has increased in human nutrition and health, edible oil, livestock feed, biofuel and many other industrial and pharmaceutical applications.

All of the sectors involved in the soybean production to consumption value chain have responded to the scientific and technological developments. As a result, soybean production and productivity have increased to comply with the demand of a globalized economy. There have been many initiatives towards soybean production systems that are ecologically and socially sound and sustainable to counteract environmental degradation and climate change.

The benefits of direct soybean use for human health have been widely explored in many countries of Asia, but not so in other regions of the globe. However, during the 1980s, 1990s and 2000s, scientific and technological developments in processing and utilization of soybean have increased exponentially towards exploiting the full benefit of this golden grain and expanding soybean nutritional advantages to a greater number of people of the world.

Soybean has now become the preferred vegetable protein for food applications due to its multiple functional properties. It is a cost-effective, high-quality ingredient that can replace dairy, egg and meat proteins as consumers search for ever-increasing variations to diet staples.

16.2 Diet and Longevity

Human beings are the highest form of life on our planet and need food for survival. Each and every person desires to live longer and would perhaps, if possible, prefer to live forever. However, as of now, everyone born on this most beautiful and dynamic planet Earth has to die one day. That said, a person can enhance his or her life span through careful living. One of the major requirements of careful living is a balanced diet based on a proper mix of plant and animal produces, in addition to a disciplined daily routine and living including physical and mental exercises.

Globally, people are becoming conscious of their longevity and searching for an appropriate diet that can lead them to a disease-free, healthy and happy life. To achieve these goals, one of the best options is to go with nature – live and eat seasonal and regional foods and adopt naturopathy and diet therapy in the case of minor ailments, if needed. In this respect, soybean has a tremendous potential to be transformed into a number of healthy foods, suiting individual requirements, across the globe.

Utilization research has shown soybean to be an excellent source of nutrition and health-promoting phytochemicals (Ali, 1991; Holt, 1998; Patricia and Newton, 1998; Gandhi *et al.*, 2008). Soy protein is the best plant protein, least expensive and, when taken along with cereals, protein quality improves by about 30%. On account of its health and economic benefits, soybean has been and continues to be a major source of good nutrition around the world. A number of soy-based healthy food products have been developed and are available on the market. Soybean may be processed and easily used in the home kitchen. A total of 30–50 g of carefully processed soybean in the daily diet helps to prevent many diseases and leads to better health, happiness and longevity.

16.3 Production and Productivity

The world production of soybean during 2007/2008 was about 223 million t (Table 16.1). The five major soybean-producing countries of the world are the USA, Brazil, Argentina, China and India. The highest soybean productivity is 2890 kg ha^{-1} in the USA and the world average is 2430 kg ha^{-1}. The total production of major oilseeds in the world during 2007/2008 was about 393 million t, with the share of soybean at around 57% (Table 16.2). The other major oilseeds are mustard/rapeseed (*Brassica* species), cottonseed (*Gossypium* species), groundnut (*Arachis hypogaea*), sunflower (*Helianthus annuus*), palm kernel (*Elaeis guineensis*) and coconut (*Cocos nucifera*).

Table 16.1. World soybean production (2007/2008) and productivity of soybean in the major producing countries of the World (2006/2007) (compiled from FAO, 2009; Soystat, 2009; USDA, 2009).

	Production			Productivity	
Country	Production (million t)	% of total production	Country	Average yield (kg ha^{-1})	Rank
USA	70.60	31.70	USA	2890	1
Brazil	62.00	27.90	Canada	2860	2
Argentina	47.00	21.10	Argentina	2680	3
China	14.40	6.50	Brazil	2670	4
India	9.50	4.30	Paraguay	2350	5
Paraguay	6.50	2.90	China	1740	6
Others	13.00	5.00	India	950	7
World total	223.00	100.00	World average yield	2430	

Table 16.2. World production of major oilseeds during 2007/2008 (compiled from FAO, 2009; USDA, 2009).

Oilseeds	Production (million t)	% of total world production
Soybean	223	57.0
Mustard/rapeseed	48	12.3
Cottonseed	45	11.7
Groundnut	32	8.0
Sunflower	28	7.0
Palm kernel	11	3.0
Coconut	6	1.0
Total	393	100

Table 16.3. Production and productivity of soybean in India during 1970/1971 to 2008/2009 (compiled from Directorate of Economics and Statistics, 2007; SOPA, 2008).

	Soybean	
Production year	Production (million t)	Productivity (kg ha^{-1})
1970/1	0.01	426
1980/1	0.44	728
1990/1	2.60	1015
1999/2000	5.05	895
2000/1	5.01	860
2001/2	5.40	900
2002/3	4.30	758
2003/4	6.93	1074
2004/5	6.12	849
2005/6	7.40	963
2006/7	7.15	950
2007/8	9.47	1070
2008/9 (est.)	10.82	1124

Table 16.4. Vegetable oil production in India (SOPA, 2008) and the world (USDA, 2009) during 2006/2007.

India			World		
Vegetable oil	Production (million t)	% of total production	Vegetable oil	Production (million t)	% of total production
Mustard	2.2	31.45	Palm	38.1	31
Soybean	1.3	18.60	Soybean	35.7	29
Cottonseed	0.9	12.85	Rapeseed	17.8	15
Rice bran	0.7	10.00	Sunflower	10.2	8
Groundnut	0.6	8.55	Groundnut	4.9	4
Sunflower	0.6	8.55	Cottonseed	4.8	4
Coconut	0.4	5.74	Palm kernel	4.3	4
Sesame	0.1	1.43	Coconut	3.3	3
Others	0.2	2.83	Olive	2.9	2
Total	7.0	100	Total	122.0	100

The production and productivity of soybean in India is growing (Table 16.3). In 2007/2008, soybean production in India was about 9.5 million t, with an average productivity of 1070 kg ha^{-1}. As per 2007/2008 data, the major soybean-producing states of India are Madhya Pradesh (4.98 million t), Maharashtra (3.24 million t) and Rajasthan (0.74 million t), accounting for about 52.4%, 34.1% and 7.8%, respectively, of total soybean production in India. It is expected that soybean production in India may reach 20 million t by 2020.

Soybean oil augments the total Indian edible oil pool by about 19% by contributing 1.3 million t (Table 16.4). The other major edible oils in India are mustard, cottonseed, rice bran (*Oryza sativa*), groundnut, sunflower and coconut. Palm and soybean oils dominate the world edible oil supply by contributing about 60% of total world vegetable oil production. The contribution of olive (*Olea europaea*) oil to global total vegetable oil production is 2%.

16.4 Nutritional and Economic Benefits

Soybean is one of the oldest food sources known to humans. On an average, it contains about 40% protein, 23% carbohydrates, 20% oil, 5% mineral, 4% fibre and 8% moisture (Gopalan *et al.*, 1974; SOPA, 2002). Soybean is recognized for its value in enhancing and protecting health. Soy protein contains all of the eight essential amino acids. The recent discovery of the value of soy-isoflavones and their role in disease prevention has created a special interest in soybean. It has boundless food potential. However, soybean also contains some antinutritional factors such as trypsin inhibitor, urease and flatulence factors. Hence, it requires careful processing before utilization.

Soybean, being rich in protein and calories, has a great potential to tackle the problem of protein–calorie malnutrition from which many people in India and other developing countries suffer. Soybean contains twice as much protein as pulses, groundnut, meat and fish; three times as much as eggs; and more than ten times that of milk. In addition, soybean is the most economical source of dietary protein in the world and is superior to other plant proteins. Soybean does not contain lactose. Hence, soy milk and other dairy analogues are best suited to lactose-intolerant people. Soybean is also a very good food source for those with diabetes. Overall, soybean is an environment-friendly crop that is needed for better global health.

India has a population of >1100 million people. The majority (65–70%) are vegetarians who derive their proteins from pulses, cereals and milk and to some extent from oilseeds such as groundnut, sesame and soybean. In general, the quality of the protein eaten by the population is poor. Better-quality proteins from egg, meat and aqua products are costly and only a small proportion of the population has access to them. About 30% of the Indian population is below the poverty line and does not have enough purchasing power for good-quality dietary proteins. It is therefore important to provide them with an alternative source of dietary protein that is financially affordable. Soybeans meet this requirement. Hence, for India, one option is to make use of soybean as a protein source to augment the conventional protein supply at a cost/price that is affordable by all, especially those with lower incomes. In India, the cost of 1 kg protein from full-fat soy flour is just US$1.5, as compared to US$4, US$5, US$8, US$10, US$12 and US$18 from split pulses (dal), egg, milk, chicken, fish and meat, respectively (Gandhi *et al.*, 2008).

Soybean is high in protein and is one of the very few plant-based foods to contain all of the essential amino acids. Therapeutically, soybeans lower cholesterol, boost the immune system and, being high in fibre, ease constipation. Soybean is extremely versatile. It contains a high proportion of polyunsaturated fat. It can be processed into a number of fermented and non-fermented food products. Research into phytochemicals has shown that soybean contains phytoestrogens, and so may help in managing irregular periods, premenstrual syndrome, menopausal hot flushes and post-menopausal problems such as osteoporosis, fatigue and vaginal dryness (Holt, 1998; Connie, 1999). It may also help guard against cancers, including prostate cancer.

The US Food and Drug Administration has approved a health claim stating that 25 g of soy protein in a daily diet low in saturated fat and cholesterol can help reduce total and low-density lipoprotein cholesterol (FDA, 1999). Various research studies undertaken the world over have indicated that the inclusion of soy foods in the daily diet not only provides good-quality protein, but also helps prevent diseases such as diabetes, breast cancer, osteoporosis, heart attack and memory loss (Holt, 1998; Patricia and Newton, 1998; Messina, 2002; SOPA, 2002). The use of whole-bean-based food provides all of the nutritional benefits that soybean offers and, when included in a diet with cereals, the food provides an excellent source of nutrition.

16.5 Processing and Products

Soybean is processed into a very wide range of products to realize its astonishing potential as food, feed, pharmaceutical and industrial products. Traditionally, soybean has been utilized mainly as fermented (e.g. sauce, *miso*, *nattō*) and non-fermented (e.g. oil, milk, tofu, flour) foods. During the 20th century, however, with the increased demand for meat and eggs, the use of soy products as feed has been extensively developed, mainly in Western countries and to a lesser extent in Asia. In addition, during the 1980s, 1900s and 2000s, there have been tremendous improvements in soybean processing and utilization technologies and significant developments in marketing. Technologies are constantly being adapted to produce better-quality milk and milk products and soy protein isolates. New findings also include the physiological functions of soybean for human nutrition and health.

The global goal for soybean processing and utilization is to strengthen the development of new food, feed, pharmaceutical, cosmetic and industrial products, including co-products and ingredients for speciality applications. As of now, soybean derivatives are gaining importance not only in nutritious food products, but also as sources of phytochemicals and nutraceuticals to reduce the risk of coronary heart disease, osteoporosis, cancer, diabetes and so on. The emphasis in the non-food or industrial products markets is on biodegradable adhesives, plastics, coatings, inks, lubricants, biodiesel and more.

Scientific advances made in characterizing the phytochemical properties of molecular components of soybean, with the aid of sophisticated research instruments and processing equipment, are leading to the discovery of new structural and functional properties of ingredients and product performances. The advances made are related to chemical and physiological properties of soybean oil and protein and relating them to changes that occur because of breeding programmes or genetic development, growing-season variations, agronomic

Table 16.5. Exploiting soybean food potential: technology and products (fermented and non-fermented) (Gandhi *et al.*, 2008; reprinted with permission).

Form of soybean	Technology	Products
Whole soybean (direct food uses)	Separation, soaking, blanching, boiling, drying, size reduction, fermentation, extrusion, packaging, storage, marketing	Full-fat soy flour, milk, paneer (tofu), curd, ice-cream, tempeh, sauce, sprouted and roasted snack, extruded snack foods, soy fortified bakery, fermented foods
Partially de-fatted soybean (oil and cake)	Mechanical expression, physical refining, enzyme, cooking, size reduction, packaging, storage, marketing	Oil, margarine, medium-fat soy flour, bakery foods
Fully de-fatted soybean (oil and meal)	Solvent extraction, refining, hydrogenation, size reduction, separation and concentration, packaging, storage, marketing	Oil, vanaspati, soy meal, de-fatted soy flour, lecithin, soy protein concentrate, isolates and hydrolysates, speciality and health foods
By-products of soybean (hull, okara and whey)	Size reduction, fermentation, separation, packaging, storage, marketing	Dietary fibre, single-cell proteins, citric acid, enzymes, alcohol

factors and processing parameters. The restructuring of separated and modified triglycerides of oils and polypeptides of proteins with unique properties and chemical-enzymatic modifications of components to enhance functionalities, such as foaming, emulsification, film and adhesive properties, are enhancing the continued development of new food, feed and industrial applications.

Soybean is generally used as a raw material in the form of whole beans or partially or fully de-fatted cake or meal for making various soy-based food products. Whole beans are used for making full-fat soy flour, dairy analogues and fermented and snack foods. Soy flour can also be made from partially or fully de-fatted beans (cake/meal) and used in making baked products, texturized soy proteins, protein isolates and concentrates, extruded snack foods and so on. A range of technologies – physical, chemical, biological or a combination of these – are used in making various soy-based fermented and non-fermented foods (Table 16.5). However, the option of technology depends on the type of product and its use.

Soy nuts

Soy nuts are whole soybeans that have been soaked in water and roasted until brown. Soy nuts, like whole soybeans, are excellent source of protein, fat and isoflavones. Soy nuts can be eaten as an alternate to peanuts, which are expensive and pose the problem of aflatoxins. Soy nuts provide higher protein at lower cost. Roasted soy nuts have been found to have at least 60% less fat than peanuts. Most conventional nuts are incredibly high in fat;

Table 16.6. Comparison of nutritive value of major nuts with soy nuts (adapted from SFPWA, 2008).

Nut	Calories (per 100 g)	Carbohydrate (%)	Fat (%)	Protein (%)
Almonds	618	26.4	53.6	15.4
Cashew nuts	607	18.2	49.3	31.8
Soy nuts	500	28.2	26.4	37.1
Peanuts	607	25.0	53.6	17.9

Fig. 16.1. Flow diagram for the manufacture of soy nuts (adapted from SFPWA, 2008).

in comparison, soy nuts have less fat and more protein (Table 16.6). Soy nuts are similar in texture and flavour to peanuts, yet far less expensive.

Many soy foods, including roasted soy nuts, are rich in calcium. Other calcium-rich soy foods include tofu, *tempeh*, textured vegetable protein and soy milk fortified with calcium. Calcium in soy foods is readily absorbed by the body.

Plant-based foods contain dietary fibre. A fibre-rich diet is very important in reducing the risk of certain types of cancer and heart disease. Soybeans, especially roasted soy nuts, are an excellent source of fibre.

The protein in soy is 'complete protein of highest quality, equal to that of meat and milk products' (SFPWA, 2008). Soybeans are exceptionally high in protein; therefore, soy is an inexpensive way to add protein to the diet. About 40% of the calories in soybeans come from protein, while other beans contain approximately 20–30% protein. A flow diagram for manufacturing soy nuts is shown in Fig. 16.1. Tables 16.6 and 16.7 give a comparison of the nutritive value of the major conventional nuts with soy nut and a nutritional composition of oil-roasted and dry-roasted soy nuts, respectively.

Table 16.7. Composition of oil-roasted and dry-roasted soy nuts (adapted from SFPWA, 2008).

Component	Dry roasted	Oil roasted
Carbohydrate (%)	32.7	33.6
Crude fibre (%)	5.4	4.6
Calcium (mg 100 g^{-1})	4.0	3.9
Energy (kcal 100 g^{-1})	450	474
Fat (%)	21.5	25.4
Moisture (%)	1	2
Protein (%)	39.6	35.2
Zinc (mg 100 g^{-1})	4.8	3.1

Soy milk and tofu

Soy milk is a creamy, milk-like product that is made by soaking and grinding soybean in water. Soy milk has been known in mainland China for centuries. Hot soy milk is used as a breakfast beverage in mainland China, Japan, Taiwan and Thailand. During the last few decades, soy milk has been introduced to other parts of the world.

Besides being rich in protein, vitamins and minerals, soy milk is a very economical, lactose free, highly digestible and nutritious alternative to a dairy and meat-centred diet. It is cholesterol-free product with a very low fat content and is rich in polyunsaturated fatty acids of phospholipids, especially lecithin and also linolenic acid. Soy milk generally contains around 7–8% total solids. Adding 3–4% sugar and about 0.05% salt brings it to a sugar, salt and total solid level that is approximately identical to 2% fat cow's milk (i.e. about 12–13% total solids). This can be consumed as such or after sweetening and diluting. Alternatively, soy milk can be made into yogurt (curd) or tofu (*paneer*) or used directly in cooking.

Tofu is the most popular soy product. Also called soy *paneer* in India and bean curd in China, it is a tasty and very nutritious product made by coagulating hot soy milk with food-grade chemicals such as calcium chloride, magnesium chloride, calcium sulphate, acetic acid and citric acid. It is a versatile food that can be converted into a variety of value-added products. Tofu is a perfect and inexpensive substitute for milk cheese or paneer. Nutritionally, its protein is as good as that derived from animal sources. Tofu is a very porous product that absorbs the flavour of the food with which it is cooked.

Tofu is an extremely perishable food. It should be kept in water under proper refrigeration and the water should be changed daily. This will keep tofu fresh for a week or more, depending on the refrigeration conditions. Preferably, tofu should be used when it is as fresh as possible. Vacuum-packed tofu is also available on the market.

The undissolved residual portion left after extracting soy milk from soybeans during the process of making soy milk is known as *okara*. *Okara* is rich in protein and fibre and is mainly used in the preparation of biscuits and other bakery items and for the thickening of soups and gravies. It is also used as a main protein source in cattle feed. It is a highly perishable product with a very limited shelf life under normal conditions. Proper refrigeration can preserve it for a week or more.

Texturized soy protein product

Texturized soy protein (TSP), also known as soy nuggets or granules, is made from edible-grade de-fatted soy flour containing about 50% protein. TSP is a versatile, economical, nutritional and convenient good-quality protein source. As a principal ingredient in daily food, TSP has not only gained momentum over the years, but has also shown commendable penetration into different markets. By its nature, the product goes exceptionally well with vegetable and non-vegetable dishes alike in India. As the product is a convenient, protein-rich and affordable substitute to fresh vegetables, it has found wide acceptance in regions and seasons where such vegetables are scarce.

TSP is made using extrusion technology. An extruder consists of a barrel in which the food material, de-fatted soy flour, is carried by a turning screw and subjected to intense frictional stress, which generates heat under high pressure. The heat and pressure reach a maximum by the time the food material reaches the terminal end, where it is ejected through the nozzle. The texture of the product undergoes considerable change during passage through the barrel, from globular to fibrous, and when the material comes out through the nozzle it expands to give a porous texture preferred for human consumption. Due to the high pressure and temperature during cooking in the barrel, heat-labile anti-nutritional factors are destroyed while nutritional properties and values are maintained. The extruder is a very flexible piece of equipment and can be used for any flour or combination of flours. The texture of the product can be varied from fully expanded to a tightly compact, meat-like product. The length of the cooking barrel can be changed. As the product emerges, it is cut using rotating cutter knives. The shape of the final product can be altered by adjusting the dies and the cutting arrangement. TSP contains about 50% protein and other nutrients.

Oil extraction

Various products, co-products and by-products that can be obtained in soybean solvent extraction processing (Cherry, 2004) are shown in Fig. 16.2. The main commercial products of soybean are oil and meal (protein). Soy meal contains 45–50% good-quality protein and can be processed into a variety of high-protein co-products such as livestock feeds, de-fatted soy flours, soy protein concentrate, soy protein isolates and TSP (Table 16.8).

Much soybean oil is used for edible applications such as for cooking, salad oil, dressing, shortening, margarine, mayonnaise and confectionery coating. Soybean oil is also being used for biodiesel, lubricants and cleaning

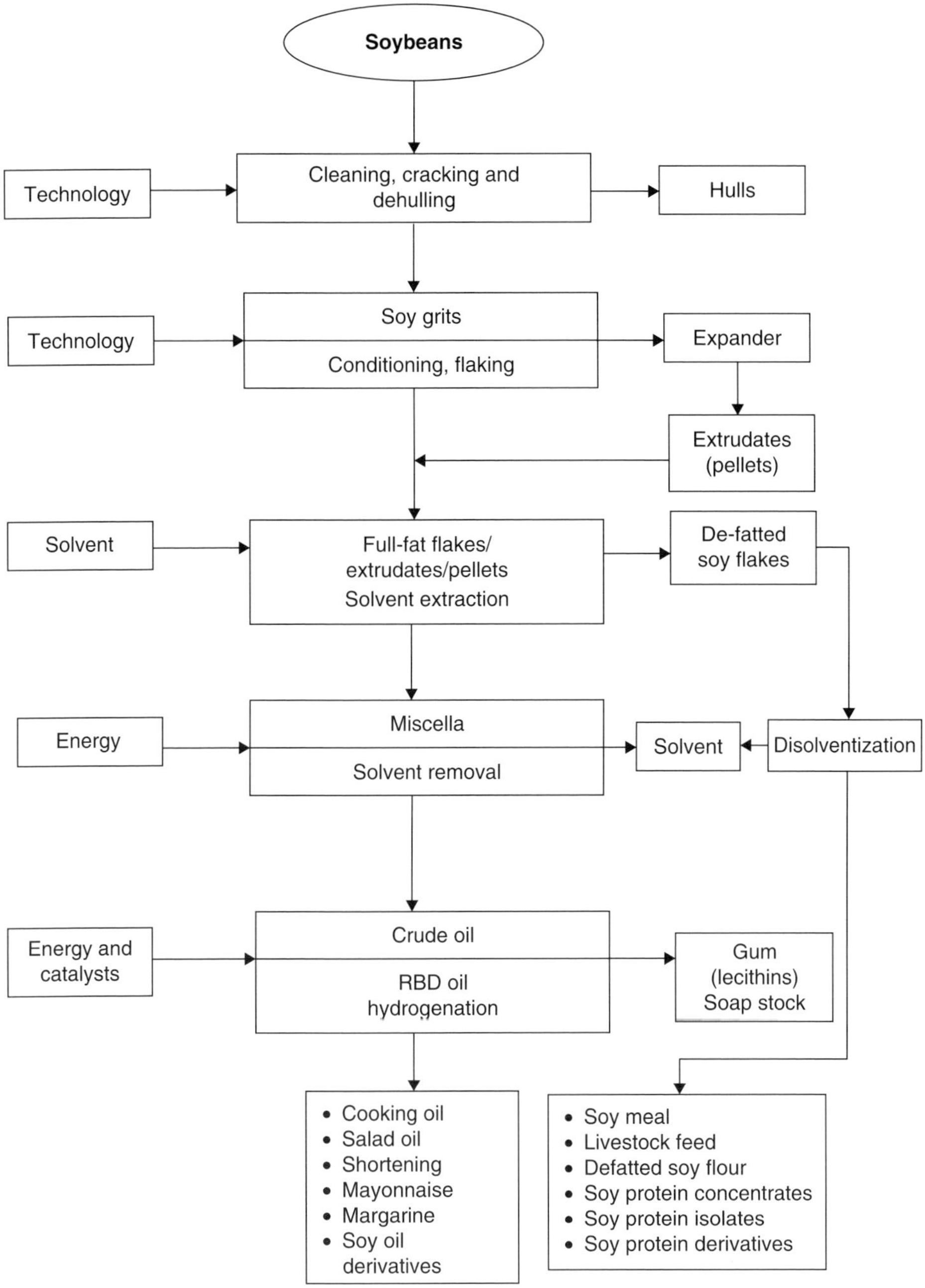

Fig. 16.2. Flow diagram for soybean solvent extraction processing and resultant products, co-products and by-products (modified from Cherry, 2004).
RBD, refined, bleached and deodorized.

Table 16.8. Products, co-products and by-products of the soybean solvent processing industry (adapted from Cherry, 2004).

Products	Co-products	By-products
Oil	Refined, bleached and deodorized oil: cooking oil, salad oil, shortening, margarine, mayonnaise, biodiesel. Lecithin, soy oil derivatives	Hull, germ, gum, feed fat, soapstock, spent catalysts, distillates/sludge, spent bleaching earth
Protein	Soy meal: livestock feed, de-fatted soy flour, soy protein concentrate, soy protein isolates, texturized soy protein, soy protein derivatives	

agents. Lecithin, once treated as a by-product of crude oil refining, has grown in use and value to become a co-product of soybean processing. It is a valuable ingredient, primarily a mixture of phospholipids including phosphatidylcholine, phosphatidylethanolamine, phosphatidylserine, phosphatidylinositol and phosphatidic acid. These components have outstanding functional properties when applied to food, feed and other products. The pharmaceutical and cosmetic industries have benefitted greatly from this ingredient. The unique lipophilic and hydrophilic properties of lecithin components make them very useful.

It is estimated that there are currently 400–500 non-food soy and soybean-containing products being manufactured and marketed by 200–300 companies all over the world. These products can be placed into three categories: consumer products, ingredient intermediates and institutional or industrial products. Examples of these products are listed in Table 16.9.

During soybean processing into various products and co-products, by-products such as hull, molasses, soap stock, deodorizing distillates, spent bleaching earth and spent catalyst are obtained. These by-products are natural compounds and their use or disposal has been a challenge to the soybean-processing industry. A number of these by-products are returned to the land as fertilizer or soil modifiers, fed to livestock as feed, burnt for energy or applied to value-added conversion. The by-products are rich in carbohydrates, protein, lipids, lignins and phytochemicals. More focused research is needed to utilize these by-products as value-added co-products, similar to as has been successfully done with the once by-product, now a co-product, lecithin. Some soybean processing by-products are briefly described in Table 16.10.

Soybean continues to earn its worldwide importance as an accepted food and feed resource. The ability to use processed soy products and ingredients in many applications as biodegradable substitutes for petroleum-based products is a rapidly growing industry. Moreover, some selected soybean by-product-based ingredients are gaining recognition as major nutritional sources of nutraceuticals and functional foods.

Soybean oil is extracted using solvent and expeller technologies. Hexane is the most common solvent used in the soybean oil industry. Expander

Table 16.9. Examples of soybean-based non-food products and ingredients (adapted from Cherry, 2004).

Consumer products	Ingredient or intermediate products	Institutional or industrial products
These products are sold by companies in packaged forms of different sizes through stores and used every day. Examples: Lubricant and greases Biodiesel Auto polish Building material composites Candles Carpets Coatings Hair care Motor oil Adhesives Engine oil Waxes Personal care	These are not finished products, but chemicals and ingredients used by manufacturers in making finished products and sold by companies in bulk. Examples: Industrial oil for paint or varnishes Industrial protein for cosmetics Industrial solvent Lubricant base stocks Surfactants/emulsifiers Waxes for specialty applications	These are used by companies in construction and by institutions such as schools, hospitals and farms. They are packaged in large quantities for multipurpose application, depending upon the needs of the users. Examples: Agricultural adjuvant Dielectric fluid Dust suppressants Fuel oil emulsifier Lubricant Industrial cleaner Metal-working fluid Odour reduction Printing ink Saw guide oil

Table 16.10. Soybean processing by-products and their potential uses (adapted from Cherry, 2004).

By-product	Description and potential uses
Hulls	A by-product of the seed-crushing process. Hulls are made up of cellulose, hemicellulose, lignin, protein and minerals, and can be converted to valuable co-products such as feed that can be used as a replacement, up to 10%, for forage fibre in dairy cow rations to increase milk yield. Hulls can also be processed and used as dietary fibre. They lower blood serum cholesterol. Ground soy hull fine particulate matter treated with citric acid works as an excellent metal absorbent, especially of copper ions, for cleaning waste streams. Another component of soy hulls is the enzyme peroxidase, which has industrial applications.
Soy molasses	Obtained while preparing soy protein concentrate from soy meal. An ethanol extract of soy molasses called phytochemical concentrate (PCC) has been shown to repress genomic DNA elastogenic damage and point mutations in mammalian cells. Isoflavones, genistein, genisten, daidzein and daidzin have been identified in PCC. These compounds express a wide range of growth suppression of selected human colon cancer cells.

(Continued)

Table 16.10. *continued*

By-product	Description and potential uses
Deodorizer distillates or sludge	Obtained during soybean oil refining, and also known as scum oil. Sludge is composed of a complex mixture of aggregated compounds, tocopherol (vitamin E), sterol esters (campestene, sitosterol), squalene and free fatty acids (FFA). Tocopherol, the basis of the demand to process this by-product, can be used by the vitamin market and as a nutraceutical, sterol or feed stock for hormone production; squalene, a high carbon source for cosmetic and cholesterol biosynthesis; and fatty acids for biodiesel. A supercritical CO_2 extraction method may be economically feasible for preparation of tocopherol from deodorized distillate/sludge.
Spent bleaching earth/clay	Contains a substantial amount of absorbed oil. The by-product is prone to spontaneous combustion and is a source of recoverable oil using steam, aqueous, solvent or pressure extraction treatments. It can also be used as a feed supplement, pellet binder and metabolizable energy source for poultry diets.
Feed fat and soapstock	Feed fat is obtained while de-gumming soybean crude oil for refining and it is used for animal feed. Further refining produces soapstock. This is a mixture of FFA, phosphatides and unsaponifiable materials. It is also used in animal feed.
Spent catalysts	Results from the use of nickel and clay support in the hydrogenation of fats and oils. The recovered material contains nickel, clay, residual oil and filtration media used to aid in separation of the physically fine spent catalyst. Incineration is being used as one way to recycle nickel into the stainless steel industry, while the released energy is used for generating electricity.

technology has improved plant capacity and reduced solvent and energy losses. The technology of supercritical (SC) fluid extraction (SCFE) is emerging as a safe and viable extraction system. Extrusion-aided mechanical extraction of soybean oil is an environment-friendly technology for the production of soybean oil and edible-grade soy cake. Membrane technology may be used for separation of oil from miscella. Physical refining offers an increased oil yield and reduced energy consumption. Refined soybean oil is used in various types of cooking and may also be converted into margarines, shortenings and other products. Soy lecithin has many usages. Soy by-products such as soy meal, with 50% protein, offer a wide range of products for food and livestock feed. Soy hull is a good source for the production of single-cell protein for animal feed.

The purpose of oil milling is to separate oil from the protein-rich residual mass/cake. Oil and cake are then used for food, feed or fertilizer. Whether the oil is to be separated by pressing, solvent or a combination of the two, the seed is usually first cooked and flaked to rupture the cell walls, reduce oil viscosity and increase the rate of diffusion.

Oil extraction technology has not changed much and both expeller and solvent extraction technologies are standard options for the oil industry throughout the world. Solvent extraction technology is normally preferred for low-oil seeds such as soybean. Better mechanical and thermal designs leading to improved processes and equipment for desolventizing and miscella distillation are offering substantial energy savings. A flash-desolventizing technology is available to produce edible-grade, high-protein meal. For miscella distillation, waste heat utilization and three-stage evaporation with a spherical liquid film system is now the state of the art technology to reduce energy consumption as well as solvent loss.

One of the new technological developments is in seed preparation using an extruder/expander. This offers better energy and solvent efficiency and an increased plant capacity. Expander technology for direct extraction plants, processing low-oil-bearing seeds such as soybean offers steam saving up to 60 kg t^{-1} of seed, lower solvent loss and about 0.3% more oil recovery. This technology may also be adapted for high-oil seeds. Membrane separation of oil from solvent offers scope for energy saving. SCFE is emerging as a safe and viable extraction system. Extrusion-aided mechanical extraction of soybean oil has emerged as an environment-friendly and decentralized small-scale technology that is suitable for low soybean-production catchment areas. It results in chemical-free, high-quality edible soybean oil and cake.

The function of the preparation process of soybean is to prepare the seed for expelling and/or extraction of the oil by mechanical methods, solvent or a combination of both. If possible, the hulls and other materials should be removed from the seed kernel or meat. The unit operations involved in the preparation of seed and their purposes are given in Table 16.11.

An innovative and new technology in pre-extraction is the addition of an expander, also known as an extruder or enhancer press, to the conventional preparation after flaking (Singh and Ali, 1992; Ali, 2004). The flakes are extruded to form pellets, which enhance the extraction and drainage properties of the flakes. The extruder consists of a worm screw in a barrel. Flakes with about 18% moisture are fed in and high temperature and pressure are generated as the flakes pass through the barrel. When the material leaves the barrel, the sudden drop in pressure causes expansion of steam, which puffs the final product to produce favourable drainage and extraction properties. Excess moisture is removed and the material is cooled to 60°C before extraction. Success has also been found by extruding soybean at 10–14% moisture (Rittener, 1984). The subsequent increase in extraction efficiency and oil quality justifies the additional cost of an extruder or expander.

Oil is removed or extracted from the flakes or pellets by an organic solvent (n-hexane) to form an oil/solvent mixture called miscella. The oil is recovered from the miscella by removing the solvent. Since hexane is explosive, flammable and expensive, efforts have been made to investigate alternate solvents such as alcohol, isopropanol and CO_2. Pressurized CO_2 (named as an SC fluid) (Freidrich and Pryde, 1984)-extracted oil contains less phospholipids than hexane-extracted oil, but they are otherwise similar. It is expensive because of equipment cost and the need to generate high pressure. The de-fatted flakes from the extractor contain about 30% hexane by weight, which is removed through a desolventizer-toaster. The process also

Table 16.11. Major unit operations in seed preparation before extraction (adapted from Ali, 2004).

Unit operation	Purpose
Cleaning	Removes foreign materials from soybean. Includes the separation of plant tissues, pebbles, dust and so on to protect the processing equipment and enable the production of high-quality soy products.
Drying	A moisture content of about 10% is needed for effective removal of the soy hull and hence drying of soybean before dehulling.
Cracking	Cracking is performed to break soybean into small pieces for dehulling and flaking. It produces 4–6 cotyledon fragments/meat per bean and the hulls are separated from the cotyledon fractions by aspiration. Fines produced in the process are included in the meat for oil extraction to maximize the extraction yield.
Conditioning	The cracked soybean grits/meats are conditioned using heat and moisture to obtain the optimum plasticity necessary for soy flake production prior to oil extraction. Steam heating raises the moisture to about 11%.
Flaking	The conditioned soy grits are flaked to a thickness of 0.25–0.37 cm. This enables better flow of solvent through the bed and improves solvent penetration to oil bodies. It also reduces the diffusion distance to which the solvent or miscella (oil and solvent mixture) moves to extract oil.

enhances the nutritional value of soy protein by inactivating trypsin inhibitors and other naturally occurring toxicants. Steam comes into contact with the flakes and the heat of vaporization released from the condensing steam vaporizes the hexane, which is subsequently condensed and recovered.

The solvent extraction process is favoured over mechanical expellers because it leaves less oil in the meal. However, the possible escape of solvent from the system is a constant air pollution and explosion hazard. Economic and social factors have revived interest in searching for cheaper and safer solvents such as ethanol and isopropanol. However, it is SC fluid technology that may be a viable alternative to the current extraction methods. SCFE is a developing technology. It is a way to strip soluble extractives from prepared plant materials without physical damage or chemical change. SCFE is the substitution of a fluid in its SC state for hexane in the conventional solvent extraction process.

Commercial uses of SCFE include decaffeination of coffee and the production of spice extracts and a few costly food components and pharmaceuticals from plant materials. Ammonia, ethylene, toluene and CO_2 all show promise for SCFE. Of these, CO_2 offers unique advantages. It is abundant, non-reactive, non-toxic and environmentally harmless. Minor leaks or accidental losses would be of little consequence. All vegetable oils are soluble in SC-CO_2 and the optimal performance of SC-CO_2 has been found to be in an easy temperature range of 35–70°C.

Full-fat soy flakes are readily extracted with SC-CO_2 at pressure of 200–700 kg per cm^2 at 50°C. Soybean oil thus extracted is lighter in colour and contains less iron and about one-tenth the phosphorus of hexane-extracted crude oil from the same beans. The successful use of SC-CO_2 could free >20 million gallons of costly hexane per year for essential energy uses. It therefore shows that a technology for oil extraction with SC-CO_2 needs to be adapted or developed. This may replace the existing technology of oil extraction in coming years.

In mechanical extraction, the oil seed is subjected to extreme heat and pressure with oil mechanically forced from the oil cell. As the material is subjected to great heat during this operation, naturally occurring urease is inactivated and protein is denatured, making the product suitable for feed purposes. The quality of mechanically pressed and filtered oil is higher than that obtained from solvent extraction as less oil-soluble impurities (e.g. phosphatides) are removed, and is suitable for direct consumption. Efficiently pressed cake will retain 4–6% residual oil. Solvent-extracted meal has <1% residual oil. However, the main disadvantage of solvent extraction is high equipment cost, and it may not be economically feasible unless the capacity is ≥50 t day^{-1}.

Mechanical extraction of soybean oil is often preferred by small enterprises. A screw press is generally used. The advantages are low initial costs and no requirement for solvents. However, a disadvantage of this extraction method is the low oil yields. Dry extrusion cooking of whole or dehulled soybean disrupts the cell structure of cotyledons. Consequently, the oil is released from the spherosomes into the matrix. When such soy-extrudate is passed through a screw-press, about 70% of the total soybean oil is recovered in a single pass. This technology could be used for processing soybean into oil and edible cake in low soybean-production catchment areas. It does not use any chemicals and, therefore, it is an environment- and worker-friendly soybean oil expression technology.

Soybean requires careful processing using heat treatment to make it edible. One such treatment is dry extrusion. The adoption of dry extrusion as a pre-treatment to soybean helps in expelling about 70% of the total oil in a single pass and results in an edible-grade soybean cake with about 50% protein and 4–6% oil. The cake can be converted into medium-fat soy flour – a good source of protein and calories for human consumption.

The cleaned soybean is dehulled and converted into grits and fed to the extruder after conditioning. During the extrusion of soy grits, the cell walls of oil globules are ruptured and oil is exposed on the surface of feed particles. This helps with quick and easy expression of oil with a relatively lower pressure application on the semi-fluid extrudate fed to the expeller.

Medium-fat soy flour may be used for the fortification of flour prepared from cereals or pulses (10–15% level of soy flour). The blended flour (soy and cereal or pulse) is excellent for baking bread, biscuits and so on. Both soy products – oil and flour – have a distinct market segment that requires better functional properties of traditional recipes in addition to high nutritional value. Oil has a ready market among edible oil consumers and can be sold to refining units. The medium-fat flour could find ready acceptability in army canteens, hostels, hotels, railway catering services and various other community kitchens, as well as in individual households through retail stores.

The extrusion-expelling plant has a great potential in low soybean-production catchment areas (3000–15,000 t year^{-1}) or where soybean is likely to be cultivated in the near future. In addition to the technical and economical advantages, the system is pollution-free and avoids the use of chemicals for the extraction of oil. It produces a natural oil with a good shelf life and nutritionally good-quality protein, as it retains most of the lysine amino acid, which, upon blending (5–15% of soy flour) with cereals or pulses, improves the protein efficiency ratio.

A 1 t h^{-1} extrusion-expelling soybean plant consisting of buildings, transport facilities and equipment such as a grain cleaner, destoner, dehuller, dry extrusion cooker and accessories, oil expeller, oil filter and grinding mill may need a capital investment of about Rs 6 million (Rs 4 million fixed investment and Rs 2 million working capital). Such a plant may produce 3600 t of medium-fat soy flour, 600 t of oil and 600 t of by-products from 4800 t of soybean by working for 16 h day^{-1} for 200 days in a year (US$1 = Rs 50). The techno-economic feasibility analysis of producing medium-fat (4–6%) soy flour and oil using extrusion-expelling technology appears to be viable from a financial perspective, in addition to providing high-quality and low-cost food items to society.

Refining crude soy oil

Crude soybean oil contains a lot of non-glyceride impurities. Some of these substances are beneficial and some are not. The most objectionable impurities are free fatty acids, gum/phosphatides, volatile/odoriferous compounds and dark-colour pigments. These are removed or reduced as far as possible by the refining process to achieve an edible oil. Crude oil can be refined by either chemical processing or physical refining. However, as of now, about 90% of the world soybean crude oil is refined by chemical methods (Table 16.12) and only a small amount is physically refined.

The shelf life of refined soy oil depends on its fatty acid composition and the type of packaging employed. Oils rich in polyunsaturated acids deteriorate faster in the presence of oxygen and light with the passage of time. Therefore, the most appropriate way of packing soy oil is to use an opaque container with zero oxygen permeability and fill the oil to the top without any air or head space. Such packaging aids and containers include metal cans, dark-glass bottles and opaque plastic containers. Some of the constituents of soybean crude and refined oils are given in Table 16.13.

Soybean oil properties and its usage

Physicochemical properties of soybean oil depend on its fatty acid composition and processing conditions. These are measured and monitored in terms of refractive index; iodine valve; smoke, flash and fire points; melting point;

Table 16.12. Operations/reactions involved in the chemical refining of soybean oil (compiled from Ali, 1995, 2004; Narula, 1997).

Operation	Detail
Degumming	Gum/phosphatides (soy lecithin) are removed using hydration and centrifugation processes.
Neutralization	Neutralization removes or reduces free fatty acids. This is done through a chemical reaction between free fatty acids and a base (NaOH). In this process, non-hydrolysable gum, trace metals and saponifiable matter are also reduced or removed. Neutralization is performed in batches as well as in a continuous operation.
Bleaching	Bleaching is done to remove or reduce the colour of oil. It is basically a physical operation involving the absorption of colour pigments in the oil into the micropores of the activated earth under vacuum and at a suitable temperature. Three basic pieces of equipment – a bleaching reactor, activated earth/clay and a filtration system – are required.
De-odourization	This is done to remove or reduce volatile and odoriferous compounds. The process is basically a vacuum steam distillation operation at an elevated temperature of 200–250°C and at 3 Torr vacuum.

Table 16.13. Some of constituents of soybean crude and refined oil (Ali, 1995; Narula, 1997).

	Present in	
Constituent	Crude oil	Refined oil
Triglycerides (%)	95–97	99
Phosphatides (%)	1.5–2.5	0.003–0.045
Unsaponifiable matter (%)	1.6	0.3
Plant sterol (%)	0.33	0.13
Tocopherol (%)	0.15–0.21	0.11–0.18
Hydrocarbon (%)	0.014	0.010
Free fatty acids (%)	0.3–1.0	≤0.05
Iron (ppm)	1–3	0.1–0.3
Copper (ppm)	0.03–0.05	0.02–0.06

crystallization behaviour; polymorphism; and a few others. The oxidative stability of soybean oil has been greatly improved because of the progress made in processing and refining technologies. Properly processed soy oil, particularly hydrogenated soy oil, is now suitable for many food applications.

A wide variety of products based on edible soybean oil are available on the market. Salad and cooking oil, shortening, margarine, mayonnaise, salad dressing and confectionery coating are some of the available products.

With improved flavour stability and stronger emulsion, soy oil has now become the oil of choice for most manufacturers. The soy oil used in mayonnaise is usually partially hydrogenated followed by winterization. Soybean oil is also used in salad dressing. Mono- and diglycerides of edible soybean oil and soybean lecithin are among the most commonly used emulsifying agents approved for food uses. Other applications of soybean oil are in canned foods, confectionery coatings, pudding mixes, pancake and waffle mixes, macaroni and cheese mixes, pasta sauces, dry breakfast cereals, frozen fried sea foods, meat patties, pizza mixes and others. Partially hydrogenated soybean oil is used in some whipped toppings.

Soybean oil is also used for industrial applications such as emulsifiers, lubricants, plasticizers, surfactant, plastics, solvent and resins. Soy oil is biodegradable, of low toxicity to humans and the ecosystem, renewable and a non-contributor to volatile organic chemicals, making it a superb material for industrial applications. Soybean oil is successfully used in printing ink, lubricants, pesticide carriers and paints and coatings.

Lecithin, an edible by-product of soybean crude oil refining, has a variety of useful functionalities. It is a mixture of phospholipids. Soy lecithin purification consists of the removal of non-lecithin components such as carbohydrates, proteins and other contaminants. A wide range of functionalities of lecithin has been applied in a variety of industries such as the food, pharmaceutical and cosmetic industries (Table 16.14).

Enhancing soybean oil quality

One of the major quality considerations of soybean oil is its oxidative stability. This refers to the resistance of soy oil to oxidative reactions during handling, storage and processing. The outcome of the reaction leads to off-flavour development. Normal soy oil contains about 4% stearic, 8% linolenic, 11% palmitic, 23% oleic and 54% linoleic acids. Because of a large proportion of polyunsaturated fatty acids (85%), particularly linolenic acid, soy oil often has a problem with lacking flavour stability during processing, storage and application. There are two ways to improve soy oil stability. One is processing and the other is through genetic/breeding means. Additional processing of refined soybean oil is through interesterification, winterization and fractionation (Table 16.15).

Plant and environment safety

Industrial and environmental safety is becoming increasingly important to oil processors. Soybean handling and its milling can present a number of hazards. Practical safety precautions must be taken when personnel work with cleaning, storage or transportation tanks. Falling and head injuries may occur on slippery surfaces in restricted spaces. Protective clothing, an air supply and breathing apparatus, a worker in attendance in case of emergency and a harness and safety line are all sound safety measures. Various

Table 16.14. Use of soybean lecithin in food and industrial products (compiled from Szuhaj, 1979; Cherry, 2004).

Products	Purpose
Margarine	As an emulsifier and anti-spattering agent.
Confection and snack foods	Crystallization control in chocolates, viscosity control in caramels and for anti-sticking properties in coatings. As a softener in chewing gum and for prevention of agglomeration in breakfast cereals and bars.
Instant foods	As a wetting and dispersing agent and emulsifier in cocoa powder, instant drinks, instant cocoa, instant coffee, protein drinks, coffee whiteners, cake mixes, puddings, instant toppings and instant soup.
Bakery products	As a starch complexing agent for crystallization control and also as an emulsifier. Wetting and release agent. Areas of application include breads, rolls, doughnuts, cookies, cakes, pasta products, pies and crusts.
Cheese products	As an emulsifier and release agent.
Meat and poultry processing	As a surfactant, sealant and browning agent. Emulsifier.
Dairy products	As an emulsifier, wetting and dispersing agent, anti-spatter agent.
Miscellaneous products	As an emulsifier and crystallization control in products such as peanut spread, salad products, flavour and colour solubilization.
Package aid	As a release agent and sealant for products such as interior coating, sausage, casing coating and stocking nets.
Processing equipment	As a release agent, lubricant, anti-corrosive agent on processing equipment such as frying surfaces, extruders, conveyors, boilers, dryers, blenders and evaporators.
Cosmetics	As an emulsifier, emollient, softening agent, texture controller, lubricant and anti-corrosive agent in shampoos and lipsticks, creams and oils for the hands and body.
Pharmaceuticals	Dietary supplement for man and animals, as an emulsifying and dispersing agent for intermuscular injections, vitamins, creams and ointments.
Plastic and rubber industry	As an internal release agent and plasticizer, injection and die moulding of polyethylene, polypropylene, nylon and rubber; tyre-manufacturing; polymer extrusion, coloration and pigmentation and toy manufacturing.
Glass and ceramics	As a release and dispersing agent.
Paper and printing	As an emulsifier, wetting and dispersing agent in typewriter ribbon and paper manufacture.
Petroleum industry	As a lubricant and detergent.
Metal processing	As a lubricant, flushing aid, anti-corrosive agent in cutting, threading, milling, turning, welding, wire-drawing, rolling, casting, polishing and so on.
Miscellaneous	As an emulsifier, wetting and dispersing agent in pesticides and adhesives. As a softening, lubricating and penetrating agent in the textile and leather industries.

Table 16.15. Addition processing of refined soy oil to enhance its quality (adapted from Ali, 1995, 2004; Narula, 1997).

Process	Purpose
Interesterification	Changes the triglyceride melting point and crystallization without altering the fatty acid composition.
Winterization	A process of removal of solids that settle out from the oil at 4–10°C. The solid portion is known as starine and is separated out so that oil does not precipitate during refrigeration of salad oil. The final oil recovery is 75–85% of the total oil winterized.
Fractionation	An alternative to winterization where triglycerides are separated by thermo-mechanical means. It produces oil fractions with distinctive properties.

emissions and waste products are produced in soy oil processing that could pose environmental problems if not properly disposed of.

The application of expander technology in solvent extraction, extrusion-aid in mechanical expression and membrane technology in soybean oil refining may rise in future. The extrusion-expelling system provides mechanically expressed soy oil and edible soy cake, facilitating the use of both oil and protein for human consumption (Nelson *et al.*, 1987). This system is an environment- and worker-friendly technology. Soybean and its constituents provide many health benefits and may therefore, in the future, be treated as a medicinal crop rather than a mere source of oil and protein. Hence, the future of soy is very bright throughout the world.

16.6 Soybean Oil Industry in India

The soybean oil industry in India is about 30 years old. It started with a few plants of about 100 t day^{-1} capacity and grew from these few plants to about a few hundred t day^{-1} of soybean. Today, the processing capacity is approximately 51,000 t day^{-1}. The largest plant in India can process about 1500 t day^{-1}, but most plants have a capacity in the range of 300–500 t day^{-1}. There are about 122 major soybean processing plants in India (Table 16.16) and most are located in Madhya Pradesh, Maharashtra and Rajasthan. The total installed processing capacity of soybean is about 15.5 million t, while the total production of soybean in India, at present, is about 10 million t.

The total world soy meal export during 2006/2007 was about 53 million t, in which India's share was about 3.5 million t (7%). Argentina and Brazil dominate the world soy meal export market, sharing about 49% and 22%, respectively (Table 16.17). Soy meal export from India has been increasing steadily, rising from 2.2 million t in 2000/2001 to 4.9 million t in 2007/2008 (Table 16.18).

Table 16.16. State-wise number of solvent extraction plants with their capacity in India (Ali, 1994; SOPA, 2008).

State	No. of plants	Capacity (t day^{-1})	Capacity (t $year^{-1}$)
Andhra Pradesh	5	1,370	411,000
Chhattisgarh	7	1,910	573,000
Gujarat	9	3,300	990,000
Haryana	1	300	90,000
Karnataka	2	800	240,000
Madhya Pradesh	48	28,175	8,452,500
Maharashtra	37	10,025	3,007,500
Punjab	2	450	135,000
Rajasthan	10	4,700	1,410,000
Tamil Nadu	1	300	90,000
Total	122	51,330	15,399,000

Table 16.17. World soy meal export during 2006/2007 (compiled from FAO, 2009; Soystat, 2009; USDA, 2009).

Soy meal exporting country	Quantity exported (million t)	% of total soy meal export
Argentina	26.5	49
Brazil	11.6	22
USA	7.9	15
India	3.5	7
Bolivia	1.0	2
Others	2.4	5
Total	52.9	100

Table 16.18. Export of soy meal from India during 2000/2001 to 2007/2008 (SOPA, 2008; reprinted with permission).

	Export (quantity and value)	
Year	Million t	Million Rs
2000/2001	2.215	18,900
2001/2002	2.681	23,481
2002/2003	1.554	14,763
2003/2004	3.582	46,566
2004/2005	1.904	18,078
2005/2006	4.183	37,228
2006/2007	3.526	36,190
2007/2008	4.888	73,320

16.7 Soy Food Industry and Market in India

The primary interest in soybean in India has been the oil. However, increasing attention is now being paid to the protein potential that soybean offers. Today, about 15% of the total soybean production in India goes to direct food and feed uses, 10% is used for seed and 75% is processed for oil and protein. Crop residues such as leaves and fine straw are used as feed and the hard and woody stem for fuel in rural areas.

The major food uses of soybean in India are of oil, protein and lecithin. The present soybean oil processing capacity is about 15.5 million t of soybean, while that for food is about 2 million t. The Indian soybean industry has the capability to process soybean for food, feed, pharmaceutical and industrial applications. The present soy food industry has about 500 units.

The major food uses of soybean in India are currently edible oils, TSP, flours, bakery products, milk/*paneer* (tofu), soy protein concentrate/isolates/hydrolysates, lecithin and others. There is a need is to create an awareness about soy products and their benefits and make such products available on the market through small-scale decentralized soy food processing enterprises. Domestic-level processing and utilization of soybean for food and feed needs to be given priority, especially in the rural sector. Central and state developmental agencies may come forward to form an implementation plan. The hardware and technology are available to create a number of soy foods that match with Indian food recipes and food habits.

Acceptance of soybean foods in India is increasing, but at a slow pace because it is a new introduction to the food baskets of Indian people. In order to accelerate the process of promotion of soy foods, it will be necessary to create awareness of their economic and health benefits, transfer presently available technology and develop new and diversified products and human resources.

Processed soybean in the form of full-fat soy flour would cost Rs 30 kg^{-1} in the current retail market, with about 40% protein and other nutrients. This compares with the average cost of 1 kg split pulse (dal), which is about Rs 50 with a protein content of about 25%. Quality-wise, soy protein is better than pulse protein. In fact, soy protein is of the best quality among all plant proteins. Soy-based food items, such as full-fat soy flour containing nutrients as well as phytochemicals, is healthy and economically affordable for all sections of the Indian population, especially those living below the poverty line. The cost of 1 kg soy protein in the form of full-fat soy flour is Rs 75, whereas 1 kg protein in the form of dal (split pulses) is Rs 200. The protein efficiency ratio of soy protein increases considerably when combined with cereal and other legume proteins. Health benefits that may be derived from the regular use of soybean in the daily diet are given in Table 16.19.

Indian snack foods such as cookies, *sev*, *laddu*, *halwa* and roasted or fried soybean have been prepared by replacing 15–30% of the traditional ingredients with soy products and found to be acceptable by the people (Gandhi *et al.*, 2008). Soy-fortified wheat flour gives good preparation of chapatis, *puri* and *paratha*. Chickpea flour (*besan*) mixed with soy flour

Table 16.19. Major health benefits from the regular use of soybean in the daily diet (compiled from Patricia and Newton, 1998; Liu, 1999; Danji, 2000).

Nutrients	Health benefits
Protein	Lowers blood cholesterol
Carbohydrates	Relaxes constipation. Good for diabetics
Minimum saturated fats	May help to prevent cardiovascular diseases
Minerals	Overall health promotion
Vitamins	Overall health promotion
Phytochemicals	Prevents cancer, helpful in menopause and osteoporosis

Table 16.20. Traditional and non-traditional soy products being used in India and those with high potential for use in the near future (Ali, 1996; reprinted with permission from the Food and Agriculture Organization of the United Nations).

Traditional products		
Non-fermented	Fermented	Non-traditional products
Cooking oil	Yogurt	Texturized soy protein
Hydrogenated fat	*Tempeh*	Health foods
Flours, flakes and grits	Sauce	Lecithin
Roasted/fried nuts		Protein concentrate
Sprouted beans		Protein isolate
Cooked beans		Protein hydrolysate
Milk		Liquid protein
Paneer (tofu)		
Bakery products		
Splits (dal)		
Vadi		

results into a good-quality fried snack foods such as *sev* and *pakoda* (Gandhi *et al.*, 2008). Soy flour incorporated into *idli/dosa* batter gives good and acceptable products. Soy-cereal extruded snacks have been found quite acceptable by consumers. Green soybean can also be used as a vegetable. Soy products can fit very well into various Indian food recipes, provided they are properly processed and blended. Technologies and hardware are necessary to develop entrepreneurs in soy food manufacturing and marketing. Traditional and non-traditional soy products are listed in Table 16.20.

Many small-scale entrepreneurs are interested in soy-based food manufacture. However, they wish to see the working of a complete plant to know the actual investment cost and the facts about plant capacity and product quality, to judge the market potential and to be sure of training facilities available for their production staff, before deciding to establish a soy food plant. Project reports along with economic profiles and analysis-based actual commercial-scale plants are needed. Training in soy-based food production and machinery is also required for individuals, groups and entrepreneurs for domestic use and commercial manufacturing.

Despite possessing a number of good features, soybean is associated with a few constraints for food uses. These are the beany flavour of soy food products and the oxidative instability of soybean oil. So far, two major approaches have been used to overcome these constraints: innovative processing and plant breeding. Now, other means such as marketing efforts, medical discoveries, consumer education and changes in dietary habits are also needed. The acceptance of soy foods in India has been rather slow. However, more and more people are now tilting towards soy food because of its economic, nutritional and health benefits.

16.8 Emerging Trends

People are becoming health conscious and the demand for speciality foods is therefore increasing. Soybean has a tremendous potential to be transformed into a number of such foods suited to people's requirements. The daily use of soybean in the diet can provide balanced nutrition at a low cost, in addition to health benefits. Awareness of these aspects is now spreading. The likely utilization pattern of soybean in the 21st century is in direct food uses, mechanically expressed and physically refined soy oil, livestock and aqua feed and pharmaceutical and other industrial products. The strategy would be for the complete utilization of soybean constituents for food, feed and pharmaceutical products. This requires needs-based and high-quality research and development in the areas of soybean processing and utilization.

Soybean was originally processed to improve its shelf life, inactivate or remove anti-nutritional factors, improve safety, make desirable sensory changes, produce more convenient products and add value. Today, however, some of these reasons have changed, with a focus on maintaining or saving the components of soybean that have health benefits. For example, instead of removing oligosaccharides that cause flatulence in some people, these are now looked upon favourably because they increase the bifidobacteria population in the colon, giving a protective effect against pathogenic organisms. Isoflavones and trypsin inhibitor (Bowman-Birk inhibitor) are believed to provide protection from cancer (Danji, 2000; Messina, 2002). With increasing interest in the use of soybean in food, due to its health benefits, the demand for direct food uses of soybean may increase in the future.

Soybean is one of the nature's wonderful nutritional gifts of plant origin and it provides a high-quality protein with minimum saturated fat. Soybeans help people feel better and live longer, with an enhanced quality of life. Soybeans contain all of the three macronutrients required for good nutrition, as well as fibre, vitamins and minerals. Soybean protein provides all of the essential amino acids in the amounts needed for human health. Almost 40% of the calories from soybeans are derived from protein, making soybeans higher in protein than any other legume and many animal products. The quality of protein in soybean is remarkable. Health professionals consider soy protein to be a superior protein. The amino acid pattern of soy protein is virtually equivalent in quality to that of milk and egg protein. During the 1990s, the Food and Agriculture Organization/World Health Organization Protein Evaluation Committee put soy protein on a par with egg and milk protein (FAO/WHO, 1990).

Unlike many other good sources of protein, soybean not only has a higher percentage of oil, but also a good-quality fatty acid profile. It has a low saturated fat content with a high amount of polyunsaturated fatty acids, and is a readily available source of essential fatty acids. Compared with other legumes, soybean contains more than double the amount of most of minerals, especially calcium, iron, phosphorus and zinc, and a very low sodium content.

Highlighting research findings with respect to the nutritional and economic benefits of using soybean through the print and electronic media has a great potential to increase the demand for soy foods. The industry must respond to such a demand with an array of soy-based and isoflavone-fortified conventional foods. For soy foods to become truly mainstream, a variety of convenient, user-friendly products is needed and efforts should be made to make them available on the market at an affordable price. The industry will have to take soy foods to the consumer, rather than relying on the consumer to seek them out. Conventional breads, snacks, crackers and breakfast cereals to which soy has been added are likely to be particularly attractive. A breakfast cereal that combines oats or corn with soy flakes represents a convenient way for consumers to incorporate soy into their diet and it does not require lifestyle modification. Nearly half the soy protein needed to lower cholesterol could easily be consumed at one sitting if such a cereal provides 5–6 g of soy protein and is eaten in combination with soy milk.

The need is to create an awareness about soy products and their benefits and to make such products available on the market through small-scale decentralized soy food processing enterprises. Domestic-level processing and utilization of soybean for food and feed need to be given priority, especially in the rural sector. Central and state development agencies may come forward to make implementation plans. The hardware and technologies for a number of soy foods that match traditional food recipes and food habits are available. The need is to make use of such indigenous facility for the benefit of the people. The following measures are suggested to accelerate soybean food uses in the world:

- Create awareness among the masses about the economic, nutritional and health benefits of soybean and its products using print and electronic media.
- Train individuals, groups and entrepreneurs in the manufacturing and marketing of soy-based food products and machinery.
- Make technical support available to potential entrepreneurs in the form of project reports, consultancy and services.
- Link research and industry to refine the product and modify technologies, with time, for better efficiency and the output of a high-quality product.

- Forge a strong political will and positive government policies to encourage the production and domestic utilization of soybean.

16.9 The Future of Soybean

Soybean has been and will continue to be a major source of well-being for the people of the world in both developing and developed countries. Scientific and technological advances have increased soybean production and productivity and thereby the global economy. The benefits of soybean use on human nutrition and health have been widely explored during the last 10–15 years and, as a result, the direct food use of soybean is increasing at a very fast rate throughout the world.

The global population is increasing, and with it the demand for food (of plant and/or animal origin). Soybean is a very rich and economical source of food and nutrition for humans as well as for livestock. Hence, more soybean will be needed in the future. The average income of people in Asia and Africa is increasing, and with it the demand for animal products. Soy meal is a major livestock protein source around the world. Hence, the demand for soy meal will rise. New industrial applications of soybean will increase the demand for soybean and its products. Soybean enriches the soil and also protects the environment. Hence, for sustainable agricultural productivity soybean has to be used in the cropping system. New findings on the role of soybean in human health will boost the direct food uses of soybean globally and almost a majority of people in the world will eat soybean by 2025.

In the near future, soybean production and utilization efforts should be concentrated on varietal development for higher productivity and specific end uses; mechanization of soybean production and post-production operations for higher yields, minimum post-production losses and better-quality grains; and direct food uses of soybean and the development of decentralized, small-scale soybean processing units in rural and urban areas.

For all of this to happen, there must be a commitment to and missionary zeal from all of those involved in soybean production and the consumption value chain. The backward link of industry with farmers will ensure a regular supply of good-quality soybean that suits specific products and needs. Through such an arrangement, all of the partners involved in the soybean value chain – namely farmers, processors and consumers – will benefit to a great extent.

References

Ali, N. (1991) Final Report of Indo-US Sub-Project on Soybean Processing and Utilization. Technical Report No. CIAE-SPU/7(85-91)/91/140. CIAE, Bhopal, India.

Ali, N. (1994) Small scale soybean processing industry in Sri Lanka – a case study. *Agricultural Engineering Today* 18, 25–31.

Ali, N. (1995) Milling of oilseeds. In: *Treatise on Fats, Fatty Acids and Oleochemicals*. Industrial Consultants (India), Karampura, New Delhi, India, pp. D1/1–D1/6.

Ali, N. (1996) Soybean processing and utilization in India. The Country Report presented in the Regional Expert Consultation on Soybean Processing and Utilization in Asia. FAO Publication, 1996/2, Bangkok, Thailand, pp. 327–328.
Ali, N. (2000) Soybean processing technology and promotional efforts for its food uses. In: *Proceedings of the International Soybean Processing and Utilization Conference (ISPUC-III)*, Tsukuba, Tokyo, Japan, 15–20 October 2000, pp. 327–328.
Ali, N. (2004) Soybean oil milling technology and modified usage of oil and its by-products. In: *Proceedings of the VII World Soybean Research Conference. IV International Soybean Processing and Utilization Conference and III Brazilian Soybean Congress*, Foz do Iguassu, Brazil, 29 February–5 March 2004, pp. 852–862.
Ali, N. (2008) *Soybean – The Medicinal Crop of the Future*. Soybean Processing and Utilization Centre, Central Institute of Agricultural Engineering, Bhopal, India.
Cherry, J.P. (2004) Utilization-application of by-products from processing soybean. In: *Proceedings of the VII World Soybean Research Conference, IV International Soybean Processing and Utilization Conference and III Brazilian Soybean Congress*, Foz do Iguassu, Brazil, 29 February–5 March 2004, pp. 1111–1119.
Connie, W. (1999) Soy and osteoporosis. In: *Proceedings of the World Soybean Research Conference–VI*. University of Illinois, Urbana-Champaign, IL, USA, p. 428.
Danji, F. (2000) Recent progress in research and technology for processing and utilization of soybean. In: *Proceedings of III International Soybean Processing and Utilization Conference*, Tsukuba, Tokyo, Japan, 15–20 October 2000, pp. 11–16.
Directorate of Economics and Statistics (2007) *Agricultural Statistics at a Glance*. Directorate of Economics and Statistics, Department of Agriculture and Cooperation, Ministry of Agriculture, Government of India, Krishi Bhavan, New Delhi, India.
FAO (2009) Food outlook – global market analysis. Available at: http://www.fao.org/docrep/011/ai482e/ai482e06.htm (accessed 2 June 2009).
FAO/WHO (1990) Protein Quality Evaluation FAP/WHO Nutrition Meetings Report series 51. FAO/WHO, Rome, Italy.
FDA (1999) Food labeling, health claim, soyprotein and coronary heart diseases. *Federal Register* 57, 699–733.
Freidrich, J.P. and Pryde, E.H. (1984) Supercritical CO_2 extraction of lipid bearing materials and characterization of the products. *Journal of American Oil Chemists' Society* 61, 223–228.
Gandhi, A.P., Kulkarni, S.D. and Jha, K. (eds) (2008) Highlights of Achievements of Soybean Processing and Utilization Centre. Technical Bulletin No. CIAE/2008/142. Central Institute of Agricultural Engineering, Bhopal, India.
Gopalan, C., Rama Sastri, B.V. and Balasubramanian, S.C. (1974) *Nutritive Value of Indian Food*. National Institute of Nutrition, Hyderabad, India.
Holt, S. (1998) *Soy Revolution – The Food of the Next Millennium*. M. Evans and Company, New York, USA.
Liu, K. (1999) Current constraints in soybean food utilization and efforts to overcome them. In: *Proceedings of the World Soybean Research Conference VI*. University of Illinois, Urbana-Champaign, IL, USA, pp. 409–418.
Messina, M. (2002) *Soya and Health*. Extension Pamphlet, American Soybean Association and United Soybean Board, New Delhi, India.
Narula, O.P. (Editor-in-chief) (1997) *Treatise on Fats, Fatty Acids and Oleochemicals*. Industrial Consultants (India), Karampura, New Delhi, India.
Nelson, A.I., Wijayratane, W.B., Yeh, S.W. and Wei, L.S. (1987) Dry extrusion as an aid to mechanical expelling of oil from soybean. *Journal of American Oil Chemists' Society* 64, 1341–1347.

Patricia, G. and Newton, H.H. (1998) *The Whole Soy Cookbook*. Three River Press, New York, USA.

Rittener, H. (1984) Conditioning of oil bearing materials for solvent extraction by extrusion. *Journal of American Oil Chemists' Society* 6, 1200–1203.

SFPWA (2008) Project report on soy food products. Soy Food Promotion and Welfare Association (SFPWA), New Delhi, India.

Singh, J. and Ali, N. (1992) Recent developments in edible oil milling technology. *Journal of Institution of Engineers (India)* 72, 23–27.

SOPA (2002) *Soybean for Health and Nutritional Security in India*. The Soybean Processors Association of India, Indore, India.

SOPA (2008) *Soybean Industry Scenario in India*. The Soybean Processors Association of India, Indore, India.

Soystat (2009) World statistics. Available at: http://www.soystats.com/2009/page_30.htm (accessed 7 September 2009).

Szuhaj, B.F. (1979) Food and industrial uses of soybean lecithin. In: *Proceedings of the World Soybean Research Conference–II*. WSRC–II and NCS University, Raleigh, NC, USA, pp. 681–692.

USDA (2009) *World Oilseed Production – Oil Crops Year Book – 2009*. USDA, Washington, DC, USA.

17 Nutritional Value of Soybean

Vineet Kumar, Anita Rani and G.S. Chauhan
Directorate of Soybean Research, Indore, Madhya Pradesh, India

17.1 Introduction

Soybean (*Glycine max* (L.) Merrill) seed has a unique confluence of properties of legumes as well as oilseeds. The protein content, which is approximately 40%, far exceeds that of the other legumes except for lupin (*Lupinus* species). Although the oil content of soybean seed (20%) is far less than that of other oilseeds such as peanut (*Arachis hypogaea*), mustard (*Brassica* species) and sunflower (*Helianthus annuus*), soybean oil is the second largest in terms of production and use. Beyond protein and oil content, soybean seed has been reported to contain biological nutraceutical ingredients including isoflavonoids, tocopherols, lecithin, biopeptides such as lunasin and Bowman-Birk factor, which have been reported to provide protective effects against hormone-dependent cancers, cardiovascular disease (CVD), diabetes, osteoporosis and menopausal problems. Although soybean been used for the treatment of ailments in southeastern countries for many years, recent scientific investigations confirming the clinical significance of these health-promoting components of soybean has helped in the acceptance of soy as the functional food of the century. Despite these virtues, soybean does suffer from the presence of protease inhibitors, flatulence factors, nutrient-binding phytic acid and off-flavour-producing lipoxygenases in higher concentrations than most other legumes. Fermented soy products (i.e. fermented tofu, *miso*, *tempeh* and *nattō* from the traditional bastion of Southeast Asia), which have reduced levels of some of these undesirable components, have not evoked much response in other parts of the world due to their unique processing and taste, while non-fermented soy products have not been adopted due to the off-flavour associated with them.

Soy meal is the major protein source in animal feed for livestock and poultry production. However, poor metabolizable energy due to oligosaccharides, deficiency of sulphur-containing amino acids and the costs involved in degrading the high concentration of phytic acid to enhance bioavailability of nutrients and avoid phosphorus pollution are some of the serious issues associated with the use of soy meal in the poultry and swine industries.

This chapter discusses the nutritional value of the major biological components of soybean, supported by established studies and recent findings. Special attention is given to the efforts made through conventional breeding, mutation and biotechnological approaches to develop specialty soybeans to address the needs of the soy food and feed industries.

17.2 Basic Nutrients

Soybean protein is as good as animal protein

Soybean is the second largest source of protein, after lupin, for the vegetarian population of the world. On average, the protein content in commercial cultivars is approximately 40%, ranging from 34% to 48% depending upon the genotype, growing environment and cultural practices of the crop. Most of the soybean produce is processed for extraction of the oil and the resultant meal contains approximately 48% protein. Soy protein concentrate and soy protein isolate are two widely used soy products in the food and pharmaceutical industries. The former is soy flour devoid of soluble carbohydrates and contains about 70% protein, while the latter is soy flour devoid of both carbohydrates and dietary fibre and contains as high as 90% protein. Until 1990, according to protein quality evaluation methods, the protein efficiency ratio (based upon the requirement of young growing rats) of soy protein was considered inferior to that of animal protein. Subsequently, in 1991 the World Health Organization adopted a new method for protein quality evaluation called the protein digestibility corrected amino acid score (PDCAAS) (Schaafsma, 2005). This compares the amino acid profile of a food protein to the requirements of 2- to 5-year-old child. The amino acid requirement profile of this age group in particular was chosen as the criterion because it is higher in this age group than in adolescents and adults. PDCAAS is computed by dividing the actual requirement of the most limiting amino acid (i.e. cysteine and methionine in soybean) for the 2- to 5-year-old human body by the value present in the food protein, multiplied by the digestibility of the protein source. Based upon this new method, the amino acid score of soybean was found to be equivalent to that of animal protein, as the sulphur-containing amino acid concentration in soy protein was not significantly less than that required by the human body. Therefore, the global vegetarian population can address its daily protein requirement ($0.8\,g\,kg^{-1}$ of body weight) through soy products. The requirement of poultry and pigs for sulphur-containing amino acids is higher than the concentrations present in soy meal. Globally, however, soybean meal accounts for 63% of all the protein sources used in animal feed because being a rich source of lysine, tryptophan, threonine, isoleucine and valine, it complements well with other cereal grains, which are deficient in these amino acids.

Proteins present in soybean are globulin storage proteins. Based upon sedimentation coefficients, soy globulins have been categorized as 2S, 7S, 11S and 15 S. Of these storage proteins, 11S (glycinin) and 7S (β-conglycinin) together constitute about 65–80% of the total seed protein. The formation of bean curd (tofu) is attributed to their tendency to precipitate under high salt concentrations. Although tofu has long been in vogue in the traditional diet of Southeast Asia, it has more recently become increasingly popular among health-conscious people in other regions of the globe.

Glycinin and β-conglycinin differ in their amino acid profiles, functionality and health implications. The glycinin fraction is relatively high in methionine and cysteine content (Kitamura, 1995). Furthermore, it not only aggregates faster but also results in larger clusters than β-conglycinin (Tay *et al.*, 2005). For tofu preparation, in addition to high protein content, soybean genotypes with high levels of 11S fraction are desirable. Soybean accessions with >50% protein content have been reported (USDA, 2001); however, due to the inverse relationship between protein content and seed yield, globally there are very few cultivars that have not only an ultra-high protein content, but also deliver a high yield. Concerning the health benefits of these storage proteins, allergenicity against soy products that is found in some individuals, especially children (Ahn *et al.*, 2003), has been ascribed to the 7S fraction (Bittencourt *et al.*, 2007).

The bean with an ideal ratio of n-6 to n-3

Apart from possessing 40% protein, soybean seed also contains 20% oil. Globally, soybean is second only to palm (*Elaeis guineensis*) oil in production and use. Furthermore, it is a good source of essential fatty acids – linoleic acid (53%) and α-linolenic acid (8%). Linoleic and linolenic acids are also termed as omega-6 (n-6) and omega-3 (n-3) fatty acids because of the presence of a double-bond 6 carbon and 3 carbon away from the last carbon (omega), respectively. Collectively, they are referred as polyunsaturated fatty acids (PUFA) because of the presence of more than one unsaturated bond in the structure. They are essential because the human body cannot synthesize them due to its inability to introduce double bonds between the terminal methyl group and the first double bond present in the carbon chain of the respective fatty acid. More importantly, these fatty acids are involved in the biosynthesis of prostaglandins, the key component of brain nerve, retinal and reproductive tissues.

Apart from rapeseed (*Brassica* species) and canola (*Brassica campestris*) oil, soybean oil is the important source of α-linolenic acid (n-3), an omega-3 fatty acid, for vegetarians, among various vegetable oils available on the global market. The dietary intake of linoleic and linolenic acid needs to be well balanced and the ratio of n-6:n-3 should be around 5:1; this is near to human cell membranes, as indicated in a clinical study (Chan *et al.*, 1993). An imbalance in the n-6:n-3 ratio has been suggested as a cause of many chronic diseases such as diabetes, CVD and osteoporosis (Simopoulos *et al.*,

Table 17.1. Comparative account of linoleic acid, linolenic acid and the n-6:n-3 ratio of several vegetable oils.

Oilseed	C18:2 (ω6 or n-6) (%)	C18:3 (ω3 or n-3) (%)	(ω6 or n-6):(ω3 or n-3)	Reference
Soybean	53	7–9	6–7	Rani *et al.* (2005)
Peanut	32.5	–	n-6 only	Nagaraju and Belur (2008)
Sunflower	65.2	0.3	n-6 only	Zadak *et al.* (2006)
Safflower	81.4	0.4	n-6 only	Lee *et al.* (2004)
Mustard	25.3	11.3	2.0	Chowdhury *et al.* (2007)
Canola	18.8	6.2	3.0	Huang *et al.* (2008)
Olive	8.7	1.0	8.7	Zadak *et al.* (2006)
Palm	10.2	0.5	20	Zadak *et al.* (2006)
Corn	48.7	0.8	60	Rodrigues and Gioielli (2003)

1999). Therefore, the hype that arose around the total dietary intake of PUFA during 1980s has subsided and the type of PUFA, rather than total PUFA, is currently being emphasized. The Paleolithic diet of *Homo sapiens* (i.e. green plants, fruits, vegetables and grains) had equal amounts of n-6 and n-3 fatty acids (Eaton and Konner, 1985). With the opening of oilseed processing units at the turn of previous century, however, an over-dependence on vegetable oils has disturbed the ratio of n-6:n-3 in many populations across the globe due to changing dietary patterns (Gebre-Egziaber *et al.*, 2008). The ratio of n-6 (linoleic acid):n-3 (α-linolenic acid) fatty acids of various vegetable oils available on the global market, worked out based upon their fatty acid compositions reported in the literature, is presented in Table 17.1. This shows that the ratio of n-6 (linoleic acid):n-3 (α-linolenic acid) fatty acid in soybean oil is in the proximity of the ideal ratio.

Minerals and vitamins

Soybean contains about 5–6% ash content, which is an index of its mineral concentration. Potassium, generally recommended for treating hypertension, is found in the highest concentration (2.3%) in soybeans. In addition, other major minerals – calcium (0.2%), magnesium (0.3%) and phosphorus (0.6%) – are also found. Silicon, zinc, iron, manganese, copper, molybdenum, boron, chromium and lead are the important minor minerals present in soy flour. The iron content in soybean varieties is about 8 mg 100 g^{-1} on a dry weight basis. Most of these minerals are retained with meal instead of following the oil fraction.

Soybean contains both water-soluble and oil-soluble vitamins. The water-soluble vitamins (thiamin, riboflavins, pantothenic acid and niacin) are not lost during oil extraction. A kilogram of soy flour contains approximately 3.25, 3.11, 16.9 and 29.7 mg of vitamin B_1 (thiamin), vitamin B_2 (riboflavin), vitamin B_5 (pantothenic acid) and vitamin B_6 (niacin), respectively. Pantothenic acid and niacin are generally prescribed for controlling high blood pressure. Mature soybean contains almost negligible amount of vitamin C (ascorbic acid); however, every 100 g of immature green seeds and an equal amount of soy sprout contains about 16 and 30 mg of this vitamin, respectively. Above all, the newly discovered vitamin pyrroloquinoline quinone, a water-soluble vitamin that is being judged as a new member of the vitamin B family and plays a major role in the metabolism of lysine, has been reported to be present in some soy foods such as tofu and *nattō*.

17.3 Functional Components

'Functional food', a major buzzword of food industries across the globe, is the term assigned to foods that contain biological components that deliver special health benefits to the consumer. The demand for functional foods may rise in the decades or even centuries to come as people in both developed and developing countries become more aware of the relationship between diet and health. Although the Koreans and Chinese have been acquainted with the therapeutic value of soybean for centuries, it is only recent research findings that have highlighted the presence of special bio-ingredients in soybean in other countries across the globe. The biological ingredients present in soybean that deliver special health benefits are summarized in Table 17.2 and some are discussed below.

Table 17.2. Biological components of nutraceutical significance in soybean.

Biological component	Function	Health benefits	Concentration in regular soybean
Soy peptides:			
Lunasin	Inhibition of acetylation of histones (Galvez *et al.*, 2001)	Anti-cancer	0.1–1.4% of the seed
Bowman-Birk	–	Anti-oral, head and neck cancer (Meyskens, 2001), skin smoothener (Wallo *et al.*, 2007)	About 20% of the total trypsin inhibitor activity
Soy-peptides with antihypertensive activity	Inhibition of angiotensin-converting enzyme	Anti-hypertension (Gibbs *et al.*, 2004)	Fermented soy products are a rich source
Special peptides in soy hydrolysates	Stimulation of low-density lipoprotein cholesterol receptor	Hypocholesterolemic effect (Cho *et al.*, 2007) Anti-obesity (Rho *et al.*, 2007)	Black soybean is rich in anti-obesity peptides

(*Continued*)

Table 17.2. *continued*

Biological component	Function	Health benefits	Concentration in regular soybean
Isoflavones	Estrogen-like activity	Anti-breast cancer (Atkinson and Bingham, 2002), anti-osteoporosis (Song *et al.*, 2008), alleviates menopausal depression	0.3% of the seed
	Antioxidant activity	Reduces cardiovascular diseases (Zhuo *et al.*, 2004; McVeigh *et al.*, 2006)	–
Tocopherols (α, β, χ, δ-isomers)	Free-radical scavenger (strong antioxidant)	May help to prevent Alzheimer's disease and Parkinson's disease, improves the immune system (Bramley *et al.*, 2000)	1.5 mg g^{-1} oil
Lecithin	Component of the cell membrane	May help to prevent Alzheimer's disease and Parkinson's disease	0.5–1.5% of the seed
	As methyl group inhibits the formation of homocysteine	Reduces cardiovascular diseases	–
	Synthesis of VLDL which transport dietary alcohol from liver	Reduces formation of gall bladder stones	–
	Soy liposomes	Skin moisturizer (Betz *et al.*, 2005)	–
Saponins (Group A, Group B)	Inhibits sialyl transferase activity (Chang *et al.*, 2006)	Anti-cancer	0.5% of the seed
	Amphipathic molecule	Hypocholesterolemic	–
Sterols	Emulsifier	Reduces total and low-density lipoprotein cholesterol (Matvienko *et al.*, 2002)	0.02–0.08% of the seed
Raffinosaccharides	As a prebiotic increases the population of bifidobacteria (Tomomatsu, 1994)	Inhibits pathogenic bacteria, reduces the risk of colon cancer (Pool-Zobel *et al.*, 2002), stool bulking	6% of the seed

	Daidzein	Genistein	Glycitein
R_1	H	OH	H
R_2	H	H	OCH_3

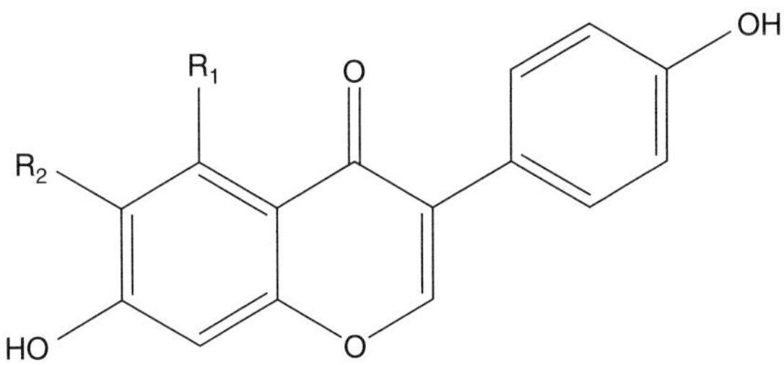

Fig. 17.1. Structure of isoflavones.

Isoflavones

Of all the functional components present in soybean, soy isoflavones are the most investigated biomolecules of the last decade (Rochfort and Panozzo, 2007). They are flavonoid compounds with two benzyl rings (C6) joined by a three-carbon chain. Major isoflavones in soybean exist in four forms: (i) as free aglycones (genistein, daidzein and glycitein) (Fig. 17.1); (ii) as β-glucosides when sugar moiety is attached to aglycone; (iii) as malonyl; and (iv) as acetylated derivatives of β-glucosides.

Twelve isomers are present in the seed, but they are converted back to their corresponding aglycones in the human gut prior to absorption (Setchell and Classidy, 1999). The concentration of total isoflavones ranges from 1 to 3 mg g^{-1}. This varies depending upon the genotype and environmental conditions (Kumar *et al.*, 2007a; Rebeiro *et al.*, 2007). Different processing methods have also been reported to influence the levels of isoflavones in soy food (Coward *et al.*, 1998). Studies also indicate that cultural practices such as increased doses of fertilizer and irrigation enhance their concentration in soybean seeds (Vyn *et al.*, 2002; Bennett *et al.*, 2004). It would not be wrong to attribute the increased interest in soy-derived foods to the increased discovery of the protective effects of isoflavones against several killer diseases of this century. Some are given in detail below.

Breast cancer

Incidences of breast cancer, a common disease in western countries, are now on the rise in other parts of the world also. Barnes *et al.* (1990) was the first to show that a soybean diet reduces the incidence of mammary tumours in rats. Subsequently, several epidemiological studies indicated the inverse relationship between breast cancer and soy food intake (Lee *et al.*, 1991; Pisani *et al.*, 2002; Wu *et al.*, 2008). Isoflavone supplements (40 mg day^{-1})

result in a decrease in breast tissue density in post-menopausal women (Atkinson and Bingham, 2002). Being structurally similar to endogenous estrogen, isoflavones bind to estrogen receptors and exert an estrogen-like effect (Beard *et al.*, 1996). Kurzer (2002) showed that consumption of 65 mg isoflavones day^{-1} favourably affected estrogen metabolism. Soy isoflavones have also been implicated in the prevention and therapy of prostate cancer (Holzbeierlein *et al.*, 2005).

Concerns were raised when isoflavones were found to stimulate cell proliferation in breast-cancer-sensitive cell lines (Zava and Duwe, 1997; Bail *et al.*, 2000). Messina and Wood (2008) addressed these concerns by highlighting the point that isoflavone exposure at levels consistent with the dose in historical Asian soy foods does not elicit adverse stimulatory effects on breast tissue. Estrogenic effects of soy isoflavones have also been implicated in the association of soy food with moderating post-menopausal symptoms such as hot flushes, fatigue and sweating.

Cardiovascular diseases

By 2020, CVD will be a leading cause of death. The number of deaths due to CVD is projected to pass 20 million $year^{-1}$. Contrary to earlier beliefs, CVD is not confined to developed countries; developing economies such as India and China together account for more deaths due to CVD than the developed countries of the world combined. Soy proteins reduce the risk of heart stroke, heart arrest, atherosclerosis and so on. Anderson *et al.* (1995) conducted a meta-analysis of the effects of soy protein intake on serum lipids. They found that an average consumption of 47 g day^{-1} soy protein led to a significant reduction in total cholesterol (9%), low-density lipoprotein (LDL) cholesterol (13%) and triglycerides (11%). Subsequently, a meta-analysis of eight randomized controlled trials in human subjects attributed these lipid-lowering effects to soy isoflavones (Zhou *et al.*, 2004).

However, contradictions persist concerning the potential mechanisms of the lipid-lowering actions of soy isoflavones. Oxidative damage to the cellular lipids is a significant contributor to the development of CVD. Lipid peroxidation of PUFA is associated with the formation of hydroperoxides, free-radical intermediates and secondary oxidation products, which are excreted in urine. Fritz *et al.* (2003) found very low concentrations of secondary lipid oxidation products – aldehydes and carbonyl compounds (the biomarkers of lipid peroxidation) – in the urinary excretions of ten healthy women who were fed dietary soy isoflavones. This study implicated the role of *in vivo* antioxidant activities of soy isoflavones in reducing the risk of CVD. Apart from LDL cholesterol, high-density lipoprotein (HDL) cholesterol, total cholesterol and apolipoproteins play no less an important role in predicting heart diseases. People with a normal LDL cholesterol level but high levels of apolipoprotein B are at high risk for CVD. Apo-lipoprotein A-1 provides a protective effect against heart attack similar to that of HDL cholesterol. Some clinical studies have speculated about the role of soy isoflavones in reducing the risk of CVD by modulating levels of apolipoproteins, but the reports were inconsistent (Psuka *et al.*, 2002; Hall *et al.*, 2006; McVeigh *et al.*, 2006).

Diabetes and the renal diseases

Type 2 diabetes is reaching epidemic proportions worldwide, with developing country such as India alone accounting for 40 million patients. Studies have suggested a role for isoflavones in reducing the risk of the disease (Jayagopal *et al.*, 2002; Ali *et al.*, 2005; Nordentoft *et al.*, 2008). Chronic kidney disease is also increasing at a rapid rate consequent to the increasing incidence of diabetes. Stephenson *et al.* (2005) reported that 40% of new cases of renal disease are related to diabetes. While reviewing studies pertaining to the beneficial effects of soy protein consumption for renal function, Anderson (2008) concluded that the role of soy isoflavones and soy peptides in improving renal function in diabetic neuropathy should be investigated.

Osteoporosis

Osteoporosis, literally meaning 'porous bone', is a metabolic disease of bone characterized by low bone mass and deterioration of bone tissues, making the individuals prone to fracture. This is a global disease affecting 150 million individuals worldwide, cutting across ethnicity and race. Although an inadequate intake of calcium and vitamin D, unhealthy lifestyles marked by excessive alcohol and tobacco consumption and lack of exercise have been cited as some of the causes of the disease, the onset of menopause with a concomitant decline in estrogen renders women prone to the disease. Worldwide, the population of postmenopausal women is expected to reach 1.2 billion by 2030, indicating the magnitude of the problem that will exist in just a few years from now. Messina *et al.* (2004) reviewed studies showing a positive effect of soy product intake on bone health. Several studies looking at the role of soy isoflavones on bone mineral density have been conducted with peri- or postmenopausal women (Nagata *et al.*, 2002; Branca, 2003). Recently, Song *et al.* (2008) showed that a 1 mg day^{-1} intake of isoflavone resulted in increases in bone mineral density of 0.26% in the femoral neck and 0.31% in Ward's triangle in young Korean women.

Biopeptides

Short-chain amino acids produced because of gastrointestinal enzymatic digestion in the human gut or by the hydrolysis of parent proteins during the processing of food with special biological activities have attracted wide attention in the recent past. These special biopeptides may possess from 2 to ≥20 amino acids (Kitts and Weiler, 2003). The presence of bioactive compounds in soy proteins has been reviewed (Elvira and De, 2006). Soy peptides present in soy protein hydrolysate have been reported to possess antihypertensive, anti-cancer and antioxidant properties (Kim *et al.*, 2000; Shin *et al.*, 2001; Wu and Ding, 2001).

Antihypertensive peptides exert their biological activity by inhibiting angiotensin-converting enzyme, which is responsible for converting decapeptide angiotensin I into vasoconstricting octapeptides. Soy foods, especially fermented preparations, are a rich source of angiotensin-converting enzyme inhibiting peptides (Okamoto *et al.*, 1995; Gibbs *et al.*, 2004). Bowman-Birk factor and lunasin are the major peptides with anti-cancer properties. The former, a protease inhibitor, has been reported to be effective in the chemoprevention of oral, head and neck cancers (Armstrong *et al.*, 2000; Meyskens, 2001). It is localized mainly in the seed-coat region of the seed (Sessa and Wolf, 2001). Lunasin, a 43-amino acid biopeptide and constituent of the 2S fraction of soybean proteins, exerts its anti-cancer properties by inhibiting the acetylation process of core histone by binding to non-acetylated H3 and H4 histones (Galvez *et al.*, 2001). Genotypic variation ranging from 0.10 to 1.33 g $100\,g^{-1}$ soy flour has been reported for the concentration of lunasin peptide (Gonzalezde *et al.*, 2004). Soy protein isolate and hydrolysate contain higher levels of lunasin than soy flour and soy concentrate. Recently, soybean protein hydrolysate has been reported to exert a hypocholesterolemic effect by influencing LDL-receptor transcription in human hepatocytes (Cho *et al.*, 2007, 2008). Circulatory bad LDL cholesterol is removed from the plasma by the highly specific LDL receptor and is internalized via receptor-mediated endocytosis. Soy peptide from black soybean has anti-obesity effects (Rho *et al.*, 2007). In a nutshell, the discovery of soy peptides with special health-promoting properties have given a new dimension to the functional-food status of soybean.

Tocopherols

Soybean oil is not only the richest source of tocopherols, but also contains all of the four isomers of tocopherols (α-, β-,γ- and δ-tocopherol). Tocopherols are exploited in pharmaceutical applications. The four isomers of tocopherols, which vary in the number and position of methyl substituents on the chroman ring (Fig. 17.2), possess antioxidative activities in biological systems in the order $\alpha > \beta > \gamma > \delta$ (100%, 50%, 10% and 3% relative activity for α-, β-, γ- and δ-tocopherol, respectively). Moreover, α-tocopherol is preferentially retained and distributed in the body. Medical evidence has indicated that an intake of 400 IU day^{-1} tocopherols results in a decreased risk for arteriosclerosis, cancers and degenerative diseases such Alzheimer's and Parkinson's disease and an improved immune system (Bramley *et al.*, 2000). Genotypic variations have been reported for all of the four isomers in soybean seeds (McCord *et al.*, 2004; Rani *et al.*, 2007). The γ-isomer is the dominant component (60%) of total tocopherols, while the β-isomer is found in the lowest concentration.

Lecithin

Crude soybean oil is a rich source of lecithin, a mixture of naturally occurring phospholipids (phosphatidylcholine [sometimes commonly called

	R_1	R_2	R_3
α	CH_3	CH_3	CH_3
β	CH_3	H	CH_3
γ	H	CH_3	CH_3
δ	H	H	CH_3

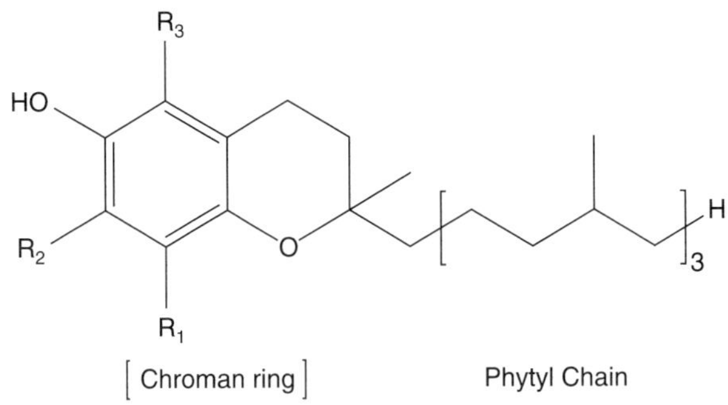

Fig. 17.2. Structure of tocopherols.

lecithin], phosphatidylethanolamine, phosphatidylinositol) extracted as a by-product during the degumming of crude soybean oil. It constitutes about 0.5–1.5% of the soybean seed or 1–3% of crude soybean oil. Commercially, lecithin is available from a dark-tan to reddish-brown colour and in a fluid state to powdered form; it constitutes about 75% phospholipids, while the rest is unrefined oil, moisture and so on. Lecithin is an important nutraceutical component of soybean. It improves liver function, cardiovascular health, fetal brain development, memory function and the reproductive system. Lecithin is an indispensable component of cell membranes, constituting about 10% of the human spinal cord and 5% of the brain; hence, its deficiency restricts the free passage of nutrients from and into the cells. Therefore, it is of great therapeutic value for patients with Alzheimer's disease who are deficient in the neurotransmitter acetylcholine. Lecithin is also recommended for relieving depression. More importantly, the choline component in lecithin is the second largest methyl group donor after methionine that keeps the levels of homocysteine (demethylated methionine) under control in the blood, thereby reducing the risk of a heart attack. Lecithin is also required for the synthesis of very low-density lipoprotein, which acts as a vehicle for exporting dietary cholesterol from the liver. In the deficiency or absence of lecithin, lipid accumulation begins in the liver, leading to the formation of gall bladder stones. As an emulsifier, soy-derived lecithin is extensively used in the food and confectionery industries.

Saponins

Soy saponins are triterpenoid compounds with one or two polysaccharide side chains. They constitute about 0.5% of soybean seeds. However, the concentration of saponins in the seed is influenced by the environmental conditions in which the plant has been raised (MacDonald *et al.*, 2005) and the processing of soy products (Anderson and Wolf, 1995; Chauhan and Chauhan, 2008). Based upon their aglycone core, saponins have been divided into: (i) group A, the acetylated saponins, which impart astringency to soy products; and (ii) group B, saponins with an aglycone structure conjugated to DDMP (2,3-dihydro-2,5 dihydro-6-methyl 4H-pyran-4-one). Group A saponins are concentrated in seed germs, while those from group B are uniformly distributed in the embryo and cotyledons (Berhow *et al.*, 2006). It is the latter group of soy saponins (group B) that has been implicated in providing several health benefits. Group B saponins exert a hypocholesterolemic effect due to their soap-like properties, which stem from the presence of both hydrophilic and hydrophobic components in their structure (Potter, 1995). They also exert anti-mutagenic effect (Berhow *et al.*, 2000) and possess antiviral activity against human immunodeficiency virus *in vitro* (Okubo *et al.*, 1994). Their anti-cancer activity in humans is attributed to their sialyl transferase-inhibiting activities (Chang *et al.*, 2006). Saponins, in general, have also been found to prevent dental caries and platelet aggregation.

Phytosterols

Phytosterols (i.e. campesterols, stigmasterol and sitosterol) in soybean (Fig. 17.3) are obtained as by-products during crude oil processing for tocopherol extraction. They share a common identical ring structure with animal cholesterol, but differ in the side chain. Campesterol and sitosterol are distinguished from each other by the presence of a methyl group for the former and an ethyl group for the latter at carbon 24. Stigmasterol is characterized by the presence of unsaturation at carbon 22. The total phytosterol content, as determined by gas liquid chromatography, in soybean seed is in the range of 0.202–0.843 mg g^{-1}. Sitosterol makes up the largest proportion of total sterols, followed by campesterol and stigmasterol (Yamaya *et al.*, 2007). Several clinical studies have shown that a diet moderately enriched with phytosterols results in a 10% reduction of the total cholesterol content and a 15% reduction in the LDL cholesterol content in human subjects (Law, 2000; Matvienko *et al.*, 2002). Phytosterols lower the cholesterol content by inhibiting its incorporation into micelle, and hence its absorption through the intestine.

Oligosaccharides (raffinose and stachyose)

Prebiotics are non-digestible food components that impart beneficial effects to health by selectively activating probiotics such as bifidobacteria.

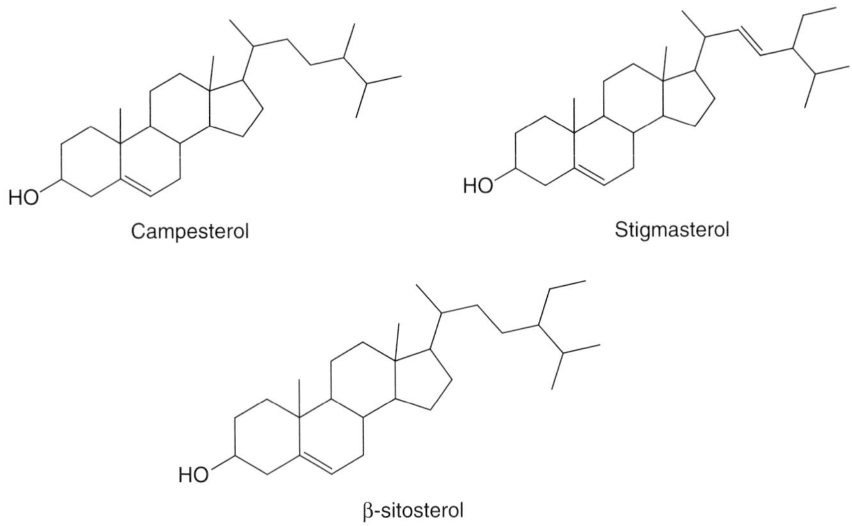

Fig. 17.3. Structures of soy sterols.

Oligosaccharides (raffinose and stachyose) are present in soybean and soy products and constitute about 0.5% and 4.0% of the seed, respectively. Although they are considered undesirable due to their flatus-inducing properties, recent studies have indicated that they also have beneficial effects. They have been reported to stimulate the growth of bifidobacteria in the colon (Tomomatsu, 1994), which provides various health effects. They also inhibit the growth of pathogenic bacteria (*Clostridia perfringens*, *Escherichia coli*, *Salmonella*, *Campylobacter* and *Listeria*) and enhance bulking of the stool, which dilutes the toxins produced by certain Gram-negative bacteria and eliminate them from the intestines. They convert sugars into lactic and acetic acid and thus reduce the colonic pH, which is beneficial for colonic mucosa. Furthermore, they reduce the risk of colon cancer (Pool-Zobel *et al.*, 2002), modulate the immune system (Bland *et al.*, 2004) and contribute to bone health (Nzeusseu *et al.*, 2006). Purified oligosaccharides in a powdered form are marketed for human consumption in Japan.

Oligomeric proanthocyanidin

Several *in vitro* studies have shown that black-seed-coated soybeans possess three to four times more antioxidative activity than yellow-seed-coated soybeans (Furuta *et al.*, 2003; Xu and Chang, 2008). Although no study has pinpointed the biological constituents responsible for the high antioxidative activity of black soybean compared to yellow soybean, oligomeric proanthocyanidin – a bioflavonoid that has been strongly linked to reductions in cancer risk and CVD – is found in high concentrations in black soybean.

17.4 Lactose-free Bean: a Boon for Lactose Intolerants

Soybean seeds are devoid of lactose sugar. Therefore, soy milk, being free from lactose – as compared to animal or human milk, which contain about 5–7% lactose – is an ideal substitute for lactose intolerants who are congenitally deficient in lactase enzymes. It is also recommended for adults who become malabsorbers of lactose with ageing. Furthermore, soy milk is nutritionally on a par with cow's milk. Globally, lactase deficiency varies across populations, ranging from 5% in northern Europe to >90% in southeastern countries. Lactose intolerance has been reported in South American Indians, South African Bantu, Nigerians, Australian Aborigines, Israeli Jews, African-Americans, northwestern Russians, Japanese, Thais and Malaysians. Tandon *et al.* (1981) reported the incidence of lactose intolerance to be in the magnitude of 66.6% for those from the south of India and 27.4% for those from the north.

17.5 Soy in Weight Management and Cosmetics

Obesity has become an important issue in some parts of the world as it leads to development of chronic diseases such as diabetes and CVD, with care costing a large part of the national health-care budget of the countries such as the UK and the USA. Studies have shown that a soy diet results in weight loss in women (Cope *et al.*, 2007; Maskarinec *et al.*, 2008). The weight-reducing property of soybean has been attributed to the low glycemic index (Blair *et al.*, 2006) and high calcium concentration (Lukaszuk *et al.*, 2007) present in soy foods. Furthermore, soy protein has been reported to regulate insulin levels by stimulating the adiponectin (Lihn *et al.*, 2005) and activating the peroxisome-proliferator activated receptors (Morifuji *et al.*, 2006). This may impact obesity, as a high concentration of insulin has been found to be a major cause of obesity.

Soy proteins have also found applications in the cosmetics industry. Lipid vesicles, commonly called liposomes, have an application in skin treatments as natural membranes of the skin and liposome both have the same bilayer structure. A liposome formulation prepared from soybean phospholipids (lecithin) has been shown to be very effective in moisturizing the skin (Betz *et al.*, 2005). Bowman-Birk factor, present in the seed coat as a protease inhibitor, has cosmeceutical properties. A soy moisturizer rich in Bowman-Birk inhibitor is being marketed for improving skin tone, skin pigmentation and other photo-ageing attributes. Soy Bowman-Birk factor inhibits melanosome phagocytosis by keratinocytes via protease-activated receptor 2 (Wallo *et al.*, 2007). It also inhibits the ornithine carboxylase required for hair growth on the skin.

17.6 Overcoming Biochemical Constraints that Affect the Nutritional Value of Soybean and Limit its Utilization

Despite myriad health benefits, soybean contains certain biological components that limit its nutritional value, affecting its utilization in food and feed uses. Some of the antinutritional factors present in soybean seed can be reduced considerably by either processing or enzymatic inactivation, but at an additional cost to the food or feed industries. Therefore, conventional plant-breeding, mutation and transgenic approaches are being followed to develop special soybean genotypes with genetically reduced or absent undesirable components. The efforts made in this direction are summarized in Table 17.3 and discussed below.

Development of Kunitz trypsin inhibitor free soybean genotypes

One of the major constraints in the acceptance of soy foods is the presence of trypsin inhibitors, namely Kunitz trypsin inhibitor (KTI) (20 kDa) and

Table 17.3. Conventional breeding, molecular and transgenic approaches for the development of specialty soybean genotypes with improved nutritional value.

Specialty genotypes	Purpose	Strategies
Null Kunitz inhibitor	Improved nutritional value; bringing down the cost of processing soy products	Conventional breeding, marker-assisted selection (Kim *et al.*, 2006)
Reduced raffinose and stachyose	Less flatulence	Mutation breeding (Sebastian *et al.*, 2000; Hitz *et al.*, 2002)
Low phytic acid	Enhanced nutrient bioavailability; to dispense with the need for fortification with metal ions and phosphorus	Mutation breeding (Sebastian *et al.*, 2000; Hitz *et al.*, 2002)
Lectin-free	Improved nutritional value	Mutation breeding (George *et al.*, 2008)
Null lipoxygenases (*Lx1*, *Lx2*, *Lx3*)	Reduced beany flavour for easy acceptance by the consumer	Mutation breeding, marker-assisted selection (Kitamura and Ujiie, 2004)
Enhanced sulphur-containing amino acids	Improved nutritional value for feed (higher ratio of 11S:7S)	Transgenic (Townsend and Thomas, 1994; Krishnan, 2005)
Enhanced tocopherols	Pharmaceutical applications	Marker-assisted selection
Low isoflavones	Less astringency in soy products and for preparation of soy infant formulae	Marker-assisted selection
High oleic and low linolenic acid	Improved oxidative stability of oil without requiring partial hydrogenation	Germplasm screening (Kumar *et al.*, 2007b), mutation, transgenic (Fehr and Curtiss, 2004; Kumar *et al.*, 2004)

Bowman-Birk factor (8 kDa) as the major antinutritional factors. These protease inhibitors, which account for up to 25% of soybean protein, have been found to be responsible for growth inhibition, pancreatic hypertrophy and hyperplasia in experimental animals (Yanatori and Fujita, 1976). KTI constitutes about 80% of the trypsin inhibitor activity and is heat labile. The heat treatment employed to inactivate KTI has its own limitations. There is always some level of residual activity of the antinutrient depending upon the temperature, time and conditions of heating (Machado *et al.*, 2007; Yuan *et al.*, 2008); more importantly, heating affects protein solubility (Machado *et al.*, 2007). Above all, the heat treatment is not cost-effective for soy-processing units. Furthermore, the presence of KTI necessitates the boiling of beans for its best possible inactivation prior to grinding of soybean with wheat (1:9) for making chapatti flour in countries such as India, which is an extra effort at the household level. KTI exists in four different forms: Ti^a, Ti^b, Ti^c and a fourth form lacking the KTI polypeptide that has been found in PI157440, PI196168 and PI542044. The absence of KTI protein is inherited as a recessive allele to Ti^a, Ti^b and Ti^c. It has been designated as 'ti' and is also referred as the null allele. Therefore, the development of KTI-free soybean varieties by introgression of the null KTI allele in high-yielding soybean varieties is one of the major breeding objectives in soybean-producing countries. Varieties devoid of KTI that have been developed are 'Kunitz soybean', 'BRM 925' and 'BRM 262'. The finding of single sequence repeat (SSR) markers (Satt 228) tightly linked to the *Ti* locus (Kim *et al.*, 2006) has expedited the development of KTI-free soybean varieties.

Development of soybean with reduced oligosaccharides (low flatulence)

One of the major reasons for people's aloofness from soy foods in many countries, especially where fermented soy products are not in vogue, is the flatulence experienced on consumption. Raffinose and stachyose are the two flatulence-inducing sugars, which constitute about 0.5% and 4% of the total soybean seed on a dry weight basis, respectively. Both raffinose and stachyose are the galactosyl derivatives of sucrose. Raffinose is one unit of galactose attached to a sucrose moiety with α 1→6 glycosidic linkage; stachyose is one unit of galactose attached to raffinose with the same glycosidic linkage. Therefore, they are collectively referred to as raffinosaccharides or raffinose family oligosaccharides (RFOs). The RFOs remain undigested in the upper intestine as *Homo sapiens* lack the α 1→6 glycosidase required for degradation of α 1→6 galactosidic linkage. They then pass on to the lower intestinal tract where they are metabolized by intestinal microflora, leading to the production of CO_2, hydrogen and methane. These gases cause abdominal discomfort. Unlike some of the other undesirable components such as trypsin inhibitor and lipoxygenases in soybean, the RFOs are not heat labile; however, they can be reduced to an extent by soaking or boiling through leaching.

The RFO content in soy meal impacts the efficiency of the poultry and swine industries as animals fed on soybean meal attain satiety early, and the weight of the animal, which is economically important, is not realized to the maximum genetic potential. The RFOs reduce the metabolizable energy of the animal feed. It has been shown that a complete removal of raffinose and stachyose from the animal feed can improve the metabolizable energy in animals by 12% (Graham *et al.*, 2002). Therefore, genetic reduction or elimination of RFOs is one of the prime plant breeding objectives. A substantial reduction in RFOs has been achieved in mutants LR28 and LR33 using N-nitroso-N-methyl urea (Sebastian *et al.*, 2000), with a total RFO concentration of 1.37% and 0.88%, respectively, in the seed. Molecular characterization of LR28 and LR33 has shown mutations at the raffinose synthase and myoinositol-p synthase levels of the RFO pathway, respectively (Hitz *et al.*, 2002). Combining the two mutations through conventional plant breeding approaches has led to a breeding line with 0.24% raffinose and 0.47% stachyose.

Breeding for low phytic acid

Phytic acid (1,2,3,4,5,6 inositol hexaphosphate) is a heat-stable antinutritional factor that constitutes about 1.0–4.6% of the seed in regular soybean cultivars (Kumar *et al.*, 2005). The concentration of phytic acid accumulated in the same genotype can vary from location to location and has been reported to be affected by various soil factors such as organic phosphorus status, pH and soil temperature that affect the uptake of phosphorus (Kumar *et al.*, 2005). The antinutritional impact of phytic acid is because of its binding properties, attributed to the presence of six PO_4^- groups, with nutritionally important minerals such as zinc, magnesium, calcium, iron, copper and manganese. Thus, in humans and animals lacking the phytase enzyme required for the hydrolysis of phytic acid, phytic acid–metal complexes formed because of the binding of the phytic acid with nutritionally important metal ions are not absorbed from the intestine. Consequently, the bioavailability of these minerals is affected. Livestock industries that utilize soybean meal as the major feed have had to invest heavily in supplementation of the soy feed with commercial preparations of phytase enzyme or inorganic phosphorus. Furthermore, in intensive animal production regions of the world where soybean meal is the major component of animal feed, the undigested phytic acid phosphorus excreted by monogastric animals has caused an accumulation of inorganic phosphorus in agricultural soils and waterways. Therefore, additional supplementation of the soy feed with inorganic phosphorus exacerbates the environmental pollution caused by undigested phosphorus.

The phytic acid content of seed has also been reported to affect the texture, consistency and yield of tofu (Toda *et al.*, 2005; Hou and Chang, 2006). Ishiguro *et al.* (2008) showed that the optimal coagulant concentration required for tofu-making is affected by the phytic acid content in seed. Therefore, the development of soybean cultivars with low phytic acid is an important plant breeding objective to address not only the requirement of the soy feed and food industries, but also the serious issues concerning the environment. Soybean genotypes that are currently available for low phytic acid content have been developed using induced mutation. Wilcox *et al.* (2000) developed a mutant soybean line that contained 0.2% phytic acid phosphorus. Sebastian *et al.* (2000) reported a mutant line (LR33) that was low in phytic acid as well as RFO content. The single recessive mutation was due to a change in the single base of the codon for an amino acid residue 396 of the peptide, which resulted in a 90% decline in myoinositol 1-phosphate synthase activity (Hitz *et al.*, 2002).

Breeding for low lectin content

Lectins are proteinaceous antinutritional factors with a concentration range of 2.2–4.0 g kg^{-1} of soybean seed. They bind to the carbohydrate moiety of glyco-conjugates, which constitute about 11% of the human body, without affecting the covalent bond. One such lectin from soybean, soybean agglutinin, causes the clumping of human erythrocytes, which have N-acetyl D-galactose as an antigen in the human gut. Lectins are known to reduce natural killer cells, decrease blood insulin levels, enlarge the pancreas and interfere in the absorption of nonheme iron. Although they can be inactivated by moist heat treatment, N-acetyl D-galactose present in the food protects them from inactivation during processing (Yukiko *et al.*, 1999). Therefore, the identification of soybean genotypes with null or a reduced lectin content is important to improve the nutritional value of soybean. Pull *et al.* (1978) identified a null allele for lectin content, which was found to have an insertion of 35 kb in the gene coding for soybean agglutinin (Goldberg *et al.*, 1983). More recently, George *et al.* (2008) reported low lectin content mutants using γ-irradiation, which ranged from 2.5×10^5 to 27.5×10^5 HAU mg^{-1}.

Development of soybean with improved amino acid composition

According to PDCAAS (Schaafsma, 2005), as per the new evaluation method for measuring protein quality, soy protein is equivalent to that of egg protein for humans. However, soybean meal is not able to meet the requirement of sulphur-containing amino acids for poultry and swine production. As a result, these industries have to bear the cost of supplementing the soybean meal with synthetic methionine. Therefore, soybean cultivars with enhanced levels of sulphur-containing amino acids are needed for improving the efficiency of poultry and swine industries. The basis for achieving such plant breeding objectives lies in the manipulation of the ratio of 11S (glycinin) and 7S (β-conglycinin) storage proteins, which account for 70% of total seed protein (Nielsen, 1996). Since the latter protein fraction is deficient in sulphur-containing amino acids, a plant breeding approach focusing on the reduction of the β-conglycinin fraction can result in soybean genotypes with sulphur-rich proteins. A Japanese soybean breeding line

Tohoku 124 (Yumemori) has been bred using null alleles for α and α' subunits of β-conglycinin. This breeding line contained 1.2 times more cysteine and methionine per gram of protein and offered an additional advantage of being devoid of allergenic protein, which is combined with α-subunits of β-conglycinin. Krishnan (2005) reviewed conventional breeding, transformation with heterologous sulphur-rich proteins and the introduction and expression of synthetic genes with a balanced amino acid composition employed for the enhancement of sulphur-containing amino acids in soybean. Transgenic soybean lines with elevated sulphur-containing amino acids (15–40%) have been developed from heterologous protein of the Brazil nut (*Bertholletia excelsa*) (Townsend and Thomas, 1994); however, commercial soybean cultivars could not be undertaken due to a reported allergy caused by the 2S albumin fraction of the Brazil nut. Dinkins *et al.* (2001) reported on a transgenic soybean expressing δ-zein from corn, with methionine and cysteine content increased by 12–20% and 15–35%, respectively.

Development of soybean with high oleic and low linolenic acid

Oil from regular soybean cultivars consists of 11% palmitic acid (16:0), 4% stearic acid (18:0), 23% oleic acid (18:1), 53% linoleic acid (18:2) and 7% linolenic acid (18:3). Linolenic acid, though an essential fatty acid for human, is also considered the main culprit for the poor shelf life of soybean oil because the rate of oxidation of linolenic acid, linoleic acid and oleic acid are in the ratio of 21.6:10.3:1. Oleic acid, being less susceptible to oxidation, would preferably be high. The partial hydrogenation employed by industries to improve oxidative stability results in the production of undesirable *trans* fatty acids, about which serious health concerns have been raised by the medical fraternity due to their atherogenic and diabetogenic properties (Lichtenstein *et al.*, 2003). Consequently, many countries have made it mandatory for processors to label *trans* fats content on food packing. To obviate the need for partial hydrogenation, soybean cultivars with a low linolenic and high oleic acid content are being searched for and developed around the world (Fehr and Curtiss, 2004; Kumar *et al.*, 2004). Two mid-oleic sources, namely FA 22 and M 23, have been developed using mutation breeding while N98-4445A has been developed through hybridization and selection. Recently, soybean germplasm screening has led to the identification of mid-oleic acid (>45%) and oleic acid content remained stable across three growing years at the same location (Kumar *et al.*, 2007b). The lowest level of linolenic acid reported in soybean is 1% for the line A29 (Ross *et al.*, 2000), and the oil obtained from it has been rated as superior to high-oleic soybean. As both oleic acid and linoelic acid are controlled by multiple genes, the use of molecular markers for rapid introgression of the traits has been suggested (Bilyeu *et al.*, 2005). A major quantitative trait locus associated with reduced linolenic acid content has been identified on LG G3-B2 (Yarmilla *et al.*, 2006). By using a transgenic approach through co-sense suppression of the fatty acid desaturase gene responsible for conversion of oleic to linoleic acid, Health Canada recently produced the high-oleic soybean lines G 94-1, G 94-19 and G 168 with as high as 80% oleic acid, and claimed that these lines have no human food safety concern (Health Canada, 2000).

Development of null lipoxygenases soybean genotypes (varieties with a reduced beany flavour)

Beany flavour in soy products is a prime deterrent for the wider acceptance of soy foods in many countries barring China, Japan and southeastern countries. The lipoxygenase enzyme (EC1.13.11.12), an iron-containing dioxygenase that constitutes about 1–2% of the seed protein, is responsible for this flavour. It catalyses the oxidation of the PUFA-containing *cis cis* 1,4 pentadiene moiety, leading to the formation of hydroperoxides. These are subsequently hydrolysed, leading to the formation of beany-flavour-producing aldehyde and ketone compounds. Therefore, lipoxygenase is not an antinutritional factor, as is sometimes inadvertently mentioned, but definitely makes people soybean-averse in some regions of the world with its beany-flavour-producing property.

In seed, lipoxygenase is present in three isozymic forms – Lox-I, Lox-II and Lox-III – which have been categorized into two classes. Class I is characterized by a high pH optima of around 9.0 and the formation of large amounts of 13-hydroperoxides such as Lox-I, while class II designates a pH optima of around 7.0 and the formation of equal amounts of 9- and 13--hydroperoxides such as Lox-II and Lox-III. The presence of each isozymes is controlled by single dominant gene: *Lx1*, *Lx2* or *Lx3*. Absence of these isozymes is due to the presence of a single null allele, which is recessive to *Lx1*, *Lx2* and *Lx3*. *Lx1* and *Lx2* are tightly linked in repulsive phase, while *Lx3* segregates independently from the other two.

Soy preparations made from genotypes lacking in lipoxygenases have a reduced beany flavour and score high on the hedonic scale (Torres-Panaranda *et al.*, 2006). The heat inactivation of the enzyme at industry level is not only cost-ineffective, but also leads to insolubilization of proteins. Therefore, the development of lipoxygenase-free varieties through genetic elimination is the key to reducing the beany flavour. Genotypic variation and the influence of growing environment on lipoxygenases in soybean seed are well documented in the literature (Marczy *et al.*, 1995; Kumar *et al.*, 2003). Japanese plant breeders have developed triple-null lipoxygenase cultivars, namely Ichihime (Suzuyutaka-derived), Kyushu 133 (Enerie-derived), Kyushu 126 (Tamahomare-derived) and L star (Fukuyutaka-derived), using mutation strategy (Kitamura and Ujiie, 2004). Exploiting a Japanese parent as the source of the triple-null lipoxygenase alleles, five triple-null lipoxygenase lines (KY15-4, KY11-59, KY11-83, KYKY10-126 and Ky23-76) have been developed in the USA by backcrossing the regular cultivar '7499' (Pfeiffer, 2008). Similarly, five triple-Lox lines (GC96-19-86, GC96019-96, GC96020-51, GC96020-54 and GC96020-57) have been developed at the Asian Vegetable Research and Development Center by backcrossing with

AGS 29. SSR markers linked to Lox-1, Lox-2 and Lox-3 can be used for expediting the development of lipoxygenase-free genotypes. Kim *et al.* (2004), based on the results of linkage analysis between Lx_2 and the SSR markers, have shown that Lx_2 is positioned on one end of LG F in the frame map, flanked by the SSR markers Satt522 and Sat074. Yarmilla *et al.* (2006) mapped Lox-1 and Lox-2 as single major genes to the same location on LGG13-F and Lox-3 on LGG11-E.

Development of low isoflavones

Although isoflavones have been implicated in providing numerous health benefits, as described earlier in this chapter, concerns have been raised about the possible adverse effects of isoflavones on fetal development and on infants fed on soy-based formulae (Mendez *et al.*, 2002; Chen and Rogan, 2004). This has led some countries to formulate recommended safe upper limits for daily isoflavone intake (Morandi *et al.*, 2005). Therefore, it is important to breed soybean cultivars with low isoflavone levels in the seed for the development of soy infant formulae with reduced levels of isoflavones. Song *et al.* (2007) reported a daily intake of 75,000 μg of soy isoflavone as the upper safe limit. The US Food and Drug Administration recommends a daily intake of 25 g soy protein to receive all of the health benefits of soybean. Therefore, assuming about 40% protein in a soybean seed, a daily intake of 62.5 g of soybean seeds is to be considered safe if it contains <1200 μg isoflavones g^{-1} of soy flour. Accordingly, soybean genotypes containing >1200 μg isoflavones g^{-1} may be considered under a high-isoflavones category.

17.7 Conclusions

The food and therapeutic value of soybean has been known to people in southeastern countries for many years. Research findings from the past 15 years have revealed the health-promoting functions of some of the biological components present in soybean. Although this has helped in creating global awareness regarding the significance of incorporating soy in the regular diet, the use of soybean for food purposes is negligible in some of the major soybean-growing countries. A beany flavour, flatulence and poor digestibility are cited some of the common reasons for non-preference for soy foods. Fermented soy foods and vegetable soybean, which have very low levels of these undesirable components, are not in vogue in many countries. Research efforts are underway to develop specialty soybeans for easy acceptance of soy foods. With the identification of null alleles for trypsin inhibitor and lipoxygenases and subsequent reports of SSR markers associated with these traits, it has become possible to breed KTI- and lipoxygenase-free soybean varieties. Although efforts using mutation strategy have helped in the development of genotypes with reduced levels of phytic acid and RFOs, commercial cultivars lacking or with reduced levels of these deterrents are not yet available.

This chapter has not only highlighted the nutritional value of each biological ingredient, but also emphasized the need for varieties with special characteristics. A brief review of the literature has shown the absence of research works focusing on the manipulation of starch content in soybean. Enhancement of this carbohydrate would definitely impact the use of soybean in food in South Asian countries such as India, Pakistan and Bangladesh, where a variety of recipes are prepared from pulses (grain legumes) with high concentrations of starch. This would also contribute to attenuating the crunch scenario of grain legumes in these regions of the world and addressing nutritional security.

References

Ahn, M., Park, H.Y., Shin, M.Y. and Lee, S.I. (2003) Prevalence of soy protein hypersensitivity in cow's milk protein sensitive children in Korea. *Journal of Korean Medical Science* 18, 473–477.

Ali, A.A, Velasquez, M.T., Hansen, C.T., Mohamed, A.I. and Bhathena, S.J., (2005) Modulation of carbohydrate metabolism and peptide hormones by soybean isoflavones and probiotics in obesity and diabetes. *Journal of Nutritional Biochemistry* 16, 693–699.

Anderson, J.W. (2008) Beneficial effects of soy protein consumption for renal function. *Asia Pacific Journal of Clinical Nutrition* 17 (Suppl. 1), 324–328.

Anderson, J.W., Johnston, B.M. and Cook Nowel, M.E. (1995) Meta analysis of the effects of soy protein intake on serum lipids. *New England Journal of Medicine* 333, 276–282.

Anderson, R.L. and Wolf, W.J. (1995) Compositional changes in trypsin inhibitors, phytic acid, saponins and isoflavones related to soybean processing. *Journal of Nutrition* 125, 581–588.

Armstrong, W.B., Kennedy, A.R., Wan, X.S., Atiba, J., McLaren, C.E. and Meyskens, F.L. (2000) Single dose administration of Bowman-Birk inhibitor concentrate in patients with oral leukoplakia. *Cancer Epidemiology, Biomarkers and Prevention* 9, 43–47.

Atkinson, C. and Bingham, S.A. (2002) Mammographic breast density as a biomarker of effects of isoflavones on the female breast. *Breast Cancer Research* 4, 1–4.

Atkinson, C., Warren, R.M.L.W., Dowsett, M., Day, N.E., Bingham, S.A. (2001) Effects of isoflavones on breast density oestradiol, and gonadotrophins: A double blind randomised placebo controlled trial. Endocrine Society Annual Meeting Program Abstracts, Endocrine Society Press, Bethesda, MD, USA, p. 433.

Bail, J.C., Champavier, Y., Chulia, A.J. and Habirinoux, G. (2000) effects of phytoestrogens on aromatose, 3 beta and 17 beta-hydroxysteroid dehydrogenase activities and human breast cancer cells. *Life Science* 66, 1281–1291.

Barnes, S., Grubbs, C., Setchell, K.D.R. and Carlson, J. (1990) Soybeans inhibit mammary tumors in models of breast cancer. *Progress in Clinical Biological Research* 347, 239–253.

Beard, J.L., Burton, J.W. and Theil, E.C. (1996) Prurified ferritin and soybean meal can be sources of iron for treating iron deficiency. *Human and Clinical Nutrition* 126, 154–160.

Bennett, J.O., Yu, O., Heatherly, L.G. and Krishnan, H.B. (2004) Accumulation of genistein and daidzein, soybean isoflavones implicated in promoting human health, is significantly elevated by irrigation. *Journal of Agricultural and Food Chemistry* 52, 7574–7579.

Berhow, M.A., Wagner, E.D., Vaughn, S.F. and Pewa, M.J. (2000) Characterization and antomutagenic activity of soybean saponins. *Mutation Research* 448, 11–22.

Berhow, M.A., Suk, B.K., Vermillion, K.E. and Duval, S.M. (2006) Complete quantification of group A and group B soyasaponins in soybeans. *Journal of Agricultural and Food Chemistry* 54, 2035–2044.

Betz, G., Aeppli, A., Menshutina, N. and Leuenberger, H. (2005) *In vivo* comparison of various liposome formulations for cosmetic application. *International Journal of Pharmaceutics* 296, 44–54.

Bilyeu, K.D., Pallavalli, L., Sleper, D.A. and Beuselikck, P.R. (2005) Mutations in soybean microsomal omega-3 fatty acid desaturase genes reduce linolenic concentration in soybean seeds. *Crop Science* 45, 1830–1836.

Bittencourt, A.L., Soares, M.F.M., Pires, R.R., Honmoto, C.S., Tanaka, M.K., Jacob, C.M.A. and Abdalla, D.S.P. (2007) Immunogenicity and allergenicity of 2S,7S and 11S soy protein factions. *Brazilian Journal of Pharmaceutical Sciences* 43, 597–606.

Blair, R.M., Henley, E.C. and Tabor, A. (2006) Soy foods have low glycemic and insulin response indices in normal weight subjects. *Nutrition Journal* 5, 35–43.

Bland, E.J., Keshavarz, T. and Bucke, C. (2004) The influence of small oligosaccharides on the immune system. *Carbohydrate Research* 339, 1673–1678.

Bramley, P.M.I., Elmadfa, I., Kafatos, A., Kelly, F.J., Manios, Y., Roxborough, H.E., Schuch, W., Sheehy, P.J.A. and Wagner, K.H. (2000) Vitamin E. *Journal of Science of Food and Agriculture* 80, 913–938.

Branca, F. (2003) Dietary phyto-estrogens and bone health. *Proceedings of the Nutrition Society* 62, 877–887.

Chan, J.K., McDonald, B.E., Gerrard, J.M., Bruce, V.M., Weaver, B.J. and Holub, B.J. (1993) Effect of dietary α-linolenic acid and its ratio to linoleic acid on platelet and plasma fatty acids and thrombogenesis. *Lipids* 28, 811–817.

Chang, W.W., Chia, Y.Y., Lin, T.W., Wang, P.H. and Tsai, Y.C. (2006) Soyasaponin I decreases the expression of alpha 2,3 linked sialic acid on the metastatic potential of B16F10 melanoma cells. *Biochemistry and Biophysics Research Communication* 341, 614–619.

Chauhan, O.P. and Chauhan, G.S. (2008) Development of antinutrients free soy beverage using microwave treated soybeans. *Journal of Food Science and Technology* 45, 496–500.

Chen, A.M. and Rogan, W.J. (2004) Isoflavones in soybean infant formula: A review of evidence for endocrine and other activities in infants. *Annual Review of Nutrition* 24, 33–54.

Cho, S., Juillerat, M.A. and Lee, C. (2007) Cholesterol lowering mechanism of soybean protein hydrolysate. *Journal of Agricultural and Food Chemistry* 55, 10599–10604.

Cho, S., Jullirat, M.A. and Lee, C.H. (2008) Identification of LDL-receptor transcription stimulating peptides from soybean hydrolysate in human erythrocytes. *Journal of Agricultural and Food Chemistry* 56, 4372–4376.

Chowdhury, K., Banu, L.A., Khan, S. and Latif, A. (2007) Studies on fatty acid composition of edible oil. *Bangladesh Journal of Scientific and Industrial Research* 42, 311–316.

Cope, M.B., Erdman, J.W. and Alison, D.B. (2007) The potential role of soy foods in weight and adiposity reduction: An evidence-based review. *Obesity Review* 9, 219–235.

Coward, L., Smith, M., Kirk, M. and Barnes, S. (1998) Chemical modification of isoflavones in soy foods during cooking and processing. *American Journal of Clinical Nutrition* 68, 1486S–1491S.

Dinkins, R.D., Reddy, M.S.S., Meurer, C.A., Yan, B., Trick, F., Thibaud-Nissen, J.J., Finer, J.J., Parrott, W.A. and Collins, G.B. (2001) Increased sulphur aminoacids in soybean plants overexpressing the maize 1.5 kDa zein protein. *In vitro Cellular & Development Biology – Plant* 37, 742–747.

Eaton, S.B. and Konner, M. (1985) Paleolithic nutrition: A consideration of its nature and current implications. *New England Journal of Medicine* 312, 283–289.

Elvira, D.M. and De, L.B. (2006) Soybean bioactive peptides: A new horizon in preventing chronic diseases. *Sexuality, Reproduction and Menopause* 4, 91–95.

Fehr, W.R. and Curtiss, C.F. (2004) Breeding for fatty acid composition. In: Moscardi, F., Hoffmann-Campo, C.B., Saraiva, O.F., Galerani, P.R., Krzyzanowski, F.C. and Carrão-Panizzi, M.C. (eds) *Proceedings of the International Soybean Processing and Utilization Conference*. Foz do Iguassu, Brazil, pp. 815–821.

Fritz, K.L., Seppanen, C.M., Kurzer, M.S. and Csallany, A.S. (2003) The *in vivo* antioxidant activity of soybean isoflavones in human subjects. *Nutrition Research* 23, 479–487.

Furuta, S., Takahashi, M., Takahata, Y., Nishiba, Y., Oki, T. and Masuda, M. (2003) Radical scavenging activities of soybean cultivars with black seed coats. *Food Science and Technology Research* 9, 73–75.

Galvez, A.F., Chen, N., Macasieb, J. and De, L.B. (2001) Chemopreventive property of a soybean peptide that binds to deacetylated histones and inhibits histone acetylation. *Cancer Research* 61, 7473–7478.

Gebre-Egziaber, F., Rabasa-Lhoret, R., Bonnet, F., Bastard, J.P., Desage, M., Skilton, M.R., Vidal, H. and Laville, M. (2008) Nutritional intervention to reduce the n-6/n-3 fatty acid ratio increases adiponectin concentration and fatty acid oxidation in healthy subjects. *European Journal of Clinical Nutrition* 62, 1287–1293.

George, M.A., Bhide, S.V., Thengane, R.J., Hosseini, G.H. and Manjaya, J.G. (2008) Identification of low lectin mutants in soybean. *Plant Breeding* 127, 150–153.

Gibbs, B.F., Zongman, A., Masse, R. and Mulligan, C. (2004) Production and characterization of bioactive peptides from soy hydrolysate and soy fermented foods. *Food Research International* 37, 123–131.

Goldberg, R.B., Hoschek, G. and Vodkin, L.O. (1983) An insertion sequence blocks the expression of a soybean lectin gene. *Cell* 33, 465–475.

Gonzalezde, M.E., Vasconez, M., De, L.B. and Nelson, R. (2004) Lunasin concentration in different soybean genotypes, commercial soy protein and isoflavones products. *Journal of Agricultural and Food Chemistry* 52, 5882–5887.

Graham, K.K., Kerley, M.S., Firman, J.D. and Allee, G.L. (2002) The effect of enzyme treatment of soybean meal on oligosaccharide disappearance and chick growth performance. *Poultry Science* 81, 1014–1019.

Hall, W.L., Vafeiadou, K., Hallund, J., Bugel, S., Reimann, M. and Koebnick, C. (2006) Soy-isoflavone-enriched foods and markers of lipid and glucose metabolism in postmenopausal women: Interactions with genotype and equol production. *American Journal of Clinical Nutrition* 83, 592–600.

Health Canada (2000) Novel Food Information – Food Biotechnology: High Oleic Soybean Lines G 94-1, G 94-19, and G 168. Food Directorate, Health Protection Branch, Health Canada, Ottawa, Canada.

Hitz, W.D., Thomas, J.C., Kerr, P.S. and Sebastain, S.A. (2002) Biochemical characterization of a mutation that confers a decreased raffinosaccharides and phytic acid phenotype on soybean seeds. *Plant Physiology* 128, 650–660.

Holzbeierlein, J.M., McIntosh, J. and Thrasher, J.B. (2005) The role of soy phytoestrogens in prostate cancer: Preventive and alternative medicine. *Current Opinion in Urology* 15, 17–22.
Hou, H.J. and Chang, K.C. (2006) Yield and textural properties of tofu as affected by the changes of phytate content during soybean storage. *Journal of Food Science* 68, 1185–1191.
Huang, S.S.Y., Fu, C.H.L., Higgs, D.A., Balfry, S.K., Schulte, P.M. and Brauner, C.J. (2008) Effects of dietary canola oil level on growth performance, fatty acid composition and ionoregulatory development of spring chinook salmon parr, *Oncorhynchus tshawytscha*. *Aquaculture* 274, 109–117.
Ishiguro, T., Ono, T. and Nakasato, K. (2008) The localization of phytate in tofu curd formation and effects of phytate on tofu texture. *Journal of Food Science* 73, C67–C71.
Jayagopal, V., Albertazzi, P., Kilpatrick, E.S. and Howarth, E.M. (2002) Beneficial effects of soy phytoestrogen intake in postmenopausal women with type-2 diabetes. *Diabetes Care* 25, 1709–1714.
Kim, M.S., Park, M.J., Jeong, W.H., Nam, K.C. and Chung, J.I. (2006) SSR marker tightly linked to the Ti locus in soybean. *Euphytica* 152, 361–366.
Kim, M.Y., Keun, H.A., Hwan, J.T., Young, H.E., Van, K., Kuk, Y.I. and Lee, S.H. (2004) Single nucleotide polymorphism discovery and linkage mapping of lipoxygenase-2 gene (Lx_2) in soybean. *Euphytica* 135, 169–177.
Kim, S.E., Lim, H.H., Lim, J.Y., Kang, Y.I., Woo, H.J. and Lee, H.J. (2000) Anticancer activity of hydrophobic peptides from soy proteins. *Biofactors* 12, 151–155.
Kitamura, K. (1995) Genetic improvement of nutritional and food processing quality in soybean. *Japanese Agricultural Research Quarterly* 29, 1–8.
Kitamura, K. and Ujiie, A. (2004) Recent progress in soybean breeding for processing and utilization in Japan. In: Moscardi, F., Hoffmann-Campo, C.B., Saraiva, O.F., Galerani, P.R., Krzyzanowski, F.C. and Carrão-Panizzi, M.C. (eds) *Proceedings of International Soybean Processing and Utilization Conference*. Foz do Iguassu, Brazil, pp. 808–814.
Kitts, D.D. and Weiler, K. (2003) Bioactive proteins and peptides from food sources: Applications of bioprocesses used in isolation and recovery. *Current Pharmaceutical Design* 9, 1309–1323.
Krishnan, H.B. (2005) Engineering soybean for enhanced sulphur amino acid content. *Crop Science* 45, 454–461.
Kumar, V., Rani, A., Tindwani, C. and Jain, M. (2003) Lipoxygenases and trypsin inhibitor activity as influenced by growing location. *Food Chemistry* 83, 79–83.
Kumar, V., Rani, A. and Joshi, O.P. (2004) Fatty acid profile of released cultivars of Indian soybean: Identification of comparatively low linolenic and high oleic acid cultivars. *Indian Journal of Agricultural Sciences* 74, 388–391.
Kumar, V., Rani, A., Rajpal, S., Srivastava, G., Ramesh, A. and Joshi, O.P. (2005) Phytic acid in Indian soybean: Genotypic variability and influence of growing location. *Journal of Science of Food and Agriculture* 85, 1523–1526.
Kumar, V., Rani, A. and Chauhan, G.S. (2007a) Isoflavones concentration in popular soybean genotypes of Malwa region in Central India. *Soybean Research* 5, 8–13.
Kumar, V., Rani, A., Shakya, A.K., Verma, S.K., Hussain, S.M. and Chauhan, G.S. (2007b) Identification of soybean germplasm for high oleic and low linolenic acid and stability of the trait across different growing years. In: *Proceedings of National Seminar on Changing Global Vegetable Oil Scenario: Issues and Challenges before India*, Directorate of Oilseeds Research, Hyderabad, India, 29–31 January 2007, pp. 126–127.

Kurzer, M.S. (2002) Hormonal effects of soy in premenopausal and post menopausal women. *Journal of Nutrition* 132, 570S–573S.
Law, M. (2000) Plant sterol and satnol margarines and health. *British Medical Journal* 320, 861–864.
Lee, H.P., Gourmey, L., Duffy, S.W., Esteve, J., Lee, J. and Day N.E. (1991) Dietary effects on breast cancer risk in Singapore. *Lancet* 337, 1197–1200.
Lee, Y.C., Oh, S.W., Chang, J. and Kim, I.H. (2004) Chemical composition and oxidative stability of safflower oil prepared from safflower seed roasted with different temperatures. *Food Chemistry* 84, 1–6.
Lichtenstein, A.H, Erkilla, A.T., Lamarche, B., Schwab, U.S., Jalbert, S.M. and Ausman, M. (2003) Influence of hydrogenated fat and butter on CVD risk factor remnant like particles, glucose, insulin, blood pressure and C reactive protein. *Atheroscelerosis* 171, 97–103.
Lihn, A.S., Pedersen, S.B. and Richelsen, B. (2005) Adinonectin: Action, regulation and association to insulin sensitivity. *Obesity Review* 6, 13–21.
Lukaszuk, J.M., Luebbers, P. and Gordon, B.A. (2007) Preliminary study: Soy milk as effective as skim milk in promoting weight loss. *Journal of American Dietetic Association* 107, 1811–1814.
MacDonald, R.S., Guo, J.Y., Copeland, J., Browning, J.D., Sleper, D. Jr, Rottinghaus, G.E. and Berhow, M.A. (2005) Environmental influences on isoflavones and saponins in soybeans and their role in colon cancer. *Journal of Nutrition* 135, 1239–1242.
Machado, F.P.P., Queiroz, J.H., Oleveira, M.G.A., Piovesan, N.D., Peluzio, M.C.G., Costa, N.M.B. and Moeira, M.A. (2007) Effects of heating on protein quality of soybean flour devoid of Kunitz inhibitor and lectin. *Food Chemistry* 107, 649–655.
Marczy, J., Sz, M.L., Simon, M. and Szajani, B. (1995) Comparative study on the lipoxygenase activities of some soybean cultivars. *Journal of Agricultural and Food Chemistry* 43, 313–315.
Maskarinec, G., Aylward, A.G., Erber, E., Takata, Y. and Kolonel, L.N. (2008) Soy intake is related to a lower body mass index in adult women. *European Journal of Nutrition* 47, 138–144.
Matvienko, O.A., Lewis, D.S., Swanson, M., Arndt, B., Rainwater, D.L., Stewart, J. and Anekel, D.L. (2002) Single daily dose of soybean phytosterols in ground beef decreases serum total cholesterol and LDL cholesterol in young, mildly hypercholesterolemic men. *American Journal of Clinical Research* 76, 57–64.
McCord, K.L., Fehr, W.R., Wang, T., Welke, G.A., Cianzio, S.R. and Schnebly, S.R. (2004) Tocopherol content of soybean lines with reduced linolenate content. *Crop Science* 44, 772–776.
McVeigh, B.L., Dillingham, B.L., Lampe, J.W. and Duncan, A.M. (2006) Effect of soy protein varying in isoflavone content on serum lipids in healthy young men. *American Journal of Clinical Nutrition* 83, 244–251.
Mendez, M.A., Anthony, M.S. and Ara, L. (2002) Soy-based formulae and infant growth and development: A review. *Journal of Nutrition* 132, 2127–2130.
Messina, M. and Wood, C.E. (2008) Soy isoflavones, estrogen therapy, and breast cancer risk: Analysis and commentary. *Nutrition Research* 7, 17.
Messina, M., Ho, S. and Alekel, D.L. (2004) skeletal benefits of soy isoflavones: A review of the clinical trial and epidemiologic data. *Current Opinion in Clinical Nutrition and Metabolism Care* 7, 649–658.
Meyskens, F.L. (2001) Development of Bowman-Birk inhibitor for chemoprevention of oral, head and neck cancer. *Annals of New York Academy of Sciences* 952, 116–123.

Morandi, S., Agostina, A.D., Ferrario, F. and Arnoldi, A. (2005) Isoflavone content of Italian soybean food products and daily intakes of some specific classes of consumers. *European Journal of Food Research and Technology* 221, 84–91.

Morifuji, M., Sanbongi, C. and Sugiura, K. (2006) Dietary soya protein intake and exercise training have an additive effect on skeletal muscle fatty acid oxidation enzyme activities and mRNA levels in rats. *The British Journal of Nutrition* 96, 469–475.

Nagaraju, A. and Belur, L.R. (2008) Rats fed blended oils containing coconut oil with groundnut oil or olive oil showed an enhanced activity of hepatic antioxidant enzymes and a reduction in LDL oxidation. *Food Chemistry* 108, 950–957.

Nagata, C., Shimizu, H., Takami, R., Takeda, N. and Yasuda, K. (2002) Soy products intake and serum isoflavonoid and estradiol concentration in relation to bone mineral density in postmenopausal Japanese women. *Osteoporosis International* 13, 200–204.

Nielsen, N.C. (1996) Soybean seed composition. In: Verma, D.P.S. and Shemaker, R.C. (eds) *Soybean: Genetics, Molecular Biology and Biotechnology*. CAB International, Wallingford, UK, pp. 127–163.

Nordentoft, I., Jeppesen, P.B., Hong, J., Abdula, R. and Hermansen, K. (2008) Increased insulin sensitivity and changes in the expression profile of key insulin regulatory genes and beta cell transcription factors in diabetic KKAy-Mice after feeding with a soybean protein rich diet high in isoflavones content. *Journal of Agricultural and Food Chemistry* 56, 4377–4385.

Nzeusseu, A., Diest, D., Haufroid, V., Deresseux, G., Devogelaer, J.P. and Manicourt, D.H. (2006) Inulin and fructo-oligosaccharides differ in their ability to enhance the density of cacellous and cortical bone in the axial and peripheral skeleton of growing rats. *Bone* 38, 394–399.

Okamoto, A., Hanagata, H., Matsumoto, E., Kawamura, Y., Koizumi, Y. and Yanadiga, F. (1995) Angiotensin I converting enzyme inhibitory activities of various fermented foods. *Bioscience, Biotechnology and Biochemistry* 59, 1147–1149.

Okubo, K., Kudou, S., Uchida, T., Yoshiki, Y., Yoshikoshi, M. and Tonomura, M. (1994) Soybean saponin and isoflavonoids, structure and antiviral against immunodeficiency virus *in vitro*. *American Chemical Society Symposium Series* 547, 332–339.

Pfeiffer, T. (2008) Breeding triple-null lipoxygenase soybean cultivars. Available at: http://www.uky.edu/ag/newcrops/soybreeding.htm (accessed 28 February 2010).

Pisani, P., Bray, F. and Parkin, D.M. (2002) Estimates of the worldwide prevalence of cancer from 25 sites in the adult population. *International Journal of Cancer* 97, 72–81.

Pool-Zobel, B., Van-Loo, J., Rowland, I. and Roberfroid, M.B. (2002) Experimental evidences on the potential of prebiotic fructans to reduce the risk of colon cancer. *British Journal of Nutrition* 87, S273–S281.

Potter, S.M. (1995) Overview of proposed mechanisms for the hypocholestemic effect of soy saponins. *Journal of Nutrition* 125, 606S–611S.

Psuka, P., Korpelainen, V., Hoie, L.H., Skovlund, E., Lahti, T. and Smerud, K.T. (2002) Soy in hypercholesterolaemia: A double blind, placebo-controlled trial. *European Journal of Clinical Nutrition* 56, 352–357.

Pull, S.P., Pueppke, S.G., Hymowitz, T. and Orf, J.H. (1978) Soybean lines lacking the 120000 dalton seed lectin. *Science* 200, 1277–1279.

Rani, A., Kumar, V., Verma, S.K., Shakya, A.K., Husain, S.M. and Chauhan, G.S. (2005) Interrelationship between oil content and fatty acid composition in Indian soybean (*Glycine max*) cultivars. *Indian Journal of Agricultural Sciences* 77, 195–198.

Rani, A., Kumar, V., Verma, S.K., Shakya, A.K. and Chauhan, G.S. (2007) Tocopherols content and profile in soybean: Genotypic variability and correlation studies. *Journal of American Oil Chemists' Society* 84, 377–383.

Rebeiro, M.L.L., Mandarino, J.M.G., Carr-Panizzi, M.C., Oliveira, M.C.N., Campo, C.B.H., Nepomuceno, A.L. and Ida, E.I. (2007) Isoflavone content and beta-glucosidase activity in soybean cultivars of different maturity groups. *Journal of Food Composition and Analysis* 20, 19–24.

Rho, S.J., Park, S., Ahn, C.W., Shin, J.K. and Lee, H.G. (2007) Dietetic and hypocholestrolemic action of black soy peptide in dietary obese rats. *Journal of Science of Food and Agriculture* 87, 908–913.

Rochfort, S. and Panozzo, J. (2007) Phytochemicals for health, the role of pulses. *Journal of Agricultural and Food Chemistry* 55, 7981–7994.

Rodrigues, J.N. and Gioielli, L.A. (2003) Chemical interesterification of milk fat and milk fat-corn oil blends. *Food Research International* 36, 149–159.

Ross, A.J., Fehr, W.R., Welke, G.A. and Cianzio, S.R. (2000) Agronomic and seed traits of 1% linolenic soybean genotypes. *Crop Science* 40, 383–386.

Schaafsma, G. (2005) The Protein Digestibility-Corrected Amino Acid Score (PDCAAS) – a concept for describing protein quality in foods and food ingredients: A critical review. *Journal of the Association of Official Analytical Chemists International* 88, 988–994.

Sebastian, S.A., Kerr, P.S., Pearlstein, R.W. and Hitz, W.D. (2000) Soybean germplasm with novel genes for improved digestibility. In: Drackely, J.K. (ed.) *Soy in Animal Nutrition*. Federation of Animal Science Society, Savoy, IL, USA, pp. 56–74.

Sessa, D.J. and Wolf, W.J. (2001) Bowman-Birk inhibitors in soybean seed coats. *Industrial Crops Products* 14, 73–83.

Setchell, K.D. and Cassidy, A. (1999) Dietary isoflavones: Biological effects and relevance to human health. *Journal of Nutrition* 129, 758S–767S.

Shin, Z., Yu, R., Park, S.A., Chung, D.K., Ahu, C.W. and Nam, H.S. (2001) His-His-Leu, an angiotensin I converting enzyme inhibitory peptide derived from Korean soy paste exerts antihypertensive activity *in vitro*. *Journal of Agricultural and Food Chemistry* 49, 3004–3009.

Simopoulos, A.P., Alexander, L. and Salem, N. Jr (1999) Essentiality of recommended dietary intakes for omega-6 and omega-3 fatty acids. *Annals of Nutrition and Metabolism* 43, 127–130.

Song, W.O., Chun, O.K., Hivang, I., Shin, H.S., Kim, B.G., Kim, K.S., Lee, S.Y., Shin, D. and Lee, S.G. (2007) Soy isoflavones as safe functional ingredients. *Journal of Medicinal Food* 10, 571–580.

Song, Y.J., Paik, H.Y. and Joung, H. (2008) Soybean and soy isoflavone intake indicate a positive change in bone mineral density for 2 years in young Korean women. *Nutrition Research* 28, 25–30.

Stephenson, T.J., Serchell, K.D., Kendall, C.W., Jenkins, D.J., Anderson, J.W. and Pantl, P. (2005) Effect of soy protein-rich diet on renal function in young adults with insulin-dependent diabetes mellitus. *Clinical Nephrology* 64, 1–11.

Tandon, R.K., Joshi, Y.K., Singh, D.S., Narendranathan, M., Balakrishnan, V. and Lal, K. (1981) Lactose intolerance in North and South Indians. *The American Journal of Clinical Nutrition* 34, 943–946.

Tay, S.L., Xu, G.Q. and Conrad, P. (2005) Aggregation profile of 11S, 7S and 2S coagulated with GDL. *Food Chemistry* 91, 457–462.

Toda, K., Takahashi, K., Ono, T., Kitamura, K. and Nakamura, Y. (2005) Variation in the phytic acid content of soybeans and its effect on consistency of tofu made from soybean varieties with high protein content. *Journal of Science of Food and Agriculture* 86, 212–219.

Tomomatsu, H. (1994) Health effects of oligosaccharides. *Food Technology* 10, 61–65.

Torres-Panaranda, V., Reitmeier, C.A., Wilson, L.A., Fehr, W.R. and Narvel, J.M. (2006) Sensory characteristics of soymilk and tofu made from lipoxygenase-free and normal soybeans. *Journal of Food Science* 63, 1084–1087.

Townsend, J. and Thomas, L.A. (1994) Factors which influence the Agrobacterium-mediated transformation of soybean. *Journal of Cellular Biochemistry Supplement* 18A, 78.

USDA (2001) Industrial uses of agricultural materials: Situation and outlook report. *IUS* 11, 6–9.

Vyn, T.J., Yin, X., Bruulsema, T.W., Jackson, J.C., Rajcan, I. and Bruder, S.M. (2002) Potassium fertilization effects on isoflavone concentrations in soybean. *Journal of Agricultural and Food Chemistry* 50, 3501–3506.

Wallo, W., Nebus, J. and Leydon, J.J. (2007) Efficacy of a soy moisturizer in photoaging: A double-blind, vehicle-controlled, 12-week study. *Journal of Dermatology* 6, 917–922.

Wilcox, J.R., Premchandra, G.S., Young, K.A. and Roboy, V. (2000) Isolation of high seed inorganic P, low phytate soybean mutant. *Crop Science* 40, 1601–1605.

Wu, A.H., Yu, M.C., Tseng, C.C. and Pike, M.C. (2008) Epidemiology of soy exposures and breast cancer risk. *British Journal of Cancer* 98, 9–14.

Wu, J. and Ding, X. (2001) Hypotensive and physiological effect of angiotensin converting enzyme inhibitory peptides derived from soy protein on spontaneously hypertensive rats. *Journal of Agricultural and Food Chemistry* 49, 501–506.

Xu, B. and Chang, K.C. (2008) Total phenolics, phenlic acids, isoflavones and anthocyanins and antioxidant properties of yellow and black soybeans as affected by thermal processing. *Journal of Agricultural and Food Chemistry* 56, 7165–7175.

Yamaya, A., Endo, Y., Fujimoto, K. and Kitamura, K. (2007) Effects of genetic variability and planting location on the phytosterol content and composition in soybean seeds. *Food Chemistry* 102, 1071–1075.

Yanatori, Y. and Fujita, T. (1976) Hypertrophy and hyperplasia in the endocrine and exocrine pancreas of rats fed soybean trypsin inhibitor or repeatedly injected with pancreozymin. *Archives of Histology, Japan* 39, 67–69.

Yarmilla, R., Poysa, V.W., Yu, K., Rajcan, I., Ablett, G.R. and Pauls, K.P. (2006) Seed and agronomic QTL in low linolenic acid, lipoxygenase-free soybean [*Glycine max* (L.) Merrill] germplasm. *Genome* 49, 1510–1527.

Yuan, S., Sam, K.C., Chang-Sam, K.C., Liu, Z. and Xu, B. (2008) Elimination of trypsin inhibitor activity and beany flavour in soy milk by consecutive blanching and ultrahigh-temperature processing. *Journal of Agricultural and Food Chemistry* 56, 7957–7963.

Yukiko, U., Yuki, S., Takako, N., Tomohisa, O. and Koji, M. (1999) Stability of soybean lectin during food processing and in digestive tract. *Soy Protein Research* 2, 59–64.

Zadak, Z., Hyspler, R., Ticha, A., Solichova, D., Blaha, V. and Melichar, B. (2006) Polyunsaturated fatty acids, phytosterols and cholesterol metabolism in Mediterranean diet. *Acta Medica* 49, 23–26.

Zava, D.T. and Duwe, G. (1997) Estrogenic and antiproliferative properties of genistein and other flavonoids in human breast cancer cells *in vitro*. *Nutrition and Cancer* 27, 31–40.

Zhuo, X.G., Melby, M.K. and Watanabe, S. (2004) Soy isoflavone intake lowers serum LDL cholesterol: A meta-analysis of 8 randomized controlled trials in humans. *Journal of Nutrition* 134, 2395–2400.

18 Uses of Soybean: Products and Preparation

Rita S. Raghuvanshi[1] and Kavita Bisht[2]
[1]Dean, College of Home Science, GB Pant University of Agriculture and Technology, Pantnagar, Uttarakhand, India; [2]SAP Kanya Mahavidyalaya, Kichha (Kumaon University, Nainital), India

18.1 Introduction

Soybean (*Glycine max* (L.) Merrill) is one of the most economical and valuable agricultural commodities in the legume group because of its unique chemical composition. Among the plant-based foods, soybean is fairly unique because the protein in soybean products is considered one of the most complete proteins. Acceptable in almost all diets, soybean products contain virtually no cholesterol and are lactose-free and relatively low in saturated fat. Easily digested by humans, these proteins may provide a number of health benefits, reduce the costs of food production and impart functional benefits to numerous food products. The increasing acceptance of soybean protein is due to its versatility and functionality in food applications. Soybean's high protein content makes it a valuable component in formulated foods.

18.2 Physicochemical and Functional Properties

Among cereal and other legume species, soybean has the highest protein content (around 40%) and second highest fat content (about 20%). The protein is of very high quality because it contains all of the essential amino acids, thus making it very important for vegans. Other valuable components found in soybean include phospholipids, vitamins and minerals (Gopalan *et al.*, 1989). Soybean is a good source of antioxidants such as lecithin and vitamin E. It is also rich in magnesium, which has important functions in the bones, heart and arteries. The approximate composition of full-fat soybean flour is 41.0% protein, 20.0% fat, 5.3% ash, 2.7–3.9% crude fibre and 25.0% carbohydrate (Mustakes, 1971; Kellor, 1971; Krishna *et al.*, 2003), whereas that of de-fatted soybean flour is approximately 50.5–52.0% protein, 1.0–1.5% fat, 3.0–3.2% crude fibre and 5.7% ash. Michael and Alison

(2003) reported the approximate composition of low-fat soybean flour is 52.05% protein, 6.31% ash and 7.39% fat.

To utilize soybean ingredients effectively, food processors should have detailed information on the methods of preparation and processing of soybean products, because these affect the composition and functional properties of the component proteins. Protein solubility is closely related to the functional properties needed for bakery food application. Heat treatment, especially moist heat, rapidly insolubilizes soybean protein. The more dispersible types of soybean flours are used in bakery and cereal products, by adding them directly to dough. Enzyme-active soybean flour has a minimum water solubility of 70% (Pingle, 1974). The ability of protein to aid in the formation and stabilization of emulsions is critical in the preparation of meat sausages and cake batters. In general, the emulsifying capacity of soybean protein products is enhanced by rising solubility. Accordingly, soybean proteins progressively reduce interfacial tension as the concentration is increased (Kinsella, 1979).

Foaming – the capacity of proteins to build stable foams with gas by forming impervious protein films – is an important property in some food applications, including beverages, angel cakes and sponge cakes. The foaming properties of various soybean protein products have been studied and soybean isolates have been found to be superior to soybean flour and concentrates (Kinsella, 1979). Processing conditions can vary the amount of water that can be absorbed. Soybean proteins differ considerably from wheat proteins in their chemical composition, as well as in physical properties (such as their total lack of elasticity). Adding soybean proteins to wheat flour thus dilutes the gluten proteins and the starch. On the other hand, soybean proteins exhibit a strong binding power that provides some resistance to dough expansion. This can be partially overcome by increasing the amount of water used in dough making and by a longer proofing time. The binding power of soybean flour is closely related to its high water-absorption capacity, which in the case of the de-fatted product is equivalent to 110% by weight. With full-fat flour, however, no measurable increase in dough absorption results from normal usage levels of the soybean product (Pyler, 1988).

Gels are characterized by relatively high viscosity, plasticity and elasticity. The ability of a gel structure to provide a matrix to hold water, fat, flavour, sugar and other food additives is useful in variety of products, such as chicken or ham analogues made from textured soy protein and fibrous soybean protein. Soybean flour and concentrates form soft, fragile gels, whereas soy isolates form firm, hard and resilient gels. The general procedure for producing a soy protein gel involves heating the protein solution at 80–90°C for 30 min followed by cooling at 4°C (Kinsella, 1979). Heating reduces the gel-forming capacity of isolated soybean protein and at >100°C there is complete loss of gel structure (Shemer, 1974).

Water-holding capacity is a measure of trapped water that includes both bound and hydrodynamic water. It affects the texture, juiciness and taste of the product. Water-holding capacities of soybean flour, concentrate and isolate have been reported as 2.6, 2.75 and 6.25 $g\,g^{-1}$ of solids, respectively (Kinsella, 1979). All soy protein concentrates, regardless of the process used, have certain fat-and water-holding characteristics. This ability of soy protein enhances the shelf life of bakery products.

18.3 Uses of Soybean

Soybean is a versatile legume grain with uses in human consumption as well as in animal consumption, industry and medicine. A general description of soy processing and its uses is shown in Fig. 18.1.

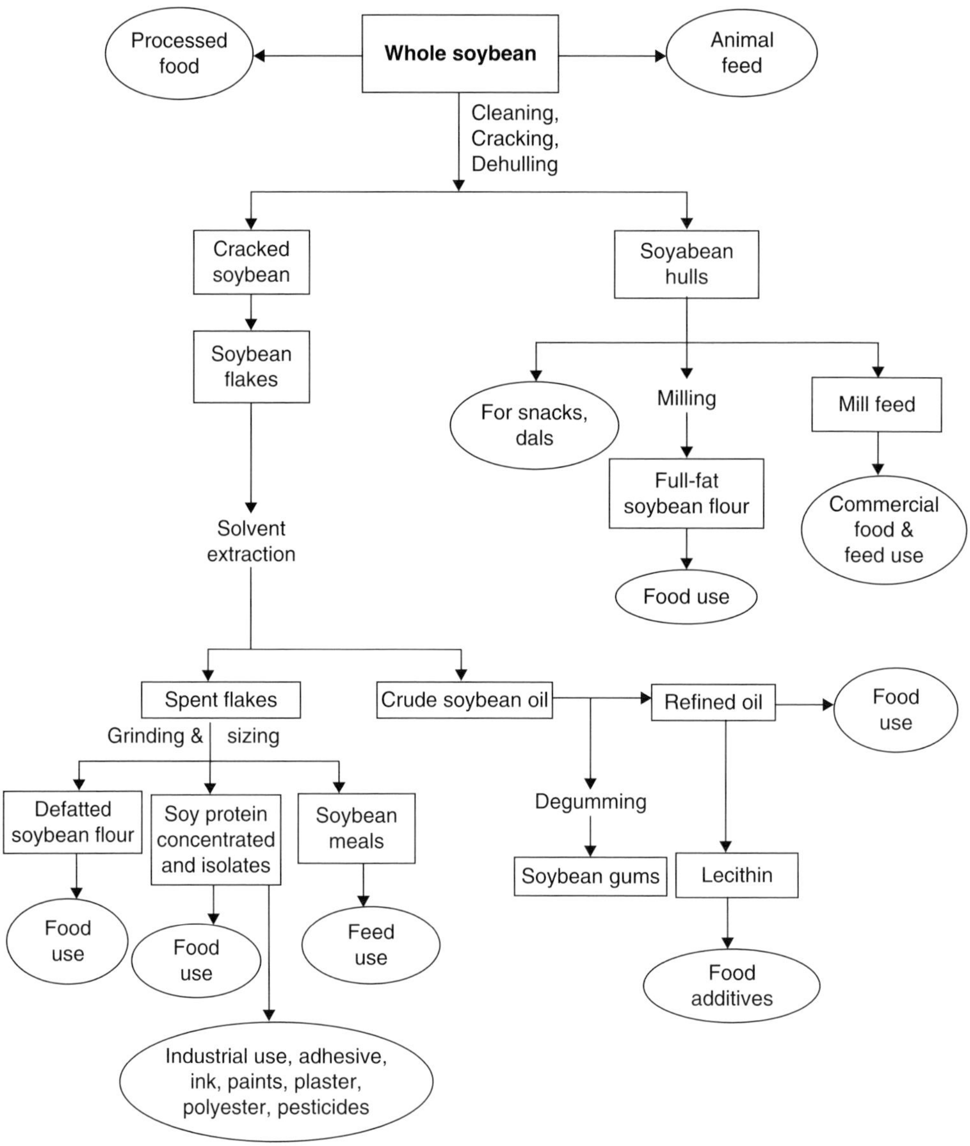

Fig. 18.1. Soybean processing and usage.

Food uses

Many types of soybean foods are available throughout the world today. Some are produced through the use of modern processing techniques in a large soybean processing plant, whereas others are produced in more traditional ways, owing their history to oriental processing techniques. These foods are usually referred to as traditional soybean foods. The main food products of soybean are full-fat soybean flour, de-fatted flour, soybean grits, soybean flakes, soybean milk, soybean-fortified bakery products, ready to eat snacks, soybean sprouts, fermented products and oil. Among these products, soy milk has great potential to supplement dairy milk and is nutritionally comparable to human and cow's milk.

Green pods

The fully developed green pods are harvested for their green seeds. The pods are removed by hand and the seeds are boiled until tender. An alternative technique is to boil the pods first and then shell the beans. The cooked beans can be eaten as they are or combined in many dishes. Their flavour is unique. Soybeans have not been readily accepted as a fresh, frozen or canned vegetable because of their peculiar odour and the difficulty in shelling the green bean. The protein of green immature soybean has been reported to be superior in nutritive value to the mature bean and, when properly cooked, the biological value of the protein compares favourably with that of casein and beef liver.

Dried seeds

Normal boiling of soybeans, as done with most kinds of dried beans, results in an off-flavour (enzyme-substrate reaction) that many people do not like. The following technique avoids this reaction by destroying the enzyme by heat and is a basic technique for several other foods. Bring two parts of water to boiling temperature. Add one part of soybeans and boil for 5 min. Meanwhile, boil 4 parts of water. Remove the seeds from the first water, rinse them and boil them in the second water for a further 5 min. Discard the water and rinse again. The product is called preboiled soybeans.

Soy flour

One of the most common forms in which soybean is used in population diets is as flour. Soybean flour may be used in the human diet as an ingredient of a wide variety of common dishes such as soup, stews, beverages and desserts; in the formulation of bakery and cereal products; as a meat extender; as a starting material for the preparation of infant formulas, protein concentrates or isolates; or as a protein supplement to cereal grains and other foods. In addition to being an excellent source of iron, calcium and B vitamins, it is also rich in high-quality protein. Soybean flour can be made with or without removal of the natural oils during processing and is thus named full-fat or de-fatted soybean flour, respectively. The general steps followed in soybean flour preparation are cleaning, cracking and dehulling whole soybean, followed by tempering and oil extraction. The remaining meal is desolventized, toasted, dried and milled to de-fatted soybean flour. For full-fat soybean flour, the cleaned, cracked and dehulled whole grains are dried and then milled (Kellor, 1971).

Soy milk

Soy milk is an aqueous extract of soybeans that is inexpensive, highly digestible and nutritious. It contains no cholesterol or lactose, and is a good source of protein and iron. It can be fortified with calcium, vitamin D and vitamin B_{12}. Because soy milk is lactose-free, it can be used as a substitute for bovine milk for lactose-intolerant people (Liu *et al.*, 1995; Gandhi, 2000). Soybean milk contains less sodium than cow's milk and is therefore better for persons with high blood pressure (Manay and Shadaksharaswamy, 2000). Parihar (1977) and Gandhi (2000) reported that 4% protein content in soy milk is comparable to 3.7% protein in cow milk. Soy milk and soybean paneer yield and quality are affected by several factors, such as the soybean cultivar (Skurray *et al.*, 1980; Wong *et al.*, 1983), growth environment (Schaefer and Love, 1992) and milk-processing methods (Wang and Chang, 1995), the nature and concentration of coagulant and soy paneer processing methods. Soybean cultivars differ in their chemical components, including proteins, lipids and minerals, that may influence the yield and quality of soy milk and tofu.

There are various methods for production of soybean milk. Soybean milk can be prepared from soybean with or without hulls/soy protein isolates/spray-dried soy milk powder or soy milk can be prepared after clarifying the insoluble fibres (Gandhi, 2000). It is prepared by grinding soaked beans with water to obtain an emulsion. The emulsion is cooked for 20 min and then margarine, sugar, salt, lime and malt are added. The cooked product is then homogenized or emulsified and may be used fresh or spray dried (Manay and Shadaksharaswamy, 2000) to give milk powder.

At household-level soy milk is prepared by removing dirt from soybean, washing and soaking overnight, draining the water and grinding in a mixer with hot water. Additional hot water is added to make slurry and then boiled for 15–20 min. It is then filtered through a muslin cloth. Flavourings such as vanilla essence or crushed cardamom and sugar are added to obtain soybean milk.

Soybean oil

Soybean oil is a natural extract from whole soybeans. Odourless and flavourless, this clear oil is excellent for stir frying as it brings out the flavours of foods. To produce soybean oil, the soybeans are cracked, adjusted for moisture content, rolled into flakes and solvent-extracted with commercial hexane. The oil is then refined, blended for different applications and sometimes hydrogenated. Due to its versatility, soybean oil is used by the food industry in a variety of food products including salad dressings, sandwich spreads, margarine, bread, mayonnaise, non-dairy coffee creamers and snack foods. The high smoke point of soybean oil allows it to be used as a frying oil. Soybean oil is often hydrogenated to increase its shelf life or to produce a more solid product.

Soybean oil is considered healthy for the heart as it is cholesterol-free and low in saturated fatty acids. The major unsaturated fatty acids in soybean oil triglycerides are linolenic acid (C18:3; 7%), linoleic acid (C18:2; 51%) and oleic acid (C18:1; 23%). It also contains saturated fatty acids (i.e. 4% stearic acid and 10% palmitic acid). Soybean oil contains natural antioxidants (vitamin E), which remain in the oil even after extraction. These antioxidants help prevent oxidative rancidity. As with fish oils, soybean oil contains omega-3 fatty acids, known to be protective against various cardiovascular diseases. In the process of hydrogenation, unhealthy *trans* fats are produced that may raise blood cholesterol levels and increase the risk of heart disease. Food manufacturers are now trying to remove *trans* fats from their products. Soybean oil has a shelf life of a year, but it may be better to store the oil for only a few months at room temperature. Soybean oil should be stored in a dry, dark location away from heat (About Soya, 2009).

Soy protein

Soy protein isolate is a highly refined or purified form of soy protein with a minimum protein content of 90% on a moisture-free basis. Edible soy protein 'isolate' is derived from de-fatted soy flour with a high solubility in water (high nitrogen solubility index). The aqueous extraction is carried out at pH <9. The extract is clarified to remove the insoluble material and the 'supernatant' is acidified to a pH range of 4–5. The precipitated protein curd is collected and separated from the whey by centrifuge. The curd is usually neutralized with alkali to form the sodium proteinate salt before drying. Soy isolates are mainly used to improve the texture of meat products, but are also used to increase protein content and enhance flavour and as an emulsifier.

Soy protein concentrate is produced by immobilizing the soy globulin proteins while allowing the soluble carbohydrates, soy whey proteins and salts to be leached from the de-fatted flakes or flour. The protein is retained by one or more of several treatments: leaching with 20–80% aqueous alcohol/solvent; leaching with aqueous acids in the isoelectric zone of minimum protein solubility (i.e. pH 4–5); leaching with chilled water (which may involve calcium or magnesium cations); and leaching with hot water of heat-treated, de-fatted soy meal or flour. Soy protein concentrate is about 70% soy protein and is basically soybean without the water-soluble carbohydrates. Soy protein concentrate is widely used as a functional or nutritional ingredient in a wide variety of food products, mainly in baked foods, breakfast cereals and in some meat products. Soy protein concentrates are available in granule, flour and spray-dried powder form (Smith and Circle, 1972).

Soy flakes

Horan (1974) has described general steps for the procedure of making de-fatted and full-fat soy flakes. These involve cleaning soybeans, followed by cracking and dehulling the grains. The dehulled grains are then conditioned, after which they are flaked and then extracted in solvent to obtain de-fatted flakes. The solvent thus obtained is used for the extraction of oil and neutral-grade lecithin, while the de-fatted flakes are further toasted and cooled.

Soy grits

Grits are produced from coarse ground flakes. The separation into various grades is achieved by mechanical shifting or a combination of air classification and shifting. Soybean flour and soy grits are differentiated on the basis of granulation. Grits have a particle size of >100 mesh, while flour has a particle size of <100 mesh (Bastiaens, 1976). In order to obtain a desired texture in some food products, grits are used rather than the soybean flour.

Medicinal uses

There is no denying that soybean has many health benefits. These benefits are mainly derived from the quality of the soybean proteins and from the isoflavones, genistein and daidzein. Soybean has been shown to be beneficial in conditions of lactose intolerance, high cholesterol, heart disease, cancer, menopausal symptoms, osteoporosis and diabetes.

Soybeans contain a variety of anti-carcinogenic phytochemicals, including lunasin and lectins. Lunasin is a polypeptide that arrests cell division and induces apoptosis in malignant cells. Lectins are glycoproteins that selectively bind carbohydrates; lectins are being used in medicine in a variety of new applications. Medical research has shown that foods rich in soybean protein may be protective against prostate cancer by helping to promote healthier prostate tissues. The components of soybean that may be helping to prevent colon cancer are called isoflavones and saponins. Soybean also contains omega-3 fatty acids that help to provide protective benefits against breast cancer.

Soybean may help to prevent heart disease by reducing total cholesterol and low-density lipoprotein cholesterol and preventing plaque build up in the arteries, which may lead to stroke or heart attack. Potential mechanisms by which soybean isoflavones might prevent atherosclerosis include a beneficial effect on plasma lipid concentrations, antioxidant effects, antiproliferative and antimigratory effects on smooth muscle cells, effects on thrombus formation and maintenance of normal vascular reactivity (Anthony *et al.*, 1998). The soybean diet is the most potent dietary tool for hypercholesterolemia.

Soy foods have a low glycemic index and are high in dietary fibre, and thereby help in the management of diabetes. In addition, soy foods can provide additional benefits for controlling heart disease, one of the most prevalent complications of diabetes.

The isoflavone genistein seems to inhibit bone breakdown and may have similar effects to estrogens in maintaining bone tissue. Diets that are high in animal protein cause more calcium to be excreted in the urine. Replacing animal protein with soy protein may help to prevent calcium loss from the bones and reduce osteoporosis risk.

Estrogens play a role in the control of body temperature. Soybean contains phytochemicals such as genistein, daidzein, and other phytoestrogens. These are the botanical equivalents of the human female hormone, but their effect is milder than that of estrogen and progesterone. However, they may ease menopausal symptoms such as hot flashes, night sweats and vaginal dryness and perhaps alleviate premenstrual difficulties such as cramping and irritability. Isoflavones in soybean may also offer some relief from the pain, swelling, nausea and bleeding of endometriosis (Lock, 1991; Cassidy *et al.*, 1995).

All soy formulas are lactose-free and are fortified with L-methionine, taurine, carnitine and iron. They are used commonly in the empirical management of acute gastroenteritis and intolerance to cow's milk protein. Furthermore, the use of soy formulas has been found to significantly reduce the prevalence of atopic diseases in the first 6 months of life, as well as in children with infantile atopic dermatitis, recurrent bronchiolitis and bronchial asthma (Quak and Tan, 1998).

Fodder and feed uses

Fodder refers particularly to food given to animals, rather than which they forage for themselves. It includes hay, straw, compressed and pelleted feeds, oils and mixed rations and sprouted grains and legumes. Soybean is also used as an animal fodder. Ruminants need a high fibre content in their feed. Soybean hulls are extensively and exclusively used for roughage in feeding livestock (United Soybean Board, 2008). This feedstuff is a source of highly digestible fibre that does not contain starch (Coverdale *et al.*, 2004). Soy protein concentrates are preferred because the absence of water-soluble carbohydrates not only increases the protein content, but also keep the flatulence problem under control. Soybean meals play an important role in the production of fish feed and pet foods. Soybean proteins and the linoleic and linolenic acids present in full-fat soybean meal may even improve the fur quality of mink. Soybean meal represents one of the major feed ingredients for cattle, especially during winter. The presence of soybean meal in poultry rations allows optimum growth of the animals. Soybean meal can be added to the mixture of pollen and honey that is provided for feeding the bee larvae.

Non-food versatile uses of soybean

Soy-based materials are gaining popularity in the construction industry. Soybean oil, soybean oil refining by-products and emulsion of soybean oil methyl esters are being used as release agents for concrete, competing against petroleum-based release agents. This soybean oil is being promoted for its environmental advantages and low toxicity and skin irritation to workers. When soy candles burn, they do not get as hot as paraffin candles and their fragrance spreads faster. Soy candles also burn more cleanly and do not leave soot like paraffin. Crayons made from soybean oil have better colour and do not rub off like other crayons. These all-natural crayons also cost less.

A product called 'soysilk', made from the residual compounds of soybeans following tofu manufacturing, is quickly becoming a yarn of choice. It is being used for clothing and for a cuddly new toy called 'Tofu Bear'. Soybean can also be found in everyday beauty products. Soybean is not as greasy as other products and it helps to protect the skin from the sun. Soybean is used to make shampoos and conditioners that provide nutrients to hair.

Lubricants made from soybean oil protect metal better than other lubricants, as they do not dry out like other oils and reduce cost of multiple applications. Soybeans are also used to make oil for hydraulic systems. Soy hydraulic oil is better for the environment and easier to clean up and recycle than standard hydraulic oils. Soybeans are also used to make paints and tough coatings for many surfaces. These are safe for the environment and safe enough for use in food packages.

Soybean oil is being used to develop toner for use in laser printers, copiers and fax machines. Paper printed with soy toner is easier to recycle and comes out cleaner and brighter (Fan *et al.*, 1999). Soy foams are currently being developed for use in coolers, refrigerators, automotive interiors and even footwear (Soybean Producers Association, 2007).

Soybean oil is being used in insect sprays for orchards and trees. Soybean resins are used to make fibreglass, which is being tested to make strong yet lightweight parts for farm equipment, cars and boats. A product called 'soapstock', made from soybean parts, forms an environmentally safe coating that protects roads and helps control dust on gravel roads. Soybean solvents remove grease, paint, oil and stains without harming materials. Unlike other solvents they can be cleaned with water, making them better for the environment.

18.4 Methods Used in Soy Product Preparation

The acceptance of soy-based fresh foods had always remained under scanner because of the beany flavour, difficulty in cooking and presence of antinutritional factors such as trypsin inhibitors, hemagglutinins, flatulence factors and phytic acid. Several methods of soybean preparation – including soaking, blanching, germination, enzyme treatment and ultrafiltration – have been useful in the removal of undesirable factors and making soy foods more palatable and acceptable (Chauhan and Chauhan, 2007). Various methods may be used for the preparation of soybean products, some of which are discussed here.

Soaking

Hydration of soybeans is the first step in soy milk manufacture. Small, hard beans imbibe less water, which results in a significantly reduced yield of soy milk. The initial moisture content of beans also influences the hydration rate of soybean (Smith *et al.*, 1960).

The rate of water absorption in soybean is regulated by the calcium content in the seed coat, the surface area and the structure of the micropyle (Saio, 1976). It has been found that soaking soybeans for 10 hours at ambient temperature, followed by blanching in 0.5% sodium bicarbonate for 30 min, resulted in the loss of 11% total bean solids and 5% protein (Nelson *et al.*, 1976). At 100°C, soybeans absorb water equal to their weight in approximately 15 min and reach a peak within 2 hours; in comparison, at 21°C soybeans absorb an equal weight of water after approximately 4 hours of steeping. Steeping soybean in 0.5% sodium bicarbonate at 40°C reduces the water uptake process (Johnson and Snyder, 1978). Therefore, temperature and time significantly affect water uptake. Soaking seeds in an alkaline medium aids in the processing of soy milk and improves its quality as it helps to reduce the undesirable beany flavour (Muhammad and Khan, 2000). Wong *et al.* (1983) reported that the bulk of solids lost during soaking comprises low molecular weight water-soluble oligosaccharides such as raffinose and stachyose, which are considered responsible for causing flatulence. The remaining quantities of carbohydrates leach out during soaking.

Germination

Sprouting is a process by which the dormant embryo in the seed wakes up and begins to grow into a seedling. During the germination process, seeds are wrapped in a wet paper towel and placed in a seed germinator set at 30°C and 85% relative humidity for 24–72 hours. After the removal of sprouts, the seeds are used to prepare different soy products such as soy beverage, soybean powder, bread and soy milk.

Pathak (2005) reported that germinated soybean seeds produced significant anti-diabetic effects by regulating blood sugar levels and were more effective than oral hypoglycaemic drugs. Singh (1978) reported that in the Bragg and Kalitur varieties of soybean, trypsin inhibition was decreased from 52.24% and 58.21% to 16.42% and 19.40%, respectively, after 96 hours of germination. Chauhan and Chauhan (2007) developed an anti-nutrient-free soybean beverage using germinated soybeans and reported that the beverage was devoid of oligosaccharides, with very low quantities of phytic phosphorus, trypsin inhibitor and saponins.

Bau and Debry (1979) prepared various protein fractions from non-germinated and germinated soybeans. Germination tended to improve the nutritional quality of protein products as measured by protein efficiency ratios. Vitamin C content increased from 0 to 25 mg $100\,g^{-1}$ during germination. Flour from germinated soybeans can be used to replace wheat flour in some formulations to improve nutritional quality. Selection of the right variety combined with a suitable germination process could provide a good source of bioactive compounds from soybean and their germinated products for nutraceutical applications (Pei and His, 2006). Sprouting induces hydrolysis of soybean polypeptides and polysaccharides, limiting the cross-linking of these macromolecules during and after heat treatment, thereby delaying the coagulation of soybean extract (Nsofor and Maduako, 1992). When soy milk is developed from sprouted soybeans, it is more digestible and nutritious than that from the unsprouted soybeans. Germination of seeds causes hydrolysis of macromolecules, which facilitates digestion.

Fermentation

Fermented foods may be defined as those that have been subjected to the action of microorganisms or enzymes so that desirable biochemical changes cause significant modifications to the food. By fermentation, the food may be made more nutritious, more digestible, and safer or have better flavour (Raghuvanshi and Singh, 2009). Fermentation is carried out directly on cooked soybeans with a selected organism under specific conditions. The various products that are prepared through soybean fermentation include *miso*, *shoyu*, *nattō*, *tempeh*, *suju* and fermented soy milk. Ara *et al.* (2002) reported that fermentation removes unpleasant tastes such as astringency; this may be attributed to the microorganisms involved in brewing in the process of fermentation having resolved the unpleasant taste element of the soybean. Several *in vitro* models are used to detect the antioxidant effect of the fermented soybean extract, which is compared to vitamin C. Fermented soybean extract can function both as an antioxidant and as a free-radical acceptor that can convert free radicals into harmless substances through an energy-decreasing procedure (Hu *et al.*, 2004).

Soybean flour fermented and steamed with chickpea (*Cicer arietinum*) flour is called *dhokla*; with rice (*Oryza sativa*) flour, it is called *idli*. Kanekar *et al.* (1992) reported that if soybean meal mixed with chickpea is fermented for 18 hours at 30°C then it results in a decrease in trypsin inhibitor activity. There is slight increase in protein quality on germination and fermentation (Khader, 1983). Replacement of pulses with soybean up to 50% can improve the palatability of the *idli*. The fermentation of soy *idli* increased amino nitrogen, free sugars, niacin and riboflavin, and significantly decreased the phytate content. The substitution of black gram (*Vigna mungo*) by soybean resulted in increased water requirements. Substitution of up to 30% resulted in an increase in batter volume during fermentation. Substitution of chickpea with soybean beyond 25% considerably decreased the palatability of *dhokla* (Tambe *et al.*, 1971). Fermented soybean products, (e.g. *miso*, *tempeh*) have been associated with a reduction in cancer and heart disease incidence.

Blanching

Blanching is a process in which soybeans are boiled in water for a definite period to make them soft and to inactivate lipoxygenase and trypsin inhibitors. Akpapunam *et al.* (1997) reported that soybean seeds are blanched in water at 90°C for 7 min. Blanching reduces soy solids yield of the extracted concentrates, indicating that blanching denatures and insolubilizes soybean proteins (Kinsella, 1985; Cheman *et al.*, 1989), thereby limiting solids extractability from the blanched cotyledons (Nsofor and Maduako, 1992). The blanched samples contain less soluble protein and less starch than unblanched samples. Nelson *et al.* (1976) observed complete destruction of trypsin inhibitor when beans were soaked in 0.5% sodium bicarbonate solution or water and blanched for 5–10 min. They also observed that beans soaked for 8 hours in 0.5% sodium bicarbonate solution and blanched for 20 min had a highly acceptable mouthfeel. Chauhan *et al.* (1998) subjected soy milk to different physical and chemical treatments and reported that atmospheric blanching resulted in considerable losses of vitamins B_1, B_2 and methionine; these losses were further increased by the use of sodium bicarbonate during blanching. They further observed that phosphorus and iron contents were not affected by sodium bicarbonate and only a slight loss in calcium content was observed.

Boiling

In the boiling process, soybeans are boiled and various products such as soy milk can be prepared. In Japanese dishes, soybeans are thoroughly softened in boiled water under pressure to increase both digestibility and flavour. Heating intact soybeans in boiling water for 10–20 min is sufficient to improve the digestibility of soybean proteins. Morinaga (2001) reported that trypsin inhibitor is inactivated by heating soybeans in boiling water. In intact soybeans, trypsin inhibitor is mostly inactivated after heating for 20 min in boiling water.

Dry roasting

Dry roasting (i.e. roasting without the presence of fat) or parching is a traditional Indian household practice for roasted and popped products. It involves initially sprinkling the grains with a little water, which may contain common salt. The pulse is then mixed with four times its own volume of preheated sand or salt of about 240–335°C. The pulse is subsequently roasted by rapid mixing in the pan using a ladle. During this process the pulse temperature increases from an appropriate initial 26°C to 132°C in a period of 2–3 min. The roasted material is separated from the sand or salt by sieving. This process brings about a light, porous texture in the pulse. These are traditionally eaten as a snack.

Dry roasting of soybeans has been followed for centuries in Japan for production of 'Kinako' (Smith and Circle, 1972). Kinako was originally made in the home by roasting soybeans over an open flame and grinding them into a powder, which was then used as a coating or ingredient in other foods. Phytic acid content decreases on roasting (Khetarpaul and Goyal, 2008). This may be attributed to the formation of insoluble complexes between phytic acid and other components (Kumar *et al.*, 1978; Chitra *et al.*, 1996). Higher protein digestibility may be attributed to an opening of the protein structure through denaturation, leading to increased accessibility of the protein to enzymatic attack (Wu *et al.*, 1994) and structural disintegration of enzyme inhibitors (Vijayakumari *et al.*, 1995).

Frying

Frying may involve deep-oil frying, shallow frying or sautéing. In deep-oil frying, the food is totally immersed in hot oil. Cooking is rapidly completed as the temperature is 180–220°C and results in an increased calorific value of the food. A salted and spiced product named 'soy nuts' is developed by deep-fat frying blanched soybean splits. It contained approximately 45% protein and 35% fat (Singh, 1978). Khetarpaul and Goyal (2008) also prepared fried soy dal, which is nutritionally rich with high protein digestibility (*in vitro*) and acceptable sensory taste.

Extrusion

Extrusion has been defined as the process by which moistened, starchy or proteinaceous materials are plasticized by a combination of high pressure and mechanical shear (Hauck, 1980). Several food products have been developed to add protein content by combining cereal and soybean flour (Cheman *et al.*, 1992; Adesina *et al.*, 1998), such as noodles, savoury snacks and textured nuts. Osundahunsi (2006) reported that extrusion cooking reduces the moisture content of products and hence prolongs shelf life.

18.5 Preparation and Recipe

Several preparations are made using soybean as a main ingredient or as a replacement for other pulses or cereal flour. Recipes for fermented soybean have been in use since ancient times in China, Thailand, Burma and north-east India. In northern parts of India, black soybean is used as a soup, dal or roasted snacks and mainly consumed during the winter months. Some of the products prepared from soybean are shown in Fig. 18.2.

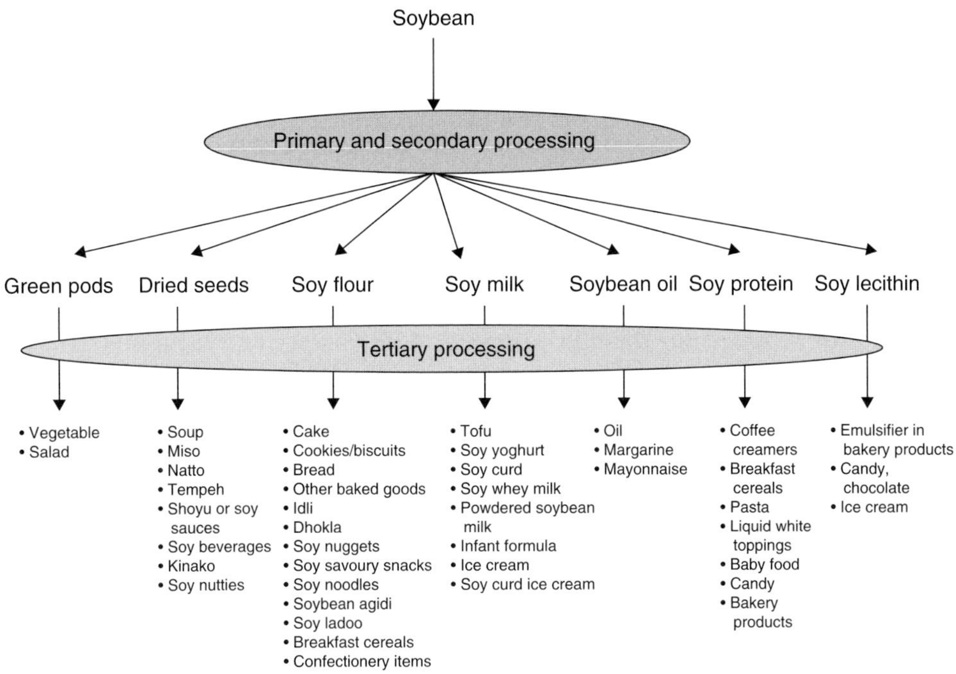

Fig. 18.2. Soybean and its preparations.

Use of green leaves and pods

Fully developed pods, while still green, are harvested for their green seeds. These are removed by hand and then boiled until tender. An alternative technique is to boil the pods first and then shell the beans. The cooked beans can be eaten as they are or in the form of a vegetable or salad/*chat*, or they can be combined in many dishes. Soy *chat* can be prepared by mixing boiled green soybeans (250 g), boiled and diced potato (75 g), chopped onion (150 g), tomato (75 g), finely chopped ginger (5 g), chopped green chillies (2 g), mango powder (¼ tsp) and salt (to taste). These green soybeans can also be eaten in the form of a vegetable preparation known as '*nimona*', which is generally made with table pea (*Pisum* species) or tender chickpea. Preparation of *nimona* involves grinding green soybeans (250 g) into a coarse paste. Potatoes (50 g) are diced, fried and kept aside. The oil is heated in a pan and tempered with asafoetida and ground spices (100 g onion, 2 g garlic, 5 g ginger, 2 g green chillies, ¾ tsp coriander powder, ½ tsp turmeric, ½ tsp spice mixture of black pepper, red cardamoms, mace, cinnamon and two cloves) are added. After a little frying, tomatoes are added and fried and then soy paste is added and further fried until it leaves oil. Fried potatoes are then added with salt (to taste) and water and cooked until a thick gravy is obtained. Nimona is served with rice (Raghuvanshi, 2003).

The green leaves of soybean can be used for the preparation of soy pakori and soy soup. To prepare soy pakori, tender soybean leaves (150 g) are washed and chopped and added along with onion (100 g), salt (to taste) and green chillies (10 g) to a chickpea flour (250 g) batter. Small portions of batter mixture are deep fried in oil until light brown and then served with sauce. Soy soup can be prepared by grinding tender soybean leaves (200 g) to a fine paste. Finely chopped onions (150 g), salt (to taste), sugar (¼ tsp), bread (two slices) and boiled potatoes (150 g) are added to 150 ml of water and boiled until the onion and bread are soft. The soybean-leaf paste is then boiled with this bread/onion mixture with added ginger (5 g), garlic (2 g) and chopped chillies (2 g). The soup is then run through a strainer after cooling and served with added crushed black pepper and a teaspoon of cream (Raghuvanshi, 2003).

Use of grains

Preboiled soybeans can be boiled soft and the resulting mashed paste used for the preparation of soup. Soy sauce is prepared by cooking preboiled soybeans for 2 hours with spices to a consistency desired to form a sauce. Soy namkeen is a roasted product of soybean in salt. It can be prepared by soaking 100 g of soybean in water for 4 hours. This is then dried in the shade on filter paper for 1 hour and roasted on preheated salt at 250°C for 15–20 s (Raghuvanshi, 2003). Another product of dried soybean seeds is fried soybean dal, which is prepared by soaking 100 g dehulled soybean in 400 ml of water for 4 hours and then spreading it on filter paper to remove any adhering water. The hydrated dal is then deep fried in oil and salt is sprinkled over it (Khetarpaul and Goyal, 2008).

Halwa is a very popular dessert in India. It is traditionally prepared from wheat semolina. The preparation of instant *halwa* mix involves roasting wheat semolina (100 g) and soy semolina (60 g) separately until golden brown. Ghee (90 g) is heated in a vessel to which wheat and soy semolina are added, mixed and cooled to 60–70°C. Powdered sugar (155 g), fried cashew nuts (1.25%) and cardamom powder (0.3%) are added and mixed thoroughly. The product can be reconstituted by adding 100 g dry mix to 130 ml of boiling water and stirring continuously for 3–4 min until it attains the desired consistency (Yadav *et al.*, 2007).

Nowadays, health-conscious people are replacing high-calorie drinks with soy beverages. These can be prepared by wrapping seeds in a wet paper towel and placing in a seed germinator set at 30°C and 85% RH for 48–72 hours. The germinated seeds are taken out and the sprouts removed. The seeds are then dehulled, blanched in 0.5% sodium bicarbonate for 30 min and washed with water four times. They are then ground in a colloid mill and the slurry is diluted 12 times with water. The slurry is then homogenized at 5000 psi and 6% sugar is added, followed by boiling for 5 min, adding flavour, packing and storing at low temperature (Chauhan and Chauhan, 2007).

Miso is a whitish-brown, brown or reddish-brown fermented soup-base paste. This thick, salty paste is high in protein and low in fat and calories.

The basic process for making *miso* is to first wash the beans, then soak them so that they absorb enough water for cooking. The soaked beans are cooked in water or steam. After cooling, the beans are mixed with salt and a *koji* starter (*Aspergillus oryzae* mould fungus) and allowed to ferment at 25–30°C for varying periods from 1 week to >2 years, depending on the final product requirements. Fermentation breaks down the protein and carbohydrate content to form palatable flavour components (Burke, 1996).

Nattō is made of fermented whole soybeans. It has a sticky, viscous coating with a cheesy texture. In Asian countries, *nattō* is traditionally served as a topping for rice, in miso soups and with vegetables. The basic process for making *nattō* is to wash, soak and steam the beans, allow them to cool to 60°C and mix in a starter of *Bacillus nattō* for an 8-hour fermentation process at 35°C (Burke, 1996).

Tempeh is a fermented soybean product originating from Indonesia (Nout *et al.*, 1993). It is made of whole cooked soybeans infused with a culture to form a dense, chewy cake used as a meat substitute. It can be marinated and grilled or used in soups. *Tempeh* is high in fibre, calcium, B vitamins, iron and protein. It is cholesterol-free and low in saturated fat (The George Mateljan Foundation, 2010). The principal steps in making *tempeh* include soaking soybeans in water until the hulls can be easily removed by hand. The dehulled soybeans are then boiled with excess water for 30 min, drained and spread for surface drying. Small pieces of *tempeh* from a previous fermentation are mixed with the soybeans. The inoculated beans are wrapped with banana leaves and allowed to ferment at room temperature for 1 day. By this time, the beans are covered with white mycelium and bound together by mycelium as a cake, which has a pleasant odour. Traditionally the cake, which is consumed within a day, is cut into thin slices, dipped into a salt solution and fried in coconut oil. Sliced tempeh can be baked or added to soup (Hesseltine and Wang, 1972).

Shoyu or soybean sauce is a dark-brown liquid made by the fermentation of a combination of soybeans and cereals, usually wheat. It has a pleasant aromatic odour and salty taste, suggesting a meat extract (Japan Federation of Soy Sauce Manufactures Cooperatives, 2008).

Use of soy flour

Soy flour is used in the preparation of cakes, biscuits, bread and other baked goods. For making soy biscuits, 110 g of fat is rubbed into the mixed dry ingredients (110 g wheat flour, 110 g soy flour, 55 g sugar, 10.8 g baking powder and 6 g common salt). Milk is added in the requisite volume and the mixture is kneaded into a stiff dough. The dough is rolled out on a sheeting board to a uniform thickness of about 0.4 cm. The sheet is stamped out in circular shapes of about 5.8 cm diameter, using a biscuit cutter. The biscuit cuts are placed on lightly greased baking trays, covered, rested for about 15 min and baked for 12 min at 185°C (Onweluzo and Iwezu, 1998). To prepare soy bread, active yeast (20 g) is added to warm water and set aside. The remaining ingredients, including ¼ cup brown sugar, 1 tsp salt, 2 tablespoons shortening, ½ cup milk, ½ cup water, 1½ cups soybean flour and 4½ cups whole wheat flour, are combined and added to the yeast mixture. The dough is kneaded hard five to six times and is placed in a greased glass bowl and left to rise for 1 hour. The bowl is covered with a clean kitchen towel. The dough is turned out onto a floured surface and kneaded twice more. The dough is returned to the bowl and allowed to rise until doubled in size. The dough is then made into two separate loaves and placed into an oiled bread pan and baked at 375°F for 55 min (Physicians Laboratories, 2009).

Snack foods have long been a part of diets both in developing and developed countries. But most snack foods, being cereal-based, are either poor sources of protein or contain low-quality protein. Singh *et al.* (2006) and Narayan *et al.* (2007) made efforts to produce extruded snack food by replacing sorghum (*Sorghum bicolor*) and kodo (*Paspalum scrobiculatum*), respectively, with 20% soy flour, improving the protein quality. Soy flour is also used for the preparation of an Indian sweet known by as *ladoo*. For this, ghee (clarified butter) is heated in a deep saucepan and wheat flour (100 g) and soybean flour (50 g) are roasted separately until brown. Both of the flours are mixed and ground sugar and cardamom powder are added. This is then mixed thoroughly and made into round balls (*ladoos*) and garnished with coconut powder or silver foil (Raghuvanshi, 2003). Ramakrishnan *et al.* (1976) found that acceptable *idli* can be prepared with a 2:1 ratio of rice and soybean. Singh (1970) prepared *idli* using soybeans (1 cup), rice (1 cup) and black gram dal (1 cup). All three were soaked and ground to a fine paste and kept for 8–10 hours for fermentation. The paste is then filled in *idli* cups and steamed in a pressure cooker for 7–10 min.

A method for the preparation of *agidi* (commonly consumed in Nigeria) supplemented with soy flour has been given by Akpapunam *et al.* (1997). For preparing soybean flour, soybeans are sorted, cleaned and blanched in water at 90°C for 7 min. The blanched beans are then soaked in 0.5% $NaHCO_3$ solution for 6 hours (Johnson and Snyder, 1978). The soaked beans are dehulled, sun-dried and finally milled into flour in a corn mill. The flour is sieved through cheese cloth to obtain fine and uniform particle-size flour. Five flour blends are prepared by mixing maize (*Zea mays*) and soybean flours in different proportions. For the preparation of *agidi*, the slurry containing 30 g of flour blends and 150 ml water is cooked in an aluminium pot for 5 min with constant stirring at about 85°C on an electric stove. The highly viscous paste formed is poured into a 250 ml glass beaker and allowed to cool for about 1 hour, during which time it solidified into a gel called *agidi*. It is eaten alone or with vegetable soup.

Use of soy milk

Flavoured soy milk to be drunk as a beverage contains an added sweetener, oil, salt and flavour. In soy milk beverage the water to bean ratio is 7:1, whereas in plain soy milk it is 5:1. When lactic acid bacteria are used for fermentation, fermented soy milk is produced. In infant formulas, soybean milk is fortified with vitamins and minerals. Soy yogurt is prepared by mixing soy milk with an equal quantity of commercial milk, followed by fermentation (Yoshimoto and Sato, 2001). Tuitemwong *et al.* (1993) reported that fermentation and flavouring significantly change the major attributes of soy milk.

Ice cream is a delicious frozen food that is prepared by using both dairy and non-dairy products (De, 1986). According to the Prevention of Food Adulteration Act and Rules of India (PFA, 1954), ice cream is a frozen product obtained from cow's or buffalo's milk or a combination thereof from cream and/or other milk products. Soybeans are an excellent and cheap source of calories, protein and fat and thus hold a great promise for substituting milk solids in functional properties (Tyagi, 1984). Patil and Jha (2008) gave a method of preparation for soy ice cream that involved adjusting soy milk to 10% solids content by adding finely ground full-fat soy flour. The optimum amount of glyceryl monophosphate and propylene glycerol alginate (3 g each l^{-1} of ice cream mix before freezing) are added to the formulation of 650 ml soy milk with 10–12% solids, 60 g milk powder, 150 g sugar and 100 g cream, which is pre-homogenized. The homogenized mix is allowed to age at 4°C for 24 hours before making ice cream.

Soy whey milk is cheaper and can be used in the formulation of frozen concentrate, soft-serve-type desserts and prepared foods (Anonymous, 1972). A method for manufacturing a packaged soybean curd with a long shelf life without the inclusion of any artificial additives, such as coagulating agents, germicides and the like, is often used in the USA. Soybean juice is subjected to lactic acid fermentation until its pH reaches a value of ≤5 and is then subjected to heating at 60–95°C for 10 min to adjust the desirable curd tension. Soy curd can be further processed to make soy curd ice cream, as illustrated in Fig. 18.3.

Tofu is a cheese-like food made from soybean milk. Tofu is cholesterol-free, low in sodium and a good source of calcium, iron and B vitamins. High quantities of available protein and oil result in high tofu yields. A high protein-to-oil ratio produces a hard (or firm) tofu; a high oil-to-protein ratio makes a soft (or silky) tofu (Burke, 1996). To prepare tofu, soybeans are washed, soaked overnight and then ground with water. The finely ground mixture is strained through a coarse cloth to separate the soybean milk from the insoluble residue. After the soybean milk is heated to boiling, calcium or magnesium sulphate is added to coagulate the proteins. The coagulated milk is then transferred into a cloth-lined wooden box and pressed with a weight on top to remove the whey. A soft but firm cake-like curd (tofu) forms. This can be consumed directly (Hesseltine and Wang, 1972) or cooked as a *paneer* curry.

Soy curd→Stirring→Blending for 10 min→Pasteurization (76°C for 10 min)→Cooling to room temperature→Ageing for 12 h at 4°C→Freezing in batch freezer→Hardening in the freezer at 0°C→Soy curd ice cream

Fig. 18.3. Flow chart for the preparation of soy curd ice cream.

In Nigeria, cow's milk is substituted with 20% soy milk to prepare *waranksi* (a soft unripe cheese) with acceptable sensory quality (Akubor *et al.*, 2006). In India, soybean *paneer* or *chenna* is prepared by blending soy milk with cow's milk in the ratio 30:70. A total of 500 ml blended milk is mixed for 2 min in a kitchen churn at a speed of 1500 rpm and then boiled for 15 min. The milk blend is then coagulated using 5% citric acid (v/w) solution at 80°C. The coagulated mass is allowed to settle for 5 min and the suspension is then filtered through double-layered muslin cloth and pressed with a weight to remove water. When set, it is cut to 1 inch squares and made into *palak paneer* curry or *paneer pakori*, a fresh snack item served with chutney or sauce.

18.6 Conclusions

Soybean has great potential as an exceptionally nutritive and very rich protein food. Soybeans are versatile and can be used in a number of different ways. The most common use of the soybean is as food for humans. The soybean has various functional properties such as good foaming capacity, high moisture-holding capacity and emulsifying capacity and can, therefore, be used in diverse processes for various food preparations. Soybean can be processed in a variety of ways. Common soybean products include green pods, dried seeds, soy flour, soy milk, soybean oil, soy protein and soy lecithin. These are further processed for the preparation of various products such as vegetables, salads, soups, *miso*, *nattō*, *tempeh*, *shoyu*, *kinako*, tofu and various baked goods. Soybean products are the main ingredients in many meat and dairy substitutes. Different processing methods such as boiling, blanching, roasting, frying, germination and fermentation increase the nutritive value of soybean either directly or by decreasing anti-nutritional factors. Soybean also has medicinal, feed and fodder and industrial uses. Soybeans contain the isoflavones genistein and daidzein, which are sources of phytoestrogens in the human diet and useful in the maintenance of good health.

References

About Soya (2009) Soya – information about soy and soya products. Available at: http://www.soya.be/soybean-oil.php (accessed 25 February 2009).

Adesina, A.A., Sowbhagya, C.M., Bhattacharya, S. and Zakiuddin, A.S. (1998) Maize soy based ready to eat extruded snack food. *Journal of Food Science and Technology* 35, 40–43.

Akpapunam, M.A., Badifu, G.I.O. and Etokudo, E.P. (1997) Production and quality characteristics of Nigerian *Agidi* supplemented with soy flour. *Journal of Food Science and Technology* 34, 143–145.

Akubor, P.I., Offonry, S.U. and Isolokwu, P.C (2006) Effect of proportion of soy milk and pH on the quality attributes of soy warankasi a Nigeria soft cheese. *Journal of Food Science and Technology* 43, 101–103.

Anonymous (1972) Soy-whey-milk offers economic substitute for frozen whole milk concentrate. *Quick Frozen Foods* 34, 141–142.

Anthony, M.S., Clarkson, T.B. and Williams, J.K. (1998) Effects of soy isoflavones on atherosclerosis: Potential mechanisms. *American Journal of Clinical Nutrition* 68, 1390S–1393S.

Ara, K., Yoshimatsu, T., Ojima, M., Kawai, S. and Okubo, K. (2002) Effect of new lactic acid fermentation on sensory taste of fermented soy milk. *Journal of the Japanese Society for Food Science and Technology* 49, 377–387.

Bastiaens, F.G. (1976) Oilseed flour for human food. *Journal of the American Oil Chemists' Society* 53, 310–315.

Bau, H.M. and Debry, G. (1979) Germinated soybean protein products: Chemical and nutritional evaluation. *Journal of the American Oil Chemists' Society* 56, 160–162.

Burke, P. (1996) Marketing food-quality soybeans in Japan. American Soybean Association, St. Louis, MO, USA.

Cassidy, A., Bingham, S. and Setchell, K. (1995) Biological effects of isoflavones in young women: Importance of the chemical composition of soybean products. *British Journal of Nutrition* 74, 587–601.

Chauhan, G.S., Singh, J.D. and Tomar, N.S. (1998) Nutritional changes in soymilk subjected to different physical and chemical treatments. *Journal of Food Science and Technology* 35, 271–273.

Chauhan, O.P. and Chauhan, G.S. (2007) Development of anti-nutrients free soy beverage using germinated soybean. *Journal of Food Science and Technology* 44, 62–65.

Cheman, Y.B., Wi, L.S. and Nelson, A.I. (1989) Acid inactivation of soybean lipoxygenase with retention of protein solubility. *Journal of Food Science* 54, 963–967.

Cheman, Y.B., Mohmad, N.B., Abdul, K. and Tan, T.K. (1992) Evaluation of soy flour and high protein rice-soy snack formation. *Journal of Food Science and Technology* 27, 715–719.

Chitra, U., Singh, U. and Rao, P.V. (1996) Phytic acid, *in vitro* protein digestibility, dietary fiber and minerals of pulses as influenced by processing methods. *Plant Foods for Human Nutrition* 49, 307–316.

Coverdale, J.A., Moore, J.A., Tyber, H.D. and Miller Auwerda, P.A. (2004) Soybean hulls as an alternative feed for horses. *Journal of Animal Science* 82, 1663–1668.

De, S. (1986) *Outlines of Dairy Technology*. Oxford University Press, New Delhi, India.

Fan, X.D., Deng, Y., Waterhouse, J., Pfromm, P. and Corr, W.W. (1999) Development of an easily deinkable copy toner using soy-based copolyamides. *Journal of Applied Polymer Science* 74, 1563–1570.

Gandhi, A.P. (2000) Soymilk – a potential supplement for dairy milk. *Indian Food Industry* 19, 392–397.

Gopalan, C., Sastri, B.V.R. and Balasubramanian, S.C. (1989) *Nutritive Value of Indian Foods*. National Institute of Nutrition, and Indian Council of Medical Research, Hyderabad, India.

Hauck, K. (1980) Marketing opportunities for extrusion cooked products. *American Association of Cereal Chemists* 25, 594–595.

Hesseltine, C.W. and Wang, H.L. (1972) Fermented soybean food products. In: Smith, A.K. and Circle, S.J. (eds) *Soybeans: Chemistry and Technology*. The AVI Publishers, Westport, CT, USA, pp. 404–405.

Horan, F.E. (1974) Soy protein products and their production. *Journal of American Oil Chemists' Society* 51, 67A–73A.

Hu, C.C., Hsiao, C.H., Huang, S.Y., Fu, S.H., Lai, C.C., Hong, T.M., Chen, H.H. and Lu, F.J. (2004) Antioxidant activity of fermented soybean extract. *Journal of Agricultural and Food Chemistry* 52, 5735–5739.

Japan Federation of Soy Sauce Manufactures Cooperatives (2008) Japanese Soy Sauce. Available at: http://midwestsgippers.com/images/2008kaneko.pdf (accessed 8 March 2010).

Johnson, K.W. and Snyder, H.E. (1978) Soymilk – a comparison of processing methods on yield and composition. *Journal of Food Science* 43, 349–353.

Kanekar, P., Joshi, N., Sarnaik, S. and Kelkar, A. (1992) Effect of fermentation by lactic acid bacteria from soybean seeds on trypsin inhibitor (TI) activity. *Food Microbiology* 9, 245–249.

Kellor, R.L. (1971) Soy flour and grits for use in food products. *Journal of the American Oil Chemists' Society* 48, 481–493.

Khader, V. (1983) Nutritional studies on fermented, germinated and baked soy bean (*Glycine max*) preparations. *Journal of Plant Foods* 5, 31–37.

Khetarpaul, N. and Goyal, R. (2008) Development of soynutties and fried soy dhal: Sensory and nutritional evaluation. *Journal of Food Science and Technology* 45, 105–107.

Kinsella, J.E. (1979) Functional properties of soy proteins. *Journal of American Oil Chemists' Society* 56, 242–258.

Kinsella, J.E. (1985) Functional criteria for expanding utilization of soy protein in foods. In: Shibles, R. (ed.) *World Soybean Research Conference III: Proceedings.* Westview Press, Boulder, CO, USA and London, UK, pp. 152–157.

Krishna, A., Singh, G., Kumar, D. and Agarwal, K. (2003) Physico-chemical characteristics of some new varieties of soybean. *Journal of Food Science and Technology* 40, 490–492.

Kumar, K.G., Venkataraman, L.V., Jaya, T.V. and Krishnamurthy, K.S. (1978) Cooking characteristics of some germinated legumes: Changes in phytins, Ca^{++}, Mg^{++} and pectins. *Journal of Food Science* 43, 85–93.

Liu, K., Orthaefer, F. and Thompson, K. (1995) The case of food grade soybean varieties. *INFORM* 6, 593–596.

Lock, M. (1991) Contested meanings of the menopause. *Lancet* 337, 1270–1272.

Manay, N.S. and Shadaksharaswamy, M. (2000) *Foods: Facts and Principles*. New Age International (P) Ltd, New Delhi, India, 540 pp.

Michael, A.P. and Alison, M.J. (2003) Variability in soy flour composition. *Journal of the American Oil Chemists' Society* 80, 551–562.

Morinaga, K. (2001) Effect of tissue destruction of soybean seeds on the inactivation of trypsin-inhibitor by heating in water. *Journal of the Japanese Society for Food Science and Technology* 48, 416–421.

Muhammad, I. and Khan, S. (2000) Optimization of soaking conditions of whole soybean seed for soymilk production. *Sarhad Journal of Agriculture* 16, 101–104.

Mustakes, G.C. (1971) Full fat and defatted soy flour for human nutrition. *Journal of the American Oil Chemists' Society* 48, 815–819.

Narayan, P.N., Siddalinga, S.M., Babu, S.S.T. and Semwal, A.D. (2007) Protein quality of sorghum-soy based extruded snack food. *Journal of Food Science and Technology* 44, 165–167.

Nelson, A.I., Steinberg, M.P. and Wie, L.S. (1976) Illinois process for preparation of soymilk. *Journal of Food Science* 41, 57–61.

Nout, M.J.R., Ruikes, M.M.W., Bouwmeestar, H.M. and Beljaars, P.R. (1993) Effect of processing conditions on the formation of biogenic amines and ethyl carbonate in soybean tempeh. *Journal of Food Safety* 13, 293–303.

Nsofor, L.M. and Maduako, O. (1992) Stabilized soymilk for ambient tropical storage: A preliminary report. *International Journal of Food Science and Technology* 27, 573–576.

Onweluzo, J.C. and Iwezu, E.N. (1998) Composition and characteristics of cassava–soybean and wheat–soybean biscuits. *Journal of Food Science and Technology* 35, 128–131.

Osundahunsi, O.F. (2006) Functional properties of extruded soybean with plantain flour blends. *Journal of Food, Agriculture and Environment* 4, 57–60.

Parihar, A.S, Mittal, M., Data, I.C., Quadri, M.A. and Kushwah, H.S. (1977) Organoleptic evaluation and nutritive value of recipes of soymilk and soy residue. *Journal of Food Science and Technology* 14, 130–132.

Pathak, M. (2005) Soaked and germinated soybean seeds for blood sugar control: A preliminary study. *Natural Product Radiance* 4, 405–409.

Patil, R.T. and Jha, K. (2008) Optimization of stabilizer and emulsifier requirement for production of soy ice cream by response surface analysis. *Journal of Food Science and Technology* 45, 267–269.

Pei, Y.L. and His, M.L. (2006) Bioactive compounds in legumes and their germinated products. *Journal of Agricultural and Food Chemistry* 54, 3807–3814.

PFA (1954) The Prevention of Food Adulteration Act & Rules. Available at: http://mohfw.nic.in/pfa%20acts%20and%20rules.pdf (accessed 26 February 2010).

Physicians Laboratories (2009) Recipes for tofu, tempeh, textured soy protein and many other soyfoods. Available at: http://soyfoods.com/recipes/index.html (accessed 26 February 2010).

Pingle, W. (1974) Full-fat soy flour. *Journal of the American Oil Chemists' Society* 51, 74A–76A.

Pyler, E.J. (1988) *Baking Science and Technology*. Sosland Publishing Co., Merriam, USA.

Quak, S.H. and Tan, S.P. (1998) Use of soy protein formulas and soy-food for feeding infants and children in Asia. *American Journal of Nutrition and Dietetics* 68, 1444S–1446S.

Raghuvanshi, R.S. (2003) *Nutritious Soy Greens Preparations for Health*. GB Pant University of Agriculture and Technology, Pantnagar, India.

Raghuvanshi, R.S. and Singh, D.P. (2009) Preparation and use. In: Erskine, W., Muehlbauer F.J., Sarker A. and Sharma, B. *The Lentil: Botany, Production and Uses*. CAB International, Wallingford, UK, pp. 408–424.

Ramakrishnan, C.V., Parekh, L.J., Akolkar, P.N., Rao, G.S. and Bhandari, S.B. (1976) Studies on soy idli fermentation. *Journal of Plant Foods* 2, 28–31.

Saio, K. (1976) Tofu relationship between texture and live structure. *Cereal Food World* 24, 342–345.

Schaefer, M.J. and Love, D.J. (1992) Relationship between soybean components and tofu texture. *Journal of Food Quality* 15, 536–539.

Shemer, M. (1974) Studies on the effect of thermal processing on the chemical, nutritional and functional properties of soybean protein. *Dissertation Abstracts International University of Illinois*, Urbana, IL, USA.

Singh, D.S., Garg, S.K., Singh, M. and Goyal, N. (2006) Effect of processing parameters on the quality of extrudates made out of soy-kodo blends. *Journal of Food Science and Technology* 43, 434–437.

Singh, R. (1970) *Soyahar: Indian Recipes of Soybean*. U.P. Agriculture University Press, Pantnagar, India.

Singh, S. (1978) Potential of soyprotein in improving Indian diet. In: *Proceedings of the International Protein Food Conference*, Singapore, 25–27 January 1978, pp. 70–73.

Skurray, G., Cunich, J. and Carter, O. (1980) The effect of different varieties of soybean and calcium ion concentration on the quality of tofu. *Food Chemistry* 6, 89–95.

Smith, A.K. and Circle, S.J. (1972) Protein products as food ingredients. In: Smith, A.K. and Circle, S.J. (eds) *Soybeans: Chemistry and Technology*, Vol I. The AVI Publishing Company, Inc., Westport, CT, USA, pp. 339–388.

Smith, A.K., Watanabe, T. and Nash, A.M. (1960) Tofu from Japanese and United States soybean. *Food Technology* 14, 332–336.

Soybean Producers Association (2007) How soybeans are used. Available at: http://www.ncsoy.org/aboutsoybeans/usesof soybeans/tabid/1012/Default.aspx (accessed 7 January 2009).

Tambe, P., Srivastava, A. and Bardhan, D. (1971) Chemical analysis and palatability of different soybean recipes based on cereal pulse combination. In: *Highlights of 1971 Soybean Research.* Annual Report, Jawaharlal Nehru Krishi Vishwavidhyalaya Jabalpur, India, pp. 147–149.

The George Mateljan Foundation (2010) Tempeh. Available at: http://www.whfoods.com/genpage.php?tname=foodspice&dbid=126 (accessed 8 March 2010).

Tuitemwong, P., Erickson, L.E., Fung, D.Y.C., Setser, C. and Perng, S.K. (1993) Sensory analysis of soy yogurt and frozen soy yogurt produced from rapid hydration hydrothermal cooked soy milk. *Journal of Food Quality* 16, 223–229.

Tyagi, A.K. (1984) Development of a process for manufacturing soy ice-cream. MSc. thesis, submitted to GB Pant University of Agriculture and Technology, Pantnagar, India.

United Soybean Board (2008) Animal Agriculture and the Soybean Checkoff. Available at: http://www.animalag.org/ (accessed 10 March 2010).

Vijayakumari, K., Siddhuraju, P. and Janardahanan, K. (1995) Effect of various water or hydrothermal treatments on certain anti-nutritional compounds in the seed of the tribal pulse. *Plant Foods for Human Nutrition* 48, 17–29.

Wang, C.C.R. and Chang, S.K.C. (1995) Physicochemical properties and tofu quality of soybean cultivar Proto. *Journal of Agricultural and Food Chemistry* 43, 3029–3034.

Wong, H.L., Swai, E.W. and Kwolek, W.F. (1983) Effect of soybean varieties on the yield and quality of tofu. *Cereal Chemistry* 60, 245–248.

Wu, W., William, W.P. and Kunkel, M.E. (1994) Thermal effects on *in vitro* protein of red kidney bean (*Phaseolus vulgaris* L.). *Journal of Food Science* 5, 1187–1191.

Yadav, D.N., Sharma, G.K. and Bawa, A.S. (2007) Optimization of soy-fortified instant sooji halwa mix using response surface methodology. *Journal of Food Science and Technology* 44, 297–300.

Yoshimoto, T. and Sato, K. (2001) Properties of soymilk and yogurt like food: Properties of soymilk and yogurt like food prepared by using defatted soybean flour. *Journal of the Japanese Society for Food Science and Technology* 48, 906–912.

19 Vegetable Soybean

S. Shanmugasundaram[1] and Miao-Rong Yan[2]
[1]New Jersey, USA; [2]AVRDC – The World Vegetable Center, Shanhua, Tainan, Taiwan

19.1 Introduction

Vegetable soybean (*Glycine max* (L.) Merrill) is called *edamame* in Japan, *mao dou* in China, *poot kong* in Korea and *tua rae* in Thailand (Lumpkin and Konovsky, 1991). Around the world, soybean is a major oil crop, used mostly for human consumption; the protein meal is used as animal feed. Processed soy products such as soy milk, tofu, soy protein nuggets and soy-enriched imitation meat patties are increasingly popular for their nutritional value. Although whole shelled green soybeans have been used as a vegetable in China, Japan, Korea, Taiwan and Thailand, in other Asian countries and in the West the history of trying to promote soybean as a vegetable is as old as the introduction of soybean itself (Lumpkin and Konovsky, 1991). For example, when soybean was introduced into the USA it was first used as a forage crop, but during World War I the US Department of Agriculture researched and selected large-seeded soybeans with a sweet taste as a source of protein-rich food. Canned green soybeans were marketed as the major protein source during World War II as well. However, as economic conditions improved after the war, soybeans were replaced with meat and meat products (Bernard, 2001). With recent increases in heart disease, obesity and stroke, health-conscious consumers are again turning to green vegetables, including vegetable soybeans (Shanmugasundaram, 2005).

Soybean originated in northeastern China. The first written record of soybean is dated 2838 BC, and the Chinese have been cultivating soybeans for thousands of years. Among the soy foods, stems with green soybean pods and soybean sprouts were mentioned in *Dong-Jin-Mong-Hua-Lu* in AD 1147 (Gai and Guo, 2001). *Edamame* – green pods attached to the stem – first appeared in Japan in 1275. These green vegetable soybeans were cooked and served in the pod as a snack. In AD 1406 during China's Ming Dynasty, green vegetable soybeans were mentioned (Shurtleff and Lumpkin,

2001). However, *mao dou* – green hairy pods – were first mentioned in AD 1620 in the *Account of the Vegetable Gardens in Runan* by Zhou Wenhua; the vegetable was also called *qing dou*, green bean.

Soybeans can be boiled or steamed in the pods and shelled, removed from the pods to be boiled, baked or steamed, or toasted over a fire (Gai and Guo, 2001; Shurtleff and Lumpkin, 2001). In the USA, green vegetable soybeans were first mentioned in 1855 and in 1915 William J. Morse introduced a number of large-seeded vegetable soybeans from Japan and Korea and found them to be comparable to Lima beans. For a more detailed account of the history of vegetable soybean see Gai and Guo (2001) and Shurtleff and Lumpkin (2001). For a bibliography and source book on vegetable soybean, see Shurtleff and Aoyagi (1994).

Vegetable soybeans are known by a variety of names: *edamame*, *mao dou*, fresh green soybeans, green soybeans, edible green soybeans, green vegetable soybeans, vegetable-type soybeans, garden soybeans, garden-type soybeans, garden soy and others (Shurtleff, 2001).

Definition

Shanmugasundaram *et al.* (1991) defined vegetable soybeans as those that are harvested after the R6 but before the R7 growth stage. Vegetable soybeans have a large seed size (>30 g 100^{-1} dry seeds). For the Japanese market the following qualifications are preferred: in addition to the seed size, a 500 g frozen pod packet should contain ≤175 pods; the pod and bean colour should be dark green; the pubescence should be grey; the number of beans per pod should be two or more and should be without blemishes; the pod length and width should be ≥5.0 cm and ≥1.4 cm, respectively; the hilum should be grey or light-coloured; and soybean should have good flavour, aroma, texture and a slightly sweet taste (≥10% sucrose). The mature seed coat colour can range from yellow or green to brown or black. Vegetable soybeans are marketed as: (i) fresh pods attached to the stem; (ii) fresh pods detached from the stem; (iii) fresh shelled green beans; (iv) frozen pods; and (iv) frozen green beans (Shanmugasundaram *et al.*, 1991). A group of brown-seeded varieties in Japan, called Dadacha-mame, have an aroma similar to fragrant rice (Fushimi and Masuda, 2001).

Distinction from grain soybean

Vegetable soybeans differ from grain soybeans mostly in seed size. Vegetable soybeans are large-seeded (≥30 g 100^{-1} dry seeds). However, in some countries such as Nepal, grain soybeans are harvested at the green pod stage and marketed as vegetable soybeans; grain soybean varieties have also been used as vegetable soybean in China, Taiwan and Thailand. Recently, the introduction of new, high-quality vegetable soybeans has changed consumer attitudes in all of these countries. Vegetable soybean is slightly sweeter compared to the grain type, which is oily and slightly bitter. Grey pubescence is desirable due to the pod appearance after cooking for vegetable soybean, but for grain soybean there is no pubescence preference. Farmers and local growers have recognized the good taste and flavour of some of the large-seeded vegetable soybeans and selected them as different from grain-type soybeans, particularly in Japan. Private seed companies have used them to develop productive and tasteful vegetable soybean varieties. The dry seed coat colour of these varies from yellow to green, brown and black (Kitamura, 2001). Among 46 *edamame* cultivars, Masuda and Harada (2001) found that 'Chakari' and 'Murasaki' (Dadacha-mame) had sucrose concentrations of 16 g 100 g^{-1} dry weight, which is more than twice that of the grain-type cultivar 'Enrei'; the same authors also found that when the green beans of starch-rich *edamame* cultivars 'Tanbakuro' and 'Koitozairai' were boiled, there was enzymatic generation of maltose from starch, a quality highly desirable in vegetable soybeans. Akazawa *et al.* (1997) found that vegetable soybeans contain a higher level of water-soluble nitrogen than grain soybean. Phytic acid levels are higher in vegetable soybean than grain soybean, which makes for tender beans that cook faster.

Nutritional value

During the second century BC, Wu mentions that vegetable soybeans enhance the yang principle and have medicinal value (Shurtleff and Aoyagi, 1994). In the 1950s in the USA, vegetable soybeans were considered distinctly superior to grain soybeans for human consumption (Weber, 1956). Compared to sweet green peas (*Pisum sativum* L.) vegetable soybeans are rich in protein, fat (cholesterol-free), phosphorus, calcium, iron, thiamine, riboflavin, vitamins A, B1, E and C, folic acid, isoflavones and dietary fibre (Table 19.1). Vegetable soybean has a lower percentage of flatulence-producing starches compared to grain soybean. Like the grain soybean, vegetable soybean also has anti-nutritional factors. Trypsin inhibitor (TI) activity is low in vegetable soybean compared to grain soybean. One third of the TI activity remains in vegetable soybean after boiling for 5 min. Vegetable soybean is highly nutritious, yet the nutritional value is not a major factor determining its market value. Pod and green bean appearance, taste, flavour, texture and nutritional value, in that order, are the five most important quality requirements for vegetable soybean (Masuda, 1991).

Soy isoflavones, genistein, daidzein, and to a limited extent glycitein and their β-glycosides, have been reported to have an anti-carcinogenic effect on breast cancer in premenopausal women (Messina, 2004) and androgen-sensitive and -insensitive prostate cancer in men (Kucuk, 2004; Zhou, 2004).

At 25 and 40 days after flowering (DAF), the α, β, γ, and δ tocopherol content is 32, trace, 1038, 148 $\mu g\ g^{-1}$ lipid and 44, 2, 1124, and 306 $\mu g\ g^{-1}$ lipid, respectively. At the mature grain stage, 75 DAF, the four tocopherol contents are 109, 62, 1109, and 402 $\mu g\ g^{-1}$ lipid, respectively. At the vegetable

Table 19.1. Nutritional composition (values 100 g^{-1}) of vegetable soybean, grain soybean and green pea.

	Vegetable soybean		Grain soybean		Green pea	
Nutrient composition	Raw	Cooked	Raw	Cooked	Raw	Cooked
Energy (kcal)	147	141	446	173	81	84
Moisture (g)	67.5	68.8	8.54	62.55	78.86	77.87
Protein (g)	12.95	12.35	36.49	16.64	5.42	5.36
Fat (g)	6.8	6.4	19.94	8.97	0.4	0.22
Total carbohydrate (g)	11.05	11.05	30.16	9.93	14.45	15.63
Crude fibre (g)	4.2	4.2	9.3	6	5.1	5.5
Ash (g)	1.7	1.6	4.87	1.91	0.87	0.92
Phosphorus (mg)	194	158	704	245	108	117
Calcium (mg)	197	145	277	102	25	27
Iron (mg)	3.55	2.5	15.7	5.14	1.47	1.54
Vitamin A (μg RAE)	9	8	1	0	38	40
Vitamin B1 (mg)	0.435	0.26	0.874	0.155	0.266	0.259
Vitamin B2 (mg)	0.175	0.155	0.87	0.285	0.132	0.149
Vitamin C (mg)	29	17	6	1.7	40	14.2
Vitamin E (mg)	(1476)[a]	–	0.85	0.35	0.13	0.14
Folate (μg)	165	111	375	54	65	63
Isoflavones (mg)	20.42	13.79	128.35	54.66	–	–

RAE, retinol activity equivalent.
Values of isoflavones obtained from US Department of Agriculture (USDA)–Iowa State University Database on the Isoflavone Content of Foods (1999). Other values obtained from USDA National Nutrient Database for Standard Reference, Release 20 (2007). –: data not available in the USDA database.
[a]Value in parenthesis is total tocopherol content (μg g^{-1} lipid) at 40 days after flowering (Masuda, 1991).

soybean stage, the total tocopherol content is around 12 μg per seed (40 DAF). At 75 DAF, the total tocopherol content is 83 μg per seed (Masuda, 1991). Thus, tocopherol, protein and TI activity increases with maturity. However, sucrose increases up to the vegetable soybean stage and then starts to decrease. Therefore, vegetable soybean should be harvested at the optimum stage to get the sweet taste. Frozen soybean has more sucrose (e.g. 1.7% versus 1.1%) and amino acids (e.g. alanine at 30 mg 100 g^{-1} fresh weight versus 16 mg 100 g^{-1} fresh weight) than fresh. Delaying the harvest, after the appropriate stage, results in low sucrose content.

19.2 Production

Historically, vegetable soybeans were produced in the Yangtze River valley, Jiangsu, Zhejiang, Anhui, Jiangxi, Hunan and Hubei in China. Recently, vegetable soybeans have been produced in Shandong, Henan, Tianjin, Beijing and north, south and southeast China (Gai and Guo, 2001). At present nearly 90% of China's vegetable soybeans are produced in Zhejiang, Jiangsu, Fujian, Guangdong, Hunan and Shanghai (Wu, 2004).

Table 19.2. Vegetable soybean area and production in selected countries.

Country	Area (ha)	Production (t)	Year	Reference
China	284,000	1,704,000	2003	Wu (2004)
Japan	12,200	71,000	2006	TFVMA (2008, personal communication)
Taiwan	8,200	57,300	2006	TFVMA (2008, personal communication)
Thailand	3,200	20,000	2007	S. Daruphan (Chiang Mai, Thailand, 2008, personal communication)
Indonesia	600	3,000	2006	TFVMA (2008, personal communication)
Vietnam	140	700	2006	TFVMA (2008, personal communication)

TFVMA, Taiwan Frozen Vegetable Manufacturers' Association, Kaohsiung, Taiwan.

In Japan, the government regulates the supply and price of 14 major vegetables; vegetable soybean is not one of them. It is one of the 29 other vegetables, the lowest price of which is controlled by the government. The area planted to vegetable soybean increased from about 11,000 ha in 1971 to about 14,700 ha in 1981; thereafter, the area has either stagnated or decreased (Iwamida and Ohmi, 1991; Nakano, 1991). In 2005 Japan produced about 77,000 t vegetable soybean from around 13,000 ha, while in 2006 the area was only 12,200 ha with production of 71,000 t (Table 19.2). The per capita supply of vegetable soybean was about 0.29 kg.

Traditionally vegetable soybeans were shelled and marketed as fresh green beans in Taiwan. A multipurpose cultivar, 'Jikkoku', from Japan (called 'Shih Shih' in Taiwan) was introduced for both grain and vegetable production (Cheng, 1991). Prior to 1975, the total area and production of vegetable soybean in Taiwan was negligible, but both increased from 6500 ha and 40,000 t in 1980 to nearly 10,000 ha and around 65,000 t in 1990, respectively (Cheng, 1991; Lin and Cheng, 2001). However, due to the increasing value of land and high cost of labour, vegetable soybean production slowly decreased from 1999 in Taiwan, and its share was taken by China, Thailand and Indonesia, where production costs are cheaper. In 2003, 2004 and 2005 the area and production of vegetable soybean in Taiwan was 9600, 10,300, and 8800 ha and 77,000, 80,000 and 61,000 t, respectively. The area decreased even further in 2006 (Table 19.2).

Since 1900, the pods from grain soybean have been used as vegetable soybean in Thailand. Green pods attached to the stem are boiled and sold in rural markets. In 2007, Thailand produced approximately 20,000 t, and Indonesia produced approximately 3000 t from about 500–600 ha (Table 19.2). In South Korea, the production of vegetable soybean has gained in popularity since the mid-1990s (Park *et al.*, 2001).

In the USA, vegetable soybeans are grown in California, South Carolina, Ohio, Illinois, Kentucky, Virginia, Indiana, Iowa, Oregon, Washington and Hawaii. Exact statistics for the area and production are unavailable.

The Asian Vegetable and Research and Development Center (AVRDC) – The World Vegetable Center has promoted the vegetable soybean in Africa. As a result, vegetable soybeans are now produced and marketed in

Zimbabwe, Mauritius, Uganda, Tanzania, Zambia, Sudan and Mozambique (Chadha and Oluoch, 2004). A total of 15 countries, including Bangladesh (GC 83005-9), India (GC 98009-1-1-2), Pakistan (AGS 190) and Sri Lanka (AGS 190), have released more than 38 vegetable soybean cultivars from AVRDC breeding materials (Shanmugasundaram and Yan, 2004).

19.3 Cultural Practices

Location, season and cropping system

Vegetable soybean is commonly cultivated in a rice-based cropping system or on bunds of rice fields. The cropping season varies with location, cultivar and environmental conditions. In almost all Asian countries, vegetable soybean is planted in the spring, summer and autumn in open fields; planting dates differ with season and location, depending upon temperature and day length. Forcing and semi-forcing is performed in heated glasshouses, greenhouses and vinyl tunnels, mainly in China, Japan and Korea. The diversity of cropping methods ensures a continuous supply of vegetable soybean throughout the year. In northern and central China and Japan, vegetable soybean is planted in the spring and summer; in the south of China and Kyushu, vegetable soybean is also cultivated in the autumn (Kokubun, 1991; Wu, 2004). For example, in Hokkaido, vegetable soybean is seeded in May and harvested in August, while in Kyushu, seeding is from March to May and harvesting is from June to August. Forcing and semi-forcing sowings are carried out in December to January in heated glasshouses or vinyl tunnels, and harvest runs from February to June. Forcing is usually expensive, but it makes fresh vegetable soybean available well before the normal season crop. Early-maturing cultivars are planted to bring the crop to the market early and fetch a higher price. Late-maturing cultivars are used in the normal planting season in open fields in cold regions to extend the harvesting until September to October (Kokubun, 1991).

In China, vegetable soybean is cultivated in Jiangsu, Shanghai, Zhejiang, Anhui, Jiangxi, Hunan and Hubei during spring and summer. The spring crop yields 4.5–6.0 t ha^{-1} while the summer crop yields 6.0–7.5 t ha^{-1}. A small area is planted during autumn. In southeast China, vegetable soybean is planted in all three seasons in paddy fields; productivity ranges from 4.5 to 9.0 t ha^{-1}. In Taiwan, vegetable soybean is planted in the spring, summer and autumn following the paddy crop as well as vegetables, including potato, maize or peanut (Tsay *et al.*, 1991). In Thailand, the crop is planted in the rainy season in the central plains and during the dry, cool season in the north around Chiang Mai. Vegetable soybean is cultivated as a monocrop and can be grown continuously where there are no endemic virus or root diseases. Overall, there are limited areas where it is cultivated on paddy field bunds or as an intercrop. It is also rotated with other vegetable crops, as in Mauritius.

Temperature, soil type, land preparation and sowing

The optimum temperature for growing vegetable soybean is the same as that for grain soybean. The soil temperature at sowing is very important for good germination and good seedling development. The minimum soil temperature for germination is between 13 and 18°C. A temperature of 21–32°C during the growing period is best for good crop development. In glasshouses or vinyl tunnels the recommended temperature is <21–32°C during the day and >7°C at night (Kokubun, 1991). In a study conducted by Xuan and Chang (2003) the yield of vegetable soybean was highest when there was >95 h of sunshine and 90 mm rainfall at the pod-filling stage and 26°C diurnal temperature at the pod-maturing stage.

Although soybean can be grown in a variety of soil conditions, a highly fertile and healthy soil with good drainage conditions is preferred. The optimum soil pH for vegetable soybean is 6.0. However, vegetable soybean can be grown in soils with a pH of 5.8–7. The land is usually ploughed and harrowed to break the clods and to bring the soil to a good tilth. Good-quality seed with at least 85% germination should be used. Under Asian conditions the seeds are usually treated with Arasan (bis (dimethylthiocarbamyl) disulphide) or Ceresan (ethylmercurichloride) 75% WP at the rate of 3 g a.i. kg^{-1} of seed. Treating seeds with *Bradyrhizobium* inoculum, especially in areas where soybean has not been grown in the past, also helps to increase production (Chen *et al.*, 1991). In forcing culture, the seeds are sown in rows to obtain 20–30 plants per m^2, while for normal-season cultivation in the field 5–10 plants per m^2 are used under temperate conditions in Japan (Kokubun, 1991). The spacing between rows can be 66–91 cm and within the row the seeds are sown at 7.5 cm apart. The normal plant population density in Japan and the USA is around 170,000–370,000 plants ha^{-1} (Kokubun, 1991; NSRL, 2008). Increasing plant density increases plant height, but decreases stem thickness, number of nodes, number of branches, pods per plant and dry matter accumulation. Ning *et al.* (2006) found the weight of 100 green beans and the quality of vegetable soybean to be unaffected by differences in plant density.

In forcing, semi-forcing and early-maturity cultivation in Japan and China, raising seedlings and transplanting them is a common practice to ensure rapid and uniform growth and development and high yield. Transplanting is occasionally done for normal-season and late-season crops in open fields to avoid missing plants due to poor germination and bird attacks on emerging seedlings (Kokubun, 1991; Zhang, 2004). The seedlings are raised in soil beds inside the glasshouse or vinyl tunnel or in nursery boxes. Paper or plastic pots should be used to avoid damaging the root system while transplanting the seedlings. In Japan, transplanting takes place 15–20 days after emergence when the primary leaf has expanded. In forcing culture the seedlings are transplanted twice, first to pots and then from pots to the soil inside the glasshouse or vinyl tunnel (Kokubun, 1991). In China, 10- to 15-day-old seedlings are used for transplanting (Zhang, 2004).

Until the mid-1990s, vegetable soybean was planted in Taiwan in the paddy field during the autumn season (September to October sowing and

December to January harvesting). After the harvest of paddy, the vegetable soybean seeds are dibbled close to the rice stubble without any tillage. The spacing between and within the rows is similar to the spacing for rice (25 × 25 cm). After sowing the seeds, rice straw is used as a mulch to cover the seed and prevent weeds. Super phosphate and potash fertilizers are broadcast before mulching. In some low-lying areas where there is excess moisture after the harvest of the paddy following mulching, the rice straw is burnt. The seedlings germinate in 5–7 days. The operation is very labour intensive and is acceptable where cheap labour is available. In fact, this method spread from Taiwan to Thailand, Indonesia and Vietnam. However, due to economic conditions and a labour shortage for agriculture, Taiwan has mechanized vegetable soybean production. The development of a zone-tillage pneumatic precision seeding machine marked the beginning of a highly mechanized system of vegetable soybean production. Different kinds of planting machines have been developed that make ridges and furrows and plant and cover the seeds with soil in one operation. The planter is attached to a small tractor. Spacing between rows is 40 cm and within the row seeds are sown to get about 400,000 plants ha^{-1}. For a cultivar with 30 g 100^{-1} seeds, the seed rate is around 120–150 kg ha^{-1}. The seed rate should be adjusted depending upon the seed size and expected germination rate. For detailed information with illustrations on the mechanization, see Shanmugasundaram and Yan (2001).

Soon after emergence, field seedlings should be covered with nets to protect them from birds (Kamiyama, 1991). In Taiwan, shiny vinyl ribbons are tied around the field to frighten birds. Rabbits, deer and other animals can also damage the young seedlings. A fence around the field is recommended to protect the seedlings.

Fertilization, weeding and irrigation

Vegetable soybean, being a legume, fixes atmospheric nitrogen in the soil through the *Bradyrhizobium* bacteria, and normally it does not require nitrogen fertilizer application. However, depending upon a soil test and the amount of fertilizer applied to the previous crop, a starter nitrogen fertilizer of 25–30 kg N ha^{-1} can be applied at the time of sowing. Based on soil tests in Taiwan, the recommended fertilizer application is about 10 t ha^{-1} of compost, 60 kg N ha^{-1}, 30 kg P ha^{-1} and 50 kg K ha^{-1}. Half of the nitrogen fertilizer is applied as a basal dressing and the other half as a top dressing at the flowering and pod-formation stage. To ensure optimum size and good quality of seed, another dose of 20 kg N ha^{-1} can be given at the seed-filling stage (Chen *et al.*, 1991). In Japan, compost is applied at 10–150 t ha^{-1}. The nitrogen, phosphorus and potassium fertilizer rate is 30–40, 150, and 80–100 kg ha^{-1}, respectively (Kamiyama, 1991). Lime is also applied at the rate of 1000 kg ha^{-1} (Kokubun, 1991). In trials conducted in Taiwan, the results showed that a basal dressing of 50 kg N ha^{-1}, 30 kg P ha^{-1} and 20 kg K ha^{-1} at sowing, a top dressing of 70 kg N ha^{-1} and 10 kg P ha^{-1} at 15 days after sowing and a top dressing of 50 kg N ha^{-1} and 50 kg P ha^{-1} at the pod-initiation stage gave the highest yield and good-quality pods. *Bradyrhizobium* inoculation along with 20 kg N ha^{-1} increased both the number and weight of nodules (Hung *et al.*, 1991). A fertilizer rate of 25 kg N ha^{-1} plus *Bradyrhizobium* inoculation was excellent for alluvial soil in Vietnam's Mekong Delta (Diep *et al.*, 2002). If there is no rain following fertilizer application, irrigation is necessary for proper absorption of the nutrients. Potassium sulphate is better than potassium chloride. If the soil is deficient in micronutrients such as boron, zinc or molybdenum, these should be provided as chelates (NSRL, 2008).

For weed control, a pre-emergence herbicide such as Lasso (alachlor) or Pursuit (imazethapyr) is sprayed at 1.5 kg a.i. ha^{-1}. Intercultivation is performed once or twice during the crop season to control weeds. Hand weeding is also performed to eliminate weeds when necessary. Weed control up to the R1 growth stage is extremely important, because the crop does not cover the ground at that time; after the R1 growth stage, the crop canopy covers the ground well and suppresses weeds. No Roundup Ready soybean gene has been found in the vegetable soybean cultivars grown in Taiwan (Cherng and Tay, 2002).

Optimum soil moisture (50% of the soil) is essential for good germination. Under optimum soil moisture and temperature the seed germinates and the seedling emerges in about 5–10 days. After the rice harvest, if the field is dry then it is irrigated, and when the soil moisture comes to the right stage, ploughing, ridging, furrowing and sowing are carried out. After seedling emergence, irrigation is given at 15- to 20-day intervals until the pods are well developed. The frequency and amount of irrigation depends on the type of soil, rainfall, drainage, season and crop duration to maintain proper soil moisture. On heavy clay soils with good water-holding capacity, usually three to four irrigations are sufficient; loamy and sandy loam soils require more frequent irrigations. Irrigation during initial flowering at the R1–R2 growth stages (Fehr *et al.*, 1971) and early podding period accelerates pod filling and seed filling and increases the yield (Zhang, 2004). Insufficient moisture at the R1–R2 and R3–R5 growth stages induces flower and pod drop. Therefore, optimum moisture should be maintained during these critical stages to achieve a high yield and good quality (Kokubun, 1991).

Insect pest and disease control

The insect pests and diseases that attack vegetable soybean are the same as those that attack grain soybean. For detailed information, readers are referred to chapters on insect pests (Chapter 14) and diseases (Chapter 13) in this book.

Different groups of insect pests attack the vegetable soybean at different growth stages. Bean flies or bean stem miners, *Melanagromyza sojae*, *M. phaseoli*, *Ophiomyia centrosematis* and *Dolichostigma* species attack the emerging seedlings by laying their eggs on the unifoliolate leaves. The young larvae tunnel through the veins and feed inside the stem, causing seedlings to wilt and die prematurely. The recommended chemicals at label rates are (i) Hostathion 40% emulsifiable concentrate (EC) (triazophos) and (ii) Azodrin (monocrotophos) 60% water-soluble emulsion (WSE) (Khadkao, 1992). Prior to the R1 growth stage, bean leaf rollers (*Lamprosema indicata*, *L. diamenalis*), whiteflies (*Bemisia tabaci*) and common cut worm (*Spodoptera litura*) attack the plant. Azodrin can effectively control the leaf rollers. Dimethoate (phosphorodithioic acid or O,O dimethyl S-(2-(methylamino)-2 oxoethyl) dithiophosphate) can control most of the leaf-eating caterpillars and sucking insects. Neem seed kernel extract effectively controls *B. tabaci* (Abdullah *et al.*, 2001b). During the R1–R6 growth stages, stink bugs (*Nezara viridula*, *Piezodorus hybneri* and *Riptortus* species) can attack and suck the nutrients from the plant. The infestation can result in flower and pod drop and ill-filled or empty pods. The field should be scouted for stink bugs and if three to four stink bugs are seen in a metre of the row then it is time to spray (Hostathion 40% EC or Azodrin 60% WSE). The puncture mark of stink bugs can result in undesirable dark spots and discoloration of the pods. The larvae of pod borer (*Heliothis armigera*) feed on the leaves, flower buds and flowers, pierce into the pods, eat the seeds and pupate inside the pod, resulting in unmarketable pods. Azodrin or Hostathion can effectively control this pest. It is preferable to use the insecticide at the R1–R2 growth stage.

In China, pest outbreaks are more common in the autumn than in spring. Growing a cyst-nematode-resistant variety for 3 months significantly reduces the number of nematode eggs; as a result, the following vegetable soybean crop produces a higher yield compared to that from a cyst-nematode-infested field where a susceptible cultivar was grown prior to vegetable soybean (Uragami *et al.*, 2005).

Because vegetable soybean is harvested for fresh use, farmers should avoid using systemic insecticides and fungicides. Application of insecticides should be discontinued 10–15 days prior to harvesting the pods (Abdullah *et al.*, 2001a).

Various bacterial, fungal and viral diseases and nematodes causing damage to soybean are described in the latest compendium of soybean diseases (Hartman *et al.*, 1999). A few selected diseases that reduce yields and affect the quality of the pods and seeds are mentioned here. Bacterial pustule (*Xanthomonas axonopodis* pv. *glycines*) can defoliate the leaves prematurely. The characteristic symptoms include a yellow halo around the brown spot; the lesions may coalesce to become large and irregular. In the early morning the spots have a clear bacterial ooze. Choosing vegetable soybean cultivars resistant to the disease is the best control option.

Phytophthora sojae causes root rot and it is a major disease in the USA, northern China and Japan. It causes pre- and post-emergence damping off and stem and root rot. Race-specific resistant cultivars have been developed and should be used. Cultivars with partial resistance and tolerance have been identified in grain soybean (Ferro *et al.*, 2004) and it would be useful to incorporate them in vegetable soybean.

Soybean rust (*Phakopsora pachyrhizi* and *P. meibomiae*) is one of the most devastating diseases of soybean. Initial symptoms include a water-soaked chlorotic polygonal lesion, which eventually develops into a tan or reddish-brown lesion. Once the urediniospores develop at the site of infection, the dispersal of spore dust can clearly be seen in the morning on windless days. The leaves abruptly turn yellow and premature leaf abscission occurs. Petioles and the young stem may also become infected. When the inoculum and environment are favourable, infection can occur at the cotyledon and primary leaf stage. The lesions are abundant on older leaves, especially on the abaxial (underside) surface of the leaf. The lesions may contain one to several erumpent, globose, ostiolate uredia. The number of uredia per lesion increases with lesion age. Urediniospores are exuded through the central pore in the uredium. The spores are in clumps. From infection to urediniospore production takes 11–12 days and reinfection occurs. Soybean rust symptoms appear similar to bacterial pustules. Through a hand lens or microscope, soybean rust will reveal the uredium and urediniospores. However, in a bacterial pustule, a fissure and necrotic tissue alone is visible (Shanmugasundaram, 1998). The use of tolerant cultivars is recommended for the management of this disease. In their absence, Dithane M-45 or Mancozeb (ethylene bisdithiocarbamate) at the rate of 2.0 kg ha^{-1} once every 3 weeks can be used as a prophylactic. This keeps the disease under control and reduces yield loss. In the USA and Latin America there are a number of approved chemicals for use against Asian soybean rust.

Downy mildew caused by *Peronospora manshurica* is a cool-weather disease. It begins with yellow-green spots that gradually become grey or greyish purple. Tufts of grey fungal growth may be visible on the abaxial surface of the leaves. In addition to yield reduction, the disease causes poor-quality pods and seeds. Resistant cultivars are available. In the case of susceptible cultivars, the crop should be sprayed at the onset of the first symptom with 30 g Ridomil MZ ((R,S)-2-(2,6-dimethylphenyl)-methoxyacetylamino)-propionic acid methyl ester) 20 l^{-1} of water three times at 10-day intervals (Nantapan, 1992).

Anthracnose is caused by *Colletotrichum truncatum* and *C. gloeosporioides*. Symptoms appear as black lesion on cotyledons, leaves, stems and pods. For early detection of anthracnose infection on vegetable soybean plants, Chen *et al.* (2006) designed two species-specific primer pairs – Colg 1/Colg 2 (expected size of 443 bp) and Colg 1/CT 2 (375 bp) – that allow differentiation of *C. gloeosporioides* and *C. truncatum* in multiplex polymerase chain reaction. A Benomyl (Benlate; benzimidazole) spray at 30 g 20 l^{-1} water at 10-day intervals from the R3 to R5 growth stages, as needed, gives effective control.

Purple seed stain is caused by *Cercospora sojina*. Since the disease specifically discolours the seed, the quality of the seed is drastically reduced. Resistant cultivars are available in grain soybean and the trait should be transferred to vegetable soybean. In the absence of resistant cultivars, a fungicide should be sprayed at the pod-filling stage. Spraying should be stopped 15 days prior to harvest to avoid undesirable residues (Kokubun, 1991).

Soybean mosaic virus can cause yellowish green mottled or crinkled leaves and stunting of leaves and plants. Diseased pods may be stunted and curved. Seeds from diseased pods may be discoloured (hilum bleeding), which reduces the marketability of the produce. The yield is reduced. The use of resistant cultivars is recommended. Aphids or other vectors should be controlled with Hostathion spray to prevent the transmission of the disease.

Harvesting

The time of harvesting vegetable soybean is critical to marketing quality. The pod should be examined against sunlight: if the seed cavity is full while the pod is still fresh green then it is time to harvest. The harvest window for vegetable soybean is very short. Harvesting earlier than optimum results in soft beans with excessive moisture; delayed harvesting results in increased hardness of the beans and suboptimal pod colour (Kokubun, 1991).

Technically, the vegetable soybean is ready for harvest when the moisture content of the beans is 65–70%. At this time the pods are still fresh green and the leaves are just beginning to turn yellow. For the Japanese market, vegetable soybean appearance is the highest priority, followed by taste (Chiba, 1991). The taste of vegetable soybean is highly correlated to the sucrose and glutamic acid content of the seed (Masuda, 1991). Pod colour is evaluated using a colorimeter and expressed according to a Lab colour system, which was recommended by a Committee of International Illumination (Chiba, 1991). In Japan, vegetable soybean is harvested approximately 33–38 DAF, depending upon the pod colour. The harvesting time varies with different cultivars and location. For example, for cultivar 'Sapporomidori' it is 36–39 DAF, for 'Fukura' it is 42 to 45 DAF and for 'Kinsyu' it is 48 DAF (Chiba, 1991). In Taiwan, the vegetable soybean harvest time depends on the season. The time taken to harvest is longer in the spring and summer than in the autumn. In spring and summer, depending upon the cultivar, the time to harvest is about 35–45 DAF; in autumn it is about 30–35 DAF. In Thailand and Indonesia, the crop is ready for harvest in 28–30 DAF or 65–76 days after sowing (Sitani, 1992). In Thailand, Indonesia and Vietnam, harvesting is by uprooting the whole plant or cutting the plants at ground level. In other cases, labourers are employed to collect the mature pods directly from the plant.

Traditionally, vegetable soybean is harvested manually. Because the quality of vegetable soybean is affected from the time of harvest to processing, harvesting is done at midnight. Whole plants are pulled from the ground and transported to de-podding centres. Before dawn the pods are stripped under shade and the pods are transported to the factory within 2h. The pods are kept moist during transport to prevent quality deterioration; normally the vegetable soybean production area is very close to the processing factory. Pods are stripped in the shade to maintain pod quality (Liu and Shanmugasundaram, 1982; Shanmugasundaram and Yan, 2001). This practice is still followed in Indonesia, Thailand and Vietnam. Production in

Taiwan, however, is mechanized. In 1994, a tractor-mounted 95 HP FMC 1647 bean harvester was introduced from France and modified to suit the harvesting of vegetable soybean under Taiwan rice stubble conditions. The harvester could harvest about 0.22–0.29 ha h^{-1}, but losses due to different factors were 18.4–23.8%, which was unacceptable. To reduce the yield loss an improved version – a tractor-driven 190 HP FMC 7100 harvester operating at 3400 rpm – was introduced. The actual loss was kept at around 5%. The total cost of production could be reduced by about 20%. The harvested and threshed pods are collected in the bin behind the harvester. When the bin is full it is immediately taken to the factory. The turnaround time from harvest to factory is kept to a minimum, retaining good quality (Shanmugasundaram and Yan, 2001). For very small farms, a bundle thresher with a hard rubber tooth rasp bar is used. The optimum angular velocity for the threshing cylinder is 400 rpm and the damage rate is <10%. The bundle thresher can thresh about 60 kg h^{-1}, which is six to eight times faster than manually stripping pods. In China, there are mini-manual and electric threshers. The manual thresher weighs 3 kg and is easy to transport. It can thresh 5–8 kg h^{-1}. The electric thresher can thresh about 5–30 kg h^{-1} (Wu, 2004). In general, manually harvested and picked vegetable soybean pods are superior in quality compared to machine-harvested pods. The exception is the FMC 7100 harvester, with which quality is comparable to that of manually harvested pods.

Seed production

As mentioned earlier, the vegetable soybean has a large seed size, ≥30 g 100^{-1} seeds. Quality seed is a prerequisite for producing a good quality crop. Factors such as growing season, location, management inputs, diseases and insect pests, postharvest handling and storage affect seed production. Locations with a high temperature, high relative humidity and frequent and prolonged rainfall are unfavourable for seed production. Cool, dry conditions and moderate temperatures help to produce good-quality seed. Seed of early- to medium-maturing cultivars (the majority of vegetable soybean cultivars) are produced mainly in Hokkaido, where weather conditions are favourable. Seed of medium-late- to late-maturing cultivars are produced primarily in Tohoku and Hokuriku in Japan. A number of private seed companies produce their own seed as well as seed of different land races and market the seed domestically and internationally (Iwamida and Ohmi, 1991; Kamiyama, 1991; Ohmi, 2001). In Taiwan, the autumn season is excellent for seed production (Yan and Shanmugasundaram, 2001). In Thailand, seed production sowing is in the dry season from December to January 15 in northern Thailand; in the rainy season, sowing is in August in the central plains. The quality of the dry-season crop is better than that of the rainy-season crop. Taiwan imported vegetable soybean seed from Japan until the mid-1980s. With the development of AGS 292 (AVRDC line), released as 'Kaohsiung No. 1' by the Kaohsiung District Agricultural Research and Extension Station (KDARES) in 1987, Taiwan began to produce its own seed and also exported 'Kaohsiung No. 1' and other cultivars to Thailand and Vietnam.

The cultural practices for growing a vegetable soybean seed crop are the same as those described earlier for the vegetable soybean pod crop. The major difference is in pest and disease control. Moisture content of the seed is an important factor influencing seed quality. In locations where there are significant differences in day/night temperatures, the seeds of many vegetable soybean cultivars shatter upon maturity. To avoid shattering, the plants are harvested at the first sign of shattering, dried on a rack or stacked in the field or greenhouse (Kamiyama, 1991) or spread on a vinyl sheet in the field or on a concrete drying floor for slow drying in partial shade. Seeds are frequently turned for uniform drying. The moisture content of the seed at the time of harvest may be 25–28%. At this moisture content, threshing is difficult. If the moisture content is too low, the seeds crack when threshed. Results have shown that the optimum moisture content for threshing with least damage to the seeds is 18–20% (in Japan it is 16–18%) and the cylinder speed of the threshing machine should be adjusted to 10–11 rpm s^{-1} (Chen *et al.*, 1991). After threshing, the seed should be dried at 30–35°C for 24–48 h. The seed moisture content is brought down to about 9–10%. If the seed readily splits and cracks when crushed then the moisture content is about 10%; if the seeds are crushed on the drying floor and the floor becomes sticky, the moisture content is >10%. Moisture content can also be checked in the laboratory. Good-quality seed should have 98% purity. Seed of other species should be <0.2% and inert matter should be <2%. The germination rate should be >85%. The dried seeds can be stored in 0.03-mm thick polyethylene bags, sealed airtight and stored in a cool, dry place (Chen *et al.*, 1991). Farmers also store their seeds in metal bins, adding ash to prevent moisture fluctuations, and seal the bins tight with a lid with a rubber gasket. Recently, farmers in Taiwan have begun storing their seeds in air-conditioned rooms to maintain quality. At 25°C and 65% relative humidity, seeds with 10–11% moisture content retain their initial germination rate and vigour for 1 year in storage. At the AVRDC – The World Vegetable Center, seeds with 8–10% moisture content stored in airtight containers or hermitically sealed high-density polyethylene bags may maintain high seed viability for 2 years at 30°C (Yan and Shanmugasundaram, 2001).

The largest seed size reported for a vegetable soybean cultivar with a black seed coat is for 'Tanbaguro' in Japan, with 80 g 100^{-1} seeds dry weight at 11% moisture content (Hikino, 2000). The price of grain soybean seed is around US\$0.63–1.56 kg^{-1}, with an average price of around \$1.25 kg^{-1}. However, the current price of vegetable soybean seed in Taiwan is around US\$2.5–6.0 kg^{-1} depending upon the season and cultivar. The price of vegetable soybean seed in Japan is around US\$32–52 l^{-1}, depending on the cultivar (Takii, 2010). 'Tanbaguro' seed costs from US\$25–50 kg^{-1} (Hatsanaka *et al.*, 2004). Seed yields of up to 4 t ha^{-1} have been obtained for vegetable soybean under experimental conditions (Chadha and Oluoch, 2001). Under farmers' field conditions, the average yield of vegetable soybean seed is around 2.0–2.5 t ha^{-1}.

19.4 Genetic Improvement

It is likely that the selection of soybean landraces for good vegetable qualities has occurred gradually over centuries. In the middle of Edo period (1603–1868), farmers' wives began marketing boiled vegetable soybean pods attached to the stem in the streets of Edo (early Tokyo) (Lumpkin and Konovsky, 1991). Fresh vegetable soybeans with pods attached to the stem are sold even today; they are very popular and preferred over frozen pods. The increasing demand for vegetable soybean catalysed research to develop improved varieties for the discriminating consumer. In Taiwan, a Japanese cultivar called 'Jikkoku' ('Shih Shih' in Taiwan) was released in 1957 and cultivated as a dual-purpose soybean until the early 1970s (Shanmugasundaram, 1979). In Taiwan, vegetable soybeans are usually shelled and marketed as fresh green beans (Chen *et al.*, 1991). The early research on chemical composition, especially the protein, starch and sugar content and isozyme variations, laid the foundation for breeders to develop good-quality vegetable soybean for consumers (Lumpkin and Konovsky, 1991).

As early as 1930, an agricultural experiment station in Kungchuling, China, developed cultivars for special purposes, including vegetable soybeans. Most of the cultivars in China are either landraces or varieties selected by public research institutions for vegetable purposes. The main objectives in selection were large seed size, good taste and eating quality as a vegetable. Some of the older cultivars were 'Wuyuewu', 'Wuyueba' (US maturity group [MG] I and II), 'Baishulou', 'Liuyueba' and 'Baimaoliuyuewang' (MG III and IV) and 'Jiangyoudou' and 'Deqingdou' (MG V and VI). 'Baishulou', 'Baimaoliuyuewang', 'Jiangyoudou' and 'Deqingdou' were reported to be superior in quality to newer cultivars (Konovsky *et al.*, 1994). Formal breeding of vegetable soybean in China began in 1990, directed towards the international market, especially Japan. Cultivars 'Xinliuqiong', 'Zoudou 30', 'Jiaoxuan 1' and 'Jiaoxuan 2' were developed, but their quality could not meet international standards (Wu, 2004). In Zhejiang and Jiangsu provinces, popular landraces were 'Shanghai Liuyuebai', 'Wuyueba' and 'Suzhouwuyuehuang'; popular summer vegetable soybean landraces were 'Lanzi Daqingdou', 'Niutabian' and 'Zhejiang Bayueba' (Wu, 2004). Recent cultivars that have been developed and released in China are listed in Table 19.3.

In 1915, PI 34702, an introduction from Shantung province in China, was found to be good for shelled green beans in California. In 1929, Morse coined the term 'vegetable soybean' and tried to introduce a large number of vegetable soybean cultivars from Japan and China (Shurtleff, 2001). From 1929 to 1931, Dorsett and Morse collected germplasm to develop 49 *edamame* cultivars (Hymowitz, 1984). Due to protein shortages during the 1930s and 1940s, research on vegetable soybean continued (Smith and Van Duyne, 1951). With the initiative of Rodale in organic agriculture in the 1970s, interest in vegetable soybean was revived in the USA (Hass *et al.*, 1982). Twenty-two breeders from 17 North American locations, including eight north-central states, four southeastern states, Washington and Canada, conduct research on specialty soybeans. Interest in vegetable soybeans has increased in the

Table 19.3. New vegetable soybean cultivars released from different institutions in China.

Cultivar name	Releasing institution	Reference
Chuxiu	Huaiyin Agricultural Science Institute, Zhejiang	Jianfeng and Zhengwen (2002)
Xiangchun 18	Crop Research Institute, Hunan	Jianfeng and Zhengwen (2002)
Xinliuqing and Andou 3	Agricultural Science Institute, Anhui	Dai *et al.* (2001)
Liaoxuan 1	Liaoning Academy of Agricultural Sciences, Liaoning	Wu (2004)
Jiaoxuan 1 and 2	Jiaotong University, Shanghai	Wu (2004)
Qingdali 1 and 2	Fujian Academy of Agricultural Sciences, Fujian	Xu *et al.* (1999)
Shennong 951	Shanghai Academy of Agricultural Sciences, Shanghai	Zhihao *et al.* (2000)
Tezao 1	University of Anhui, Anhui	Zhihao *et al.* (2000)
Huijia 2	Jiangsu Academy of Agricultural Sciences, Jiangsu	Xin *et al.* (1997)
Ningshu 60	Nanjing Vegetable Institute, Nanjing	Wu (2004)
Taiwan 75	Introduced from Taiwan	Wu (2004)

USA as consumers become more health conscious and commercial growers seek new market niches. Breeders seek to produce large-seeded cultivars. Because grain soybeans from the USA are only 10–12 g 100^{-1} seeds, a seed size of 20–25 g 100^{-1} seeds is considered large. Pod colour, seed colour and taste have not been given that much attention in the development of cultivars. Until 1949, a total of 49 vegetable soybean cultivars were introduced or developed (two were released prior to 1920, 30 in the 1930s and five in the 1940s); they are maintained in the USDA germplasm collection. Among them, 'Banes' and 'Jorgen' are still marketed for home gardeners. In the1940s, 38 farmers grew 17 cultivars on 4000 ha. 'Bansei' was the most popular, followed by 'Etum'. The other popular cultivars were 'Aoda', 'Easycook', 'Funk Delicious', 'Giant Green', 'Hokkaido', 'Jogun', 'Rokusun', 'Sac', 'Chusei', 'Higan', 'Imperial', 'Kanro', 'Mendota', 'Sanga' and 'Sousei' (Bernard, 2001). Bernard (2001) stated 'The main overall breeding objective for the American Edamame breeder is to combine the desirable yield and plant traits of our commercial grain types with the desirable seed traits of the East Asian Edamame types.' US breeders consider 20 g 100^{-1} seeds as a minimum, 25 g 100^{-1} seeds more desirable and 30 g 100^{-1} seeds a long-range goal. Several genes appear to govern large seed size; therefore, to recover large seed size, it is necessary to back-cross to the large-seeded parent. Desirable genes for vegetable soybeans include *t*, the gene for grey pubescence and *w*, the gene for white flower colour, which also eliminates the undesirable anthocyanin pigment from green pods. The stay-green seed coat and seed embryo genes *G1* and *G2* are present in some North American and Asian vegetable soybeans (Bernard, 2001). From 1950 to 1990, 20 vegetable soybean cultivars were released in the USA (see Bernard, 2001 for the list).

Martin Weiss and Robert Weber started breeding for the large-seeded food-type soybean in Iowa; the programme was continued under the direction of Walter Fehr. From 1991 to 2000 Iowa, along with the Puerto Rico Agricultural Experiment Station, developed 41 cultivars with large seeds, high protein or null lipoxygenase (Bernard, 2001). From 1990 to 2001 several institutions in the USA and Canada released 19 large-seeded cultivars, but only the following seven have a seed size of ≥25 g 100^{-1} seeds: 'AC Onrei' (Canada), 'Gardensoy 11, 21, 31, 41' (Illinois), 'Satum' (Nebraska),and 'Ohio FG 2' (Ohio). The Gardensoy cultivars have a range of US MGs from I to IV and, therefore, can be harvested early to late over a month-long period. However, the cultivar 'BeSweet 292' from the Rupp Seed Company is relatively insensitive to photoperiod (it is duration-bound rather than season-bound) and can therefore be planted at staggered intervals to prolong the harvesting period.

Shattering is a problem in producing the seed of vegetable soybean; the transfer of non-shattering qualities from the grain soybean parent is another breeding objective. Bernard (2001) also identified 'shellability' – the ability to easily remove cooked seeds from the pod – as yet another breeding objective. From 1999 to 2002 a private seed company in Molokai Island, Hawaii, evaluated the breeding lines from the AVRDC and released 12 cultivars adapted to Molokai, South Carolina and Ohio. Mimura *et al.* (2007) identified 17 single sequence repeats (SSRs) and detected polymorphisms to differentiate 99 of the 130 vegetable soybean accessions from Japan and other countries. The study concluded that the Japanese vegetable soybean has a narrow genetic base and SSRs can describe the patterns of diversity for MG and test colour among vegetable soybeans.

The results of a four-year study conducted in Georgia, USA, to evaluate the yield potential of 10 Japanese cultivars/introductions, two Chinese cultivars and two US cultivars revealed that most of the Japanese cultivars/introductions produced green pod and green seed yield of >20 t ha^{-1} and 10 t ha^{-1}, respectively. The yields were extrapolated from a sample area of 0.38 m^2 (Rao *et al.*, 2002a, 2002b). The authors concluded that 'PI 181565', 'Tanbaguro', 'Wan Guingsi' and 'PI 200506' have potential for use in Georgia (Rao *et al.*, 2002a,b). Between 2001 and 2008, Virginia released 'Asmara', 'Randolph', (both US MG VI) and 'Owens' (MG V) cultivars (Mebrahtu *et al.*, 2005a,b, 2007). Devine *et al.* (2006) released an indeterminate vegetable soybean cultivar, 'Moon Cake', in 2003. 'Moon Cake' is good for home gardeners and can also be a good forage crop.

In Japan, various vegetable soybean cultivars can be broadly grouped into summer types ('Okuhara', 'Sapporo-midori', 'Osedefuri', 'Shiroge', 'Fukura', 'Mikawashima' and 'Yukimusume') and fall/autumn types ('Kinshu', 'Tzurunoko' and 'Yuzuru'). All have white flowers with the exception of 'Okuhara', 'Osedefuri' and 'Shiroge'. 'Fukura' is known for its sweetness, 'Kinshu' has dark pods, 'Yukimusume' has good pod colour, 'Mikawashima' has more three-seeded pods, 'Osedefuri' has good flavour and 'Tzurunoko' has large seeds (Konovsky *et al.*, 1994). Japan has laid the foundation for the quality requirements of vegetable soybean. Japan's discriminating and quality-conscious consumers demand and define the breeding objectives for vegetable soybean. A pleasant green pod colour with a clean appearance is the primary requirement, followed by grey pubescence. The length and width of the pod should be ≥5.0 and ≥1.4 cm, respectively. A 500 g plant should have ≤175 pods. The weight of 100 fresh green beans should be ≥70 g. The taste should be sweet with a sugar content of ≥10%. A bitter or astringent taste is undesirable. There should be no blemished, damaged or malformed pods (<1%). They should have a nice flavour and be easy to blanch or cook. The number of seeds per pod should be ≥2. The seed weight (dry weight basis) should be ≥30 g 100^{-1} seeds. Private and public vegetable soybean breeders in Japan and Taiwan and at the AVRDC have the above breeding objectives for developing improved vegetable soybean cultivars. In Japan, private seed companies conduct most of the applied vegetable soybean research and cultivar development, while the public sector undertakes some of the basic research on quality improvement. Lumpkin and Konovsky (1991) reviewed in detail the Japanese vegetable soybean cultivars.

From the screening of the USDA germplasm collection during 1960 in Taiwan, 13 large-seeded soybeans were selected; among them, 'PI 153210' and 'PI 179823' were considered as the best for vegetable soybean (Shanmugasundaram, 1979). 'Tzurunoko' and 'Ryokkoh' are the two vegetable soybean cultivars introduced from Japan for production in Taiwan and export of frozen pods to Japan.

The AVRDC began its vegetable soybean research in 1976. In 1981, the AVRDC joined with KDARES in breeding vegetable soybean for export to Japan. In 1985, the Council of Agriculture and the Provincial Department of Agriculture and Forestry provided financial support for vegetable soybean improvement. From 1980 to 1983, 51 vegetable soybean cultivars from 10 Japanese seed companies were screened and compared with 'Tzurunoko' and 'Ryokkoh'. Cultivars 'Tancho', 'Ryokukou', 'Nakate kaori' and a pure line selection from 'Taisho Shiroge' were selected and further evaluated. In spring, summer and autumn trials, AVRDC Glycine Selection (AGS) 292 (a pure line selection from 'Taisho Shiroge') gave a higher yield (7.7 t ha^{-1}) than 'Tzurunoko' (6.9 t ha^{-1}) and 'Ryokkoh' (6.0 t ha^{-1}). In 1987, KDARES released AGS 292 as the first vegetable soybean cultivar 'Kaohsiung No. 1' (KS No. 1). Farmers, processors and trading companies from Japan readily accepted KS No. 1. In 1990, KS No. 1 occupied 84% of the total vegetable soybean area in Taiwan (Chen *et al.*, 1991). The AVRDC collaborated with the Tainan District Agricultural Research and Extension Station and released additional improved cultivars (Table 19.4). AGS 292 is relatively less sensitive to photoperiod and temperature (Roberts *et al.*, 1996) and has been crossed with large-seeded cultivars such as 'Tanbaguro', 'Ryokkoh', 'Niutabian' (also spelled as Neu Ta Pien), 'Mikawashima', 'Blue Side', 'Setuzu' and 'Yukinoshita'. The AVRDC has developed a large number of improved pure line selections and distributed them worldwide to interested cooperators for evaluation. A total of 38 vegetable soybean cultivars from 15 countries have been released using AVRDC selections (Table 19.4).

As shown in Table 19.4, AGS 292 has been released and cultivated in seven countries around the world, indicating its wide adaptation. It has a large seed size, a large number of pods with two or more seeds, good pod

Table 19.4. Vegetable soybean cultivars released by AVRDC cooperators in different countries until 2008 (updated from Shanmugasundaram, 2001b; reprinted with permission from AVRDC).

Sr. no.	Local name	AVRDC ID no.	Year	Country	Remarks
1	AGATA[a]	GC 84123-13-1-2	2000	Argentina	–
2	GC 83005-9	GC 83005-9	1995	Bangladesh	HY, suitable for homestead cultivation
3	AGS 292	AGS 292	1990	China	–
4	AGS 337	AGS 337	1996	India	Pan and Rai, 1996
5	Swarna Vasundhara	GC 89009-1-1-2	2008	India	AVRDC, 2008
6	MKS 1	AGS 190	1995	Malaysia	HY
7	VSS 1	AGS 292	1999	Mauritius	–
8	VSS 2	AGS 339	1999	Mauritius	–
9	Rawal-1	AGS 190	1994	Pakistan	HY
10	PSB-VS 1	AGS 191	1997	Philippines	HY
11	PSB-VS 2	AGS 190	1997	Philippines	HY
12	PSB-VS 3	AGS 186	1997	Philippines	HY
13	AGS 190	AGS 190	1992	Sri Lanka	HY, suitable for soy milk, ice cream and soy nuts, less beany flavour
14	AGS 292	AGS 292	2002	Sudan	–
15	Kaohsiung No. 1	AGS 292	1987	Taiwan	HY, MH, DM, EM
16	Kaohsiung No. 2	Ryokkoh × KS 8	1991	Taiwan	HY, MH
17	Kaohsiung No. 3	PI 157424 × KS 8	1991	Taiwan	HY, MH
18	Kaohsiung No. 6[a]	AGS 292 × Nakadei Kaori	2001	Taiwan	–
19	Kaohsiung No. 7[a]	AGS 292 × Tanbaguro	2001	Taiwan	–
20	Tainan-ASVEG 2	GC 94016-10-1	2005	Taiwan	HY
21	KPS 292	AGS 292	1992	Thailand	HY
22	CM 1	AGS 190	1995	Thailand	HY, suitable for domestic consumption
23	VRQ 46	AGS 346	1999	Vietnam	EM (65–85 days), HY (11–14 $t\ ha^{-1}$), 3 crops $year^{-1}$
24	Mana	AGS 292	1999	Hawaii, USA	–
25	Makani	AGS 334	1999	Hawaii, USA	–

(*Continued*)

Table 19.4. *continued*

Sr. no.	Local name	AVRDC ID no.	Year	Country	Remarks
26	Momona	AGS 337	1999	Hawaii, USA	–
27	Nui	AGS 346	1999	Hawaii, USA	–
28	Buker's Favorite	AGS 292	1995	USA	Oregon and Washington
29	BeSweet 292	AGS 292	2002	Ohio, USA	Rupp Seed Co.
30	Koapaka	GC 97002 F3	2002	Hawaii, USA	HY, adaptation MKK, Ohio
31	Hiluhilu	GC 97022 F3	2002	Hawaii, USA	HY, adaptation MKK, Ohio, SC
32	Kanaloa	GC 97002 F3	2002	Hawaii, USA	HY, adaptation Ohio
33	Kila	GC 97022 F3	2002	Hawaii, USA	HY, adaptation SC
34	Onaona	GC 97002 F3	2002	Hawaii, USA	HY, adaptation SC
35	Mimiki	GC 97022 F3	2002	Hawaii, USA	HY, adaptation SC
36	Palanehu	GC 97002 F3	2002	Hawaii, USA	HY, adaptation MKK
37	Akua	GC 97029 F3	2002	Hawaii, USA	HY, adaptation MKK
38	Edamame 1	AGS 292	2006	Zimbabwe	Marketing started in 2007 by SeedCo (Zimbabwe and Malawi)
	Total	38		15	

DM, resistant to downy mildew; EM, early maturing; HY, high yielding; MH, suitable for mechanical harvesting; MKK, Molokai, Hawaii; SC, South Carolina, USA.
[a]Cross between AVRDC line and another cultivar.

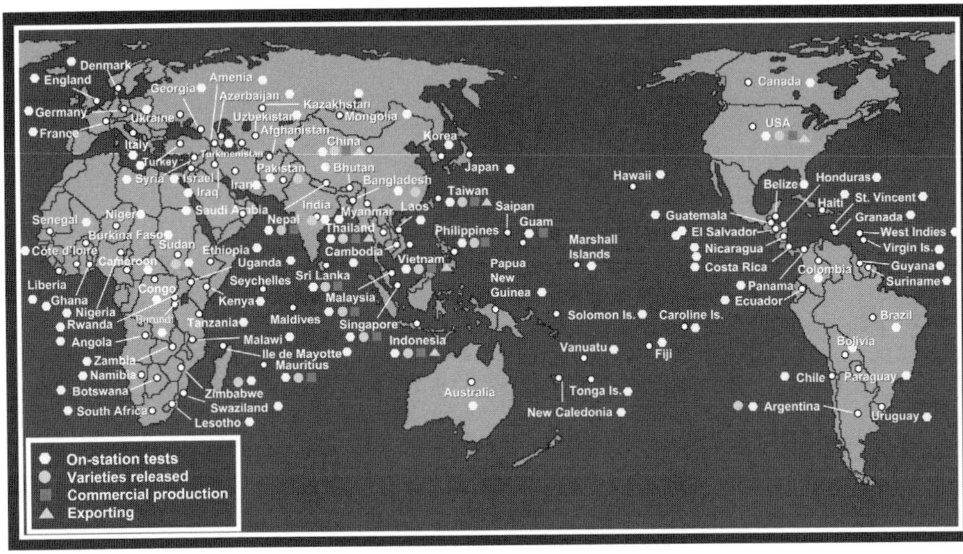

Fig. 19.1. AVRDC vegetable soybeans: evaluation, commercial production and export around the world as of 2008 (updated from Shanmugasundaram, 2001b; reprinted with permission from AVRDC).

colour, sweet taste and flavour and cooks easily. From 1979 to 1983, very few countries knew about vegetable soybean (Shanmugasundaram, 2001a). Cultivars released from AVRDC materials and in commercial production around the world as of 2008 are shown in Fig. 19.1.

At present, AVRDC vegetable soybean breeding focuses on selecting for high-graded pod yield, protein, fat, sugar, pod colour and seed size. Because there is a good correlation between pod length and pod width of two-seeded pods with seed size (Bravo *et al.*, 1980; Frank and Fehr, 1981; Shanmugasundaram *et al.*, 1991) they are used as selection criteria for seed size. Because narrow leaflets are associated with a higher number of three- and four-seeded pods (Bernard and Weiss, 1973), narrow leaflets are being introduced into vegetable soybean cultivars. At the AVRDC, both the pedigree and single seed descent methods of breeding are followed. Three generations are advanced per year, shortening the time for cultivar development. In Japan, glabrous cultivars have been reported to be resistant to soybean pod borer, but they are susceptible to soybean leaf hopper (Bernard and Weiss, 1973). The AVRDC has developed vegetable soybean breeding lines with glabrous gene, *P1*, from lines D62-7812 (G2030) and D62-7815 (G12495) from Stoneville, Mississippi, USA (Shanmugasundaram and Yan, 2001).

In Japan, there is an emerging interest in vegetable soybean with taro flavour. The cultivar 'Dadacha-mame' has a taro flavour after blanching and a brown seed coat. Fushimi and Masuda (2001) reported that the flavour of 'Dadacha-mame' is similar to that of aromatic rice. The aroma in rice is due to 2-acetyl-1-pyrroline (AP). They found the concentration of AP in 'Dadacha-mame' was high at the vegetable soybean stage, but as it matured to grain stage the concentration decreased significantly and could not be detected. Preliminary studies indicate that the basmati flavour may be governed by a single recessive gene, which can be identified in the F_2 generation by cooking and tasting the seeds (Shanmugasundaram and Yan, unpublished data). Because fragrant rice (jasmine or basmati rice) is very popular, the AVRDC has decided to incorporate the fragrance gene into selected vegetable soybean cultivars. A number of these selections are currently being evaluated in different countries. Such cultivars have potential in South Asia, where basmati rice is very popular. The name 'Dadacha-mame' has been patented in Japan. The AVRDC is also exploring the possibility of improving the isoflavones, tocopherol and antioxidant activities of vegetable soybean. AGS 292 is high in isoflavone content, at 1490 mg g^{-1} (Shanmugasundaram and Yan, 2001).

Masuda (2004) reported that the starch content of 'Tanbaguro' is higher than the sucrose content at the vegetable soybean stage. However, upon cooking the starch is converted into maltose and imparts sweetness to the vegetable soybean, even though the sucrose content is low. Such high starch at the vegetable soybean stage and the ability to convert starch to maltose after cooking may be an excellent trait to prolong the harvest window of vegetable soybean; harvested vegetable soybeans can be kept for a longer time without refrigeration and without fear of quality deterioration (Shanmugasundaram and Yan, 2001).

Although the area planted to vegetable soybean in South Korea is small, the country has an active vegetable soybean breeding programme. Korean research focuses on vegetable soybeans adapted to Korea with resistance to soybean mosaic virus and brown stem rot, and cultivars that can be used to cook with rice. The cultivars released in Korea are listed in Table 19.5.

Table 19.5. Vegetable soybean cultivars released in Korea (Kyeong-ho Chung, Korea, 2009, personal communication).

Cultivar	Year of release
Saealkong	1984
Keunolkong	1991
Hwasongputkong	1993
Hwaumputkong[a]	1993
Seockryangputkong[a]	1994
Geomjeongolkong	1996
Saeolkong	1998
Seonnogkong	2000
Danmiputkong	2002
Dajinputkong	2003
Geomjeongsaeolkong	2004
Mirang	2004
Danmi 2	2005
Nogwon	2006

[a]Japanese cv.

Lipoxygenase imparts an undesirable flavour in vegetable soybean. Using germplasm with triple null for *Lx* genes, vegetable selections with lipoxygenase nulls have been developed at the AVRDC. 'Danmi 2' developed in Korea is null for lipoxygenase 1 and 2 (Baek *et al.*, 2006). Srisombun *et al.* (2009) have identified GC 96025-43 (AGS 408) from AVRDC as triple null for *Lx* genes and GC 96026-10 (AGS 415) to be null for *Lx1* and 2. These cultivars have excellent qualities for soybean milk production in Thailand and will be released to farmers.

19.5 Processing and Utilization

Vegetable soybean requires minimal processing. In Japan, the whole plant with fresh green pods is harvested. After removing the leaves the plants are tied in a bundle and placed in wooden crates. The whole plant is boiled with pods and the cooked green beans from the pods are consumed. This is the preferred way of enjoying *edamame*. Fresh green pods detached from the stem are packed in plastic net bags (300–500 g bag^{-1}) and sold on the market using refrigeration (Iwamida and Ohmi, 1991). Shelled green beans are sold fresh every day in markets in China and Taiwan. Shelling the pods is time-consuming, labour-intensive and rather difficult because the soybean pod is hard to shell. Therefore, the pods usually are cooked for 2–3 min to soften the shell. Prior to shelling, the green pods are processed through a mechanical sieve or grading machine, which sorts abnormal, damaged and off-size pods (Shanmugasundaram and Yan, 2001). In countries where cheap labour is available, shelling is done manually. Shelling machines are used in Taiwan. Three new shelling machines have recently been developed in Japan. The large automatic P-360 can shell 100 kg h^{-1}, the small P-78 can shell boiled pods at 30 kg h^{-1} and the P-78 for raw pods can shell 20 kg h^{-1} (Atsumi, 2008). Fresh green pods are exported packed in ice through airfreight for immediate sale in supermarkets and retail outlets.

Due to a shortage of domestically produced vegetable soybean, instant quick-frozen green vegetable soybean has found a market in Japan. Taiwan began exporting frozen vegetable soybean in the mid-1980s. Taiwan soon had 27 frozen-food manufacturing companies and had captured 90% of the frozen vegetable soybean imported into Japan (Lin, 2001; Shanmugasundaram and Yan, 2004). Eleven companies have survived; almost all have established factories in southern China and export frozen vegetable soybean from there to Japan and other countries. Thailand, Indonesia and Vietnam currently have factories that process and export frozen vegetable soybeans (Lin, 2001).

Following harvesting of the pods, either manually or mechanically, the pods should reach the factory for freezing in 6 h to avoid quality deterioration. For timely delivery, the distance between the sorting and grading location and the processing factory should not exceed 200 km. During sorting the pods are graded. A and B grades are acceptable to the processors; C and D grades are unacceptable (AVRDC, 1992). Grading and sorting is one of the most important steps in processing. The raw material received in the factory is washed and cleaned well. Any debris, stems, leaves or other extraneous materials are removed. The cleaned pods go through a steam chamber where they are blanched for 1.5–3 min at 98–100°C. Returning from the blanching chamber, the pods are quickly passed through water at 0°C to retain their green colour. The pods then pass through a conveyor belt where undesirable pods are sorted out. The pods go through an instant quick-frozen chamber where individual pods are quick-frozen at –40°C. The frozen pods are then packed in 0.5–1 kg packages or bulk containers and stored at –18°C until they are ready for shipping in refrigerated containers (Liu and Shanmugasundaram, 1982; AVRDC, 1992). Shelled green beans are frozen in a similar fashion for the domestic and export markets. In processing, the sanitation quality is extremely important. The total count of bacteria should be <3 million g^{-1}. The sample should be free from *Escherichia coli* and *Salmonella*.

In China, vegetable soybeans are processed as pickles, saline green vegetable soybean, bamboo-shoot vegetable soybean and smoke-dried green soybean, which has a long storage life. Smoke-dried vegetable soybeans are cooked in brine and roasted over a fire. Similarly, bamboo-shoot vegetable soybean is cooked with bamboo shoots and Lima beans in brine and dried with smoke (Wu, 2004).

The Taiwan Frozen Food Manufacturer's Association has actively explored various vegetable soybean products to diversify the market and attract customers. In the past, frozen vegetable soybeans have been produced with or without salt. Now they are also available with different combinations of garlic, pepper, *wasabi*, stockfish, *perilla* (a leafy vegetable) or Chinese spice added after blanching. They are marketed as seasoned vegetable soybean in the supermarket. For South Asia, curry-flavoured vegetable soybean may be attractive. Green vegetable soybean is used to make milk, ice cream, yogurt, tofu, soy nuts, soy flour, noodles and meat balls. It is also combined with meat or seafood or processed as ready-made frozen food and marketed to urban consumers (Shanmugasundaram, 2001a; Shanmugasundaram and Yan, 2004).

19.6 International Trade

Taiwan exported 142 t of frozen vegetable soybean to Japan in 1971. In 1978, the quantity exported increased to 9500 t and in 1990, with about 40,000 t, the value of exports reached nearly US$80 million (Benziger and Shanmugasundaram, 1995). Japan imports about 70,000 t of fresh and frozen vegetable soybeans annually from different countries (Table 19.6). The quality of vegetable soybean dictates the price. Vegetable soybeans from Taiwan and Thailand receive a premium price from Japan compared to other countries due to their better quality.

The establishment of a vegetable soybean factory for the export of frozen vegetable soybean in Chiang Mai, a Thai/Japanese joint venture, stimulated the commercial production of vegetable soybean in Thailand. In 1990, only a small quantity of soybean was exported to Japan, but in 1999 the volume

Table 19.6. Vegetable soybean export to Japan (data obtained from Trade Statistics of the Ministry of Finance of Japan through personal communication with the Taiwan Frozen Vegetable Manufacturer's Association, 2009).

	2005			2006		
Country	Quantity (t)	Price (US$ kg^{-1})	Total value (million US$)	Quantity (t)	Price (US$ kg^{-1})	Total value (million US$)
China	31,086	1.25	38.73	29,702	1.38	40.99
Taiwan	23,572	1.66	39.27	22,198	1.77	39.29
Thailand	10,960	1.52	16.59	11,161	1.65	18.41
Indonesia	2,936	1.43	4.18	3,117	1.48	4.61
Vietnam	664	1.44	0.94	698	1.46	1.00
Total	69,218	1.44	99.71	66,876	1.54	104.30

exported had increased to 9000 t (Srisombun and Shanmugasundaram, 2001). The export of vegetable soybean from Thailand is slowly increasing. Indonesia commenced production for export to Japan in the late 1990s via quasi-government operations and through the private sector. Production and the factory are concentrated in Jember in central Java. From around 500 to 600 ha, Indonesia exports a modest 3,000 t annually to Japan. The USA imports around 10,000–15,000 t of frozen vegetable soybeans from Taiwan, China and Thailand (Bernard, 2001).

In Taiwan and Thailand, the retail price of vegetable soybean in the market is US$1.5 and 1.0 kg^{-1}, respectively. In the USA, both frozen pods and frozen shelled beans cost about US$3 kg^{-1} in the supermarket. In Mauritius, the green pods are marketed at US$2 kg^{-1} (Gangadurdoss and Hanoomanjee, 1999). In Japan, the price of fresh pods attached to the stem varies from US$6 to US$12 kg^{-1} depending upon the season and location. The price of imported frozen vegetable soybean pods ranges from US$4 to US$6 kg^{-1} pods. Currently, Japan is the major market for vegetable soybean, but it is a finite market with a fixed demand of about 70,000 t year^{-1}. Almost all of the countries involved in production are focusing on Japan. For quite some time to come, China and Taiwan will dominate the export of vegetable soybean to Japan. To a limited extent, Thailand, Indonesia and Vietnam will share the market. With the increase in the Asian population in the USA, the import of vegetable soybean from China and Taiwan to the USA will also gradually increase, unless niche locations for vegetable soybean production and processing are identified in the USA for the domestic market. The quality of the product is most important to the consumer, and cultivars that are acceptable to consumers should be cultivated and harvested at the right time to maintain that quality.

19.7 Potential for Developing Countries

Vegetable soybean is a new crop for many of the world's developing countries. Due to its high protein, dietary fibre, health-promoting phytochemicals and easy cooking, it is an attractive crop for alleviating protein malnutrition. Because it also contains cholesterol-free fat it is an excellent vehicle for the absorption of vitamin A. It is also a short growth-duration crop. From sowing to harvest takes about 65–75 days. Furthermore, the green or dried forage from the crop after the harvest can be fed to cattle or ploughed under as a green manure to enrich the soil. Vegetable soybeans can produce a total biomass of 40 t ha^{-1}, of which 25% is green pods. The dry matter yield of the residue of vegetable soybean can be 6.0–6.6 t ha^{-1} in about 80 days, compared to 5.0 t ha^{-1} from green manure crops such as *Crotalaria* or *Sesbania*. The total nitrogen, phosphorus and potassium in the residue (leaf plus stem) is about 120, 18 and 150 kg ha^{-1}, respectively (AVRDC, 1998; Shanmugasundaram *et al.*, 1992; Shanmugasundaram and Yan, 2004).

When the AVRDC began its vegetable soybean breeding programme, a large number of cooperators from around the world expressed interest in evaluating its vegetable soybean selections. From 1979 to 2000, the AVRDC has distributed 3200 vegetable soybean selections and 1600 germplasms to 665 cooperators in 87 countries. Utilizing these materials, cooperators from 15 countries have released 38 new vegetable soybean cultivars to their farmers (Table 19.4; Fig. 19.1). Thailand, Indonesia and Vietnam have been able to commercially produce, freeze and export the frozen pods and expand their domestic markets. Through the efforts of the AVRDC's Regional Center for Africa (RCA) in Arusha, Tanzania, from 1998 to 2003, 361 AVRDC vegetable soybean selections were distributed to cooperators in 26 African countries. Currently, vegetable soybean is commercially grown in Mauritius, Zimbabwe, Sudan and Uganda and marketed domestically (Gangadurdoss and Hanoomanjee, 1999; Chadha and Oluoch, 2001, 2004). The RCA has distributed new vegetable soybean cultivar seed to thousands of farmers in Tanzania, Sudan, Zambia and Mauritius.

In Asia, Malaysia engages in domestic production for the domestic market. Through the efforts of the AVRDC's Regional Center for South Asia, India has released a new vegetable soybean cultivar, 'Swarna Vasundhara', for Jharkhand in India and is aggressively promoting it to the public. Sri Lanka is also growing vegetable soybean for the domestic market. With the introduction of basmati-flavoured vegetable soybean cultivars there may be great potential for vegetable soybean in South Asia.

AVRDC vegetable soybeans have been evaluated in Argentina, Chile, Nicaragua and Brazil. Argentina has released a new vegetable soybean cultivar. Improved vegetable soybean cultivars are ready for release in South America, if niche markets can be developed (Benavidez *et al.*, 2001; Vello *et al.*, 2004).

19.8 Conclusions

Vegetable soybean has been recognized as a nutritionally valuable crop. The fresh green pods with fully developed green beans are harvested with the stem or detached from the stem and marketed in Japan; the fresh green shelled beans are also marketed. The green pods and green shelled beans are frozen and marketed. China, Japan and Taiwan have the largest area and production of vegetable soybean in the world. China, Taiwan, Thailand, Indonesia and Vietnam are major exporters of vegetable soybean to Japan and the USA.

Vegetable soybean is cultivated in different seasons in rotation with rice. During the off-season and to get the crop to market early, forcing cultivation in glasshouses and vinyl tunnels is also practised. Vegetable soybean is a labour-intensive crop. To reduce costs, cultivation of vegetable soybean has been mechanized. Quality seed is an essential requirement for vegetable soybean. Therefore, extra care should be taken to produce good-quality seeds.

All of the vegetable soybean cultivars grown in China are landraces. These landraces were introduced to Japan and selections were made for appearance, colour, taste and flavour. The quality requirements for the Japanese market are very demanding. These qualities form the basic objectives for vegetable soybean breeding. In Japan, private seed companies have developed a large number of adapted cultivars for different locations and seasons. Japanese cultivars have been introduced to Taiwan and the USA. In the USA, breeders have selected a number of introduced cultivars for vegetable soybean purposes. Several universities in the USA and the USDA have developed a number of improved vegetable soybean cultivars with large seed size. However, the extent of their commercial production is unknown.

The AVRDC and national researchers from Taiwan began joint vegetable soybean breeding in the 1980s. Focusing on their objectives and using several key strategies, the AVRDC developed a large number of breeding lines for evaluation in different countries. Utilizing the AVRDC selections, 38 improved cultivars have been released in 15 countries. Several countries in Asia and Africa are growing vegetable soybean commercially for the domestic and export markets.

Vegetable soybean is a short-duration crop. The shelled beans are used as human food. The crop biomass that remains after harvesting the green pods can be fed to cattle or incorporated as green manure to sustain soil fertility. Therefore, there is a great potential for this crop in developing countries. The new basmati-flavoured vegetable soybean cultivars have potential to expand the area under production and acceptance by consumers in South Asia and Africa.

References

Abdullah, M., Sarnthoy, O. and Jiwajinda, S. (2001a) Cypermethrin insecticide residues in vegetable soybean, *Glycine max* (L.) Merrill, at different days of pre-harvest intervals. *Kasetsart Journal of Natural Sciences* 35, 115–121.

Abdullah, M., Sarnthoy, O., Isichaikul, S. and Tantakon, S. (2001b) Efficacy of cypermethrin, neem extract and *Bacillus thuringiensis* for controlling insect pests of vegetable soybean. *Kasetsart Journal of Natural Sciences* 36, 14–22.

Akazawa, T., Yanagisawa, Y. and Sasahara, T. (1997) Concentrations of water-soluble nitrogen and amino acids as criteria for discriminating vegetable type and grain type cultivars. *Breeding Science* 47, 39–44.

Atsumi, H. (2008) Future plan of edamame shelling machine development. *Journal of the Japanese Society for Edamame Science* 6, 68–71.

AVRDC (1992) *Vegetable Soybean Production: Proceedings of a Training Course*. Shanmugasundaram, M. (ed.). Asian Vegetable Research and Development Center, Shanhua, Tainan, Taiwan.

AVRDC (1998) *AVRDC Report 1997*. Asian Vegetable Research and Development Center, Shanhua, Tainan, Taiwan.

AVRDC (2008) Vegetable soybean field day promotes a new crop. *AVRDC Newsletter*. Asian Vegetable Research and Development Center, Shanhua, Tainan, Taiwan.

Baek, I.Y., Ko, J.M., Kang, S.T., Han, W.Y., Shin, D.C., Oh, S.K., Chung, M.G., Kim, H.T., Kang, N.S., Oh, K.W., Shin, S.O., Park, K.Y., Suh, D.Y., Yun, H.T., Oh, Y.J., Lee, Y.S., Son, C.K., Kim, Y.D., Kim, H.Y. and Lee, S.H. (2006) A new vegetable soybean cultivar, 'Danmi 2' with free of lipoxygenase 1 and 2. *Korean Journal of Breeding* 38, 301–302.

Benavidez, R., Gosparini, C.O. and Morandi, E.N. (2001) Vegetable soybean in Argentina. In: *Proceedings of Second International Vegetable Soybean Conference*. Washington State University, Tacoma, WA, USA, pp. 17–20.

Benziger, V. and Shanmugasundaram, S. (1995) Taiwan's frozen vegetable soybean industry. *Technical Bulletin No. 22*. Asian Vegetable Research and Development Center, Shanhua, Tainan, Taiwan.

Bernard, R.L. (2001) Breeding vegetable soybeans in the Midwest. In: *Proceedings of Second International Vegetable Soybean Conference*. Washington State University, Tacoma, WA, USA, p. 21.

Bernard, R.L. and Weiss, M.G. (1973) Qualitative genetics. In: Caldwell, B.E. (ed.) *Soybeans: Improvement, Production, and Uses*. Agronomy Monograph 16. American Society of Agronomy, Madison, WI, USA, pp. 117–154.

Bravo, J.A., Fehr, W.R. and Cianzio, S.R. (1980) Use of pod width for indirect selection of seed weight in soybeans. *Crop Science* 20, 507–509.

Chadha, M.L. and Oluoch, M.O. (2001) Adaptation and utilization of AVRDC vegetable soybean lines in eastern and southern Africa. In: *Proceedings of Second International Vegetable Soybean Conference*. Washington State University, Tacoma, WA, USA, pp. 29–33.

Chadha, M.L. and Oluoch, M.O. (2004) Vegetable soybean research and development in Africa In: Moscardi, F., Hoffmann-Campo, C.B., Saraiva, O.F., Galerani, P.R., Krzyzanowski, F.C. and Carrão-Panizzi, M.C. (eds) *Proceedings of VII World Soybean Research Conference and VI International Soybean Processing and Utilization Conference*. Foz do Iguassu, Brazil, pp. 921–928.

Chen, K.F., Lai, S.H., Cheng, S.T. and Shanmugasundaram, S. (1991) Vegetable soybean seed production technology in Taiwan. In: Shanmugasundaram, S. (ed.) *Proceedings of Vegetable Soybean – Research Needs for Production and Quality Improvement Workshop*. Asian Vegetable Research and Development Center, Shanhua, Tainan, Taiwan, pp. 45–52.

Chen, L.S., Chu, C., Liu, C.D., Chen, R.S. and Tsay, J.G. (2006) PCR-based detection and differentiation of anthracnose pathogens, *Colletotrichum gloeosporioides* and *C. truncatum*, from vegetable soybean in Taiwan. *Journal of Phytopathology* 154, 654–662.

Cheng, S.H. (1991) Vegetable soybean area, production, foreign and domestic trade in Taiwan. In: Shanmugasundaram, S. (ed.) *Proceedings of Vegetable Soybean – Research*

Needs for Production and Quality Improvement Workshop. Asian Vegetable Research and Development Center, Shanhua, Tainan, Taiwan, pp. 17–21.

Cherng, W.L. and Tay, S.Y. (2002) Detection of Roundup Ready gene in soybean processed food and vegetable soybean by multiplex polymerase chain reaction. *Journal of Chinese Society of Horticultural Science* 68, 275–282.

Chiba, Y. (1991) Post harvest processing, marketing and quality degradation of vegetable soybean in Japan. In: Shanmugasundaram, S. (ed.) *Proceedings of Vegetable Soybean – Research Needs for Production and Quality Improvement Workshop*. Asian Vegetable Research and Development Center, Shanhua, Tainan, Taiwan, pp. 108–112.

Dai, O., Zhang, L. and Huang, Z.P. (2001) Introduction of a kind of big grain, autumn sown vegetable soybean cultivar, 'Shiyueqing'. *Seeds in China* 1, 21.

Devine, T.E., McMurtrey, J.E., Mebrahtu, T., Abney, T.S., Donald, P., Starney, D.E, Hashem, F.M. and Dadson, R.B. (2006) Registration of 'Moon Cake' vegetable soybean. *Crop Science* 46, 1810–1811.

Diep, C.N., Dang, V.H., Ngau, N.V., Son, M.T. and Duong, T.P. (2002) Effects of rhizobial inoculation and inorganic nitrogen fertilizer of vegetable soybean [*Glycine max* (L.) Merrill] cultivated on alluvial soil of Cantho province (Mekong Delta) using ^{15}N isotope dilution technique. In: Herridge, D. (ed.) *Inoculants and Nitrogen Fixation of Legumes in Vietnam. ACIAR Proceedings 109e*, Australian Centre for International Agricultural Research, Canberra, ACT, Australia, pp. 81–85.

Fehr, W.R., Caviness, C.C., Burmood, D.T. and Pennington, J.S. (1971) Stages of development descriptions for soybeans, *Glycine max* (L.) Merrill. *Crop Science* 11, 929–931.

Ferro, C.R., Hill, C.B., Miles, M.R. and Hartman, G.L. (2004) Evaluation of soybean cultivars for partial resistance and tolerance to *Phytophthora sojae*. In: Moscardi, F., Hoffmann-Campo, C.B., Saraiva, O.F., Galerani, P.R., Krzyzanowski, F.C. and Carrão-Panizzi, M.C. (eds) *Abstracts of Contributed Papers and Posters. VII World Soybean Research Conference and VI International Soybean Processing and Utilization Conference*. Foz do Iguassu, Brazil, p. 54.

Frank, S.J. and Fehr, W.R. (1981) Associations among pod dimensions and seed weight in soybeans. *Crop Science* 21, 547–550.

Fushimi, T. and Masuda, R. (2001) 2-Acetyl-1-pyrroline concentration of the aromatic vegetabe soybean 'Dadacha-Mame'. In: *Proceedings of Second International Vegetable Soybean Conference*. Washington State University, Tacoma, WA, USA, p. 39.

Gai, J. and Guo, W. (2001) History of Maodou production in China. In: *Proceedings of Second International Vegetable Soybean Conference*. Washington State University, Tacoma, WA, USA, pp. 41–47.

Gangadurdoss, M. and Hanoomanjee, P. (1999) Prospects and potential of growing vegetable soybean [*Glycine max* (L.) Merrill] in Mauritius. In: Lalouette, J.A. (ed.) *Proceedings of the Third Meeting of Agricultural Scientists*. Food and Agricultural Research Council, Reduit, Mauritius, pp. 159–166.

Hartman, G.L., Sinclair, J.B. and Rupe, J.C. (1999) *Compendium of Soybean Diseases*, 4th edn. The American Phytopathological Society, St Paul, MN, USA.

Hass, P.W., Gilbert, L.C. and Edwards, A.D. (1982) *Fresh Green Soybeans: Analysis of Field Performance and Sensory Qualities*. Rodale Press, Emmaus, PA, USA.

Hatanaka, T., Yoshihara, S., Imoto, S., Uchida, N. and Tsugawa, H. (2004) Tanbaguro: A new model genotype of soybean for tissue culture study. In: Fischer, T., Turner, N., Angus, J., McIntyre, L., Robertson, M., Borrell, A. and Lloyd, D. (eds) New *Directions for a Diverse Planet*. Proceedings of the 4th International Crop Science

Congress, Brisbane, Australia, 26 September–1 October 2004. Available at: www.cropscience.org.au (accessed 1 March 2010).

Hikino, I. (2000) Super premium soybean variety 'Tanbaguro'. In: *Proceedings of the Third International Soybean Processing and Utilization Conference*. Korin Publishing Co., Japan, pp. 35–36.

Hung, A.T., Cheng, J.H., Ma, C.H. and Kobayashi, H. (1991) Effect of fertilizer management and rhizobia inoculation on yield and quality of vegetable soybean. In: Shanmugasundaram, S. (ed.) *Proceedings of Vegetable Soybean – Research Needs for Production and Quality Improvement Workshop*. Asian Vegetable Research and Development Center, Shanhua, Tainan, Taiwan, pp. 73–84.

Hymowitz, T. (1984) Dorsett-Morse soybean collection trip to East Asia: A 50 year perspective. *Economic Botany* 38, 378–388.

Iwamida, S. and Ohmi, H. (1991) Communication links between vegetable soybean producers, processors, trading companies and seed companies in Japan. In: Shanmugasundaram, S. (ed.) *Proceedings of Vegetable Soybean – Research Needs for Production and Quality Improvement Workshop*. Asian Vegetable Research and Development Center, Shanhua, Tainan, Taiwan, pp. 22–25.

Jianfeng, M. and Zhengwen, Z. (2002) Research on the harvesting time of vegetable soybean. *Hunan Agricultural Science* 6, 28–30.

Kamiyama, Y. (1991) Vegetable soybean seed production technologies in Japan. In: Shanmugasundaram, S. (ed.) *Proceedings of Vegetable Soybean – Research Needs for Production and Quality Improvement Workshop*. Asian Vegetable Research and Development Center, Shanhua, Tainan, Taiwan, pp. 43–44.

Khadkao, S. (1992) Insect pests of soybean. In: Shanmugasundaram, M. (ed.) *Proceedings of Vegetable Soybean Training Workshop*. Asian Vegetable Research and Development Center, Shanhua, Tainan, Taiwan, pp. 27–37.

Kitamura, K. (2001) Recent edamame production information and research advances in Japan. In: *Proceedings of Second International Vegetable Soybean Conference*. Washington State University, Tacoma, WA, USA, pp. 71–74.

Kokubun, M. (1991) Cultural practices and cropping systems for vegetable soybean in Japan. In: Shanmugasundaram, S. (ed.) *Proceedings of Vegetable Soybean – Research Needs for Production and Quality Improvement Workshop*. Asian Vegetable Research and Development Center, Shanhua, Tainan, Taiwan, pp. 53–60.

Konovsky, J., Lumpkin, T.A. and McClary, D. (1994) Edamame: The vegetable soybean. In: O'Rourke (ed.) *Understanding the Japanese Food and Agrimarket: A Multifaceted Opportunity*. Haworth Press, Binghamton, NY, USA, pp. 173–181.

Kucuk, O. (2004) Soy isoflavones in the treatment of prostate cancer. In: Moscardi, F., Hoffmann-Campo, C.B., Saraiva, O.F., Galerani, P.R., Krzyzanowski, F.C. and Carrão-Panizzi, M.C. (eds) *Proceedings VII World Soybean Research Conference and VI International Soybean Processing and Utilization Conference*. Foz do Iguassu, Brazil, pp. 695–700.

Lin, C.C. (2001) Frozen edamame: Global market conditions. In: *Proceedings of Second International Vegetable Soybean Conference*. Washington State University, Tacoma, WA, USA, pp. 93–96.

Lin, F.H. and Cheng, S.T. (2001) Vegetable soybean development for export to Japan – a historical and technical perspective. In: *Proceedings of Second International Vegetable Soybean Conference*. Washington State University, Tacoma, WA, USA, pp. 87–91.

Liu, C.P. and Shanmugasundaram, S. (1982) Frozen vegetable soybean industry in Taiwan. In: *Proceedings of Symposium on Vegetables and Ornamentals in the Tropics*. University Pertanian, Serdang, Malaysia, pp. 199–212.

Lumpkin, T.A. and Konovsky, J. (1991) A critical analysis of vegetable soybean production, demand and research in Japan. In: Shanmugasundaram, S. (ed.) *Proceedings of Vegetable Soybean – Research Needs for Production and Quality Improvement Workshop*. Asian Vegetable Research and Development Center, Shanhua, Tainan, Taiwan, pp. 120–140.

Masuda, R. (1991) Quality requirement and improvement of vegetable soybean. In: Shanmugasundaram, S. (ed.) *Proceedings of Vegetable Soybean – Research Needs for Production and Quality Improvement Workshop*. Asian Vegetable Research and Development Center, Shanhua, Tainan, Taiwan, pp. 92–102.

Masuda, R. (2004) The strategy of sweetness increase of vegetable soybeans: Maltose another sugar in boiled seeds. In: Moscardi, F., Hoffmann-Campo, C.B., Saraiva, O.F., Galerani, P.R., Krzyzanowski, F.C. and Carrão-Panizzi, M.C. (eds) *Proceedings of VII World Soybean Research Conference and VI International Soybean Processing and Utilization Conference*. Foz do Iguassu, Brazil, pp. 839–844.

Masuda, R. and Harada, K. (2001) Enhancement of sweet components in vegetable soybean seeds: Starch degradation during cooking enhances flavour of immature seeds. In: *Proceedings of Second International Vegetable Soybean Conference*. Washington State University, Tacoma, WA, USA, pp. 105–108.

Mebrahtu, T., Devine, T.E., Donald, P. and Abney, T.S. (2005a) Registration of 'Randolph' soybean. *Crop Science* 45, 2644–2645.

Mebrahtu, T., Devine, T.E., Donald, P. and Abney, T.S. (2005b) Registration of 'Asmara' vegetable soybean. *Crop Science* 45, 408–409.

Mebrahtu, T., Devine, T.E., Donald, P. and Abney, T.S. (2007) Registration of 'Owens' vegetable soybean. *Journal of Plant Registrations* 1, 95–96.

Messina, M. (2004) The science behind soyfoods. In: Moscardi, F., Hoffmann-Campo, C.B., Saraiva, O.F., Galerani, P.R., Krzyzanowski, F.C. and Carrão-Panizzi, M.C. (eds) *Proceedings of VII World Soybean Research Conference and VI International Soybean Processing and Utilization Conference*. Foz do Iguassu, Brazil, pp. 73–82.

Mimura, M., Coyne, C.J., Bambuck, M.W. and Lumpkin, T.A. (2007) SSR diversity of vegetable soybean [*Glycine max* (L.) Merr.]. *Genetic Resources and Crop Evolution* 54, 497–508.

Nakano, H. (1991) Vegetable soybean area, production, demand, supply, domestic and foreign trade in Japan. In: Shanmugasundaram, S. (ed.) *Proceedings of Vegetable Soybean – Research Needs for Production and Quality Improvement Workshop*. Asian Vegetable Research and Development Center, Shanhua, Tainan, Taiwan, pp. 8–16.

Nantapan, M. (1992) Diseases and their control. In: Shanmugasundaram, M. (ed.) *Proceedings of Vegetable Soybean Training Workshop*. Asian Vegetable Research and Development Center, Shanhua, Tainan, Taiwan, pp. 38–44.

Ning, L., Zhao, Z.H., Zhang, J.X. and Yan, G.F. (2006) Effect of density on the agronomic character and yield of vegetable soybean. *Xinjiang Agricultural Sciences* 43, 271–274.

NSRL (2008) Edamame: The new garden-type soybean. Available at: http://www.nsrl.uiuc.edu/aboutsoy/edamame.html (accessed 1 March 2010).

Ohmi, H. (2001) Breeding, seed production and marketing of vegetable soybean in Japan by Snow Brand Seed Co. Ltd. In: *Proceedings of Second International Vegetable Soybean Conference*. Washington State University, Tacoma, WA, USA, pp. 139–141.

Pan, R.S. and Rai, M. (1996) Evaluation of vegetable soybean lines in Chotanagpur Plateau, Bihar, India. *TVIS Newsletter* 1. Asian Vegetable Research and Development Center, Shanhua, Tainan, Taiwan.

Park, H.K., Kim, K.H., Oh, Y.J. and Kim, S.H. (2001) Selection of vegetable soybean and development of year-round production system in Korea. In: *Proceedings of Second International Vegetable Soybean Conference*. Washington State University, Tacoma, WA, USA, pp. 143–146.

Rao, M.S.S., Bhagsari, A.S. and Mohamed, A.I. (2002a) Fresh green seed yield and seed nutritional traits of vegetable soybean genotypes. *Crop Science* 42, 1950–1958.

Rao, M.S.S., Mulliniz, B.G., Rangappa, M., Cebert, E., Bhagsari, A.S., Sapra, V.T. and J.M. Joshi (2002b) Genotype × environmental interactions and yield stability of food grade soybean genotypes. *Agronomy Journal* 94, 72–80.

Roberts, E.H., Qi, A., Ellis, R.H., Summerfield, R.J., Lawn, R.J. and Shanmugasundaram, S. (1996) Use of field observations to characterize genotypic flowering responses to photoperiod and temperature: A soybean exemplar. *Theoretical and Applied Genetics* 93, 519–533.

Shanmugasundaram, S. (1979) *Varietal Development and Germplasm Utilization in Soybeans*. AVRDC Technical Bulletin No. 13 (76-102). Shanhua, Tainan, Taiwan.

Shanmugasundaram, S. (1998) *Epidemiology and Control Strategies for Soybean Rust. Proceedings of Workshop*, Pootchefstroom, ARCI, GCI, South Africa.

Shanmugasundaram, S. (2001a) Global extension and diversification of fresh and frozen vegetable soybean. In: *Proceedings of Second International Vegetable Soybean Conference*. Washington State University, Tacoma, WA, USA, pp. 161–165.

Shanmugasundaram, S. (2001b) Vegetable soybean varietal improvement at AVRDC. In: *Proceedings of Second International Vegetable Soybean Conference*. Washington State University, Tacoma, WA, USA, pp. 173–177.

Shanmugasundaram, S. (2005) Designing vegetables with improved nutritional contents. In: Ramasamy, C., Ramanathan, S. and Dakshinamoorthy, M. (eds) *Perspectives of Agricultural Research and Development*. Tamil Nadu Agricultural University, Coimbatore, India, pp. 541–561.

Shanmugasundaram, S. and Yan, M.R. (2001) Mechanization of vegetable soybean production in Taiwan. In: *Proceedings of Second International Vegetable Soybean Conference*. Washington State University, Tacoma, WA, USA, pp. 167–172.

Shanmugasundaram, S. and Yan, M.R. (2004) Global extension of high value vegetable soybean. In: Moscardi, F., Hoffmann-Campo, C.B., Saraiva, O.F., Galerani, P.R., Krzyzanowski, F.C. and Carrão-Panizzi, M.C. (eds) *Proceedings VII World Soybean Research Conference and VI International Soybean Processing and Utilization Conference*. Foz do Iguassu, Brazil, pp. 915–920.

Shanmugasundaram, S., Cheng, S.T., Huang, M.T. and Yan, M.R. (1991) Varietal improvement of vegetable soybean in Taiwan. In: Shanmugasundaram, S. (ed.) *Proceedings of Vegetable Soybean – Research Needs for Production and Quality Improvement Workshop*. Asian Vegetable Research and Development Center, Shanhua, Tainan, Taiwan, pp. 30–42.

Shanmugasundaram, S., Tsou, S.C.S. and Yan, M.R. (1992) Vegetable soybean for sustainable agriculture. In: Huang, S.C., Hsieh S.C. and Liu D.J. (eds) *Proceedings of SABRAO International Symposium on the Impact of Biological Research on Agricultural Productivity*. Taichung District Agricultural Improvement Station, Taichung, Taiwan, pp. 379–385.

Shurtleff, W. (2001) On the English language etymology of 'green vegetable soybeans' 'edamame' 'vegetable type soybeans' and 'food-grade soybeans'. In: *Proceedings of Second International Vegetable Soybean Conference*. Washington State University, Tacoma, WA, USA, pp. 179–181.

Shurtleff, W. and Aoyagi, A. (1994) *Green Vegetable Soybeans (Edamame) and Vegetable Type Soybeans. Bibliography and Source Book, 3rd century* AD *to 1994*. Soyfoods Center, Lafayette, CA, USA.

Shurtleff, W. and Lumpkin, T.A. (2001) Chronology of green vegetable soybeans and vegetable-type soybeans. In: *Proceedings of Second International Vegetable Soybean Conference*. Washington State University, Tacoma, WA, USA, pp. 97–103.

Sitani, K. (1992) Cultivation practices for vegetable soybean. In: Shanmugasundaram, M. (ed.) *Vegetable Soybean Production: Proceedings of a Training Course*. Asian Vegetable Research and Development Center, Shanhua, Tainan, Taiwan, pp. 19–21.

Smith, J.M. and Van Duyne, F.O. (1951) Other soybean products. In: Markley, K.S. (ed.) *Soybeans and Soybean Products*, Vol. 2. Interscience, New York, USA, pp. 1055–1078.

Srisombun, S. and Shanmugasundaram, S. (2001) The history of vegetable soybean development, current status and future development in Thailand. In: *Proceedings of Second International Vegetable Soybean Conference*. Washington State University, Tacoma, WA, USA, pp. 183–186.

Srisombun, S., Malipan, A., Pa-Oblek, S., Chinchest, A., Sritussaniy, S. and Shanmugasundaram, S. (2009) Soybean variety improvement for soymilk production in Thailand. In: Qiu, L., Guan, R., Jin, J., Song, Q., Guo, S., Li, W., Wang, Y., Han, T., Liu, X., Yu, D., Jiang, L. and Peng, D. (eds) *Developing a Global Soy Blueprint for a Safe Secure and Sustainable Supply. Abstracts for oral presentations and posters of the World Soybean Research Conference VIII*, Beijing, China, 10–15 August 2009. The Chinese Academy of Agricultural Sciences, Beijing, China, p. 173.

Takii (2010) Takii Net Shop. Available at: https://www0.takii.co.jp/CGI/netshop/catalog/products/search.cgi?search_category=00000084&keyword (accessed March 3 2010).

Tsay, J.S., Lai, S.H. and Tsai, C.L. (1991) Present and potential cropping systems for soybean in Taiwan. In: Shanmugasundaram, S. (ed.) *Proceedings of Vegetable Soybean – Research Needs for Production and Quality Improvement Workshop*. Asian Vegetable Research and Development Center, Shanhua, Tainan, Taiwan, pp. 65–72.

Uragami, A., Morishita, M., Hirokane, H., Sato, F., Tokuda, S. and Hidashio, H. (2005) Control of soybean cyst nematode (*Heterdera glycines*) in vegetable soybean cultivation using a high soil temperature and cultivation of the resistant cultivar, 'Peking'. *Horticultural Research, Japan* 4, 219–223.

USDA (2007) National Nutrient Database for Standard Reference, Release 20. Food and Nutrition Information Center (FNIC), National Agricultural Library (NAL), USDA. Available at: http://www.ars.usda.gov/main/site_main.htm?modecode=12-35-45-00 (accessed 5 October 2008).

USDA–Iowa State University (1999) Database on the Isoflavone Content of Foods. Available at: http://www.nal.usda.gov/fnic/foodcomp/Data/isoflav/isoflav.html (accessed 5 October 2008).

Vello, N.A., Bernardi, W.F., Pinto, J.I.H.V., Nekatschalow, M.C., Didone, C.A., Congo, A.R. (2004) Vegetable soybean research on cultivar development and future market potential in South America with emphasis on Brazil. In: Moscardi, F., Hoffmann-Campo, C.B., Saraiva, O.F., Galerani, P.R., Krzyzanowski, F.C. and Carrão-Panizzi, M.C. (eds) *Proceedings VII World Soybean Research Conference and VI International Soybean Processing and Utilization Conference*. Foz do Iguassu, Brazil, pp. 936–941.

Weber, C.R. (1956) New and better vegetable soybeans. *Farm Science* 10, 1314.

Wu, T.L. (2004) Production and researches on vegetable soybean [*Glycine max* (L.) Merrill] in Southern China. In: Moscardi, F., Hoffmann-Campo, C.B., Saraiva, O.F., Galerani, P.R., Krzyzanowski, F.C. and Carrão-Panizzi, M.C. (eds) *Proceedings VII World Soybean Research Conference and VI International Soybean Processing and Utilization Conference*. Foz do Iguassu, Brazil, pp. 942–949.

Xin, C., Kexin, C. and Heping, G. (1997) 'Huijia' introduction of an early mature vegetable soybean. *Seed Science* 2, 10.

Xu, S.C., Huang, J.C. and Liu, D.J. (1999) High-yielding cultivation techniques of vegetable soybean. In: Xu, S.C., Huang, J.C. and Liu, D.J. (eds) *Theory and Practice of High Yield Cultivation of Vegetable Soybean in the South of China*. Fujian Science and Technology Press, Fuzhou, China, pp. 120–142.

Xuan, Z. and Chang, H.J. (2003) Effects of some meteorological factors on the yield of vegetable soybean variety 75. *Acta Agriculturare Universitatis Jiangxiensis* 25, 698–700.

Yan, M.R. and Shanmugasundaram, S. (2001) Vegetable soybean seed production – recommendations. In: *Proceedings of Second International Vegetable Soybean Conference*. Washington State University, Tacoma, WA, USA, pp. 191–194.

Zhang, W.Q. (2004) Relationship between podding traits and water requirement in vegetable soybean variety Madou 292. *Journal of Fujian Agriculture and Forestry University Natural Sciences* 33, 423–425.

Zhihao, W., Shekuan, W. and Shidong, Z. (2000) Breeding of the vegetable 'Tezao 1'. *Seeds in China* 4, 40.

Zhou, J.R. (2004) Soy and isoflavone intake inhibits experimentally induced prostate cancer. In: Moscardi, F., Hoffmann-Campo, C.B., Saraiva, O.F., Galerani, P.R., Krzyzanowski, F.C. and Carrão-Panizzi, M.C. (eds) *Proceedings VII World Soybean Research Conference and VI International Soybean Processing and Utilization Conference*. Foz do Iguassu, Brazil, pp. 686–694.

20 Global Soybean Marketing and Trade: a Situation and Outlook Analysis

Jonas N. Chianu[1], Edilegnaw W. Zegeye[2] and Ephraim M. Nkonya[3]
[1]Agriculture 2 Division (OSAN.2), Agriculture & Agro Industry Department (OSAN), African Development Bank, Tunis, Tunisia; [2]Department of Agricultural Economics, School of Agricultural Sciences and Agribusiness, University of KwaZulu-Natal, Scottsville, Pietermaritzburg, South Africa; [3]International Food Policy Research Institute (IFPRI), Washington, District of Columbia, USA

20.1 Introduction

Global soybean production and trade has changed dramatically in the past 23 years. These changes have been driven by the increasing demand for soybean meal, which accounts for 65% of animal feeds (Ash *et al.*, 2006). Until 1985, production and export of soybean was dominated by the USA. The combined soybean production of Brazil and Argentina, however, now surpasses that of the USA (Ash *et al.*, 2006). The growing economies of China, India and other developing countries have dramatically increased the demand for livestock products, which, in turn, has increased the demand for soybean meal (Delgado *et al.*, 1999). Soybean has also taken a key role in the emerging biofuel sector. Brazil and Argentina produce biodiesel, and respectively derive 66% and 100% of feedstock from soybean (Trostle, 2008).

On the trading side, high-income countries accounted for 46% of the total demand (165 million t) for soybean in 2000, 90% of which was for non-food use – mainly animal feeds and more recently biodiesel production. The European Union (EU) has been the leading importer of soybean products (USDA, 2005). However, in 2002, China's imports surpassed those of the EU, and China thus became the leading global importer of soybean. This was achieved after China eased trade restrictions and joined the World Trade Organization (WTO) in 2002 (USDA, 2005). These rapid changes in the production and trade of soybean have prompted a renewed interest in examining the drivers of these changes and forecasting future scenarios. This chapter discusses global soybean trade and examines the factors that will determine future major dynamics.

Table 20.1. Top ten oilseeds exporting countries by export percent share in 2005 (adapted with permission from Perez *et al.*, 2008).

Country	2005 share (%)	% change (1995–2005)
USA	32.0	−19.1
Brazil	25.7	19.0
Argentina	11.6	4.0
Canada	6.8	−4.0
China	3.2	−1.1
France	2.9	−2.0
Paraguay	2.9	1.3
The Netherlands	2.1	0.1
Australia	1.4	0.7
India	1.4	0.2
Total	90.0	–

Three soybean products (grains, meal and oil) are often traded. Together, these three products represent the most important agricultural exports of Brazil, accounting for 8% of Brazil's total exports in 2006 (Perez *et al.*, 2008). Producing about 80% of world soybean (Hamamoto *et al.*, 2002; Andino *et al.*, 2005), Brazil, the USA and Argentina (in that order) are also the three major exporters in the international protein meal markets. In recent years, these countries accounted for about 85% of the global trade on soybean meal, a share that is projected to increase to >90% in the future (USDA, 2005). While the USA dominates the soybean grains export market, Argentina is the principal exporter of soybean meal and soybean oil, followed by Brazil and the USA (Andino *et al.*, 2005). The major soybean importing countries and regions are India, Germany, Indonesia, Japan, Mexico, the Netherlands, Korea, Spain, Thailand, North Africa, the Middle East, Central America and the Caribbean (USDA, 2005).

The top ten oilseeds exporting countries are given in Table 20.1.

The EU remains the world's leading importer of soybean meal, as import prices for meal relative to soybean grains put pressure on crush margins, curtailing soybean (grains) imports in favour of soybean meal (USDA, 2005). However, increases in grain and rapeseed (*Brassica* species) meal feeding are expected to continue to slow the growth in EU soybean meal and soybean grain imports. Latin America, North Africa, the Middle East, South-east Asia and the former Soviet Union constitute important markets for soybean meal (USDA, 2005).

20.2 Present Situation of Export and Import

Repeatedly in the limelight

Soybean is repeatedly in the limelight these days, being a booming crop in Brazil, with an export value equivalent to US$10billion (>10% of the value

Table 20.2. Global soybean production in 2005 (adapted from FAO, 2007).

Country/region	Quantity produced (t)	%
USA	83,368,000	39.0
Brazil	53,053,000	24.8
Asia	25,746,286	12.0
Europe	3,050,403	1.4
Africa	1,238,443	0.6
Others	47,520,152	22.2
World	213,976,284	100.0

of the total Brazilian exports) in 2004 (Smaling *et al.*, 2008). By 2003/2004, Brazil was the world's largest soybean exporter and the second largest producer after the USA (see Table 20.2) (Perez *et al.*, 2008). Three-quarters of the total soybean production in Brazil is exported, mainly to China and the EU (Smaling *et al.*, 2008).

Global soybean import demands

Global soybean imports have been rapidly increasing. There has been a growing demand for soybean in Asia. The demand surge (with a nine-fold increase in soybean imports in the 10 years from 1994 to 2004) largely stems from China, which has insignificant domestic production (Smaling *et al.*, 2008). The demand surge was triggered by China's 2002 WTO membership, which ended border tariffs and in turn boosted trade. The increasing global demand for soybean has been met through a strong supply response from Brazil and Argentina. Soybean cultivation in Brazil is expected to expand further in the coming decades, mainly in response to the growing demand in Asia (Smaling *et al.*, 2008). Country statistics show that in 2004, Brazilian production was >50 million t, twice the amount realized in 1997. Area and production increases were particularly strong in the period 2001–2005, following a favourable devaluation of the Brazilian currency. Soybean exports in 2004 earned Brazil >US$10 billion, against US$4.2 billion in 2000.

Import demand for soybean oil is rising in nearly all countries and regions except for the former Soviet Union. Countries with the largest projected gains are China and India. In China, growing demand for high-quality vegetable oils outpaces domestic oil production and fuels, expanding soybean oil imports (USDA, 2005). Land-use competition from other crops constrains the area planted to vegetable oil crops in China. In India, relatively lower tariffs on soybean oil (held in check by WTO tariff-binding commitments) compared with those for other vegetable oils will favour continued strong imports of soybean oil. India accounts for an increasing share of world soybean oil imports due to burgeoning domestic demand for vegetable oils and limitations in domestic oilseed production. Low yields associated with erratic rainfed growing conditions and low input use limit oilseed production in India (USDA, 2005). In North Africa, the Middle East region and Latin America (particularly Central America and the Caribbean), income and population growth drive strong gains in soybean oil imports (USDA, 2005).

Global soybean grains and oil exports

Between 1990 and 2007, the world soybean export was dominated by three countries (the USA, Brazil and Argentina) that presently account for >80% of world soybean exports. Projections indicate that these three countries will account for >90% of the world soybean trade by 2014 (USDA, 2005). With continuing area gains, Brazil maintains its position as the world's leading exporter of soybean and soybean products in the projection. Over the projection period, ending in 2014, while the overall percentage share of soybean export by Argentina remains more or less stagnant, that of Brazil shows continuous growth, while the USA's share is set to decline. During the same projection period, Argentina's soybean grains exports hold steady at about 7 million t.

Argentina is the leading exporter of soybean oil, reflecting the country's large crush capacity, its small domestic market for soybean oil and an export tax structure that favours the exports of products rather than soybean grains (Smaling *et al.*, 2008). Increases in crush and soybean oil exports are supported by gains in Argentine soybean production due to extensive double-cropping, further adjustments to crop-pasture rotations and the addition of marginal lands in the northwest part of the country (USDA, 2005). Brazil's expansion of soybean production into new areas of cultivation enables it to increase both its volume of soybean oil exports and its share of world trade. A strong emphasis on exporting soybean products will push the combined share of Argentina and Brazil in world soybean oil exports from about 80% in 2004 to a projected proportion of about 86% in 2014 (USDA, 2005). The USA remains the world's third largest soybean oil exporter after Argentina and Brazil. However, the USA's share of the world soybean oil trade is expected to continue on a downward trend to <5% by 2014 (USDA, 2005). Similarly, the EU remains a small exporter of soybean oil, although its soybean oil export volume and share of the world soybean oil trade continues to decline.

20.3 Past Trends (Export and Import) in Global Soybean Marketing and Trade

Rising unabated since the early 1990s, the global trade in soybean grains and soybean products has surpassed that of wheat – the traditional leader in agricultural commodity trade (USDA, 2005). The world's soybean grain trade grows at an average annual rate of 3.8% compared to 2.9% for soybean oil and 2.3% for soybean meal (USDA, 2005).

Following World War II, the USA, followed by Canada, dominated the soybean industry. Soybean attained global significance shortly after World War II, when the USA made soybean exports part of its negotiated assistance packages in the reconstruction of Europe. This allowed the USA to establish a dominant position for this emerging commodity and to rule global soybean markets for two decades as the crop's sole exporter. As late as 1970, the USA accounted for two-thirds of the world's 44 million t of soybean. Canada was the second largest producer, followed by a number of European countries (Perez *et al.*, 2008).

Brazil entered the soybean industry in the early 20th century with an initial focus on family farms and domestic use. Commercial production began in the 1960s and by 1976 Brazil had gained a significant share (16%) of the world market. After a brief decline in the 1980s, soybean production in Brazil again grew dramatically (Perez *et al.*, 2008). Bolivia started growing soybean in the 1950s with production steadily growing, then taking off in the early 1990s. The area planted expanded nearly six-fold from 1985 to 1995, with exports rising from US$20 million to US$143 million (Perez *et al.*, 2008). Like other countries in the region, Bolivia's soybean boom coincided with trade liberalization.

Increases in income have differential effects on the consumption of traditional soybean foods, fats and oil, as well as livestock products. The results of income elasticity of demand, estimated in Japan for certain soybean products, shows that *miso* (a processed soybean product) has had a negative income elasticity of less than –1.0 (meaning that with a 1% increase in income, the quantity purchased decreases by >1%). *Shoyu* (another soybean product) has had an income elasticity ranging between 0 and –0.5. The income elasticity of demand ranges for other soybean products have been estimated at 0 to +0.5 for tofu, +0.5 to +1 for soybean as a whole and above +1 for fats, oils and livestock products. These estimates indicate that as the income of the Japanese consumer increases, consumption of traditional soybean foods decreases, while consumption of fats, processed oils and livestock products increases more rapidly than the rate of income (Nakamura, 1961).

Buyers sometimes indicate a preference for soybean from particular sources or countries. For instance, some European mills have preferred US soybean from the Manchurian (Primmer, 1939). Soybean oil imports normally stand high above the imports of soybean grains or other soybean products. For dependable delivery in the 1950s, Japan also continued to import most of its soybean from the USA instead of then-potential sources such as mainland China (Nakamura, 1961). Most of the domestic consumption of soybean oil was in the form of food products, such as cooking and salad oils, and most of the soybean meal used domestically was in the high-protein portion of feed rations for poultry and livestock (Rausser and Carter, 1983).

The dramatic growth in Brazilian production and export of soybean grain and soybean products during 1973–1983 eroded US dominance of the world market (Williams and Thompson, 1984) in the mid 1980s. From <1% in the early 1960s, Brazilian soybean output has grown to >24% of world production. In Brazil, soybean has replaced other arable crops (Smaling *et al.*, 2008).

In India, the expansion in area under oilseed has occurred mainly through an increase in the cropped area, but also through displacement of low-yielding coarse cereals (Acharya, 1993).

20.4 Drivers of Global Soybean Market and Trade

The key driver of soybean market growth is the macro-economy of the suppliers and consumers of soybean and its products (Informa Economics, 2005). Market demand for soybean and its value is derived from soybean meal and soybean oil (Rausser and Carter, 1983). Global movements towards biofuels, functional foods and the increasing replacement of protein sourced from fishmeal with that sourced from soybean meal in livestock feed formulations have been driving global soybean marketing and trade, leading to drastic price increases that are not likely to fall or stabilize soon. This may benefit large-scale soybean producers and increase economic incentives for emerging soybean-producing countries such as Argentina (Perez *et al.*, 2008). Some of these driving factors are further discussed below.

Biodiesel

Soybean is one of the major booming oil crops in the world and is one of the products presently being used as biodiesel, with an increasing trend in response to the growing demand for biofuels, especially in the USA (Smaling *et al.*, 2008). This is being driven by high and unstable fossil-fuel-derived energy prices. Several countries (e.g. Brazil and the USA) have created programmes for biodiesel development. As a result, a new market opened for soybean oil in Brazil in 2006/2007. In Brazil, the biodiesel programme has allowed the inclusion of 2% of biodiesel in diesel from petroleum since 2006. This proportion will become compulsory in 2008 and increase to 5% by 2013 (Smaling *et al.*, 2008). In 2008, demand for vegetable oil for biodiesel was estimated at about 500,000 t (Smaling *et al.*, 2008). It has also been estimated that the energy sector will absorb about 1.5–3.0 million t of vegetable oil by 2013 (Smaling *et al.*, 2008).

Surging demand for soybean foods, vegetable oil and animal products

The increasing global demand for animal products and vegetable oil in developed countries (e.g. Japan, the USA) and emerging markets (e.g. China, India, Brazil) has continued to exert pressure on the soybean value chain, resulting in drastic price increases (Smaling *et al.*, 2008). In China, the growing demand for high-quality vegetable oils outpaces domestic oil production and is stimulating expanding soybean oil imports (USDA, 2005). There has also been a steady change in tastes and preferences towards the greater consumption of vegetable oils and reduced consumption of animal fats (Vandenborre, 1966). There is an ever-growing demand for soybean products in China. For instance, overall demand for vegetable oil has been estimated at about 1 million t in 2008 and a projected 2.5 million t in 2013. The various and increasing uses of soybean and its products have helped it to gain a key place in the American industrial scene. In response, increasing numbers of manufacturers in the USA (and elsewhere) now primarily engage in the production of soybean oil, soybean cake, soybean meal and other soybean products.

Continuous investment and expansion in oilseed crushing capacity

Many countries (e.g. China, some countries in North Africa, the Middle East and South Asia) with limited opportunity to expand oilseed production have continued to invest in crushing capacity (USDA, 2005). Expansion in oilseed crushing facilities in importing countries has also accelerated the expansion of the soybean industries in Brazil and Argentina (Uri *et al.*, 1993), causing oilseed import demand to be maintained above the import demand for protein meal (USDA, 2005). China has a policy of expanding its crushing capacity instead of importing protein meal and vegetable oil. This policy influences the composition of world trade by raising international import demand for soybean grains and other oilseeds, rather than import demand for processed products.

Increased demand for livestock feed protein from soybean

The increased sourcing of livestock feed protein from soybean has been associated with an increased commercialization of pork and poultry production that demands a higher minimum quality of feedstuffs in terms of energy and protein content (USDA, 2005). As the livestock industry grows to meet the increasing demand for livestock products, the use of soybean meal in feed (especially for pig, chicken and rabbit production) is also becoming more important in response to changes in dietary habits and shifts in tastes and consumer preferences (Nakamura, 1961; USDA, 2005). Protein for animal feed manufacturing is increasingly sourced from soybean meal instead of fish meal, as has been the case in the past (Nakamura, 1961; Mwasha, 2006; Zulu, 2006). The problems with sourcing livestock feed proteins from fishmeal include high levels of bacterial (e.g. *Salmonella*) contamination, which causes serious production problems in poultry (diseases can lead to 100% mortality in poultry farms, about 50% reduction in egg production and about 30% reduction in hatchability), the use of drugs that have residual effects, huge livestock medical bills, a fishy smell in eggs and meat, short shelf life due to high moisture, depressed growth in broilers (due to disease) (Mwasha, 2006; Zulu, 2006) and a likelihood of mercury contamination.

Economic growth, rising per capita income and urbanization in developing countries

The changing world food demand for high-value agricultural products (including livestock products) and processed foods has mainly been attributed to strong economic growth and rising per capita income in developing countries (USDA, 2005; ASARECA, 2008). With economic growth, urbanization and changing diets, the world demand for plant-derived oils and their derivatives has soared (Smaling *et al.*, 2008). In North Africa, the Middle East region, Central America and the Caribbean, income and population growth are driving strong gains in soybean oil imports (USDA, 2005). North Africa and the Middle East are projected to experience a continued growth in import demand for grain and high-protein meals through 2014, as rising populations and incomes sustain a strong growth in the demand for animal products (USDA, 2005). Strong income and population growth in developing countries generate increasing demand for vegetable oils for human food and high-protein meals are used in livestock production. In India, cereal consumption remained unchanged between 1990 and 2005, while the consumption of oil almost doubled (ASARECA, 2008).

Profit and high price

Profit is the single most important driver of soybean production in Brazil where, although combating soybean rust disease increases the costs of producing soybean, soybean remains more profitable than other crops (USDA, 2005). Similarly, although generally profitable for direct food and feed uses, soybean for those purposes has expanded into such highly competitive markets that more profit seems likely through its utilization as a factory raw material (Primmer, 1939). The higher price for soybean in the early 1970s was the result of substantial increases in the demands for soybean grains and soybean products in the world markets (Uri *et al.*, 1993).

Global increase in human and livestock population and expansion of trade

Population, a demand shifter, is a significant factor driving the overall growth in demand for agricultural products (Vandenborre, 1966). Ever-increasing global human and animal populations, especially in developing countries, will likely lead to future increases in the demand for soybean. A major factor in the oilseed sector for the past several years has been China's large soybean imports due to its huge population (Plato and Chambers, 2004). The surge in demand from China was further triggered by its WTO membership in 2002 (Smaling *et al.*, 2008). Increases in domestic demand due to marginal increases in China's population will drive world demand and prices. An increasing global demand for animal products also increase demand for soybean due to its desired feed traits. Demand for food and feed is expected to double in the next 50 years (Gowing and Palmer, 2008), which, in turn, will push the demand for soybean and its products.

Technological development, trade liberalization and involvement of multinationals

Due to technological improvements and favourable price policies, the production of oilseeds recorded a jump during the 1980s (Acharya, 1993), bringing about land use changes. Scientists have continued to find new uses for the versatile soybean (Greenberg and Hartung, 1998). In Argentina, advanced research capacities have greatly contributed to increases in soybean production, marketing and trade (Perez *et al.*, 2008).

The adoption of broad commercial and financial liberalization measures has been instrumental to the development of soybean market and trade in South America (e.g. Brazil, Argentina, Bolivia). Such a policy environment made soybean take off in Brazil, leading to annual growth of 4.8%, especially on large properties that dominate Brazil's soybean sector (Perez *et al.*, 2008). With the capital-intensive and technologically advanced farming of vast areas of land, agribusiness has been driving growth in the commercial production of soybean in Brazil, where 85% of farms are >1000 ha in the largest soybean-producing municipality. Multinational agro-food firms and companies have begun to displace the state as the principal financiers of soybean production. In 2005, just four firms accounted for 59% of soybean processing and 61% of soybean-based exports, showing how agricultural trade liberalization has brought about growth for the large farms that dominate the soybean subsector in Brazil (Perez *et al.*, 2008). Similarly, Bolivia's soybean boom has coincided with trade liberalization in the country, assisted by significant state investment in infrastructure, subsidization of fuels used by the soybean sector, debt relief and large tax exemptions to attract investment and promote exports (Perez *et al.*, 2008).

Transgenic soybean boom and policy support

The transgenic soybean boom has pushed Argentina towards specializing in the production and export of a small number of primary products. For most of the last century, Argentina was one of the world's most important producers of meat and cereal grains and was nearly self-sufficient in food production for its population. Now, the country has lost that self-reliance as it has moved decisively towards soybean monoculture (Perez *et al.*, 2008). Argentina's double-harvest of wheat and soybean in rotation has replaced cattle ranching and other important food crops, negatively affecting food security. Nearly half of all land under cereals and oilseeds (46%) was under soybean in 2002/2003, up dramatically from 9% in 1980/1981. While soybean production increased to 20 million t from 1997–2004, production of fruits and cotton declined, as did the production of rice by 500,000 t (Perez *et al.*, 2008). The Argentine soybean model has caused the near disappearance of small-scale and family farming. Between 1988 and 2002, Argentina lost 87,000 farms, 86% of them <200 ha. Argentina's agricultural sector has become one of 'farms without farmers' (Perez *et al.*, 2008). Strong backing for policies supporting the industrial soybean sector has been amply demonstrated in South American countries (Brazil, Argentina and Bolivia). In Argentina, there is widespread legalization and adoption of transgenic soybean. In Brazil, soybean-producing states have been offered tax breaks to stimulate production, while in Bolivia the state has subsidized energy costs. In Brazil and Argentina, public funds have been used for research that benefits the private sector.

20.5 The Changing Profile of the Global Soybean Trade

At the global level, soybean is the largest source of protein for livestock feed and the second largest source of vegetable oil in the world after oil palm, accounting for nearly 65% of the worlds' livestock protein supply and 20–30% of the world's vegetable oil supply (Ash *et al.*, 2006; J.M. Mahasi, Kenya, 2008, personal communication).

Changing pattern of soybean production and export as demand increases

As the demand for soybean increases, the pattern of its production, export and import has been changing. This change has been the result of reactions of producers, exporters and importers to market incentives and domestic demand situations. Brazil and Argentina are emerging as the new world players in soybean production and export. Brazil is the second largest exporter of soybean after the USA. Assuming no subsidies, Brazil has a comparative advantage over the USA in producing soybean due to its relatively fertile virgin land and generally lower production costs (Ash *et al.*, 2006). China's significant role as an importer of soybean only came recently when the country relaxed a number of trade restrictions in compliance with the WTO, which it joined in 2002. The other emerging exporters of soybean are Paraguay, China and Canada. The export of China has, however, been declining, mainly due to the fast increasing domestic demand. Overall, China is a net importer and its import accounts for 40% of the world soybean trade (Ash *et al.*, 2006).

During recent years, the export of soybean by the USA has also been declining due to the alternative uses of soybean as a biodiesel and the recent policy changes that have given tax breaks or subsidies for the production of biodiesel from soybean. Competition (with maize) for land for ethanol production is also likely to decrease US soybean production and export (Elobeid *et al.*, 2006).

Changing demand for soybean

Based on demand and importation, the developed countries are the major consumers of soybean produced in the world. In 2000, high-income countries accounted for 46% of total soybean demand (165 million t), 90% of which was for non-food use – mainly animal feeds and, more recently, biodiesel.

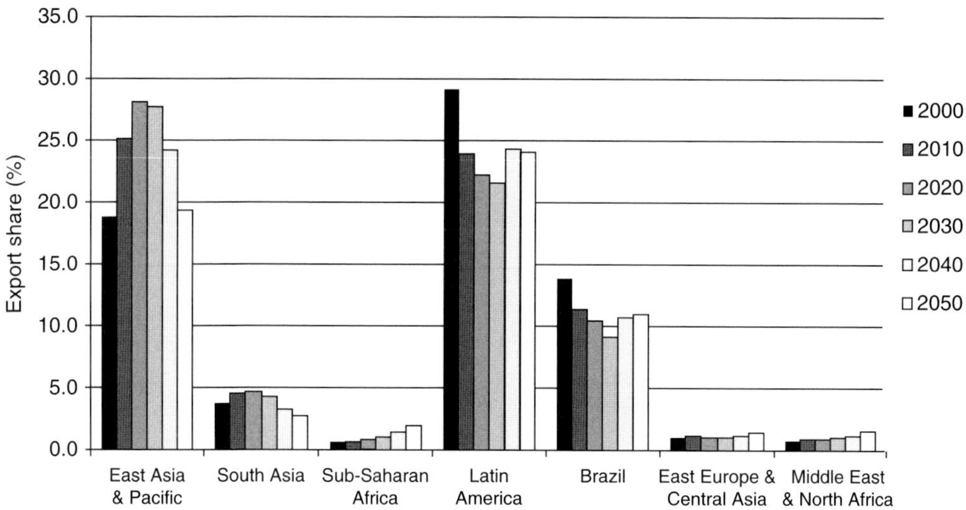

Fig. 20.1. Share of total demand for soybean in developing countries (authors' projections using the IMPACT model).

Using the International Model for Policy Analysis of Agricultural Commodities and Trade (IMPACT), developed by the International Food Policy Research Institute (IFPRI) (Rosegrant *et al.*, 2002), it is estimated that the demand for soybean in developing countries is expected to increase significantly by the year 2050. Such an increase will mainly come from the demand for livestock products in developing countries.

Similar to the USA and Brazil, the demand for soybean biodiesel is increasing in the EU. Europe has set targets for biofuels in order to reduce petroleum carbon emissions and dependence on fossil fuels (Ash *et al.*, 2006). The demand for soybean products in East Asia and Latin America presents an interesting trend (Fig. 20.1). The East Asia and Pacific share of the world demand for soybean will increase from about 0.2 (or 20%) in 2000 to about 0.25 (or 25%) in 2020, driven mainly by China. The region's share of the world demand for soybean will then start falling and reach about 0.2 (or 20%) again in 2050. Latin America and Brazil show a decreasing share of the global soybean demand. India presents a unique case in its soybean trade policies. The country's production does not meet its large demand, yet it has set prohibitive taxes to limit the importation of oilseeds (Ash *et al.*, 2006). With its large population and rising incomes, India is unable to meet its vegetable oil needs. Despite the increasing demand for animal products and the consequent demand for soybean meal, India is a minor importer of soybean meal.

Fundamental shift in Africa

There have been fundamental shifts in soybean enterprise in Africa – from being a net exporter of oilseeds until the mid 1970s to net importer after

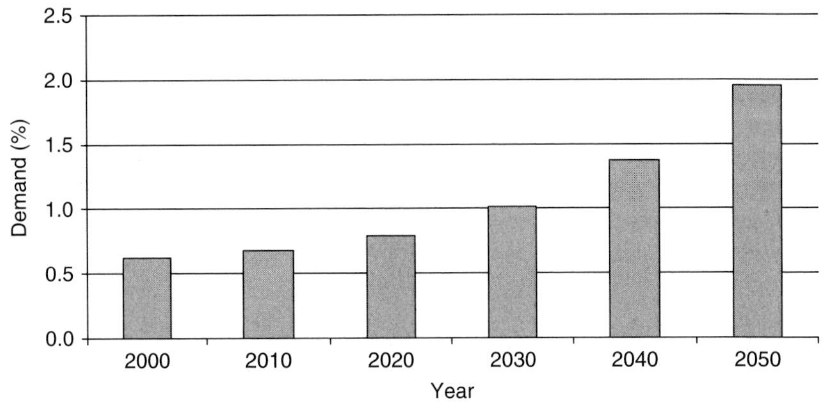

Fig. 20.2. Demand for soybean in sub-Saharan Africa (authors' projections using the IMPACT model).

(Diaz-Bonilla and Reca, 2000). There has been a dramatic decline in the world market share for African exports of processed products from oilseed. Less developed countries overtook developed countries in oilseed exports in the mid 1970s (Diaz-Bonilla and Reca, 2000). Sub-Saharan Africa (SSA) accounts for <1% of the total demand and import of soybean. Our projection shows that SSA's demand for soybean will increase significantly and consistently to reach about 2% of the world demand (Fig. 20.2). Figure 20.2 is an amplification of the crowded Fig. 20.1, although based on similar data just for SSA. It shows an increase from <1% to slightly >2%. As in other regions, this increase is mainly determined by the increasing demand for non-food soybean products.

South American soybean boom

The South American soybean boom began in earnest in the early 1990s, to the point where Brazil and Argentina began to take market share from the USA and Canada. By 2007, the USA accounted for 37% of global production, with Brazil and Argentina accounting for 24% and 20% of the market share, respectively (Perez *et al.*, 2008). In Argentina, transnational firms also control soybean processing and marketing, and even have extensive connections with the financial sector to form 'planting pools'.

20.6 Expected Future Trends (Export and Import) in Global Soybean Marketing and Trade

Several factors will work simultaneously to determine future trends in the global soybean trade. An example is domestic agricultural and trade policies in individual countries that have a role in soybean trade. Some selected details are given below.

All projections show that Brazil will soon surpass the USA as the world's largest producer and that South America will dominate the growing soybean market. South America has already replaced the USA as the largest soybean exporter, with Brazil as the largest exporter of raw soybean grains and Argentina as the largest exporter of soybean oil (Perez *et al.*, 2008). South America is also expected to surpass the USA in the predominance of transgenic soybean (Perez *et al.*, 2008). Argentina produces virtually 100% transgenic soybean, while other producers in the region now grow at least half of their soybean from transgenic seeds (Perez *et al.*, 2008).

As scientists continue to find new uses for soybean, it is expected that per capita and total consumption of soybean (oil, meal and so on) will continue to increase. This will be accelerated by economic progress, a rise in consumer incomes that will create a stronger demand for fats, oils and livestock products. In addition, the ratio of consumption by the crushing and food industries is expected to further increase in order to meet the increasing demand for soybean oil and soybean meal.

Various analysts have given their judgement regarding future developments related to the soybean industry. Some analysts have noted that continued strong growth in the global demand for vegetable oil and protein meal is expected to maintain soybean and soybean-product trade well above wheat and coarse grains trade throughout the next decade (2010–2020) (USDA, 2005). Strong competition in the international protein meal markets is expected to influence oilseed crushing margins and shift some of the import demand for oilseeds to cheaper meals. The steady competitive pressure of new oilseed crushing capacity is expected to result in many inefficient crushers being edged out.

EU soybean meal and soybean grains import has exhibited some slow growth. Increases in grain and rapeseed meal feeding are expected to continue to slow the growth in EU soybean meal and soybean imports. This is because the increase in grain and rapeseed meal replaces soybean meal, given that the two products (soybean meal and rapeseed meal) are substitutes (USDA, 2005). Abundant EU grain stocks, lower internal EU grain prices due to Agenda 2000 price cuts, increased barley production due to Common Agricultural Policy 2003 reforms, greater supplies of coarse grains from acceding countries and more rapeseed meal available as a result of the biofuels initiative are combining to slow the growth of soybean meal consumption. These factors are partially offset by increases in dairy quota and increased feeding of soybean meal.

Significant investments in oilseed crushing infrastructure by China are driving strong gains in soybean imports as China seeks to capture the value added from processing oilseeds into protein meal and vegetable oil (USDA, 2005). East Asia's trade outlook is dominated by a continuing shift from importing feedstuffs to importing meat and other livestock products. As a result, this region's import demand for protein meal and oilseeds is expected to slow. This process is occurring most noticeably in Japan (USDA, 2005).

There is intense within-region competition among the South American soybean-producing countries. Although all of the soybean-producing countries in South America (including Paraguay and Uruguay) generally follow a similar production model and belong to the same regional integration association, each government follows policies to compete with its neighbours. Paradoxically, the region that is coming to dominate global production and export of soybean and its various products is engaged more in within-region competition than in coordinating national policies to maximize synergistic benefits (Perez *et al.*, 2008).

Genetically modified (GM) and Roundup Ready (RR) soybeans are increasingly becoming important in the major soybean-producing nations of the world. RR soybean and GM seeds have been in use in the USA and Argentina since 1996 (Qaim and Traxler, 2005). Most soybean varieties presently grown in Brazil are also GM (Smaling *et al.*, 2008). In the USA, GM soybean accounts for about 87% of the total area under soybean (Andino *et al.*, 2005). The advent of RR and GM soybean is expected to become more important in the future – a situation that will demand institutional innovations to deal with potential health (safety) and environmental risks.

A significant expansion in domestic crushing in China and large imports of oilseeds will result in China exporting about 1 million t of soybean meal in the future (USDA, 2005). Increasing soybean meal exports from other South American countries (especially Paraguay) is expected to contribute to keeping the international protein meal markets very competitive. The EU continues to be a small but steady exporter of soybean meal. India also remains an exporter of soybean meal, although this has been projected to decline in the future (USDA, 2005). According to USDA (2005), Brazil's rapidly increasing area planted to soybean is enabling it to gain a larger share of world's soybean grains and soybean meal exports, despite increasing domestic feed use. Its share of world exports of soybean grains plus the soybean equivalent of soybean meal exports is expected to rise from about 35% in recent years to 45% by 2014 (USDA, 2005).

20.7 Global Soybean Price Trends, Trade and Marketing Policies

The pricing system for soybean is complex because it involves interactions between the markets for soybean grain, soybean meal and soybean oil (Uri *et al.*, 1993). Since these are different products, their prices, although not completely disconnected, are determined differently. Soybean oil is a more basic and much more speculative commodity than soybean meal (Vandenborre, 1966). Since there are numerous substitutes for soybean oil on the world market, its price is determined residually (Rausser and Carter, 1983).

Reasons for increases in the price of soybean products

As a result of the increasing demand for livestock products and the derived demand for soybean (from demand for animal products), prices of soybean

Table 20.3. Contribution of soybean to production of biodiesel in 2006/2007 by major producing countries (reprinted with permission from Trostle, 2008).

Country	Amount produced (million l)	% of total global production	% produced from soy feedstock
Argentina	443	5	100
Brazil	398	5	66
EU-27	5004	65	16[a]
USA	1927	22	74
Total	8583		

[a]About two thirds of EU biodiesel production is derived from rapeseed.

meal have increased dramatically over the last decade (i.e. between 2000 and 2010). Prices of soybean oil and soybean-based foods have also increased, due mainly to increasing dietary health in high-income countries. The increasing soybean prices have also been driven by rising incomes and the consequent higher demand for livestock products and dietary health concerns that increase demand for vegetable-based diets and fibre-rich foods.

Increasing fossil fuel prices have prompted efforts by Brazil, Argentina, the USA and EU countries to develop alternative energy sources, including soybean-based biodiesel. Increasing fossil fuel prices have also increased soybean production costs, contributing to the increasing prices of soybean and other foods (Benson *et al.*, 2008). Policies in these countries have been designed to give incentives for the production of biofuels. This has also led to increasing production of ethanol, which is maize-based. Import and export restrictions, in response to the increasing food prices, have also had the net effect of increasing world food prices (Trostle, 2008). Farmers in the USA and other countries have also substituted production of soybean for maize, leading to an increase in the soybean price due to reduced supply (Tyner, 2008). As shown in Table 20.3, soybean is the major crop used to produce biodiesel in Brazil, Argentina and the USA.

Soybean remains a favourable source of biodiesel for the major soybean-producing countries and this has contributed to the increasing soybean price in the global market. However, production of biodiesel from soybean accounts for only a small share of the total consumption of petroleum diesel, although its demand is likely to increase given its environmental advantage over fossil fuel. Compared to fossil fuel, biodiesel reduces gas emissions by 41% and produces less air pollutants (Hill *et al.*, 2006). As will be discussed below, these environmental advantages have appealed to soybean-producing countries to subsidize production of soybean and biodiesel.

Demand for soybean is also being driven by changing health concerns, as people in developing countries switch to eating more vegetables and fibre-rich foods. While the demand for animal products in Asian countries has sharply risen in response to increasing incomes, livestock product consumption in developed countries has remained stable and in some cases declined slightly in response to increasing dietary health and livestock-related food safety concerns. Soybean is perceived as having health benefits that address these concerns. In the USA, the Federal Food and Drug Administration allows foods containing 5 g of soybean protein per serving to be labelled as reducing heart disease (Ash *et al.*, 2006). The use of high amounts of soy protein (soy isoflavones) in fortified foods and supplements for the prevention of osteoporosis is growing rapidly.

Effect of policy on pattern of soybean trade

The policies of the major players in the world's soybean production and trade play a key role in determining the pattern of soybean trade. To help understand global policies and their impact on soybean trade, we here review the policies of the USA, Brazil and Argentina as exporters of soybean and of the EU as an importer.

The USA has now enacted the Energy Policy Act of 2005, which set a goal of producing 28.5 billion l of biofuel by the year 2012. More ambitious is the Energy Independence and Security Act of 2007, which sets a goal of producing 136 billion l of biofuel (of which 4 billion l will be biodiesel) by 2022 (Tyner, 2008). This has led to a robust growth in biodiesel production from soybean. The Acts are accompanied by subsidies and other incentives for farmers, developers of technologies and biofuel producers (see Tyner, 2008, for a complete review). Efforts to increase biofuel production have been driven by the need to reduce CO_2 and sulphur emissions. The USA enacted Clean Air Act amendments in 1990, seeking to lower sulphur emissions (Tyner, 2008). Assuming that only soybean-based biodiesel is blended with fossil diesel to achieve the Clean Air Act, it will require 15% of USA soybean production (Ash *et al.*, 2006).

Brazil and Argentina have also formulated policies to reduce dependence on foreign fossil fuels (Pousa *et al.*, 2007). The 2005 Brazilian Federal Law 11097 and the Argentine Federal Law 26093, enacted in 2006, both seek to increase the blend of biodiesel in petroleum diesel (Nardi *et al.*, 2008). Brazilian law has set a target for a 2% blend of biodiesel starting in 2008, increasing the requirement to a 5% blend by 2013; Argentina has set a goal of achieving a 5% blend of biodiesel in petroleum diesel, starting in 2010 (Nardi *et al.*, 2008). These policies have contributed to increased soybean production in both of these countries and to the development of new technologies for the production of biodiesel from soybean. Unlike the USA, however, policies in these countries do not give large subsidies to support soybean producers.

Unlike in South America and the USA, the production of biodiesel in the EU exceeds that of ethanol. In 2003, the region produced 1.5 million t of biodiesel, which was 82% of biofuel production in the region (Bozbas, 2008). Such a policy has undoubtedly increased the demand for soybean in the region, contributing to increasing soybean prices in the global market. Such concerted efforts to produce biodiesel are also driven by the need to reduce dependence on petroleum diesel and reduce CO_2 and sulphur emissions. Given that the EU has a lower comparative advantage (or relatively higher costs) in producing soybean, the region has reduced soybean import taxes and maintained restrictions on GM soybean (Zepeda, 2006).

Practical use of trade and marketing policies (including subsidy policies)

Policies have commonly been used to engender agricultural commodity price support (to domestic producers) and promote exports by several countries, especially Western nations. It has also been a common practice in Europe to provide export subsidies for selected commodities. Between 1986 and 1997, export subsidies in the EU amounted to 13% of the combined value of all agricultural exports from Africa, Latin America, the Caribbean and Asia (excluding China) (Diaz-Bonilla and Reca, 2000). Trade and marketing policies have also been used to support and artificially increase the competitiveness of soybean production, processing, marketing and trade in Brazil and Argentina (Williams and Thompson, 1984; Uri *et al.*, 1993) and were instrumental to the take-off in soybean production in Brazil. In Japan, farmers who plant soybean on fields diverted from rice receive extra subsidies (Hamamoto *et al.*, 2002). Japan also has soybean price support, known as 'deficiency payments' (Hamamoto *et al.*, 2002). The oilseed sector generally has lower trade protection in Organization for Economic Co-operation and Development countries than in developing countries, with higher trade barriers and tariffs in this sector (Elbehri *et al.*, 2001). The good production and export performance of oilseed products by Least Developed Countries (LDCs) is due, in part, to the fact that oilseed production has been relatively less distorted by support policies in developed countries, allowing its expansion in LDCs (Diaz-Bonilla and Reca, 2000). But still, the West engages in import tariff escalation and stringent sanitary/phytosanitary measures, playing against market participation by developing countries. Added to this, African oilseed policies have largely been restrictive and characterized by export tax (Diaz-Bonilla and Reca, 2000). In addition, overvalued local currencies, weak domestic demand for oilseeds and derived products as well as political instability have negatively affected the African oilseed sector.

20.8 Constraints to Global Soybean Marketing and Trade

In the long run, the most important constraints to global soybean production, marketing and trade emanate from the negative externalities of the expansion of the soybean sector itself. These can be summarized under plant and biodiversity loss, carbon emissions, changing air circulation and increased frequency of drought, unlawful land acquisition and inequity, forced migration of forest inhabitants (e.g. in Brazil), deforestation in order to expand the land area under soybean (Smaling *et al.*, 2008) and threats to staple food production. Some specific constraints are further discussed below.

Pest and disease problems still hamper soybean production in many countries, while the extensive use of glyphosate pesticides in large commercial soybean monocrops in Argentina has led to polluted soils. Within the last decade (the 2000s), Argentina's agricultural production system became dominated by 14 million ha of soybean monocrop, creating a production system that could be highly vulnerable to pests or diseases. The massive production of RR soybean in Argentina also has social consequences, such as displacement of rural labour, unemployment, gender inequity and the perpetuation of rural poverty (TWN, 2005). Furthermore, low soil fertility is still a major problem, especially in SSA. For instance, phosphorus, which is particularly important for the growth performance of soybean, is limiting in most soils of SSA. Due to climate change, drought is increasingly posing a natural threat to agricultural production and productivity, especially in SSA. This is especially serious given SSA farmers' lack of access to irrigation facilities. The problem of drought is compounded by SSA farmers' limited access to improved soybean seeds.

Limited crushing capacity in developing countries and GM-related constraints also pose some problems. This deprives farmers of adequate net returns and creates disincentives for them to continue to process soybean. This implies an incomplete value-chain development. GM soybeans are increasingly becoming more important, but continue to generate issues with regards to environmental and food safety concerns. This poses a challenge to the production, marketing and trade of soybean. For instance, increased adoption of GM soybeans in the USA has been one factor in the decreased performance of US soybean exports, due to lack of consumer acceptance (Andino *et al.*, 2005).

The South American soybean sector model has inherent limitations. The three cases of Brazil, Argentina and Bolivia show the limitations of the South American soybean sector model. While liberalization and an agro-export orientation have benefitted some producers, the strategy is based on undervalued natural resources and foreign enterprises dominate all parts of the industry except the farming – financing, input supply, processing, marketing and export operations. Despite dynamic growth in productivity and output, the soybean sector in South America has seen a significant drop in employment (Perez *et al.*, 2008). This fall in employment, in addition to foreign dominance and the capital-intensive nature of this sector in South America, raises questions as to the role of this crop in the lives of the poor and equity implications. In this context, the governments of soybean-producing South American countries have adopted various policies to support the industrial soybean sector, which might improve the poverty and inequity or worsen it depending on who benefits from these technologies. In Argentina, we see the widespread legalization and adoption of transgenic soybean. The increased use of agrochemicals on soybean is an environmental challenge. In Brazil, soybean-producing states have been offered tax breaks (which can help to reduce poverty) to stimulate production. Bolivia has subsidized energy costs (which can also help to reduce poverty by increasing savings) (Perez *et al.*, 2008). Public funds in Brazil and Argentina have also gone to research that has benefitted the private sector (Perez *et al.*, 2008).

Growth in the export of soybean meal and soybean grains is experiencing some constraints in selected countries. For instance, further growth in the soybean meal exports of Brazil is constrained by strong growth in the domestic consumption of soybean meal due to rapid expansion of the poultry and pork sectors (USDA, 2005). In the USA, projected declines in acreage planted to soybean and increased domestic crush limit exportable supplies (USDA, 2005).

20.9 Conclusions and Implications (Policy and Research) for the Soybean Sector

This chapter has examined the current situation and future outlook of global soybean marketing and trade, based on the available evidence. To this effect, its methodology has been a review of the literature, compiling the publicly accessible data and supplementing this information with predictions from the IMPACT model developed by IFPRI.

Soybean is an extensively processed and traded commodity with versatile uses. Because soybean is mostly used in its processed form and because it is an input to manufacture various food and non-food products, crushers and manufacturing industries are the key users of this crop. Prediction of prices and understanding the soybean pricing system is a complex undertaking because it involves various products: soybean grain, soybean meal and soybean oil.

Results from the situation analysis indicate that the USA, Brazil and Argentina are the three most important players in soybean export market. For soybean oil, the key exporters are Argentina, Brazil, the USA and the EU, in the decreasing order. China, the EU, and Japan are the major players in soybean import. As a block, the EU remains the world's leading importer of soybean meal. The increasing global demand for soybean has traditionally been met through a strong supply response from Latin America, namely Brazil and Argentina. The prospects for soybean grain export continues to be highest in Brazil, and projections show that Brazil will soon surpass the USA as the world's largest soybean producer and that South America will dominate the growing soybean market.

Results from the outlook analysis (up to 2050) based on the IMPACT model show that demand for soybean in developing countries is expected to increase significantly, mainly driven by increasing demand for non-food soybean products. Global trade in soybean products has been on the rise since 1985 and is projected to continue to increase. This increase in demand will push up soybean world market prices, which will put Africa in general and SSA in particular at a disadvantage as a net importer region whose soybean demand is predicted to increase. Several factors drive global soybean trade. Global soybean imports have rapidly increased, due mainly to the growing demand for soybean in Asia, especially China since joining the

WTO. Global population growth, shifts in tastes, preferences and food habits and the supply of alternative animal feeds are other important factors affecting the global trade in soybean and its products.

Two main challenges and opportunities are worth mentioning as far as the future of soybean is concerned. First is the growing global energy demand and soaring energy prices. These continue to increase the demand for alternative energy sources, including soybean oil as biodiesel. The second is China's ever-growing demand for soybean products. This is also true of the industrial uses of soybean oil and by-products to meet the expansion of manufacturers. These expected trends may benefit big soybean producers, but will also prove to be a challenge unless there is a technological breakthrough in efficiently producing and marketing soybean internationally. There are, however, certain constraints in the soybean sector at the global level. One of these is the incomplete soybean value chain. There is a need to develop institutions to address the missing links along the soybean value chain and reduce the transaction costs of doing business in soybean and its products. New and emerging uses of soybean and enhancing competitiveness throughout the soybean value chain will call for technological changes to meet those new uses and the growing demand. The advent of GM soybeans and their use will demand institutional innovations to deal with potential environmental and public health (safety) risks. To address the ever-increasing demand for non-grain soybean, continuous investment in oilseed crushing capacity will be needed. Research is also needed to enhance efficiency in soybean production and utilization in the livestock sector and to enhance general soybean system linkages.

Import tariff escalation and stringent sanitary/phytosanitary measures by the West are still huge bottlenecks in Latin American developing countries reaping the benefits of this sector. For LDCs to benefit from their exports, they will have to enhance their capacity in trade negotiations against export subsidies and tariff escalation. In some cases, exporting countries will have to deal with their overvalued local currencies.

The review performed in this chapter has finally generated some unanswered questions for future research. Due to its strategic importance in terms of foreign currency earning and its versatile uses, governments of the soybean-producing countries (especially in South America) invest a great deal in supporting the soybean sector. Are all the policy support interventions (e.g. tax breaks, subsidies on energy costs, commodity price support to domestic producers, export subsidies) economically justified and worth implementing? Are they sustainable? Should they be replicated in other soybean countries or regions? The ever-changing climate is expected to influence soybean production and thereby processing and marketing. Is the change in favour or against soybean production, and how? What is the environmental impact of the expansion of soybean production?

Increased unemployment as a result of soybean expansion, relocation of large local populations to make way for soybean production, forest clearance, foreign dominance and the capital-intensive nature of this sector have been experienced in South America. Meanwhile, the poor cannot easily afford livestock diets. This raises a question as to the role of this crop for the poor and the marginalized segments of the population. In the SSA, which is a net soybean importer, small-scale farmers' lack access to local soybean markets favours large stakeholders and importers. Approaches are needed to enhance local production and consumption, while enhancing smallholder access to markets.

Soaring energy costs have huge implications for the production, processing, marketing, import and export of soybean crops. The need for alternative energy sources and the competition of soybean with other crops (such as maize and jatropha) for ethanol production has many ramifications that will be revealed through further investigations. There is also a need to examine the demand and supply conditions for soybean substitutes to better understand the price trends and incentives for soybean producers. All of the above questions deserve critical examination through forward-looking research.

References

Acharya, S.S. (1993) Oilseeds and pulses – price policy and production performance. *Indian Journal of Agricultural Economics* 48, 317–333.

Andino, J., Mulik, K. and Koo, W.W. (2005) The impact of Brazil and Argentina's currency devaluation on U.S. soybean trade. Agricultural & Applied Economics Report No. 574. Center for Agricultural and Trade Studies, Department of Agribusiness and Applied Economics, North Dakota State University, Fargo, ND, USA.

ASARECA (Association for Strengthening Agricultural Research in Eastern and Central Africa) (2008) The world food situation: New driving forces and required actions. *PAAP (Policy Analysis and Advocacy Programme) Electronic Newsletter*, 11.

Ash, M., Livezey, J. and Dohlman, E. (2006) Soybean backgrounder. ERS report # OCS-2006-01, USDA, Washington, DC, USA.

Benson, T., Minot, N., Pender, J., Robles, M. and von Braun, J. (2008) *Global Food Crises: Monitoring and Assessing Impact to Inform Policy Responses. Food Policy Report No. 19*. International Food Policy Research Institute, Washington, DC, USA.

Bozbas, K. (2008) Biodiesel as an alternative motor fuel: Production and policies in the European Union. *Renewable and Sustainable Energy Reviews* 12, 542–552.

Delgado, C., Rosegrant, M., Steinfeld, H., Ehui, S. and Courbois, C. (1999) Livestock to 2020: The next food revolution. Food, Agriculture, and the Environment Discussion Paper 28. Washington, DC, USA.

Diaz-Bonilla, E. and Reca, L. (2000) Trade and agro-industrialization in developing countries: Trends and policy impacts. *Agricultural Economics* 23, 219–229.

Elbehri, A., Hoffman, L., Ash, M. and Dohlman, E. (2001) Global impacts of zero-for-zero trade policy in the world oilseed market: A quantitative assessment. Paper presented at the 4th conference on global economic analysis, Purdue University, West Lafayette, IN, USA, 26–29 June 2001.

Elobeid, A.S., Dermot, T., Hayes, J., Babcock, B. and Hart, C. (2006) *The Long-Run Impact of Corn-Based Ethanol on the Grain, Oilseed, and Livestock Sectors: A Preliminary Assessment*. Center for Agricultural and Rural Development, Iowa State University, Briefing Paper 06-BP 49.

FAO (2007) Statistical Yearbook 2005–2006 Volume 2, Issue 1, Web Edition. Available at: http://www.fao.org/ES/ESS/yearbook/vol_1_1/index.asp (accessed 20 September 2008).

Gowing, J.W. and Palmer, M.W (2008) Sustainable agricultural development in sub-Saharan Africa: The case for a paradigm shift in land husbandry. *Soil Use and Management* 24, 92–99.

Greenberg, P. and Hartung, H.N. (1998) *The Whole Soy Cookbook: 175 Delicious, Nutritious, Easy-to-prepare Recipes Featuring Tofu, Tempeh, and Various Forms of Nature's Healthiest Bean*. Three Rivers Press, New York, USA.

Hamamoto, T., Dyck, J. and Stout, J. (2002) Oilseed Policies in Japan. Electronic Outlook Report from the Economic Research Service, USDA, OCS-1102-01, Washington, DC, USA.

Hill, J., Nelson, E., Tilman, D., Polasky, S. and Tiffany, D. (2006) Environmental, economic, and energetic costs and benefits of biodiesel and ethanol biofuels. *Proceedings of the National Academy of Sciences* 103, 11206–11210.

Informa Economics (2005) China's oilseed market: Where do we go from here? A multi-client analysis and outlook. Informa Economics, Memphis, TN, USA.

Mwasha, E.S. (2006) Soybean in feed formulation in Tanzania: The past, current, and future should be done. In: Myaka, F.A., Kirenga, G. and Malema, B. (eds) *Proceedings of the First National Soybean Stakeholders Workshop,* 10–11 November 2005, Morogoro, Tanzania, DCD/MAFC, Tanzania, pp. 39–46.

Nakamura, H. (1961) The Japanese soybean market. *Illinois Agricultural Economics* 1, 7–13.

Nardi, M., Carpio, C., Davis, T. and Ward, W. (2008) Biodiesel Mandate Laws in Argentina and Brazil: An Estimation of Soybean Oil Foregone Export Revenues. Paper presented at the Southern Agricultural Economics Association Annual Meeting at Dallas, TX, USA, 2–6 February 2008.

Perez, M., Schlesinger, S. and Wise, T.A (2008) The promise and perils of agricultural trade liberalization: Lessons from Latin America. Washington Office on Latin America (WOLA), Washington, DC, and the Global Development and Environment Institute at Tufts University, Medfor, MA, USA.

Plato, G. and Chambers, W. (2004) How Does Structural Change in the Global Soybean Market Affect the U.S. Price? Report OCS-04D-01, U.S. Department of Agriculture, Economic Research Service. Available at: http://www.ers.usda.gov/publications/ocs/apr04/ocs04d01/ocs04d01.pdf (accessed 20 September 2008).

Pousa, G., Santos, A. and Suarez, P. (2007) History and policy of biodiesel in Brazil. *Energy Policy* 35, 5393–5398.

Primmer, G.H. (1939) United States soybean industry. *Economic Geography* 15, 205–211.

Qaim, M. and Traxler, G. (2005) Roundup Ready® soybeans in Argentina: Farm level and aggregate welfare effects. *Agricultural Economics* 32, 73–86.

Rausser, G.C. and Carter, C. (1983) Futures market efficiency in the soybean complex. *The Review of Economics and Statistics* 65, 469–478.

Rosegrant, M., Paisner, W., Meijer, M. and Witcover, S. (2002) *Global Food Projection to 2020: Dealing with Scarcity*. International Food Policy Research Institute, Washington, DC, USA.

Smaling, E.M.A., Roscoe, R., Lesschen, J.P., Bouwman, A.F. and Comunello, F. (2008) From forest to waste: Assessment of the Brazilian soybean chain, using nitrogen as a marker. *Agriculture, Ecosystems and Environment* 128, 185–197.

Trostle, R. (2008) Global Agricultural Supply and Demand: Factors Contributing to the Recent Increase in Food Commodity Prices. Economic Research Services (ERS) Working Paper # WRS-080, USDA, Washington, DC, USA.

TWN (Third World Network Biosafety Information Service) (2005) Re: Soybeans producers in Argentina face problems (Letter). Available at: http://www.twnside.org.sg/title2/service162.htm (accessed 2 October 2008).
Tyner, W. (2008) The US ethanol and biofuels boom: Its origin, current status, and future prospects *BioScience* 58, 646–653.
Uri, N.D., Chomo, G.V., Hoskin, R. and Hyberg, B. (1993) The integration of the market for soybeans and soybean products. *Food Policy* 18, 200–213.
USDA (2005) Agricultural Trade. USDA Baseline Projections, February 2005, USDA, Washington, DC, USA.
Vandenborre, R.J. (1966) Demand analysis of the markets for soybean oil and soybean meal. *Journal of Farm Economics* 48, 920–934.
Williams, G.W. and Thompson, R.L. (1984) Brazilian soybean policy: The international effects of intervention. *American Journal of Agricultural Economics* 66, 488–498.
Zepeda, J.F. (2006) Coexistence, genetically modified biotechnologies and biosafety: Implications for developing countries. *American Journal of Agricultural Economics* 88, 1200–1208.
Zulu, W. (2006) Use of soybean in poultry feed: Perspective from Zambia, Malawi and Botswana. In: Myaka, F.A., Kirenga, G. and Malema, B. (eds) *Proceedings of the First National Soybean Stakeholders Workshop*, 10–11 November 2005, Morogoro, Tanzania, DCD/MAFC, Tanzania, pp. 47–51.

Index

Note: Page references in *italic* refer to tables in the text; those in **bold** refer to figures